S0-CFW-923

THE FIBROBLAST GROWTH FACTOR FAMILY

ANNALS OF THE NEW YORK ACADEMY OF SCIENCES
Volume 638

THE FIBROBLAST GROWTH FACTOR FAMILY

Edited by Andrew Baird and Michael Klagsbrun

The New York Academy of Sciences
New York, New York
1991

Copyright © 1991 by the New York Academy of Sciences. All rights reserved. Under the provisions of the United States Copyright Act of 1976, individual readers of the Annals are permitted to make fair use of them for teaching or research. Permission is granted to quote from the Annals provided that the customary acknowledgment is made of the source. Material in the Annals may be republished only by permission of the Academy. Address inquiries to the Executive Editor at the New York Academy of Sciences.

Copying fees: For each copy of an article made beyond the free copying permitted under Section 107 or 108 of the 1976 Copyright Act, a fee should be paid through the Copyright Clearance Center Inc., 27 Congress St., Salem, MA 01970. For articles of more than 3 pages, the copying fee is $1.75.

∞ The paper used in this publication meets the minimum requirements of American National Standard for Information Sciences—Permanence of Paper for Printed Library Materials, ANSI Z39.48-1984.

Library of Congress Cataloging-in Publication Data

The Fibroblast growth factor family / edited by Andrew Baird and
    Michael Klagsbrun.
        p.   cm. — (Annals of the New York Academy of Sciences; v.
    638)
        Results of a conference held from Jan. 16 to Jan. 18, 1991 in La
    Jolla, Calif. and sponsored by the New York Academy of Sciences.
        Includes bibliographical references and index.
        ISBN 0-89766-689-5 (cloth : alk. paper). — ISBN 0-89766-690-9
    (paper : alk. paper)
        1. Fibroblast growth factors—Congresses.   I. Baird, Andrew,
    1954-   . II. Klagsbrun, Michael.   III. New York Academy of
    Sciences.   IV. Series.
        [DNLM: 1. Fibroblast Growth Factor—congresses.   W1 AN626YL
    v. 638]
    Q11.N5 vol. 638
    [QP552.F5]
    500 s—dc20
    [611'.0181]
    DNLM/DLC
    for Library of Congress

                                                                91-41797
                                                                CIP

SP
Printed in the United States of America
ISBN 0-89766-689-5 (cloth)
ISBN 0-89766-690-9 (paper)
ISSN 0077-8923

ANNALS OF THE NEW YORK ACADEMY OF SCIENCES

Volume 638
December 20, 1991

THE FIBROBLAST GROWTH FACTOR FAMILY[a]

Editors and Conference Organizers
ANDREW BAIRD and MICHAEL KLAGSBRUN

CONTENTS

[a]This volume is the result of a conference entitled The Fibroblast Growth Factor Family held from January 16 to January 18, 1991 in La Jolla, California and sponsored by the New York Academy of Sciences.

Part V. Developmental Biology

Part VI. From the Bench to the Bedside

Selected Oral Presentations

Poster Papers

Financial assistance was received from:

Supporters

- AMERICAN CYANAMID COMPANY
- FARMITALIA CARLO ERBA
- HOFFMANN-LA ROCHE INC.
- ICI PHARMACEUTICALS GROUP
- JOHNSON & JOHNSON/ETHICON, INC.
- SCHERING-PLOUGH RESEARCH
- TAKEDA CHEMICAL INDUSTRIES, LTD.
- THE WHITTIER INSITUTE

Contributors

- ALLERGAN INC.
- BRISTOL-MYERS SQUIBB COMPANY
- BIOGEN
- BURROUGHS WELLCOME CO.
- CALIFORNIA BIOTECHNOLOGY INC.
- CHIRON CORPORATION
- DUPONT MEDICAL PRODUCTS
- KAKEN PHARMACEUTICAL CO. LTD.
- LILLY RESEARCH
- MERCK SHARP & DOHME RESEARCH LABORATORIES
- MONSANTO COMPANY
- NATIONAL CANCER INSTITUTE/NIH
- PHARMACIA LKB BIOTECHNOLOGY INC.
- STERLING DRUG INC.
- SYNERGEN, INC.
- UPJOHN AND COMPANY

The New York Academy of Sciences believes it has a responsibility to provide an open forum for discussion of scientific questions. The positions taken by the participants in the reported conferences are their own and not necessarily those of the Academy. The Academy has no intent to influence legislation by providing such forums.

The Fibroblast Growth Factor Family

An Overview

ANDREW BAIRD[a] AND MICHAEL KLAGSBRUN[b]

[a]Department of Molecular and Cellular Growth Biology
The Whittier Institute for Diabetes and Endocrinology
9894 Genesee Avenue
La Jolla, California 92037

[b]Departments of Surgery, Biological Chemistry, and Molecular
Pharmacology
Harvard Medical School
Children's Hospital
300 Longwood Avenue
Boston, Massachusetts 02115

The fibroblast growth factors (FGFs) constitute a family of at least seven structurally related polypeptides of which basic FGF (bFGF) and acidic FGF (aFGF) are the best characterized. They are potent modulators of cell proliferation, motility, differentiation, and survival. They play an important role *in vivo* in normal physiological processes such as embryonic development, angiogenesis, nervous cell system differentiation, and wound repair. A possible role for the FGF family in pathological processes such as cancer has become evident with the identification of oncogenes that encode proteins having a 40–50% sequence homology to bFGF and aFGF. These oncogenes include int-2, hst, K-FGF, and FGF-5. Another member of the FGF family, keratinocyte growth factor (KGF), a specific mitogen for epithelial cells, has also been described recently.

The last five years have witnessed important progress in delineating the biochemical and biological properties of the FGF family. For example: (1) The endothelial cell growth factor field has been greatly clarified by the determination that over 20 growth factors previously isolated from various sources have been identified to be either aFGF or bFGF, both of which have been purified, sequenced, and cloned. (2) Five additional members of the FGF family have been identified and cloned; int-2, hst/K-FGF, FGF-5 and FGF.6, and KGF. (3) At least four FGF high-affinity receptors have been identified and cloned. (4) Heparin and cell surface heparan sulfate proteoglycan have been shown to modulate FGF activity. (5) aFGF and bFGF have been identified as components of extracellular matrix. (6) An important role for bFGF as an inducer of mesoderm in oocyte development has been ascertained. (7) A number of important biological activities of FGF have been demonstrated. These include mesodermal cell mitogenesis, angiogenesis, neurite extension, neuronal cell survival, and inhibition of myoblast differentiation. These biological properties suggest that members of the FGF family have important physiological roles in the development of the vascular, nervous, and skeletal systems, in promoting the maintenance and survival of certain tissues, and in stimulating wound healing and tissue repair.

Above all, in the last few years a number of important reagents such as recombinant growth factor, cDNA probes, and highly specific anti-FGF antibodies have become readily available to the scientific community. Thus, many investigators representing a wide variety of disciplines including embryology, cell biology, molecular biology, crystallography, and clinical medicine have entered the field and added new perspectives to our understanding of the FGFs. The result has been an explosive increase in FGF research, which in turn made a meeting devoted to the recent developments in the field highly opportune. This meeting entitled the Fibroblast Growth Factor Family held under the auspices of the New York Academy of Sciences is the first such gathering devoted solely to describing and discussing the structural and biological properties of the FGF family.

At the meeting a broad number of topics were discussed detailing FGF structure, biological activity, physiological roles, and clinical application. A proposal was made to introduce a unifying nomenclature to the FGF family in which aFGF, bFGF, int-2, hst/K-FGF, FGF-5, FGF.6, and KGF are to be renamed as FGF-1, FGF-2, FGF-3, FGF-4, FGF-5, FGF-6, and FGF-7 respectively. This was deemed necessary since the term fibroblast growth factor is too narrow to describe these multifunctional growth factors. Among the highlights of data presented during the meeting were the following. A description of the three-dimensional crystal structure of bFGF showed that the tertiary fold of bFGF was topologically identical to that of interleukin I and that a putative heparin binding site close to a receptor binding site could be identified. A comprehensive description of the structure and function of the seven FGF family members was clearly delineated as was the realization that not only is there a family of FGF ligands but of FGF receptors as well. At least four distinct genes encoding FGF receptors have been cloned, and there is selective specificity in the binding of different FGFs. The importance of cell surface heparan sulfate proteoglycans as low-affinity receptors that modulate bFGF activity was first revealed. It was demonstrated that bFGF, despite its lack of a signal sequence, gets exported by some nonclassical mechanism and can participate in the autocrine stimulation of cell motility. Further discussions of biological properties demonstrated the important roles for bFGF in stimulating the smooth muscle cell proliferation that occurs after injury to large blood vessels; the role of both aFGF and bFGF in the eye as a modulator of retinal cell differentiation, lens formation, and photoreceptor development; the role of bFGF in the brain as a neuronal differentiation and survival agent; and the influence of bFGF and other FGF family members in oocyte differentiation and embryonic development. Finally, the growing confidence that FGFs have important clinical potential was discussed in light of *in vivo* applications that included preventing degenerative diseases of the retina, healing damaged cornea, promoting neuronal survival in stroke and Alzheimer's disease, inhibiting melanoma growth, promoting wound healing in diabetics, accelerating dermal wound healing, and promoting healing of duodenal ulcers.

By the end of the conference, it was clear that FGFs are distributed ubiquitously, are multifunctional, and have clinical applications. Above all, it seems that the significance of the FGF family to biological processes of growth and differentiation is only beginning to be explored and at the current pace of FGF research, another meeting will be necessary in the future.

Nomenclature Meeting
Report and Recommendations
January 17, 1991[a]

Chairpersons: Andrew Baird and Michael Klagsbrun

Participants: Stuart Aaronson, Judith Abraham, Claudio Basilico, Daniel Birnbaum, Peter Böhlen, Wilson Burgess, Clive Dickson, John Fiddes, Mitch Goldfarb, Thomas Maciag, Gail Martin, Gordon Peters, Jeffrey Rubin, Ken Thomas, Masaaki Terada, Teruhiko Yoshida

As described extensively throughout this volume, there are at least 7 distinct gene products in the fibroblast growth factor (FGF) family which are each structurally related and which probably derive from a common ancestral gene.[1-8] It is only because the two original FGFs identified were first detected by their ability to stimulate fibroblast proliferation that they became known as acidic and basic fibroblast growth factors.[1,2,9,10] Since 1985, over three dozen activities described as chondrocyte growth factor, kidney angiogenic factor, adrenal growth factor, pituitary-derived growth factor, mesotropin, prostatropin, heparin-binding growth factor-1, heparin-binding growth factor-2, eye-derived growth factor, eye-derived growth factor 2, a-retina-derived growth factor, β-retina-derived growth factor, brain-derived growth factor, melanocyte growth factor, seminiferous growth factor, corpus luteum angiogenic factor and various other names were shown to be either one or another member of the "FGF" family. Because recombinant DNA techniques have identified another 5 related but distinct members of the family (int-2,[3] hst-1[11] also called HSTF1,[12] ks-FGF,[5] and K-FGF,[13] FGF-5,[6] FGF-6,[7] sometimes called hst-2[14] and most recently KGF[8]), a number of problems in nomenclature have arisen. Even interleukins 1α and 1β[15] are structurally related to the FGFs, though not actually considered members.

The following recommendations have been made:

1. A consensus was reached that, for historical reasons, these molecules should be described as "FGFs." It was concluded that there is a lack of any nomenclature that can accurately reflect the common structural and/or biological features of the known (and those possibly unknown) FGFs. Even the use of "heparin-binding growth factor" (HBGF) as described by Lobb *et al.*[16] was discouraged in view of the large number of unrelated molecules that bind heparin with quite high affinity. It is thus extremely important to emphasize that the use of the initials "FGF" does not imply that the protein has "fibroblast" growth stimulating activity. Indeed although several members of the FGF family can stimulate fibroblast proliferation *in vitro,* it is unlikely that this is their primary function *in vivo.* One member of the family (i.e., keratinocyte growth factor, KGF/FGF-7) has no known ability to stimulate fibroblast growth. Any functional association of the name "FGF" with "fibroblast" growth stimulating activity should thus be discouraged.

2. A consensus was reached that all the genes in this growth factor family be designated with the letters FGF and with a number (TABLE 1). It should be

[a]The initial meeting of this committee was held during the conference on January 16–18, 1991. The number of participants was subsequently expanded to develop a consensus.

xiii

TABLE 1

Common Historical Names	Generally Accepted Acronym	Proposed Names	Chromosome (Human)	Reference
Acidic fibroblast growth factor	Acidic FGF/aFGF	FGF-1	5	1,21
Endothelial cell growth factor	ECGF			
Heparin-binding growth factor-1	HBGF-1			
Basic fibroblast growth factor	Basic FGF	FGF-2	4q25	21,22
Heparin-binding growth factor-2	bFGF HBGF-2			
int-2	int-2	FGF-3	11q13	12,23,24
Kaposi sarcoma FGF Human stomach cancer transforming gene-1	ks FGF/K-FGF hst-1/HSTF-1	FGF-4	11q13	12,23,26
Fibroblast growth factor-5	hst-1 FGF-5 HSTF1	FGF-5	4q21	25
Fibroblast growth factor-6 Hst-1 related gene	FGF-6 hst-2	FGF-6	12p13	7
Keratinocyte growth factor	KGF	FGF-7		

noted that if additional members of the family are identified, they should be named sequentially. It is thus recommended that when cited in publications, a molecule like basic fibroblast growth factor be described as FGF-2.

3. It was also noted that individual genes in the FGF family can encode several molecular forms of their respective protein products.[17,18] For example, there are at least four different translation start sites in the gene for human basic FGF (*FGF-2*). Although the translation usually initiated at the AUG codon leads to the production of a 154-amino-acid protein, a 155-amino-acid protein is sometimes produced in recombinant expression systems and proteins with N-terminal amino acid extensions are produced when upstream CUG codons in their mRNAs are used as translation initiation sites.[17,18] This complexity is further exacerbated by the fact that FGFs isolated from tissues are sometimes truncated by enzymatic degradation. For example, basic FGF (FGF-2) has been characterized as a 131-amino-acid protein from ovary, kidney, and adrenal tissue, as a 146-amino-acid protein from brain and pituitary extracts, and as a 157-amino-acid protein from the placenta.[19,20] It is difficult to propose a specific nomenclature to designate the different forms of these FGFs, and it is emphasized that investigators should clearly define the specific sequence of the FGF used in their studies.

In conclusion, it is important to emphasize that the fibroblast growth factors belong to a common family because they are structurally and not necessarily biologically related. The use of this proposed nomenclature is recommended for immediate use recognizing that both historical and systematic names may be used during an appropriate interim.

REFERENCES

1. JAVE, M., R. HOWK, W. BURGESS et al. 1986. Human endothelial cell growth factor: cloning, nucleotide sequence, and chromosome localization. Science 233: 541–545.
2. ABRAHAM, J. A., A. MERGIA, J. L. WHANG, et al. 1986. Nucleotide sequence of a bovine clone encoding the angiogenic protein, basic fibroblast growth factor. Science 233: 545–548.
3. DICKSON, C. & G. PETERS. 1987. Potential oncogene product related to growth factors. Nature 326: 833.
4. TAIRA, M., T. YOSHIDA, K. MIYAGAWA, H. SAKAMOTO, M. TERADA & T. SUGIMURA. 1987. cDNA sequence of human transforming gene hst and identification of the coding sequence required for transforming activity. Proc. Nat. Acad. Sci. USA 84: 2980–2984.
5. DELLI BOVI, P., A. M. CURATOLA, F. G. KERN, A. GRECO, M. ITTMAN & C. BASILICO. 1987. An oncogene isolated by transfection of Kaposi's sarcoma DNA encodes a growth factor that is a member of the FGF family. Cell 50: 729–737.
6. ZHAN, X., B. BATES, X. HU & M. GOLDFARB. 1988. The human FGF-5 oncogene encodes a novel protein related to fibroblast growth factors. Mol. Cell. Biol. 8: 3487–3497.
7. MARICS, I., J. ADELAIDE, F. RAYBAUD, et al. 1989. Characterization of the HST-related FGF.6 gene, a new member of the fibroblast growth factor gene family. Oncogene 4: 335–340.
8. FINCH, P. W., J. S. RUBIN, T. MIKI, D. RON & S. A. AARONSON. 1989. Human KGF is FGF-related with properties of a paracrine effector of epithelial cell growth. Science 245: 752–755.
9. ESCH, F., A. BAIRD, N. LING, et al. 1985. Primary structure of bovine pituitary basic fibroblast growth factor (FGF) and comparison with the amino-terminal sequence of bovine acidic FGF. Proc. Nat. Acad. Sci. USA 82: 6507–6511.
10. GIMENEZ-GALLEGO, G., K. RODKEY, C. BENNETT, M. RIOS-CANDELORE, J. DISALVO & K. A. THOMAS. 1985. Brain-derived acidic fibroblast growth factor: complete amino acid sequence and homologies. Science 230: 1385–1388.
11. SAKAMOTO, H., M. MORI, M. TAIRA, T. YOSHIDA, S. MATSUKAWA, K. SHIMUZU, M. SEKIGUCHI, M. TERADA & S. SUGIMURA. 1986. Transforming gene from human stomach cancers and a non-cancerous portion of stomach mucosa. Proc. Nat. Acad. Sci. USA 83: 3997–4001.
12. YOSHIDA, M., M. WADA, H. SATOH, T. YOSHIDA, H. SAKAMOTO, K. MIYAGAWA, J. YOKOTA, T. KODA, M. KAKIMUNA, T. SUGIMURA & M. TERADA. 1988. Human HST1 (HSTF-1) maps to chromosome band 11q13 and coamplifies with the INT2 gene in human cancer. Proc. Nat. Acad. Sci. USA 85: 4861–4864.
13. BASILICO, C., K. M. NEWMAN, A. M. CURATOLA, et al. 1989. Expression and activation of the K-fgf oncogene. Ann. N.Y. Acad. Sci. 567: 95–103.
14. YOSHIDA, T., H. SAKAMOTO, K. MIYAGAWA, T. SUGIMURA & M. TERADA. 1991. Characterization of the hst-1 gene and its product. Ann. N.Y. Acad. Sci. (This volume).
15. SCHMIDT, J. A. & M. J. TOCCI. 1990. Interleukin-1. In Peptide Growth Factors. M. B. Sporn, A. B. Roberts, Eds.: 473–521. Springer-Verlag. New York, N.Y.
16. LOBB, R. R., J. W. HARPER & J.W. FETT. 1986. Purification of heparin-binding growth factors. Anal. Biochem. 154: 1–14.
17. FLORKIEWICZ, R. Z. & A. SOMMER. 1989. Human basic fibroblast growth factor gene encodes four polypeptides: three initiate translation from non-AUG codons. Proc. Nat. Acad. Sci. USA. 86: 3978–3981.
18. PRATS, H., M. KAGHAD, A. C. PRATS, et al. 1989. High molecular mass forms of basic fibroblast growth factor are initiated by alternative CUG codons. Proc. Nat. Acad. Sci. USA. 86: 1836–1840.
19. BURGESS, W. H. & T. MACIAG. 1989. The heparin-binding (fibroblast) growth factor family of proteins. Annu. Rev. Biochem. 58: 575–606.
20. BAIRD, A. & P. BOHLEN. Fibroblast growth factors. 1990. In Peptide Growth Factors. M. B. Sporn & A. B. Roberts, Eds.: 369–417. Springer-Verlag. New York, N.Y.
21. MERGIA, A., R. EDDY, J. A. ABRAHAM, J. C. FIDDES & T. B. SHOWS. 1986. The genes for

basic and acidic fibroblast growth factors are on different human chromosomes. Biochem. Biophys. Res. Commun. **138:** 644–651.

22. FUKUSHIMA, Y., M. G. BYERS, J. C. FIDDES & T. B. SHOWS. 1990. The human basic fibroblast growth factor gene (FGFB) is assigned to chromosome 4q25. Cytogenet. Cell Genet. **54:** 159–160.

23. CASEY, G., D. SMITH, D. MCGILLIVRAY, G. PETERS, G. DICKSON & C. DICKSON. 1986. Characterisation and chromosome assignment of the human homolog of int-2, a potential proto-oncogene. Mol. Cell. Biol. **6:** 502–510.

24. ADELAIDE, J., M-G. MATTEI, F. MARICS, J. RAYBAUD, J. PLANCHE, O. DE LAPEYRIERE & D. BIRNBAUM. 1988. Chromosomal localisation of the *hst* oncogene and its co-amplification with *int*.2 oncogene in a human melanoma. Oncogene **2:** 413–416.

25. NGUYEN, C., D. ROUX, M-G MATTEI, *et al.* 1988. The FGF-related oncogenes *hst* and *int*.2, and the *bcl*.1 locus are contained within one megabase in band q13 of chromosome 11, while the *fgf*.5 oncogene maps to 4q21. Oncogene **3:** 703–708.

26. HUEBNER, K., A. C. FERRARI, P. DELLI BOVI, C. CROCE & C. BASILICO. 1988. The FGF-related oncogene K-FGF maps to human chromosome region 11q13, possibly near *int*-2. Oncogene Res. **3:** 263–270.

Biological Activities of Fibroblast Growth Factors

DENIS GOSPODAROWICZ

University of California Medical Center
Cancer Research Institute
San Francisco, California 94143-1028

INTRODUCTION

The fibroblast growth factor family primarily consists of two closely related isoforms (basic and acidic FGF, bFGF and aFGF) whose primary roles are that of embryonic inducers.[1-3] Through their ability to interact with common receptors,[4] they exert similar effects on a wide variety of tissues, and have been implicated in controlling both cell proliferation and differentiation. As one would expect from embryonic inducers which are short range effectors,[5] FGFs are not humoral factors, but are integrated within the basement membrane of producing cells[6] where they function locally. In this review we primarily focused on early studies that led to the recognition that bFGF could affect the development of various tissues for which it acts both as a mitogen and a morphogen.

FGF AND CELL PROLIFERATION

FGFs have been shown to stimulate the proliferation of all mesoderm-derived cells tested to date.[8] Their activity was first believed to be restricted to tissues derived from that germ cell layer. Later studies have shown that they could also stimulate the proliferation of neuroectodermal cells (corneal endothelial cells) and ectodermal cells (lens cells).[9] Since thyroid cells, prostatic cells, and pancreatic cells which are derived from the endoderm also respond to FGFs, it is likely that tissues derived from the three embryonic germ layers (mesoderm, ectoderm, and endoderm) are FGF sensitive (reviewed in Reference 10). The cells' sensitivity to the FGF isoforms differ, with in most cases bFGF being 10- to 30-fold more active than aFGF.[11] However, this is not a general rule since their difference in potency may depend on the cell types. In the case of Balb/c 3T3 cells both mitogens have been reported to be nearly equipotent,[12] while in the case of the MK keratinocyte cell line, aFGF is more efficient than bFGF as a mitogen.[13] In addition, agents unrelated to FGF such as heparin and heparan sulfate have been reported, depending on the target cells, to potentiate the bioactivity of aFGF. Differences in bioactivity between two related forms of a given growth factor have already been observed. In the case of PDGF or TGFβ, their isoforms, depending on the target cells, do differ in activity.[14,15]

FGF alone can induce morphological cell transformation. This effect is likely to reflect the FGF's ability to modulate the synthesis of extracellular matrix (ECM) components and to alter the cytoskeleton organization. The transforming effect of bFGF, best seen with the Balb/c 3T3 cell line,[16] is not permanent and disappears once bFGF is removed. Consistent with its effect on cell transformation, bFGF can also support the growth of cells in the anchorage-independent configuration.[17,18] In

1

addition to being a mitogen, bFGF has also been shown to have two other properties that may be related to its mitogenic effects. It can act as a survival factor for cells seeded at clonal density, in particular for endothelial cells,[19] and does prevent cell lysis when cells are maintained in serum-free conditions under stressful conditions. This effect is best seen with vascular smooth muscle or granulosa cells which, when exposed to defined medium supplemented with transferrin and insulin alone, do lyse unless bFGF is present. bFGF also increases the life span of cultured cells, the effect being best seen with vascular endothelial cells and lens cells.[9,20]

One practical effect of the growth-promoting effect of bFGF has been the development of clonal cell strains derived from various tissue that are FGF sensitive.[21] This was particularly important in cases where those cells did not respond to serum or plasma factors but did depend uniquely on bFGF in order to proliferate *in vitro* (Reviewed in Reference 21). Another practical aspect has been the elucidation of the plasma factors cooperating with bFGF, to support dividing cells, through the cell cycle.

In the case of vascular endothelial cells it has been reported that these factors consist of transferrin and high-density lipoproteins (HDLs).[22] In their presence, bovine vascular endothelial cells exposed to FGF can be serially passaged in defined medium and the cells exhibit all of the phenotypic properties of counterpart cultures passaged in the presence of serum and FGF.[23] The requirement for transferrin is related to the fact that iron is an essential requirement for the enzyme ribonucleotide reductase whose activity in turn is strongly related with the rate of DNA synthesis and is greatly increased during the S phase.[24] The requirement for HDLs is related to their abilities to stimulate the cellular enzyme HMG CoA reductase.[25,26] This results in increased mevalonate synthesis which is further used primarily for cholesterol synthesis and secondarily for the synthesis of isopentenyl adenosine, ubiquinone, dolichols, and farnesyl pyrophosphate.

Recent studies have demonstrated that posttranslational modifications, by farnesyl pyrophosphate, of proteins involved in growth control are important for their correct insertion in plasma membrane, where they play the role of G proteins.[27–29] Schafer *et al.* have shown that in the yeast, farnesylation of the α mating factors is required both for their secretion as well as their activity[27] while in the case of the p21ras, modification by farnesyl isoprenoid is required for biological activity.[28] Other G proteins as well have by now been shown to be farnesylated.[29] This raises the possibility that with vascular endothelial cells as well as other cell types the effect of HDL on cell growth and shape[25,26] could be mediated through their ability to trigger the formation of a critical pool of farnesyl pyrophosphate belonging to the nonsterol branching isoprene pathway.

FGF AND CELL DIFFERENTIATION

It was with bFGF that it was demonstrated, for the first time, that a growth factor could act both as a mitogen and a morphogen, two properties previously thought to be mutually exclusive.[30] The most popular models for the study of FGF on cells differentiation have been myoblasts, chondrocytes, osteoblasts, nerve cells, glial cells, and vascular endothelial cells (reviewed in Reference 10). It is with vascular endothelial cells seeded at low density and passaged repeatedly in the presence of serum-supplemented medium with or without FGF[31,32] that the effect of FGF on cell differentiation was first examined. After a few passages, cultures maintained in the absence of FGF lost their ability to form at confluence a contact-inhibited cell monolayer with a nonthrombogenic apical surface. Cells instead did acquire a fibroblast phenotype and grew in multiple cell layers with a thrombogenic apical surface. The addition of FGF to such subconfluent cultures did revert their apparent dedifferentiation.[32] This resulted from the ability of FGF to modulate the synthesis as well as the vectorial deposition of various ECM components, including collagen types I and III, fibronectin, and laminin.[32-35] bFGF is known to repress the synthesis of collagen type I, a type of collagen normally produced by fibroblasts, as well as that of fibronectin.[32,33] It has also been shown to direct the deposition of ECM components toward the basal cell surface.[34] The bFGF ability to stimulate the activity of various proteases and collagenase which are cell surface associated[36] could have led to the disappearance of the ECM covering the apical cell surface of the dedifferentiated cells. This resulted within 24 to 48 hours into a reorganization of the fibroblastic looking cells into a cell monolayer configuration with a nonthrombogenic apical cell surface, characteristic of vascular endothelial cells.

In the case of chondrocytes, similar effects of FGF on cell differentiation have been reported.[37,38] In contrast to vascular endothelial cells, bFGF needs to be present during the whole period of their growth phase in order for chondrocytes to express their differentiated phenotype. Although the reasons for this are presently unknown, the effect of bFGF on chondrocyte differentiation is best explained by its ability to inhibit fibronectin and collagen type I synthesis. Both of these ECM components, if produced by chondrocytes, would induce a fibroblast phenotype, mostly because they would promote the flattening of chondrocytes on the plastic surface and would make them proliferate in a multilayered configuration. In addition bFGF does promote the synthesis of collagen type II and that of chondroitin sulfate proteoglycan to the expense of small ubiquitous proteoglycans.[37,39] This results in the formation of an ECM that has a composition and organization like that seen *in vivo* with embedded chondrocytes, exhibiting the rounded phenotype characteristics of chondrocytes *in vivo*.[37,39]

The effect of bFGF on cell differentiation has in most cases been shown to be a positive effect, although in some cases it may be paradoxal. By example, in the case of myoblasts, bFGF triggers their proliferation but does delay their fusion into myotubes.[40] This can be seen morphologically and is reflected in a delay in the appearance of acetylcholine binding sites which would signal the formation of myotubes.[40] Although this could be interpreted as indicating that bFGF blocks myoblast differentiation, such bFGF properties would be advantageous *in vivo*. Since myoblast fusion in the presence of bFGF would take place at a much higher cell density than in its absence, it would result in the formation of much larger multinucleated myotubes (or muscle fibers) in the presence of bFGF than in its absence.[40] (Other aspects of the effect of bFGF on cell differentiation are discussed in other papers.)

FGF IN EARLY EMBRYOGENESIS

The pleiotropic effects of FGF on multiple cell types of mesodermal origin could reflect the ability of FGF to act as a mesoderm inducer. Both FGF isoforms can induce the formation of mesodermal tissues when applied to ectoderm explants isolated from late blastulae of *Xenopus laevis*.[1-3] At low concentration the induced differentiation consists of concentric arrangements of loose mesenchyme and mesothelium within an epidermal jacket, while at high concentration most of the explants contain a significant number of muscle blocks.[1] When the effect of bFGF is tested in *Xenopus* gastrula ectoderm, it induces both blood islands and muscle formation. In some cases the muscle had the appearance of skeletal muscle, while in others it has the appearance of heart muscle.[41] In the case of ectoderm of *Triturus alpestris* gastrula,[42] mesodermal tissues are induced at a lower percentage whereas the induction of neural tissues is considerably increased. bFGF is known to be present in oocytes.[43] While bFGF mRNA levels are high in oocytes, they drop by 25-fold during oocyte maturation and abruptly increase at the midblastulae transition, a stage during which expression of the vegetalizing factor would be expected. In addition the concentration of bFGF is sufficiently high for it to be active as an endogenous inducer of ventral mesoderm.[44] This coupled with the appearance of FGF receptors in ectodermal tissues which closely parallels the developmental competence of ectodermal cells to respond to bFGF would tend to demonstrate that it can act as an embryonic inducer responsible for the formation of ventral mesoderm.[45] Although the mechanism by which bFGF can induce mesoderm formation is unknown, it is intriguing to note that exposure of animal pole explants to bFGF rapidly induces the expression of the homeobox-containing gene Xhox-3[46] and exposure to bFGF in conjunction with TGFβ-1 or -2 induces expression of another homeobox-containing gene, Mix.1.[47] These genes are putative transcription factors, and there is evidence to suggest that Xhox-3 may be involved in anterior-posterior axial specification.[46]

FGF AND THE EXTRACELLULAR MATRIX

Embryonic induction is defined as developmentally significant interaction between closely associated but dissimilarly derived tissue masses.[48] Its "mechanism" as revealed by tissue isolation technique involves the ECM, components of which could bind morphogenetic factors that promote or stabilize differentation of embryonic tissues.[5] Recent studies suggest that FGF could be among such morphogenetic factors.

In 1980 it was first reported that for cells sensitive to FGF, their growth factor requirement could be satisfied by growing them on an ECM produced by bovine corneal endothelial cells.[48] When FGF-sensitive cells were maintained on such a substrate, not only did they proliferate at an optimal rate but they also exhibited their proper phenotype.[49] In fact it was probably with that substrate that the neurotrophic effects of FGF were first recognized.[50] The nature of the factor(s) involved and their relationship to bFGF were then unknown, but the evidence presented did link the activity to a heparan sulfate proteoglycan complex with a small polypeptide, presumably bFGF.[51]

The putative ECM component(s) responsible for mimicking the effect of bFGF has recently been reported to be bFGF–heparan sulfate proteoglycan complexes.[52-53] Such complexes are an integral part of the ECM produced by either vascular or corneal endothelial cell ECM.[53,54] Indirect evidence for the integration of bFGF into an insoluble substrate such as the ECM can be derived from the observation that media conditioned by cell types such as capillary or corneal endothelial cells, which have been shown to produce bFGF,[54-56] have no significant impact on their proliferation.[54-56] In contrast, seeding those cells on their own denuded ECM will induce them to proliferate rapidly and to express their correct phenotype once confluent.[56] This suggests that bFGF, in contrast to other conventional growth factors, is not released in a soluble form but becomes an integral part of their insoluble substratum. Vlodavsky et al. have shown that corneal endothelial cell ECM contains a growth factor that, on the basis of its immunoreactivity and heparin affinity, seems to be closely related to bFGF,[53] and it has been reported that endothelial-cell-derived heparan sulfate can bind bFGF and protects it from proteolytic degradation.[57] Recent studies using neutralizing antibodies directed against bFGF have provided evidence that ECM-derived bFGF was solely responsible for its mitogenic activity on mesoderm-derived cells.[58] FGF, through integration into the ECM produced by these cells, could act as a local growth regulator and induce the regeneration of these tissues following wounding. Hydrolysis of the ECM does result in the liberation of heparan sulfate–FGF complexes that are biologically active.[57] In this context, it is interesting to note that during morphogenesis, the areas of greatest mitotic activity are located where the hydrolysis of ECM occurs.[59] Similarly, in the case of the embryonic kidney, angiogenesis correlates with the hydrolysis of the kidney mesenchymal stroma.[60] In the adult phase of development, heparan sulfate glycosaminoglycans present in an insoluble form in the ECM could be solubilized by a proteoglycosidase such as heparanase. This is an inducible enzyme produced either by platelets, when they attach to the subendothelium, or among other cell types by macrophages once they are activated.[61] This could ultimately lead to the solubilization of heparan sulfate–FGF complexes that would be biologically active, and could participate in various repair processes, including wound healing. The localization of FGF in the ECM, where it could be considered as an integral part of that structure and where it is protected from both proteolytic degradation and inactivation, is in agreement with its early role as an embryonic inducer.[1-3]

CONCLUSION

Since their first identification in the midseventies,[62,63] both isoforms of FGF have had a significant impact on our understanding of developmental processes taking place during early embryogenesis or tissue remodeling. A typical example could be the impact of FGF on the field of vasculogenesis and angiogenesis where bFGF was the first identified growth factor for vascular endothelial cells and for which it serves as an autocrine growth factor, controlling the proliferation, migration, and differentation of that cell type (reviewed in Reference 10). Following the cloning of the genes for the two FGF isoforms and the recognition that bFGF and aFGF were structurally related, the literature related to FGF has expanded enormously. First came the realization that FGF was part of a larger gene family now composed of seven members,[36] and aspects of FGF biology have become intertwined with such diverse areas as tumor virus biology, the biology of early embryonic events, the biology of the central and peripheric nervous systems as well as that of sensory organs, and the pathology of atherosclerosis.

Those unexpected and totally new aspects of the biology of FGF have expanded our basic understanding of the role of FGF in the control of development, and it is likely that in the years to come, other unexpected findings on the biology of FGF could expand our horizons in regard to the complexities involved in the control of cell proliferation and differentiation.

REFERENCES

1. SLACK, J. M. W., B. G. DARLINGTON, J. K. HEATH & S. F. GODSAVE. 1987. Mesoderm induction in early *Xenopus* embryos by heparin binding growth factors. Nature London **326:** 197–200.
2. KIMMELMAN, D. & M. KIRSCHNER. 1987. Synergistic induction of mesoderm by FGF and TGFβ and the identification of an mRNA coding for FGF in the early *Xenopus* embryo. Cell **51:** 869–877.
3. KNOCHEL, W. & H. TIEDEMANN. 1989. Embryonic inducers, growth factors, transcription factors and oncogenes. Cell Differ. Dev. **26:** 163–171.

4. NEUFELD, G. & D. GOSPODAROWICZ. 1986. Basic and acidic fibroblast growth factor interacts with the same cell surface receptor. J. Biol. Chem. **261:** 5631–5637.
5. HAY, E. D. 1981. Extracellular matrix. J. Cell Biol. **91:** 205–223.
6. GONZALEZ, A. M., M. BUSCAGLIA, M. ONG & A. BAIRD. 1990. Distribution of basic fibroblast growth factor in the 18 day rat fetus: localization in the basement membranes of diverse tissues. J. Cell Biol. **110:** 753–765.
7. KALCHEIM, C. & G. NEUFELD. 1990. Expression of basic fibroblast growth factor in the nervous system of early avian embryos. Development **109:** 203–215.
8. GOSPODAROWICZ, D., P. RUDLAND, J. LINDSTROM & K. BENIRSCHKE. 1975. Fibroblast growth factor: localization, purification, mode of action, and physiological significance. Nobel Symposium on Growth Factors. Adv. Metab. Dis. **8:** 302–335.
9. GOSPODAROWICZ, D., G. GREENBURG, H. BIALECKI & B. ZETTER. 1978. Factors involved in the modulation of cell proliferation in vivo and in vitro: the role of fibroblast and epidermal growth factors in the proliferative response of mammalian cells. In Vitro **14:** 85–118.
10. GOSPODAROWICZ, D. 1990. Fibroblast growth factor and its involvement in developmental processes. Curr. Top. Dev. Biol. **24:** 57–93.
11. GOSPODAROWICZ, D. 1987. Purification of brain and pituitary FGF. Methods Enzymol. **147B:** 106–119.
12. LOBB, R. R., J. W. HARPER & J. W. FETT. 1986. Purification of heparin binding growth factors. Anal. Biochem. **154:** 1–14.
13. GOSPODAROWICZ, D., J. PLOUËT & B. MALERSTEIN. 1990. Comparison of the ability of basic and acidic fibroblast growth factor to stimulate the proliferation of an established keratinocyte cell lines: modulation of their biological effects by heparin, transforming growth factor β and epidermal growth factor. J. Cell. Physiol. **142:** 325–333.
14. BECKMANN, M. P., C. BETSHOLZ, C. H. HELDIN, B. WESTERMARK, E. DiMARCO, F. P. DiFORE, K. C. ROBBINS & S. A. AARONSON. 1988. Comparison of biological properties and transforming potential of human PDGFA and PDGFB chains. Science **241:** 1346–1349.
15. ROBERTS, A., P. KONDAIAH, F. ROSA, S. WATANABE, P. GOOD, D. DANIELPACE, N. S. ROCHE, M. ROBERT, I. DAVID & M. SPORN. 1990. Mesoderm induction in *Xenopus laevis* distinguishes between the various TGFβ isoforms. Growth Factors **3:** 277–286.
16. GOSPODAROWICZ, D. & J. MORAN. 1974. Effect of a fibroblast growth factor, insulin, dexamethasone, and serum on the morphology of BALB/c 3T3 cells. Proc. Nat. Acad. Sci. USA **71:** 4648–4652.
17. NEUFELD, G., R. MITCHELL, P. PONTE & D. GOSPODAROWICZ. 1988. Expression of human basic fibroblast growth factor cDNA in baby hamster kidney-derived cells results in autonomous cell growth. J. Cell Biol. **106:** 1385–1396.
18. RIZZINO, A. & E. RUFF. 1986. Fibroblast growth factor induces the soft agar growth of 2 non-transformed cell lines. In Vitro **22:** 749–755.
19. GOSPODAROWICZ, D., J. MORAN, D. BRAUN & C. R. BIRDWELL. 1976. Clonal growth of bovine endothelial cells in tissue cultures: fibroblast growth factor as a survival agent. Proc. Nat. Acad. Sci. USA **73:** 4120–4124.
20. GOSPODAROWICZ, D. & S. L. MASSOGLIA. 1982. Plasma factors involved in the in vitro control of proliferation of bovine lens cells grown in defined medium. Effect of fibroblast growth factor on cell longevity. Exp. Eye Res. **35:** 259–270.
21. GOSPODAROWICZ, D. & B. ZETTER. 1977. The use of fibroblast and epidermal growth factors to lower the serum requirement for growth of normal diploid cells in early passage: a new method for cloning. Joint WHO/IABS Symposium on the Standardization of Cell Substrates for the Production of Virus Vaccine, Geneva: 109–130. S. Karger. Basel, Switzerland.
22. TAUBER, J.-P., J. CHENG & D. GOSPODAROWICZ. 1980. The effect of high and low density lipoproteins on the proliferation of vascular endothelial cells. J. Clin. Inves. **66:** 696–708.
23. TAUBER, J.-P., J. CHENG, S. MASSOGLIA & D. GOSPODAROWICZ. 1981. High density lipoproteins and the growth of vascular endothelial cells in serum-free medium. In Vitro **17:** 519–530.
24. THELANDER, L. & P. REICHARD. 1979. Transferrin. Annu. Rev. Biochem. **48:** 133–153.

25. COHEN, D. C., S. L. MASSOGLIA & D. GOSPODAROWICZ. 1982. Correlation between two effects of high density lipoproteins on vascular endothelial cells: the induction of 3-hydroxy-3-methylglutaryl coenzyme A reductase activity and the support of cellular proliferation. J. Biol. Chem. **257:** 9429–9437.

26. COHEN, D. C., S. MASSOGLIA & D. GOSPODAROWICZ. 1982. Feedback regulation of 3-hydroxy-3-methylglutaryl coenzyme A reductase in vascular endothelial cells: separate sterol and non-sterol components. J. Biol. Chem. **257:** 11106–11112.

27. SCHAFER, W. R., R. KIM, R. STERN, J. THORNER, S. H. KIME & J. KINE. 1989. Genetic and pharmacological suppression of oncogenic mutations in ras gene of yeast and human. Science **245:** 379–384.

28. CASEY, P. J., P. A. SALSKI, C. J. DER & J. E. BUSS. 1989. P21ras is modified by a farnesyl isoprenoid. Proc. Nat. Acad. Sci. USA **86:** 8323–8327.

29. FUKADA, Y. T., T. TAKAO, H. OHGURO, T. YOSHIZAWA, T. AKINO & Y. SIMONISHI. 1990. Farnesylated j subunit of photoreceptor G protein indispensable for GTP binding. Nature **346:** 658–660.

30. GOSPODAROWICZ, D., I. VLODAVSKY, N. SAVION & J.-P. TAUBER. 1980. The control of proliferation and differentiation of vascular endothelial cells by fibroblast growth factor. *In* Peptides: Integration of Cells and Tissue Functions. Society of General Physiologists Series. F. Bloom, Ed. **35:** 1–38. Raven Press. New York, N.Y.

31. GOSPODAROWICZ, D. 1976. Humoral control of cell proliferation: the role of fibroblast growth factor in regeneration. Angiogenesis, wound healing, and neoplastic growth. *In* Progress in Clinical and Biological Research. Membranes and Neoplasia: New Approaches and Strategies. V. T. Marchesi, Ed.: 1–19. A.R. Liss, Inc. New York, N.Y.

32. VLODAVSKY, I. & D. GOSPODAROWICZ. 1979. Structural and functional alterations in the surface of vascular endothelial cells associated with the formation of a confluent cell monolayer and with the withdrawal of fibroblast growth factor. J. Supramol. Struct. **12:** 73–114.

33. TSENG, S. C. G., N. SAVION, R. STERN & D. GOSPODAROWICZ. 1982. Fibroblast growth factor modulates synthesis of collagen in cultured vascular endothelial cells. Eur. J. Biochem. **122:** 355–360.

34. GREENBURG, G., I. VLODAVSKY, J. M. FOIDART & D. GOSPODAROWICZ. 1980. Conditioned medium from endothelial cell cultures can restore the normal phenotypic expression of vascular endothelium maintained in vitro in the absence of fibroblast growth factor. J. Cell. Physiol. **103:** 333–347.

35. GOSPODAROWICZ, D., G. GREENBURG, J. M. FOIDART & N. SAVION. 1981. The production and localization of laminin in cultured vascular and corneal endothelial cells. J. Cell. Physiol. **107:** 173–183.

36. RIFKIN, D. B. & D. MOSCATELLI. 1980. Recent developments in the cell biology of basic fibroblast growth factor. J. Cell Biol. **109:** 1–6.

37. KATO, Y. & D. GOSPODAROWICZ. 1985. Sulfated proteoglycan synthesis by rabbit costal chondrocytes grown in the presence and absence of fibroblast growth factor. J. Cell Biol. **100:** 477–485.

38. KATO, Y. & D. GOSPODAROWICZ. 1985. Effect of extracellular matrices on proteoglycan synthesis by cultured rabbit costal chondrocytes. J. Cell Biol. **100:** 486–495.

39. KATO, Y. & D. GOSPODAROWICZ. 1986. Comparison of the effect in vitro of fibroblast growth factor and exogenous extracellular matrices on the proliferation and phenotypic expression of chondrocytes in vitro. *In* Coordinated Regulation of Gene Expression R. M. Clayton & D. E. S. Truman, Eds.: 249–267. Plenum Press. New York, N.Y.

40. GOSPODAROWICZ, D., J. WESEMAN, J. MORAN & J. LINDSTROM. 1976. Effect of fibroblast growth factor on the division and fusion of bovine myoblasts. J. Cell Biol. **70:** 395–405.

41. KNOCHEL, W., H. GRUNZ, B. LOPPNOW-BLINDE, H. TIEDEMANN & H. TIEDEMANN. 1989. Mesoderm induction and blood island formation by angiogenic growth factors and embryonic inducing factors. Blut **59:** 207–213.

42. KNOCHEL, W., W. GRUNZ, B. LOPPNOW-BLINDE, A. POTING, H. TIEDMANN & H. TIEDEMANN. Mesoderm induction by FGFs and TGFs. Naturwissen-chaften **28.** (In press.)

43. KIMMELMAN, D., J. A. ABRAHAM, T. HAAPARANTA, T. M. PALISI & M. KIRSCHNER. 1988. The presence of fibroblast growth factor in the frog egg: its role as a natural mesoderm inducer. Science (Washington, D.C.) **242:** 1053–1056.

44. SLACK, J. M. W. & H. V. ISAACS. 1989. Presence of basic fibroblast growth factor in the early *Xenopus* embryo. Development **105:** 147–153.

45. GILLESPIE, L. L., G. D. PATERNO & J. H. W. SLACK. 1989. Analysis of competence: receptors for basic fibroblast growth factor in early *Xenopus* embryo. Development **106:** 203–208.

46. RUIZ, I., A. ALTABA & D. A. MELTON. 1989. Interaction between peptide growth factors and homeobox genes in the establishment of anterio posterior polarity of frog embryos. Nature London **341:** 33–38.

47. ROSA, F. 1989. Mix. 1 a homeobox mRNA inducible by mesoderm inducers is expressed mostly in the presumptive endodermal cells of *Xenopus* embryos. Cell Cambridge, Mass. **57:** 965–974.

48. GOSPODAROWICZ, D., D. DELGADO & I. VLODAVSKY. 1980. Permissive effect of the extracellular matrix on cell proliferation in vitro. Proc. Nat. Acad. Sci. USA **77:** 4094–4098.

49. GOSPODAROWICZ, D. & G.-M. LUI. 1981. Effect of substrata and fibroblast growth factor on the proliferation in vitro of bovine aortic endothelial cells. J. Cell. Physiol. **109:** 69–81.

50. FUJII, D. K., S. L. MASSOGLIA, N. SAVION & D. GOSPODAROWICZ. 1982. Neurite outgrowth and protein synthesis by PC-12 cells as a function of substratum and nerve growth factor. J. Neurosci. **2:** 1157–1175.

51. LANDER, A. D., D. K. FUJII, D. GOSPODAROWICZ & L. F. REICHARDT. 1982. Characterization of a factor that promotes neurite outgrowth: evidence linking the activity to a heparan sulfate proteoglycan. J. Cell Biol. **94:** 574–584.

52. BAIRD, A. & A. LING. 1987. Fibroblast growth factors are present in the extracellular matrix produced by endothelial cells in vitro: implications for a role of heparinase-like enzymes in the neovascular response. Biochem. Biophys. Res. Commun. **142:** 428–435.

53. VLODAVSKY, I., J. FOLKMAN, R. SULLIVAN, R. FRIDMAN, R. ISHAI-MICHAELI, J. SASSE & M. KLAGSBRUN. 1987. Endothelial cell derived basic fibroblast growth factor: synthesis and deposition into subendothelial extracellular matrix. Proc. Nat. Acad. Sci. USA **84:** 2292–2296.

54. SCHWEIGERER, L., G. NEUFELD, J. FRIEDMAN, J. A. ABRAHAM, J. C. FIDDES & D. GOSPODAROWICZ. 1987. Basic fibroblast growth factor: production and growth stimulation in cultured adrenal cortex cells. Endocrinology **120:** 796–800.

55. SCHWEIGERER, L., N. FERRARA, G. NEUFELD & D. GOSPODAROWICZ. 1988. Basic fibroblast growth factor: expression in cultured cells derived from corneal endothelium and lens epithelium. Exp. Eye Res. **46:** 71–80.

56. GOSPODAROWICZ, D. & J.-P. TAUBER. 1980. Growth factors and extracellular matrix. Endocr. Rev. **1:** 201–227.

57. SAKSELA, O., D. MOSCATELLI, A. SOMMER & D. B. RIFKIN. 1988. Endothelial cell–derived heparan sulfate binds basic fibroblast growth factor and protects it from proteolytic degradation. J. Cell Biol. **107:** 743–751.

58. GLOBUS, R., J. PLOUËT & D. GOSPODAROWICZ. 1989. Cultured bovine bone cells synthesize fibroblast growth factor and store it in their extracellular matrix. Endocrinology **124:** 1539–1547.

59. BERNFIELD, M., S. D. BANERJEE, J. E. KODA & A. C. RAPRAEGER. 1984. Remodelling of the basement membrane as a mechanism of morphogenetic interaction. *In* The Role of Extracellular Matrix in Development. R. L. Trelstad, Ed.: 545–572. Allan R. Liss, Inc. New York, N.Y.

60. EKBLOM, P. 1984. Basement membrane proteins and growth factors in kidney differentiation. *In* The Role of Extracellular Matrix in Development. R. L. Trelstad, Ed.: 173–206. Alan R. Liss, Inc. New York, N.Y.

61. MATZNER, Y., M. BAR NER, J. YAHALOM, R. ISHAI-MICHALES, Z. FUKS & I. VLODAVSKY. 1985. Degradation of heparan sulfate in the subendothelial matrix by readily released heparinase from human neutrophils. J. Clin. Invest. **76:** 1306–1316.

62. GOSPODAROWICZ, D. 1974. Localization of a fibroblast growth factor and its effect alone and with hydrocortisone on 3T3 cell growth. Nature **249:** 123–127.

63. GOSPODAROWICZ, D., J. WESEMAN & J. MORAN. 1975. Presence in the brain of a mitogenic agent distinct from fibroblast growth factor that promotes the proliferation of myoblasts in low density culture. Nature **256:** 216–219.

Structural Modifications of Acidic Fibroblast Growth Factor Alter Activity, Stability, and Heparin Dependence

KENNETH A. THOMAS,[a] SAGRARIO ORTEGA,
DENIS SODERMAN, MARIE-THERESE SCHAEFFER,
JERRY DiSALVO, GUILLERMO GIMENEZ-GALLEGO,
DAVID LINEMEYER, LINDA KELLY,
AND JOHN MENKE

Department of Biochemistry
Merck Research Laboratories
Post Office Box 2000
Rahway, New Jersey 07065

The seven known members of the fibroblast growth factor (FGF) family of homologous protein mitogens (aFGF, bFGF, INT-2, HST/K-FGF, FGF-5, FGF-6, and KGF) are from approximately 35% to 80% identical[1] including two conserved cysteine residues as shown in FIGURE 1. The function of these Cys residues and the selective evolutionary pressure for their conservation in all characterized FGFs are unknown. The conserved pair of Cys residues occurs at positions 16 and 83 in the 140-amino-acid numbering system of aFGF. The amino acid sequences of bovine and human aFGFs also reveal unique third Cys residues at positions 47 and 117, respectively.[2,3] We have characterized the influence of these three Cys residues on the activity, stability, and heparin dependence of human recombinant aFGF[4,5] by a combination of chemical modification and site-directed mutagenesis.[5,6]

ACTIVE aFGF DOES NOT CONTAIN AN INTRAMOLECULAR DISULFIDE BOND

Since conserved Cys residues in proteins are frequently present in stabilizing disulfide bonds, we first determined the oxidation state of the Cys residues in bovine-brain-derived and human recombinant aFGFs by a double alkylation technique. The free Cys residues of both proteins were rapidly trapped with iodoacetic acid, repurified by reversed phase high performance liquid chromatography (HPLC), reduced with dithiothreitol, and realkylated with iodo[14C]acetic acid. Bovine and human aFGFs were then digested with trypsin and *Staphylococcus aureus* V8 protease, respectively, and the resulting peptides purified by reversed phase HPLC. Each of the 3 Cys residues was present in a separate polypeptide. The Cys-containing peptides were each acid hydrolyzed, derivatized with phenylisothiocyanate, and the absolute amount and radioactivity of phenylthiocarbamoyl-carboxymethylcysteine (PTC-CMCys) determined. Only those Cys residues that became accessible to alkylation as a result of prior reduction were radiolabeled. The percent of each Cys

[a] Address correspondence to Dr. Thomas at Room 80W-243.

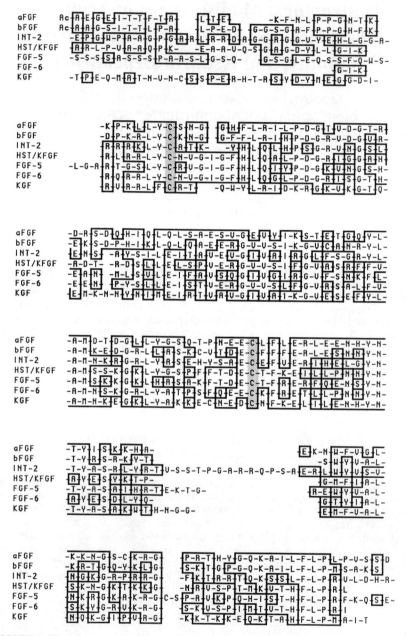

FIGURE 1. Homologies among known FGFs in region of common overlap. The seven known human FGF sequences are listed with identical residues boxed. The FGF-6 sequence is incomplete at the amino terminus. The two conserved Cys residues are denoted by stippled backgrounds.

that was radiolabeled was calculated based on the known specific radioactivity of the iodoacetic acid used for derivatization. The percent of each possible intramolecular disulfide bond could be calculated from the accessibility of the three Cys residues by solving the appropriate simultaneous equations. All three Cys residues in both bovine-brain-derived and human recombinant aFGFs are nearly fully reduced, containing only trace amounts of each possible intramolecular disulfide bond,[5] as documented in TABLE 1. This result is surprising considering that most extracellularly active proteins, including many growth factors, contain disulfide bonds that presumably contribute to their stabilization following release.

AN INTRAMOLECULAR DISULFIDE BOND IS NOT REQUIRED FOR ACTIVITY

Since aFGF has no secretory leader sequence, the material purified from bovine brain might be largely derived from the highly reducing intracellular pools. Similarly, human recombinant aFGF produced in *Escherichia coli* is synthesized and stored

TABLE 1. Chemical Analysis of Wild-Type aFGF Disulfide Bond Status[a]

Source of Mitogen	Percent Type						
	[^{14}C]PTC-CMCys			Disulfide Form			
	C_{16}	C_{83}	C_{NC}[b]	$C_{16,83}$	$C_{16,NC}$	$C_{83,NC}$	N[c]
Human[d]	4	7	5	3	1	4	92
Bovine[d]	2	2	3	1	1	1	97

[a]From Reference 5, with permission.
[b]NC denotes the nonconserved Cys at position 117 in human and 47 in bovine aFGF (140-amino-acid numbering system).
[c]N is the percent of mitogen in a nondisulfide bonded form.
[d]Human aFGF is recombinant whereas bovine aFGF is purified from brain.

under intracellular reducing conditions. A disulfide bond could be formed, however, upon oxidative activation either following release *in vivo* or during the long-duration mitogenesis assay in tissue culture. To test if a disulfide bond is required for mitogenic activity, a series of site-directed mutants of human aFGF in which any one,[5] two, or all three[6] Cys residues were substituted by serines, as depicted in FIGURE 2, was constructed, expressed, and purified to apparent homogeneity. As shown in FIGURE 3, all mutants were highly active as determined by full dose-response assays on confluent Balb/c 3T3 cells in serum-free medium. Furthermore, the forms that contain either only one or no Cys residues were the most active and least heparin dependent. Clearly, an intracellular disulfide bond is not required for mitogenic activity *in vitro*.

INTRAMOLECULAR DISULFIDE BOND FORMATION INACTIVATES aFGF

Tissue culture conditions are somewhat artificial in that multiple proteolytic inactivation mechanisms for extracellular proteins are not necessarily either func-

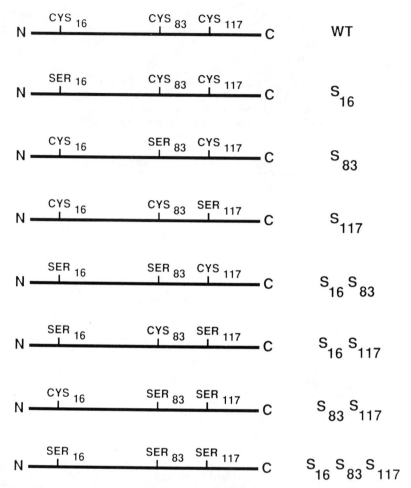

FIGURE 2. Wild-type and mutant aFGFs. The locations of the 3 Cys residues at amino acid positions 16, 83, and 117 in human WT aFGF and the Ser substitutions in the mutants are illustrated. Mutant nomenclatures denote the positions of the Ser (S) substitutions in subscripts. The 140-amino-acid-residue numbering system is used with the amino (N) and carboxyl (C) terminal ends labeled.[6]

tional or even present. Therefore, while an intramolecular disulfide bond might not be required for activity in culture, it could influence the stability of aFGF *in vivo*. If a stabilizing disulfide bond is formed then it might be expected to occur between the two evolutionarily conserved Cys residues. To test the effect of disulfide bond formation between these two Cys residues, the mutant in which the single nonconserved cysteine was converted to a serine (S_{117}) was oxidized by Cu^{2+} to force disulfide bond formation, repurified and the monomeric state confirmed by nonreducing sodium dodecyl sulphate-polyacrylamide gel electrophoresis (SDS-PAGE). Reduced and oxidized S_{117} was resolved by C_4 reversed phase HPLC, as shown in

4A, and assayed for induction of DNA synthesis in serum-free medium on confluent Balb/c 3T3 cells. Greater than 99% of the mitogenic activity was lost following oxidation but could be fully restored by subsequent reduction with dithiothreitol as shown in FIGURE 4B. Similar results were obtained with the mutants containing either of the other two pairs of Cys residues. The accessibility of the two conserved Cys residues in the S_{117} mutant, determined in both the presence and absence of the denaturant 6 M guanidinium chloride by the previously described double alkylation technique, was eliminated by oxidation and fully restored by subsequent reduction (TABLE 2).[5] Therefore, an intramolecular disulfide bond is neither required, present, nor even compatible with the mitogenic activity of aFGF.

CYSTEINE RESIDUES INFLUENCE THE RATE OF INACTIVATION

Wild-type (WT) human aFGF and the full set of seven Cys to Ser mutants were further characterized with respect to stability and heparin dependence. Inactivation rates in serum-free tissue culture medium at physiological temperature and pH were monitored by full dose-response curves assayed as a function of incubation time on confluent Balb/c 3T3 cells, also in serum-free medium. As shown in FIGURE 5, inactivation for all mutants in either the presence or absence of heparin appears to follow exponential decay kinetics. In all cases, the rates of inactivation are decreased (TABLE 3) by the presence of 50 μg/ml heparin, a glycosaminoglycan to which aFGF is known to avidly bind.[6]

In the absence of heparin, WT aFGF has a half-life of only 15 minutes. Addition of heparin extends this half-life 100-fold to 26 hours. In general, the mutants retaining either one or no Cys residues have longer half-lives than those containing

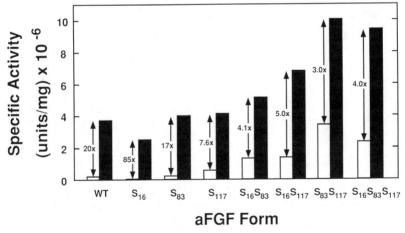

FIGURE 3. Mitogenic activities of WT and mutant aFGFs with and without heparin. The activities on confluent Balb/c 3T3 cells in serum-free media in the absence (open bars) and presence (solid bars) of 50 μg/ml of heparin were determined using a DNA synthesis mitogenic assay. Specific activity is expressed in units/milligram of pure mitogen with one unit defined as the amount of protein/milliliter required to elicit a half-maximal response (ED$_{50}$). The fold-stimulation by heparin is denoted between the double-headed arrows for each form of aFGF.[5,6]

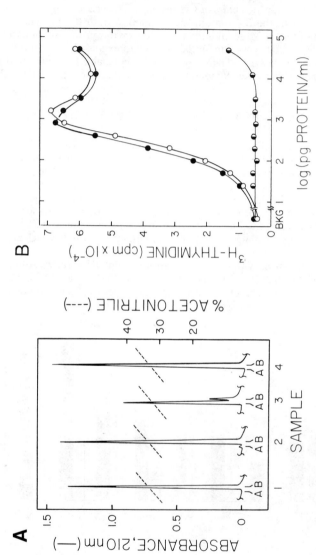

FIGURE 4. Effect of Cu^{2+} oxidation and subsequent reduction on activity of the S_{117} mutant. **A:** The untreated control S_{117} mutant protein (sample 1), after 10 min of air oxidation (sample 2), following treatment with Cu^{2+} (sample 3), and after subsequent reduction with dithiothreitol (sample 4) were chromatographed on a C_4 HPLC column and eluted with a linear gradient of acetonitrile in 10 mM trifluoroacetic acid. **B:** The samples of S_{117}, repurified by C_4 HPLC chromatography after various treatments, were assayed for the induction of DNA synthesis using confluent Balb/c 3T3 cells in serum-free medium in the presence of 50 μg/ml heparin. The dose-response curves of the nonoxidized mutant (sample 2, peak B, ○), the Cu^{2+}-oxidized mutant (sample 3, peak A, ●), and the Cu^{2+}-oxidized mutant following subsequent reduction with dithiothreitol (sample 4, ●) are shown.[5]

TABLE 2. Chemical Analysis of S_{117} Mutant Disulfide Bond Status[a]

	Untreated	Oxidized	Rereduced
Percent containing disul- fide bond	4	103 (97)[b]	5

[a]From Reference 5, with permission.
[b]First alkylation done in the presence of 6 M guanidinium chloride.

any two or all three cysteine residues. One notable exception, however, is S_{117}, the mutant retaining the two conserved Cys residues. Although rapidly inactivated in the absence of heparin, this mutant is one of the most stable forms in its presence. Thus the two conserved Cys residues appear uniquely compatible with heparin-mediated stabilization. One possible function associated with their conservation in aFGF might be to provide selective stabilization by heparin and heparan proteoglycans *in vivo* thereby restricting the activity of aFGF to the vicinity of mast-cell-derived heparin and both cell surface and basement membrane heparan proteoglycans.

RAPID INACTIVATION IS NOT CAUSED BY PROTEOLYSIS

Heparin and heparan proteoglycans have been reported to protect aFGF and bFGF *in vitro* from inactivation[7-9] and degradation by some proteases[10-12] and might

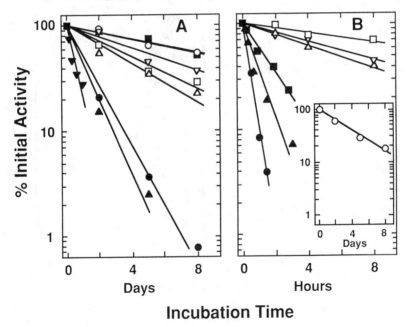

FIGURE 5. Inactivation rates of WT aFGF and all permutations of Cys to Ser mutants. The log of the percent initial mitogenic activity is plotted as a function of incubation time in serum-free medium, pH 7.4, 37°C in the presence (A) and absence (B) of heparin. Insert to B: The more stable $S_{16}S_{83}$ S_{117} mutant is presented on a longer time scale in the absence of heparin. The symbols denote WT aFGF (●), S_{16} (▼), S_{83} (▲), S_{117} (■), S_{83} S_{117} (▽), $S_{16}S_{117}$ (△), $S_{16}S_{83}$ (□), and $S_{16}S_{83}S_{117}$ (○).[6]

TABLE 3. Half-lives of Mitogenic Activities[a]

	WT	S_{16}	S_{83}	S_{117}	$S_{16}S_{83}$	$S_{16}S_{117}$	$S_{83}S_{117}$	$S_{16}S_{83}S_{117}$
+ heparin	26	12	17	240	92	79	130	240
− heparin	0.26	<0.08[b]	0.62	1.4	13	5.6	7.0	73
+/− heparin	100	>140[b]	27	170	7.1	14	19	3.3

[a]Full dose-response assays of wild-type and mutant aFGFs in either the presence or absence of 50 μg/ml heparin were determined as a function of incubation time at 37°C and pH 7.4 in serum-free Dulbecco's modified Eagle's medium/Ham's F-12 (3:1) tissue culture medium containing 1 mg/ml bovine serum albumin. The time in hours required for loss of one-half of the mitogenic activity (FIGURE 5) and the ratio of this inactivation time in the presence to the absence of heparin are listed. (From Reference 6, with permission.)

[b]The half-life was not determined because activity is lost in less than 5 minutes of incubation in the absence of heparin.

function similarly *in vivo*. In the presence of heparin, only the most rapidly inactivating mutant, S_{16}, inactivates faster than it disappears from solution as monitored by scanning densitometry of reducing SDS polyacrylamide electrophoretic gels (FIGURE 6A). However, in the absence of heparin not only WT aFGF but also the mutants retaining any two Cys residues inactivate more rapidly than they disappear from solution (FIGURE 6B).[6] Therefore, heparin is not simply protecting these forms

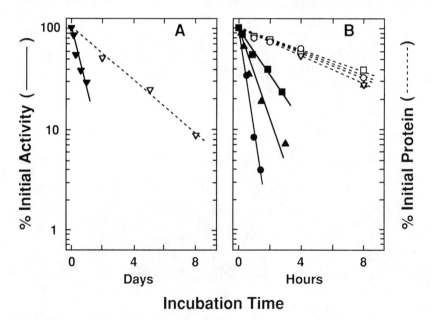

FIGURE 6. Inactivation rates compared to protein loss for rapidly inactivating WT and mutant aFGFs. The log of the percent initial mitogenic activity (closed symbols, solid lines) and the corresponding 16 kDa protein band in reducing SDS-polyacrylamide electrophoretic gels (open symbols, dashed lines) are plotted as a function of incubation time in serum-free medium, pH 7.4, 37°C in the presence (A) and absence (B) of heparin. Only the forms that inactivate faster than they disappear, WT (circles), S_{117} (squares), S_{83} (triangles), and S_{16} (inverted triangles), are shown. Other forms inactivate and disappear with similar kinetics.[6]

from proteolytic degradation. The mechanism of this rapidly nonproteolytic inactivation is unknown but could be coupled either to aggregation, oxidation of Cys and Met residues, or some other undefined process.

Moreover, both the conserved and nonconserved Cys residues clearly influence the activity, stability, and heparin dependence of aFGF and probably other homologous FGF family members. Coupling of aFGF activity to the redox potential could conceivably be a means of controlling the mitogenic and angiogenic activities of this protein *in vivo*.

REFERENCES

1. THOMAS, K. A. Biochemistry and molecular biology of fibroblast growth factors. *In* Neurotrophic Factors. J. H. Fallon & S. E. Loughlin, Eds. Academic Press. Orlando, Fla. (In press.)
2. GIMENEZ-GALLEGO, G., J. RODKEY, C. BENNETT, M. RIOS-CANDELORE, J. DISALVO & K. A. THOMAS. 1985. Brain-derived acidic fibroblast growth factor: complete amino acid sequence and homologies. Science **230:** 1385–1388.
3. GIMENEZ-GALLEGO, G., G. CONN, V. HATCHER & K. A. THOMAS. 1986. The complete amino acid sequence of human brain-derived acidic fibroblast growth factor. Biochem. Biophys. Res. Commun. **138:** 611–617.
4. LINEMEYER, D. L., L. J. KELLY, J. G. MENKE, G. GIMENEZ-GALLEGO, J. DISALVO & K. A. THOMAS. 1987. Expression in *Escherichia coli* of a chemically synthesized gene for biologically active bovine acidic fibroblast growth factor. Biotechnology **5:** 960–965.
5. LINEMEYER, D. L., J. G. MENKE, L. J. KELLY, J. DISALVO, D. SODERMAN, M.-T. SCHAEFFER, S. ORTEGA, G. GIMENEZ-GALLEGO & K. A. THOMAS. 1990. Disulfide bonds are neither required, present, nor compatible with full activity of human recombinant acidic fibroblast growth factor. Growth Factors **3:** 287–298.
6. ORTEGA, S., M.-T. SCHAEFFER, D. SODERMAN, J. DISALVO, D. L. LINEMEYER, G. GIMENEZ-GALLEGO & K. A. THOMAS. 1991. Conversion of cysteine to serine residues alters the activity, stability and heparin dependence of acidic fibroblast growth factor. J. Biol. Chem. **266:** 5842–5846.
7. GOSPODAROWICZ, D. & J. CHENG. 1986. Heparin protects basic and acidic FGF from inactivation. J. Cell. Physiol. **128:** 475–484.
8. DAMON, D. H., R. R. LOBB, P. A. D'AMORE & J. A. WAGNER. 1989. Heparin potentiates the action of acidic fibroblast growth factor by prolonging its biological half-life. J. Cell. Physiol. **138:** 221–226.
9. MUELLER, S. N., K. A. THOMAS, J. DISALVO & E. M. LEVINE. 1989. Stabilization by heparin of acidic fibroblast growth factor mitogenicity for human endothelial cells in vitro. J. Cell. Physiol. **140:** 439–448.
10. ROSENGART, T. K., W. V. JOHNSON, R. FRIESEL, R. CLARK & T. MACIAG. 1988. Heparin protects heparin-binding growth factor-I from proteolytic inactivation *in vitro*. Biochem. Biophys. Res. Commun. **152:** 432–440.
11. SALSELA, O., D. MOSCATELLI, A. SOMMER & D. B. RIFKIN. 1988. Endothelial cell-derived heparan sulfate binds basic fibroblast growth factor and protects it from proteolytic degradation. J. Cell Biol. **107:** 743–751.
12. SOMMER, A. & D. B. RIFKIN. 1989. Interaction of heparin with human basic fibroblast growth factor: protection of the angiogenic protein from proteolytic degradation by a glycosaminoglycan. J. Cell. Physiol. **138:** 215–220.

Expression, Processing, and Properties of int-2

CLIVE DICKSON, FRANCES FULLER-PACE, PAUL
KIEFER, PIERS ACLAND, DAVID MacALLAN,[a]
AND GORDON PETERS[a]

Laboratory of Viral Carcinogenesis
[a]*Laboratory of Molecular Oncology*
Imperial Cancer Research Fund
London WC2A 3PX, United Kingdom

INTRODUCTION

The *int-2* gene was originally discovered through its transcriptional activation by mouse mammary tumor virus (MMTV) and is thus classed as a protooncogene.[1,2] More recently, its ability to participate in mammary tumorigenesis has been confirmed by reintroduction as a transgene under the control of the MMTV promoter.[3] However, the gene is not detectably expressed in normal mammary epithelium suggesting that it is the virus that confers mammary tropism. Normal sites of *int-2* expression are found in developing embryos, at a variety of stages, and there are traces in adult brain and testes, but it is clear that *int-2* must function in specialized cell types rather than in a general capacity.[4-6] An obvious possibility would be as a cell to cell signaling molecule since the sequence of *int-2* predicts a polypeptide with significant homology to members of the fibroblast growth factor (FGF) family.[7,8]

These structural similarities to the FGFs have understandably influenced the functional characterization of the *int-2* product, and this paper deals with several of the obvious assays, such as mitogenicity and cell transformation. However, the situation is complicated by the fact that translation of *int-2* can initiate at two different positions, an AUG codon in a relatively favorable context for initiation, and a CUG codon located 87 nucleotides upstream.[9] Initiation at the AUG codon yields a 28.5 kilodalton (kD) polypeptide which is cleaved and glycosylated during transit through the secretory pathway.[10] In this regard it resembles the secreted FGF-related proteins, such as kFGF, FGF-5, FGF-6, and KGF.[11-15] The longer CUG-initiated product, on the other hand, partitions almost equally between the secretory pathway and the cell nucleus. Nuclear localization has also been reported for aminoterminally extended forms of basic FGF, but their significance remains unclear.[16-21] In an attempt to address such issues, we have been comparing the activities of the different int-2 products with those of other FGFs in a variety of functional assays.

RESULTS

Synthesis, Processing, and Secretion of int-2

We have previously described the synthesis of int-2 proteins in COS-1 cells following transfection with plasmids in which different *int-2* cDNAs are transcribed from the SV40 early promoter.[10] With plasmid KC3.2, which was optimized for translation from the first in-frame AUG codon, four major int-2 species were detected in cell lysates by immunoblotting with an int-2-specific antiserum. The smaller pair of proteins, around 27.5 kD and 28.5 kD in size, corresponded to the primary translation product with and without removal of an amino-terminal signal peptide, while the larger pair, at 30.5 kD and 31.5 kD, were the glycosylated equivalents. Although these data implied secretion of int-2, and antigens could be detected in the secretory pathway by immunofluorescence, it was not possible to demonstrate extracellular forms of the protein by direct analysis of the culture medium from COS-1 cells.

One explanation for these data would be that the processing of int-2 in transiently transfected COS-1 monkey cells is relatively inefficient. We have therefore repeated the analyses with lines of mouse NIH3T3 cells that were stably transfected with a vector expressing the same *int-2* cDNA sequences from a murine leukemia virus promoter.[22] In the experiment shown in FIGURE 1, cell extracts and culture media were recovered from confluent monolayers of DMI-3 cells, a cloned line of *int-2*-transformed NIH3T3 cells, and from COS-1 cells 60 hours after transfection with the KC3.2 plasmid. Immunoprecipitates of the cells and culture medium were fractionated on sodium dodecyl sulfate-polyacrylamide (SDS-PAGE) gels and the *int-2* proteins were subsequently visualized by immunoblotting. The transfected COS-1 cells contained the four intracellular forms of int-2 described previously, with very little appearing in the culture medium. In earlier work, we had failed to detect this secreted fraction by direct immunoblotting of concentrated medium, but the introduction of an immunoprecipitation step has increased the sensitivity of the assay. Nevertheless, we estimate that less than 5% of the total int-2 protein expressed in COS-1 cells appears in a recognizable form in the conditioned medium. This is in sharp contrast to the results obtained with the DMI-3 cells where the majority of int-2 proteins are found in the medium and comparatively little remains associated with the cells (FIGURE 1). In both cell types, the secreted int-2 was slightly larger in size (~ 33 kD) than the intracellular forms, consistent with the notion that further processing and addition of carbohydrate may occur in the Golgi complex prior to secretion.

Mitogenic Properties of int-2 and Other FGFs

The original identification of acidic and basic FGF relied on their ability to promote the growth of a variety of cell types and further members of the family have also been shown to be broad spectrum mitogens.[23,24] We therefore assessed the mitogenic potential of int-2 by asking whether the conditioned medium from DMI-3 cells or transiently transfected COS-1 cells could induce DNA synthesis in quiescent cells. Given the links between *int-2* and mammary tumorigenesis, we chose the C57MG mammary epithelial line as the recipient cells. The cells were rendered quiescent by placing them in 0.1% serum for at least 48 hours prior to the ^{3}H-thymidine incorporation assays. Conditioned medium was collected from cells that were also maintained in 0.1% serum for 3 days. In the case of COS-1 cells, this entailed recovering medium between 24 and 72 hours following transfection with either the KC3.2 plasmid, containing *int-2* cDNA sequences, or the parental vector KC3. With the NIH3T3 cells the comparison was between the *int-2*-transformed DMI-3 line and the nontransfected parental cells. The results were complicated by the fact that the untreated NIH3T3 cells were already secreting a mitogen as judged by the basal level of thymidine incorporation observed ($>50\%$ of the maximum response). However, in neither case did the presence of int-2 in the medium cause any stimulation over background (not shown).

As a more direct comparison between int-2 and other FGFs, we also tested the products obtained by cell-free translation. Plasmids containing cDNAs for human aFGF and FGF-5 were obtained from J. Abraham[25] and M. Goldfarb[13] respectively, and the coding regions were transferred to vectors that permitted synthesis of sense RNA from the bacteriophage T7 promoter. An analogous vector was already

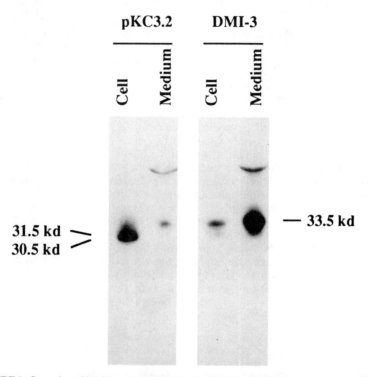

FIGURE 1. Secretion of int-2 from transfected cells. COS-1 cells were transfected with plasmid KC3.2 (pKC3.2) and the medium changed at 12 hours. After a further 48 hours cell extract and culture medium were immunoprecipitated with anti-int-2 serum and the resulting precipitates analyzed by immunoblotting using the same antiserum. An extract from a parallel culture of int-2-transformed NIH3T3 cells (DMI-3) was processed in a similar manner. COS-1 cells transfected with the parental vector and untransfected NIH3T3 cells were processed as controls, but showed no detectable int-2 products (not shown). The size of the detected int-2 proteins are indicated in kilodaltons (kD).

available for the expression of mouse *int-2,* based on the cDNA sequences contained in KC3.2, and the coding sequences for mouse kFGF and KGF were isolated using the polymerase chain reaction with primers based on the published sequences for these genes.[15,26-28] The relevant plasmid DNAs were linearized at appropriate restriction sites, and sense RNA synthesized *in vitro* with T7 RNA polymerase. Approximately 1 μg of each RNA was then used to program cell-free translation in a rabbit reticulocyte lysate, and the yield of each product was estimated by the incorporation of ^{35}S-methionine.

Equivalent amounts of the synthetic FGFs were tested for their ability to induce DNA synthesis in quiescent C57MG cells. As depicted in FIGURE 2A, lysates containing kFGF, aFGF, and FGF-5 clearly stimulated ^{3}H-thymidine incorporation in this system. For example, the addition of 5 μl of kFGF lysate gave an essentially maximum response, equivalent to the addition of 10% serum. In contrast, the lysates programmed by int-2 or KGF RNA gave only background levels of ^{3}H-thymidine incorporation, comparable to that obtained with a control lysate containing no exogenous RNA.

Analogous results were obtained with NIH3T3 cells (not shown), indicating that the negative results were not peculiar to C57MG cells. However, it was hardly surprising that KGF proved inactive since the factor was originally identified by its specificity for keratinocytes rather than fibroblasts. To confirm this specificity, the assays were repeated using the Balb/MK mouse keratinocyte cell line as the target.[15] As shown in FIGURE 2B, the lysate containing KGF was definitely mitogenic for these cells. Although kFGF was also able to elicit a modest response, confirming its broad activity, int-2 again scored negative.

Cell Transformation by int-2

Despite the lack of mitogenic activity on cells to which it is added, it is clear that *int-2* can indeed influence the growth characteristics of cells in which it is expressed. For example, some clones of int-2-expressing NIH3T3 cells adopt a transformed phenotype, including the ability to grow in semisolid medium.[22] However, int-2 appears much less effective than other secreted FGFs in eliciting transformation and, unlike kFGF and FGF-5, does not score positive in a conventional focus assay. As typified by the DMI-3 cell line, transformation seems to depend on achieving relatively high levels of int-2 expression.

It is also likely that it is the extracellular concentration of the factor that is crucial. As discussed earlier, the *int-2* gene has the capacity to specify two products, with different subcellular fates. Since all of the previous assays were conducted with the construct that was optimized for initiation at the AUG and encoding an exclusively secreted form of the protein, we considered it important to assess whether the nuclear form of int-2 could function in these assays. To this end, three plasmids were constructed in which different mutated *int-2* cDNAs were inserted into the retrovirus-based vector pDO-BS.[29] The first, designated pDO-AUG, contained the coding sequences derived from KC3.2, beginning at the AUG codon (FIGURE 3). In pDO-EXT, the sequences 5' of the AUG codon were included, encompassing the alternative initiation codon. To facilitate initiation at this position, the in-frame AUG codon was mutated to an AAG and the upstream CUG changed to an AUG (FIGURE 3). The third construct was designed to preclude secretion of int-2 by removal of the signal peptide. This was achieved by an amino-terminal truncation that removed both the CUG and AUG initiation codons but reinstated an AUG adjacent to the signal peptide cleavage site (FIGURE 3). The protein specified by this plasmid, designated pDO-NSP, was found to be exclusively nuclear (not shown) suggesting that the sequences determining nuclear localization are contained within the body of the protein rather than in the amino-terminal extension.

Each DNA was introduced into NIH3T3 cells as a calcium phosphate precipitate, and the transfected cells were selected in medium containing G418. Since the pDO-BS vector carries the neomycin resistance gene, all G418-resistant cells should

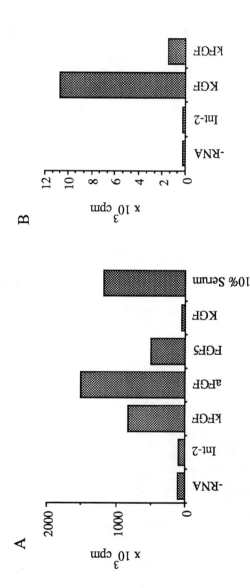

FIGURE 2. ³H-thymidine incorporation in response to synthetic FGFs. C57MG mammary epithelial cells (A), and a mouse keratinocyte line Balb/MK (B), were seeded at 5×10^4 cells/35 mm well in medium containing 10% serum, and rendered quiescent by reducing the serum concentration to 0.1% for 3 days. The cultures were then exposed to 5 μl of rabbit reticulocyte lysate containing the cell-free synthesized FGF as indicated (except for aFGF which contained 1 μl) in 0.5 ml of medium. After 16 hours the cultures were labeled for a further 3 hours with 5 μCi per well of ³H-thymidine. The cells were then lysed with 1M NaOH, 2% Na_2CO_3, and 1% SDS, and the incorporation into trichloroacetic acid (TCA) precipitable DNA determined by liquid scintillation counting.

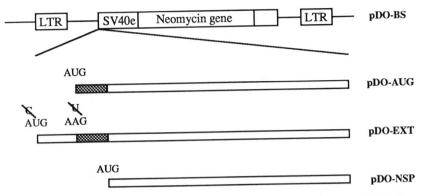

FIGURE 3. Structure of retrovirus vectors expressing *int-2* cDNAs. The vector pDO-BS[29] was used as the recipient for various *int-2* cDNAs. pDO-AUG contains *int-2* sequences beginning at the AUG codon. pDO-EXT has additional 5′ sequences to include the upstream initiation codon which was changed from CUG to an AUG. The in-frame AUG was also changed to AAG to prevent initiation at this site. pDO-NSP contains a truncated *int-2* cDNA initiating protein synthesis at an introduced AUG located at the signal peptide cleavage site.

potentially express one or other form of the int-2 protein. The pools of transfected cells were allowed to form confluent monolayers and then scored for foci of transformation as described previously. TABLE 1 summarizes the results of these experiments which confirmed that only the secreted form of int-2 encoded by pDO-AUG was capable of inducing focus formation in a dosage-dependent manner. The nuclear forms were essentially negative in this assay.

DISCUSSION

Although unified by their structural similarities, both in amino acid sequence and genomic organization, the different members of the FGF family must have evolved for specific roles. One of the current challenges in the field is therefore to identify the subtle differences that distinguish the individual factors as well as establishing the

TABLE 1. Transforming Ability of *int-2* cDNAs[a]

Plasmid	No. of Foci/Dish Concentration of Transfected DNA		
	20 ng	100 ng	500 ng
pDO-AUG	54	71	149
pDO-EXT	7	16	22
pDO-NSP	8	13	ND
pDO-BS	6	7	10

[a]NIH3T3 cells (8×10^5) were transfected with 20, 100, and 500 ng of *int-2* plasmid DNA (as depicted in FIGURE 3) in the presence of 30 μg of human placental DNA as carrier and 1 μg of pSV2neo.[22] After 24 hours the cells were trypsinized and seeded at a ratio of 1:2 in Dulbecco's modified Eagle medium with 10% newborn calf serum and 1 mg/ml G418. Nine days later the cultures were placed in medium with only 3% newborn calf serum and the foci counted after a further 7 days.

common features that underly their function. In this regard, it is already clear that there is a major distinction between FGFs that include a signal sequence for transit through the secretory pathway, such as kFGF and FGF-5, and those that do not, such as aFGF and bFGF.[8,30] We have previously suggested that int-2 may represent an intermediate case based on our inability to detect mitogenic activity or immunoreactive material in the culture medium, despite convincing evidence for a signal peptide and large amounts of protein in the endoplasmic reticulum and Golgi of expressing cells.[10] However, these ideas were based on analyses of transiently transfected COS-1 cells. We show here that COS-1 cells may be impaired in their ability to process and secrete int-2, at least when compared to NIH3T3 cells (FIGURE 1).

It is not clear why there should be such a difference between mouse cells and monkey cells, but several points are worthy of consideration. In the first place, transfected COS-1 cells contain very large amounts of int-2 protein as a result of amplification of the input plasmid. It is difficult to make a quantitative estimate, since only a proportion of the electroporated cells take up and express the input DNA in these transient assays, but the signals achieved with COS-1 cells are of the order of three- to fivefold greater than those observed in stably transfected mouse fibroblasts. However, it seems unlikely that these levels would overload the processing machinery of COS-1 cells since kFGF expressed in the same vector was found to be efficiently secreted (unpublished results of authors). Attempts to improve the secretion of int-2 by substituting the signal sequence with that of an immunoglobulin gene gave only a slight improvement (unpublished results of authors). It remains possible, therefore, that some inherent feature of the int-2 product drastically slows its release from COS-1 cells. Alternatively, monkey cells may process the secreted product such that it can no longer be easily detected with the available antisera.

The DMI-3 line, on the other hand, appeared to export int-2 very efficiently, but it is important to note that these cells were not only morphologically transformed by int-2 but subsequently selected for their ability to grow in defined medium, lacking exogenous growth factors.[22] If this phenotype depends on high levels of extracellular int-2, then clones such as DMI-3 may have been indirectly selected for efficient secretion. To pursue this issue, it will obviously be necessary to survey a wider variety of cell types, but such studies are compromised by the generally low levels of int-2 in cells or tissues in which the gene is normally expressed. Although the COS-1 and DMI-3 cells may turn out to be two anomalous extremes, they are currently the best available models in which int-2 protein can be readily detected.

In hindsight, therefore, it is surprising that the conditioned medium from transfected COS-1 cells was devoid of mitogenic activity attributable to int-2, as measured by the ability to induce DNA synthesis in the target cells tested. In the experiments presented in FIGURE 2, this conclusion was confirmed using a panel of different FGFs prepared by cell-free translation. Only int-2 scored consistently negative on C57MG, NIH3T3, and Balb/MK cells, while apart from KGF, the other products were mitogenic for all three. These findings conflict with earlier results for int-2 based on both immunological and autoradiographic assays for DNA synthesis.[10,31] The explanation is unclear, apart from possible variability in the composition of reticulocyte lysate preparations used as a source of int-2.

To date, we have only tested a limited number of recipient cells, and it will be important to survey a much wider variety of tissue types. The assumption is that int-2 will indeed be mitogenic, but given its restricted pattern of expression in the developing embryo, the true target cells may not be accessible or amenable to cell culture. On the other hand, there is no a priori reason to assume that high-affinity receptors for int-2 will be similarly restricted in their distribution. For example, at least a subpopulation of cells in the developing mammary gland must be capable of responding to int-2. A likely explanation would be that some other member of the

FGF family functions normally in mammary gland development and that in the hyperplasias observed in transgenic mice and in virally induced breast cancers, int-2 is substituting for this factor.

There is considerable scope for cross-reactivity between FGFs and their cognate receptors, but there will also be considerable variability in binding affinities. Thus, it is not clear whether the complete lack of a mitogenic response to int-2 implies the complete absence of suitable receptors. The ability of int-2 to transform NIH3T3 cells suggests that these cells must indeed have receptors that can be triggered by int-2. Perhaps the concentration of ligand achieved in a cell-free translation system is too low to elicit a response, in contrast to the local concentration achieved in cells that are actively secreting int-2. Alternatively, it may well be that mitogenicity and transformation involve different signal-transduction pathways or differ in their dependence on ligand concentration. From the comparison of transforming propensity of different *int-2* cDNAs (TABLE 1), it would seem that secretion is of prime importance to confer the transformed phenotype. It is not clear whether the inability of amino-terminally extended int-2 to induce transformation reflects its poor secretion, or whether the nuclear species modifies the cell's response to int-2. The identification of cell-surface receptors for int-2 and elucidation of the intracellular circuitry on which the secreted and nuclear forms for int-2 impinge are therefore high priorities for future research.

REFERENCES

1. DICKSON, C., R. SMITH, S. BROOKES & G. PETERS. 1984. Tumorigenesis by mouse mammary tumor virus: proviral activation of a cellular gene in the common integration region int-2. Cell **37**: 529–536.
2. PETERS, G., S. BROOKES, R. SMITH & C. DICKSON. 1983. Tumorigenesis by mouse mammary tumor virus: evidence for a common region for provirus integration in mammary tumors. Cell **33**: 369–377.
3. MULLER, W., F. LEE, C. DICKSON, G. PETERS, P. PATTENGALE & P. LEDER. 1990. The int-2 gene product acts as an epithelial growth factor in transgenic mice. EMBO J. **9**: 907–913.
4. JAKOBOVITS, A., G. SHACKLEFORD, H. VARMUS & G. MARTIN. 1986. Two proto-oncogenes implicated in mammary carcinogenesis, int-1 and int-2, are independently regulated during mouse development. Proc. Nat. Acad. Sci. USA **83**: 7806–7810.
5. WILKINSON, D., G. PETERS, C. DICKSON & A. MCMAHON. 1988. Expression of the FGF-related proto-oncogene int-2 during gastrulation and neurulation in the mouse. EMBO J. **7**: 691–695.
6. WILKINSON, D., S. BHATT & A. MCMAHON. 1989. Expression pattern of the FGF-related proto-oncogene int-2 suggests multiple roles in fetal development. Development **105**: 131–136.
7. DICKSON, C. & G. PETERS. 1987. Potential oncogene product related to growth factors. Nature **326**: 833.
8. DICKSON, C., R. DEED, M. DIXON & G. PETERS. 1989. The structure and function of the *int-2* oncogene. Prog. Growth Factor Res. **1**: 123–132.
9. ACLAND, P., M. DIXON, G. PETERS & C. DICKSON. 1990. Subcellular fate of the int-2 oncoprotein is determined by choice of initiation codon. Nature **343**: 662–665.
10. DIXON, M., R. DEED, P. ACLAND, R. MOORE, A. WHYTE, G. PETERS & C. DICKSON. 1989. Detection and characterization of the fibroblast growth factor-related oncoprotein INT-2. Mol. Cell. Biol. **9**: 4896–4920.
11. YOSHIDA, T., K. MIYAGAWA, H. ODAGIRI, H. SAKAMOTO, P. LITTLE, M. TERADA & T. SUGIMURA. 1987. Genomic sequence of hst, a transforming gene encoding a protein homologous to fibroblast growth factors and the int-2-encoded protein. Proc. Nat. Acad. Sci. USA **84**: 7305–7309.
12. DELLI-BOVI, P., A. CURATOLA, K. NEWMAN, Y. SATO, D. MOSCATELLI, R. HEWICK, D.

RIFKIN & C. BASILICO. 1988. Processing, secretion, and biological properties of a novel growth factor of the fibroblast growth factor family with oncogenic potential. Mol. Cell. Biol. **8**: 2933–2941.

13. ZHAN, X., B. BATES, X. HU & M. GOLDFARB. 1988. The human FGF-5 oncogene encodes a novel protein related to fibroblast growth factors. Mol. Cell. Biol. **8**: 3487–3495.

14. MARICS, I., J. ADELAIDE, F. RAYBAUD, M.-G. MATTEI, C. COULIER, J. PLANCHE, O. DE LAPEYRIERE & D. BIRNBAUM. 1989. Characterization of the HST-related FGF.6 gene, a new member of the fibroblast growth factor gene family. Oncogene **4**: 335–340.

15. FINCH, P., J. RUBIN, T. MIKI, D. RON & S. AARONSON. 1989. Human KGF is FGF-related with properties of a paracrine effector of epithelial cell growth. Science **245**: 752–755.

16. BUGLER, B., F. AMALRIC & H. PRATS. 1991. Alternative initiation of translation determines cytoplasmic or nuclear localization of basic fibroblast growth factor. Mol. Cell. Biol. **11**: 573–577.

17. BOUCHE, G., N. GAS, H. PRATS, V. BALDIN, J.-P. TAUBER, J. TEISSIE & F. AMALRIC. 1987. Basic fibroblast growth factor enters the nucleolus and stimulates the transcription of ribosomal genes in ABAE cells undergoing G0-G1 transition. Proc. Nat. Acad. Sci. USA **84**: 6770–6774.

18. PRATS, H., M. KAGHAD, A. PRATS, M. KLAGSBRUN, J. LELIAS, P. LIAUZUN, P. CHALON, J. TAUBER, F. AMALRIC, J. SMITH & D. CAPUT. 1989. High molecular mass forms of basic fibroblast growth factor are initiated by alternative CUG codons. Proc. Nat. Acad. Sci. USA **86**: 1836–1840.

19. FLORKIEWICZ, R. & A. SOMMERS. 1989. Human basic fibroblast growth factor gene encodes four polypeptides: three initiate translation from non-AUG codons. Proc. Nat. Acad. Sci. USA **86**: 3978–3981.

20. BALDIN, V., A.-M. ROMAN, I. BOSC-BIERNE, F. AMALRIC & G. BOUCHE. 1990. Translocation of bFGF to the nucleus is G1 phase cell cycle specific in bovine aortic endothelial cells. EMBO J. **9**: 1511–1517.

21. RENKO, M., N. QUARTO, T. MORIMOTO & D. RIFKIN. 1990. Nuclear and cytoplasmic localization of different basic fibroblast growth factor species. J. Cell. Physiol. **144**: 108–114.

22. GOLDFARB, M., R. DEED, D. MACALLAN, W. WALTHER, C. DICKSON & G. PETERS. 1991. Cell transformation by int-2—a member of the fibroblast growth factor family. Oncogene **6**: 65–71.

23. BURGESS, W. & T. MACIAG. 1989. The heparin binding (fibroblast) growth factor family proteins. Annu. Rev. Biochem. **58**: 575–606.

24. GOSPODAROWICZ, D., N. FERRARA, L. SCHWEIGERER & G. NEUFELD. 1987. Structural characterization and biological functions of fibroblast growth factor. Endocr. Rev. **8**: 95–114.

25. JAYE, M., R. HOWK, W. BURGESS, G. RICCA, I. CHIU, M. RAVERA, S. O'BRIEN, W. MODI, T. MACIAG & W. DROHAN. 1986. Human endothelial cell growth factor: cloning, nucleotide sequence, and chromosome localization. Science **233**: 541–545.

26. TAIRA, M., T. YOSHIDA, K. MIYAGAWA, H. SAKAMOTO, M. TERADA & T. SUGIMURA. 1987. cDNA sequence of human transforming gene hst and identification of the coding sequence required for transforming activity. Proc. Nat. Acad. Sci. USA **84**: 2980–2984.

27. DELLI BOVI, P., A. CURATOLA, F. KERN, A. GRECO, M. ITTMANN & C. BASILICO. 1987. An oncogene isolated by transfection of Kaposi's sarcoma DNA encodes a growth factor that is a member of the FGF family. Cell **50**: 729–737.

28. HEBERT, J., C. BASILICO, M. GOLDFARB, O. HAUB & G. MARTIN. 1990. Isolation of cDNAs encoding four mouse FGF family members and characterization of their expression patterns during embryogenesis. Dev. Biol. **138**: 454–463.

29. MORGENSTERN, J. & H. LAND. 1990. Advanced mammalian gene transfer: high titre retroviral vectors with multiple drug selection markers and a complementary helper-free packaging cell line. Nucleic Acids Res. **18**: 3587–3596.

30. THOMAS, K. 1988. Transforming potential of fibroblast growth factor genes. Trends Biochem. Sci. **13**: 141–143.

31. PATERNO, G., L. GILLESPIE, M. DIXON, J. SLACK & J. HEATH. 1989. Mesoderm-inducing properties of INT-2 and kFGF: two oncogene encoded growth factors related to FGF. Development **106**: 79–83.

Characterization of the *hst-1* Gene and Its Product

TERUHIKO YOSHIDA, HIROMI SAKAMOTO,
KIYOSHI MIYAGAWA, TAKASHI SUGIMURA,
AND MASAAKI TERADA

Genetics Division
National Cancer Center Research Institute
1-1, Tsukiji 5-chome
Chuo-ku
Tokyo 104, Japan

INTRODUCTION

The fibroblast growth factor (FGF) family comprises at least seven closely related proteins (FIGURE 1) discovered originally from diverse fields of biology: acidic and basic fibroblast growth factors (a and bFGFs) are classical peptide growth factors (Reference 1 for review); the *int-2* protein was identified as a protein coded by the cellular gene activated transcriptionally by retroviral insertion in mouse mammary tumors[2]; keratinocyte growth factor (KGF) as a mitogen specific to epithelial cells[3]; and *hst-2*/FGF6 as a product of a gene homologous to another member of the family, *hst-1*.[4,5]

The *hst-1* gene is the first transforming gene in the family and the second oncogene found to encode a growth factor. Although the gene was originally identified by the NIH3T3 focus formation assay from the genome of a human gastric cancer tissue,[6] this assay revealed that the *hst-1* gene is "activated" in DNAs of a variety of tissues including noncancerous gastrointestinal mucosae (see below). The prediction from the sequence that the gene encodes a heparin-binding growth factor was confirmed by synthesizing a recombinant *hst-1* protein and by testing its biological activity.[7] We detected the *hst-1* gene transcripts in germ cell tumors and in mouse embryos, but not in other cancerous and noncancerous tissues by northern blot analysis.[8] The *hst-1* gene is just 35 kilobase pair (kbp) downstream to the *int-2* gene in the same transcriptional orientation[9] at band q13.3 of human chromosome 11.[10] This closeness and alignment were also observed in mouse genome.[11] The *hst-1* and *int-2* genes are coamplified in several types of human cancers,[10,12,13] most notably in about 50% of human esophageal cancers,[14] although the coamplification does not appear to be accompanied by overexpression of the genes (unpublished observation).

IDENTIFICATION OF THE *hst-1* GENE FROM A HUMAN STOMACH CANCER

High-molecular-weight DNA was extracted from 58 surgical specimens obtained from 37 human gastric cancer cases. Two of the 37 cancerous and one of the 21 noncancerous gastric mucosae were found to be clearly positive in the NIH3T3 focus formation assay. A genomic library was constructed from DNA of a secondary

27

NIH3T3 transformant designated T361-2nd-1. Clusters of human genomic sequences were identified by a human repetitive sequence *Alu* as a probe, and three contiguous repetitive-free fragments were then found to hybridize to the mRNA of the T361-2nd-1 transformant but not to the mRNA of the parental NIH3T3 cells.[6] These fragments were used to screen the T361-2nd-1 cDNA library, and overlapping cDNA clones were identified.[15] The transforming activity of these clones was confirmed by placing the cDNAs under the SV40 early region promoter followed by transfection to NIH3T3 cells. Various mutated derivatives of these cDNA expression vectors revealed that one of the two possible open reading frames encodes a transforming protein (FIGURE 2). As there was no identical gene reported, we designated this novel transforming gene *hst*, which stands for human stomach.[6]

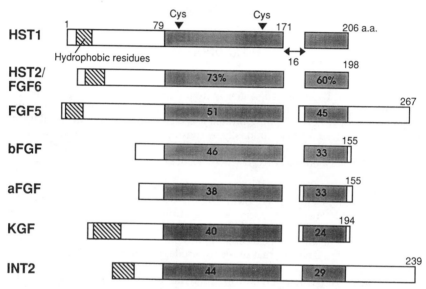

FIGURE 1. The FGF family. Amino acid sequences of currently known members of the family were aligned to that of the *hst-1* protein. Gaps were introduced to achieve the best alignment. Cross-hatched boxes indicate putative signal peptides, and two cysteine residues conserved among the family are indicated by arrowheads. Stippled areas have significant homology with the *hst-1* protein as shown by percentage.

Afterwards the gene was renamed *hst-1*, when a close homologue of *hst*, termed *hst-2*, was cloned by cross-hybridization.[4]

All the three original independent NIH3T3 foci derived from DNAs of a gastric cancer, the noncancerous portion of the same stomach, and a metastatic lymph node of another gastric cancer were found to be transformed by *hst-1*.[6] Moreover, a number of reports have followed to show the presence of the "transforming" *hst-1* gene in a variety of nongastric cancers, such as colon cancer,[16] hepatoma,[17,18] Kaposi sarcoma[19] (where *hst-1* was referred to as *KS* or *K-fgf* oncogene), melanoma,[20] and osteosarcoma.[21] These reports indicate *hst-1* is the most prevalent non-*ras* transforming gene.

```
Met-Ser-Gly-Pro-Gly-Thr-Ala-Ala-Val-Ala-Leu-Leu-Pro-Ala-Val-Leu-Leu-Ala-Leu-Leu-
Ala-Pro-Trp-Ala-Gly-Arg-Gly-Ala-Ala-Pro-Thr-Ala-Ala-Pro-Asn-Gly-Thr-Leu-Glu-

Ala-Glu-Leu-Glu-Arg-Arg-Trp-Glu-Ser-Leu-Val-Ala-Leu-Ser-Leu-Ala-Arg-Leu-Pro-Val-
Ala-Ala-Ala-Gln-Pro-Lys-Glu-Ala-Val-Gln-Ser-Gly-Ala-Gly-Asp-Tyr-Leu-Leu-Gly-Ile-
Lys-Arg-Leu-Arg-Arg-Leu-Tyr-Cys-Asn-Val-Gly-Ile-Gly-Phe-His-Leu-Gln-Ala-Leu-Pro-
Asp-Gly-Arg-Ile-Gly-Ala-His-Ala-Asp-Thr-Arg-Asp-Ser-Leu-Leu-Glu-Leu-Ser-Pro-
Val-Glu-Arg-Gly-Val-Val-Ser-Ile-Phe-Gly-Val-Ala-Ser-Arg-Phe-Phe-Val-Ala-Met-Ser-
Ser-Lys-Gly-Lys-Leu-Tyr-Gly-Ser-Pro-Phe-Phe-Thr-Asp-Glu-Cys-Thr-Phe-Lys-Glu-Ile-
Leu-Leu-Pro-Asn-Asn-Tyr-Asn-Ala-Tyr-Glu-Ser-Tyr-Lys-Tyr-Pro-Gly-Met-Phe-Ile-Ala-
Leu-Ser-Lys-Asn-Gly-Lys-Thr-Lys-Lys-Gly-Asn-Arg-Val-Ser-Pro-Thr-Met-Lys-Val-Thr-
His-Phe-Leu-Pro-Arg-Leu
```

206 a.a., 22 kDa

FIGURE 2. Amino acid sequence of the *hst-1* protein deduced from cDNA. A hydrophobic core sequence of a putative signal peptide is shadowed. The arrowhead indicates the N-terminal amino acid of the recombinant *hst-1* protein produced and processed in silkworm cells.

CLONING OF THE *hst-1* GENE FROM HUMAN GENOMIC LIBRARIES

To analyze the normal *hst-1* gene, 1×10^5 clones of a genomic library constructed from a healthy male were screened with the coding region of the *hst-1* cDNA (ORF1) at a stringent condition of hybridization.[4] Nine positive clones were classified into two groups, depending on the two different patterns of hybridization to the ORF1 probe as shown in FIGURE 3A; six of them have three *Eco*RI fragments, 5.8, 2.8, and 0.8 kbp, while each of the remaining three clones has only one 8.0 kbp *Eco*RI fragment hybridizing to the ORF1 probe. These two groups were found to represent *hst-1* and its close homologue *hst-2,* respectively. Partial physical maps are shown for these clones and also for the clone L361-Hu3, which is an *hst-1* genomic fragment cloned from the T361-2nd-1 cell, an NIH3T3 secondary transformant (FIGURE 3B).

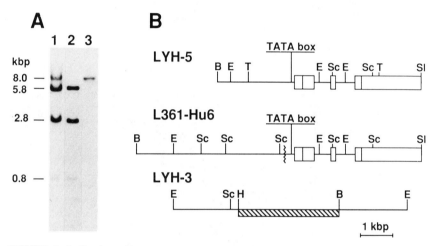

FIGURE 3. A: Southern blot analysis of *Eco*RI-digested DNAs hybridized with the probe corresponding to the coding region of the *hst-1* cDNA (ORF1). Lane 1, human placenta; lane 2, cosmid LYH-5 containing the *hst-1* gene; lane 3, cosmid LYH-3 corresponding to the *hst-2* gene. **B:** Partial physical maps of LYH-5, L361-Hu6 (*hst-1* from the NIH3T3 transformant, T361-2nd-1) and LYH-3. Boxes indicate exons of *hst-1* on LYH-5 and L361-Hu6 with the coding sequences stippled. A wavy line on the L361-Hu6 indicates the rearrangement site in the transformant genome. A region that cross-hybridizes to the ORF1 probe is cross-hatched on cosmid LYH-3. B, *Bam*HI; E, *Eco*RI; T, *Taq*I; Sc, *Sac*I; Sl, *Sal*I; H, *Hin*dIII. The *Taq*I sites are indicated only on LYH-5.

When transfected to NIH3T3 cells, the *hst-1* cosmid from the normal human library showed a transforming activity equivalent to L361-Hu3. A 4.0 kbp *Taq*I fragment, which contains a 1.3 kbp region upstream of TATA box (FIGURE 3B), was fully transforming by itself. A mouse *hst-1* fragment derived from the genome of NIH3T3 cells had a similar transforming activity.

The coding sequences were completely identical between the *hst-1* genomic fragment and the cDNA clone from the T361-2nd-1 transformant.[22] Thus, the mechanism of the activation of the transforming potential of *hst-1* is its transcriptional deregulation rather than a structural aberration. The expression of the *hst-1* seems to be tightly suppressed in normal adult cells (see below). Possible mechanisms of such transcriptional silence include gene methylation and flanking silencer sequence, both of which can be invalidated by transfection or gene cloning. This probably explains why *hst-1* is so frequently found as a transforming gene in a variety of sources of DNA.

Although the cosmids containing the *hst-2* gene transformed NIH3T3 cells only weakly in an ordinary focus formation assay, they were positive in tumorigenicity assay.[23] The *hst-2* transformants also induced distinctive foci in a defined medium lacking platelet-derived growth factor (PDGF) and FGF.[24] The *hst-2* cDNA was cloned from one of the serum-defined transformants, and the coding region was determined by the alignment with that of the *hst-1* (unpublished data). As shown in FIGURE 1, the *hst-2* protein is the closest member to the *hst-1* protein among the FGF family. Another group cloned a gene designated FGF6 also by cross-hybridization to the *hst-1* gene. The partial genomic sequence was published,[5] and we found *hst-2* is identical to FGF6.

SYNTHESIS OF THE RECOMBINANT *hst-1* PROTEIN

The biological activity of the *hst-1* protein was evaluated by synthesizing the recombinant protein in silkworm cells. Although the yield is lower and the procedure is more complicated in the baculovirus system, it is expected that posttranslational modifications occur in a way more akin to human cells, such as glycosylation, recognition, and cleavage of a signal peptide. Previous successes of synthesis of interferon-α and interleukin-3 showed that biologically active materials were secreted into the culture medium of silkworm cells.[25,26]

Briefly, the open reading frame of the *hst-1* with minimal flanking cDNA sequences (nucleotides 1 to 916 of the cDNA)[15] was cloned into an *Eco*RI site of a baculovirus transfer vector pBM030,[27] a gift from Dr. M. Furusawa (Daiichi Seiyaku Research Institute, Tokyo). Recombinant virus that produces nonfusion *hst-1* protein was generated by homologous recombination between the transfer vector and the wild-type baculovirus genome in the transfected silkworm cells. The medium of the virus-infected silkworm cells was harvested, concentrated, and purified by the simple two-step procedures of an Affi-Gel Heparin (Bio-Rad) column and reversed-phase high-performance liquid chromatography (HPLC). Growth factor activity was identified by [^{3}H]thymidine incorporation assay on NIH3T3 cells.[7] Typically, the biologically active *hst-1* fractions came through with 1.0–1.2 M NaCl fraction in the heparin-affinity chromatography (FIGURE 4). The HPLC-purified, biologically active *hst-1* protein was about 18 kilodaltons on sodium dodecyl sulfate-polyacrilamide gel electrophoresis (SDS-PAGE), and its amino acid sequencing revealed cleavage of the N-terminal 58 amino acids containing the signal peptide (FIGURE 2).

BIOLOGICAL ACTIVITIES OF THE RECOMBINANT *hst-1* PROTEIN

FIGURE 5 showed a potent mitogenic effect of the purified *hst-1* protein on NIH3T3 cells. The half-maximal stimulation was observed at 220 pg/ml in the absence of heparin and at 80 pg/ml in the presence of heparin. The potency in the presence of heparin was comparable to that of bovine pituitary bFGF (Takara Shuzo). The morphological changes of NIH3T3 cells in response to the *hst-1* protein were basically similar to the *hst-1*-gene transformants, characterized by a highly refractile, crisscross appearance with long-process formation. The changes appeared

12 hours after the addition of the protein, reached maximum after 24–48 hours, and reverted after 72 hours (data not shown).

Human umbilical vein endothelial (HUVE) cells were isolated as described[28] and cultured in the presence of 2.5 ng/ml of aFGF and 5 µg/ml of heparin. FIGURE 6 shows a typical response of HUVE cells to the purified recombinant *hst-1* protein. The half-maximal stimulation was observed at about 30 pg/ml.

The *hst-1* protein supports anchorage-independent growth of the NIH3T3 cells and NRK49F cells in a dose-dependent manner. An approximately tenfold higher concentration of the protein was required to induce the soft agar colony formation of NIH3T3 cells as compared to stimulation of DNA synthesis of the same cells.

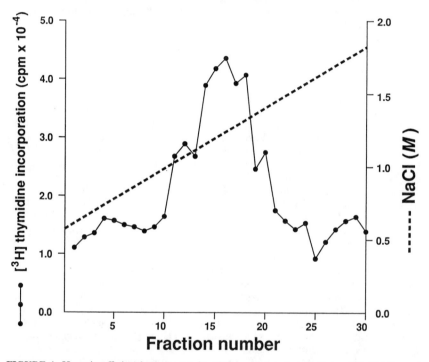

FIGURE 4. Heparin-affinity chromatography of the concentrated medium of BmN silkworm cells infected with the recombinant baculovirus. A broken line represents the NaCl gradient, and closed circles indicate mitogenic activity of each fraction as determined by [³H]thymidine incorporation assay on NIH3T3 cells.

We also observed a potent *in vivo* angiogenic activity of the recombinant *hst-1* protein using chorioallantoic membrane (CAM) assay and rat cornea assay. Furthermore, the histology of the tumors induced by subcutaneous injection of the NIH3T3 transformants of the *hst-1* gene showed a highly vascularized pattern as compared to the activated Ha-*ras* gene transformants (unpublished data). These data proved that the *hst-1* gene encodes a heparin-binding growth factor, which has *in vivo* and *in vitro* angiogenic activity comparable to FGFs.

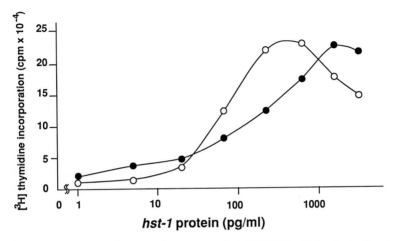

FIGURE 5. Stimulation of [³H]thymidine incorporation of NIH3T3 cells by the *hst-1* protein. The mitogenic activity was measured in the absence (filled circles) and presence (open circles) of 50 μg/ml of heparin.

EXPRESSION OF THE *hst-1* GENE

In contrast to the ubiquitous presence of FGFs, the expression of *hst-1* in adult tissues appears to be regulated tightly; our northern blot analysis failed to detect the *hst-1* message in about 80 cancerous and noncancerous cells and tissues. These samples included gastric cancers and Kaposi's sarcomas, the types of cancers in which the transforming *hst-1* gene was identified by transfection assays.[8] The expression of *hst-1* was found in germ cell tumors and in embryos (FIGURE 7). The sizes of the *hst-1* transcripts are 3.0 and 1.7 kilobases for human, and 3.0 kb for mouse.

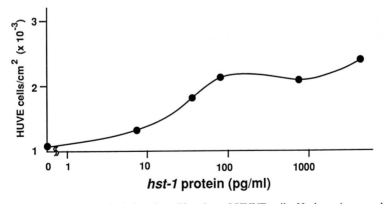

FIGURE 6. The *hst-1* protein induced proliferation of HUVE cells. No heparin was added during the assay. Cell numbers were means of triplicate determinations.

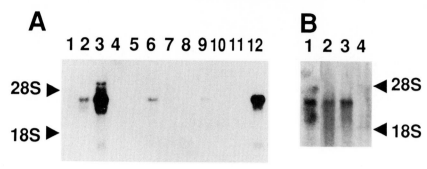

FIGURE 7. A: Northern blot analysis showing expression of *hst-1* in human germ cell tumors. Lanes 1 to 9 contained RNAs from surgical specimens of testicular germ cell tumors, lanes 10 and 11 from noncancerous portions of testes, and lane 11 from an immature teratoma cell line, NCC-IT. Three micrograms of poly(A)$^+$ RNA was hybridized with the 3' one-third of the coding region of the *hst-1* cDNA. The positions of the 28S and 18S rRNAs are indicated by arrowheads. **B:** Expression of *hst-1* in mouse embryos. Lane 1, day 11 embryos; lane 2, heads of day 14 embryos; lane 3, bodies of day 14 embryos with livers removed; lane 4, livers of day 14 embryos. Five micrograms of poly(A)$^+$ RNAs were hybridized with probe M1.8, a mouse genomic fragment of *hst-1* containing exon.[8]

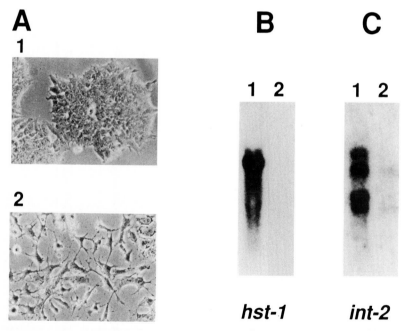

FIGURE 8. Differential expression of *hst-1* and *int-2* in the F9 cells. **A:** F9 stem cells (1) were induced to differentiate *in vitro* to parietal endodermal cells (2) as described in the text. Twenty micrograms of total RNA from each cell were hybridized with probe M1.8, an exon-containing mouse *hst-1* fragment (B), while three micrograms of poly(A)$^+$ RNA were used for probe *int-2* c (C).

Although this embryo-specific pattern of expression was also observed for *int-2* among the members of the FGF family,[29] these two genes are regulated in different ways, at least during induction of differentiation of a mouse teratocarcinoma cell line, F9.[30] This *in vitro* differentiation of F9 is thought to simulate the early stages of embryonic differentiation.[29] The *hst-1* gene was expressed preferentially in the F9 stem cells and greatly suppressed upon induction of the parietal endodermal differentiation; in contrast, transcription of *int-2* was very low in the undifferentiated F9 cells, and it was increased dramatically after differentiation (FIGURE 8). We surmise that these related oncogenes are functionally coupled and convey distinct signals during embryogenesis.

CHROMOSOMAL LOCALIZATION OF THE *hst-1* GENE

Both the human *int-2* and *hst-1* genes were mapped on the same chromosome band, 11q13.3,[10] and cosmid mapping showed[9] that they are only 35 kbp apart with *hst-1* situated 3' downstream of *int-2* in the same transcriptional orientation (FIGURE 9). This closeness (17 kbp) and the 5' *int-2-hst-1* 3' orientation is also conserved in

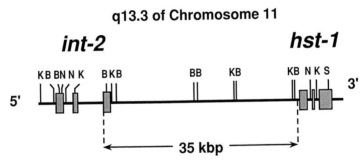

FIGURE 9. A physical map of the *int-2–hst-1* region. Exons are indicated by boxes. K, *Kpn*I; B, *Bam*HI; N, *Not*I; S, *Sal*I.

the mouse genome.[11] Considering this *int-2-hst-1* neighborhood, the following two facts were easily understood. First, *hst-1* was found to be activated transcriptionally in some of the mouse mammary tumor virus (MMTV) induced murine mammary tumors with or without *int-2* activation.[11] Second, these genes are coamplified in some human cancers.

COAMPLIFICATION OF THE *int-2* AND *hst-1* GENES IN CANCERS

The incidence of coamplification is relatively high in breast cancers (12–22%)[12,13] and remarkably high in esophageal cancers (47% in primary tumors and 100% in metastatic foci).[14] As shown in FIGURE 10, the degree of amplification is usually five- to tenfold, and in some cases amplification was noted only in metastatic foci and not in primary tumors. However, we could not detect transcripts of either of the genes in cell lines and tissues with the *int-2/hst-1* coamplification by northern blot analysis (data not shown). Although it is possible that the only low level of transcription is

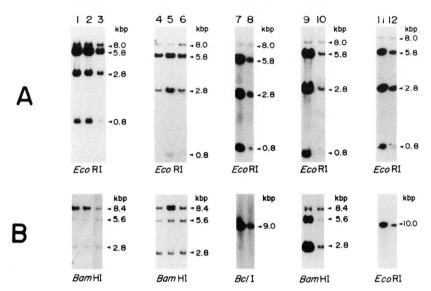

FIGURE 10. Southern blot analysis showing the coamplification of *int-2* and *hst-1* in human cancers. Genomic DNAs were digested by the restriction enzymes indicated and hybridized by the probe corresponding to the coding region of the *hst-1* cDNA (A) or BB4 or SS6, exon-containing genomic fragments of *int-2* (B). Lanes 1 to 3, a primary cancer, a lymph node metastasis, and noncancerous esophageal mucosa of a patient with esophageal cancer, respectively; lanes 4 to 6, a primary renal cell carcinoma, an esophageal cancer metastasis to the same kidney, and a noncancerous renal tissue of the same patient, respectively; lanes 7 and 8, a lymph node metastasis of a gastric cancer and noncancerous gastric mucosa of the same patient, respectively; lanes 9 and 10, a bladder cancer and a noncancerous bladder mucosa of the same patient, respectively; lane 11, A431 epidermoid cancer cell line; lane 12, COLO205 colon cancer cell line.

sufficient for the development and/or progression of the cancer, or that the transcription was turned on only at the specific period of the carcinogenesis, another plausible explanation is the presence of the unidentified cancer-related gene(s) in this *int-2/hst-1* amplification unit. The study is in progress to search for such gene(s) from the amplicon.

When the chromosomal localization of the *hst-1* gene was determined, the committee of the human gene nomenclature registered this gene officially as *HSTF1*, which stands for heparin-binding secretory transforming factor 1.[10]

REFERENCES

1. BAIRD, A., F. ESCH, P. MORMÈDE, N. UENO, N. LING, P. BÖHLEN, S.-Y. YING, W. B. WEHRENBERG & R. GUILLEMIN. 1986. Recent Prog. Horm. Res. **42:** 143–200.
2. PETERS, G., S. BROOKES, R. SMITH & C. DICKSON. 1983. Cell **33:** 369–377.
3. FINCH, P. W., J. S. RUBIN, T. MIKI, D. RON & S. A. AARONSON. 1989. Science **245:** 752–755.
4. SAKAMOTO, H., T. YOSHIDA, M. NAKAKUKI, H. ODAGIRI, K. MIYAGAWA, T. SUGIMURA & M. TERADA. 1988. Biochem. Biophys. Res. Commun. **151:** 965–972.
5. MARICS, I., J. ADELAIDE, F. RAYBAUD, M.-G. MATTEI, F. COULIER, J. PLANCHE, O. DE LAPEYRIERE & D. BIRNBAUM. 1989. Oncogene **4:** 335–340.

6. SAKAMOTO, H., M. MORI, M. TAIRA, T. YOSHIDA, S. MATSUKAWA, K. SHIMIZU, M. SEKIGUCHI, M. TERADA & T. SUGIMURA. 1986. Proc. Nat. Acad. Sci. USA **83:** 3997–4001.

7. MIYAGAWA, K., H. SAKAMOTO, T. YOSHIDA, Y. YAMASHITA, Y. MITSUI, M. FURUSAWA, S. MAEDA, F. TAKAKU, T. SUGIMURA & M. TERADA. 1988. Oncogene **3:** 383–389.

8. YOSHIDA, T., M. TSUTSUMI, H. SAKAMOTO, K. MIYAGAWA, S. TESHIMA, T. SUGIMURA & M. TERADA. 1988. Biochem. Biophys. Res. Commun. **155:** 1324–1329.

9. WADA, A., H. SAKAMOTO, O. KATOH, T. YOSHIDA, J. YOKOTA, P. F. R. LITTLE, T. SUGIMURA & M. TERADA. 1988. Biochem. Biophys. Res. Commun. **157:** 828–835.

10. YOSHIDA, M. C., M. WADA, H. SATOH, T. YOSHIDA, H. SAKAMOTO, K. MIYAGAWA, J. YOKOTA, T. KODA, M. KAKINUMA, T. SUGIMURA & M. TERADA. 1988. Proc. Nat. Acad. Sci. USA **85:** 4861–4864.

11. PETERS, G., S. BROOKES, R. SMITH, M. PLACZEK & C. DICKSON. 1989. Proc. Nat. Acad. Sci. USA **86:** 5678–5682.

12. TSUDA, H., S. HIROHASHI, Y. SHIMOSATO, T. HIROTA, S. TSUGANE, H. YAMAMOTO, N. MIYAJIMA, K. TOYOSHIMA, T. YAMAMOTO, J. YOKOTA, T. YOSHIDA, H. SAKAMOTO, M. TERADA & T. SUGIMURA. 1989. Cancer Res. **49:** 3104–3108.

13. ADNANE, J., P. GAUDRAY, M.-P. SIMON, J. SIMONY-LAFONTAINE, P. JEANTEUR & C. THEILLET. 1989. Oncogene **4:** 1389–1395.

14. TSUDA, T., E. TAHARA, G. KAJIYAMA, H. SAKAMOTO, M. TERADA & T. SUGIMURA. 1989. Cancer Res. **49:** 5505–5508.

15. TAIRA, M., T. YOSHIDA, K. MIYAGAWA, H. SAKAMOTO, M. TERADA & T. SUGIMURA. 1987. Proc. Nat. Acad. Sci. USA **84:** 2980–2984.

16. KODA, T., A. SASAKI, S. MATSUSHIMA & M. KAKINUMA. 1987. Jpn J. Cancer Res. (Gann) **78:** 325–328.

17. NAKAGAMA, H., S. OHNISHI, M. IMAWARI, H. HIRAI, F. TAKAKU, H. SAKAMOTO, M. TERADA, M. NAGAO & T. SUGIMURA. 1987. Jpn J. Cancer Res. (Gann) **78:** 651–654.

18. YUASA Y. & K. SUDO. 1987. Jpn J. Cancer Res. (Gann) **78:** 1036–1040.

19. DELLI BOVI, P. & C. BASILICO. 1987. Proc. Nat. Acad. Sci. USA **84:** 5660–5664.

20. ADELAIDE, J., M.-G. MATTEI, I. MARICS, F. RAYBAUD, J. PLANCHE, O. DE LAPEYRIERE & D. BIRNBAUM. 1988. Oncogene **2:** 413–416.

21. ZHAN, X., A. CULPEPPER, M. REDDY, J. LOVELESS & M. GOLDFARB. 1987. Oncogene **1:** 369–376.

22. YOSHIDA, T., K. MIYAGAWA, H. ODAGIRI, H. SAKAMOTO, P. F. R. LITTLE, M. TERADA & T. SUGIMURA. 1987. Proc. Nat. Acad. Sci. USA **84:** 7305–7309.

23. FASANO, O., D. BIRNBAUM, L. EDLUND, J. FOGH & M. WIGLER. 1984. Mol. Cell. Biol. **4:** 1695–1705.

24. ZHAN, X. & M. GOLDFARB. 1986. Mol. Cell. Biol. **6:** 3541–3544.

25. MAEDA, S., T. KAWAI, M. OBINATA, H. FUJIWARA, T. HORIUCHI, Y. SAEKI, Y. SATO & M. FURUSAWA. 1985. Nature London **315:** 592–594.

26. MIYAJIMA, A., J. SCHREURS, K. OTSU, A. KONDO, K. ARAI & S. MAEDA. 1987. Gene **58:** 273–281.

27. MARUMOTO, Y., Y. SATO, H. FUJIWARA, K. SAKANO, Y. SAEKI, M. AGATA, M. FURUSAWA & S. MAEDA. 1987. J. Gen. Virol. **68:** 2599–2606.

28. IMAMURA, T. & Y. MITSUI. 1987. Exp. Cell Res. **172:** 92–100.

29. JAKOBOVITS, A., G. M. SHACKLEFORD, H. E. VARMUS & G. R. MARTIN. 1986. Proc. Nat. Acad. Sci. USA **83:** 7806–7810.

30. YOSHIDA, T., H. MURAMATSU, T. MURAMATSU, H. SAKAMOTO, O. KATOH, T. SUGIMURA & M. TERADA. 1988. Biochem. Biophys. Res. Commun. **157:** 618–625.

Expression and Possible Functions of the FGF-5 Gene

MITCHELL GOLDFARB, BRIAN BATES,
BEVERLY DRUCKER, JEFF HARDIN, AND OLIVIA HAUB

Department of Biochemistry and Molecular Biophysics
Columbia University College of Physicians and Surgeons
630 West 168th Street
New York, New York 10032

INTRODUCTION

Fibroblast growth factors (FGFs) are structurally related proteins encoded by at least seven distinct genes in mammals.[a,8] FGFs are broad-spectrum mitogens and mediators of a wide range of developmental phenomena assayed *in vitro* and *in vivo* (reviewed in Reference 3). The fates of vertebrate neural cell precursors, sympatho-adrenal progenitors, skeletal muscle precursors, and amphibian undifferentiated embryonic ectoderm can all be influenced *in vitro* by prototypic FGFs (acidic FGF and basic FGF). FGF *in vivo* implants can promote angiogenesis and the formation of new nerve fiber tracts, and prevent retrograde degeneration of neurons in the central nervous system. Consistent with their potential roles in development, several FGF genes are expressed at various stages of embryogenesis (reviewed in Reference 8).

The biosynthesis of FGFs has several unusual features. First, whereas HST/K-FGF, KGF, and INT-2 proteins have hydrophobic N-terminal sequences which direct the proteins into the secretory pathway[4-6] acidic and basic FGFs lack such sequences and have more complex posttranslational fates.[3] Second, messenger RNA sequences 5' to the initiator AUG codon can impact upon FGF biosynthesis, either by providing upstream and in-frame CUG leucine codons to be utilized as alternative sites for translation initiation (bFGF and INT-2 RNAs),[1,7,20] or by containing an additional small second AUG-initiated open reading frame which might suppress FGF translation (INT-2 RNA).[5]

Fibroblast growth factor-5 (FGF-5) was discovered in our laboratory as the product of a human oncogene detected by DNA transfection assays.[31] Later cloning of the murine FGF-5 gene[9] revealed dramatic amino acid sequence conservation between mouse and human proteins, with 97% residue conservation throughout most of the protein sequences (268 amino acid residues for human FGF-5 precursor, 264 residues for murine protein). By contrast, FGF-5 is 30–50% homologous to other FGFs within a core region of 120 amino acid residues. A hydrophobic N-terminus in the predicted protein sequence suggests that FGF-5 is secreted through the classical pathway, and a second upstream AUG-initiated open reading frame in human and murine FGF-5 RNAs suggests a role for these sequences in the control of FGF-5 translation. The murine FGF-5 gene is expressed through most phases of embryogenesis,[11] and is also expressed weakly and exclusively in the adult central nervous system.[9]

---

[a]Basic FGF (bFGF), acidic FGF (aFGF), INT-2, HST/K-FGF, FGF-5, FGF-6, keratinocyte growth factor (KGF)

We have analyzed the biosynthesis of human FGF-5 at the translational and posttranslational levels, and we have characterized the sites of murine FGF-5 gene expression in embryos of various developmental stages and in the adult brain. These data were originally published as journal articles,[2,9,10] and highlights of these findings are offered in this presentation.

MATERIALS AND METHODS

All procedures are documented in our previous publications.[2,8,9]

RESULTS

Glycosylation of Secreted FGF-5

We have analyzed the apparent molecular weight of FGF-5 secreted into the culture medium of NIH 3T3 cells transformed by a human FGF-5 cDNA expression vector. Conditioned medium was shaken overnight at 4°C with a small volume of heparin-Sepharose, to which all mitogenic activity bound. The Sepharose beads were recovered by centrifugation, boiled in Laemlli sample buffer, and the eluted protein subjected to sodium dodecyl sulfate-polyacrylamide gel electrophoresis (SDS-PAGE) and immunoblotting. An FGF-5 specific, affinity purified antipeptide serum was used together with Vectastain developing reagents to detect FGF-5 protein. As shown in FIGURE 1, several species of apparent molecular weights 32,500 to 38,000 were detected by immunoblotting (lane e), while no species were detected in protein similarly recovered from HST/K-FGF transformed 3T3 cells (data not shown). All secreted FGF-5 species were larger than the 29,500 molecular weight expected of FGF-5 precursor, suggesting posttranslational modification in the form of glycosylation.

In order to test whether FGF-5 is glycosylated, aliquots of the factor were treated with neuraminidase, N-glycanase, or both enzymes combined, and were then analyzed by immunoblotting. Neuraminidase treatment of FGF-5 yielded predominantly one product of 32.5 kDa (lanes c and d), demonstrating that the larger native species (lane e) contain variable amounts of sialic acid. Further treatment with N-glycanase reduced most of the FGF-5 to 30 kDa (lane b) due to cleavage of an N-linked carbohydrate moiety presumably linked to asparagine-110, part of the sole consensus motif for N-linked sugar attachment (Asn_{110}-Ser_{112}). The major product of neuraminidase and N-glycanase treatment was still larger than the predicted polypeptide chain following N-terminal signal sequence cleavage (27 kDa). We suspect that this discrepancy is due to additional O-linked carbohydrate on FGF-5 molecules. Digestion of FGF-5 with N-glycanase alone (lane a) yielded a series of products of between 30 and 36 kDa, including a novel 31.5 kDa species not seen in the other digestions. Such a result would be obtained if O-linked carbohydrate chains bear some or all of the sialic acid residues.

ORF-1 Inhibits FGF-5 Translation

FGF-5 mRNA contains a short open reading frame (ORF-1) upstream and overlapping the FGF-5 coding sequence (FIGURE 2). We speculated that the upstream sequences containing ORF-1 suppress FGF-5 translation by one or more mechanisms: (1) The ORF-1 termination codon and the FGF-5 initiation codon overlap in the sequence AUGA, and FGF-5 synthesis may require ribosome reinitiation following ORF-1 translation. (2) FGF-5 synthesis may require leaky scanning of ribosomes past the two AUG codons in ORF-1. (3) 5' RNA secondary structure may inhibit FGF-5 translation. To distinguish among these possibilities, several site-directed point mutations were constructed in FGF-5 cDNA, as shown in FIGURE 2. The wild-type (wt) 1.1 kilobase pair (kbp) cDNA starts 22 nucleotides 3' to the mRNA 5' end.[31] In mutants C29, T111, and C29/T111, one or both of the ATG

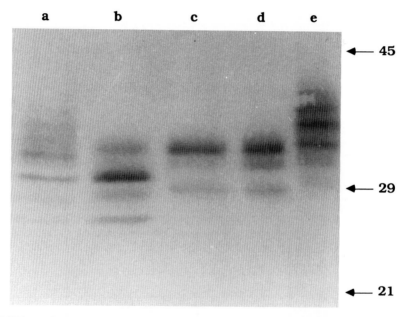

FIGURE 1. Immunoblot analysis of FGF-5 treated with glycosidases. FGF-5 in conditioned medium from 3T3 cells transformed with the FGF-5/δ1-122 vector was concentrated using heparin-Sepharose, and aliquots of eluted proteins were digested overnight with neuraminidase and/or with N-glycanase. The FGF-5 digestion products were detected by SDS-PAGE and immunoblotting. Treatment with N-glycanase alone (lane a), N-glycanase plus neuraminidase (lane b), neuraminidase alone (lanes c and d) under two different reaction conditions, or mock-digested (lane e). Positions of molecular weight standards (in kDa) are indicated.

codons in ORF-1 have been eliminated. In mutant G143, the ORF-1 TGA terminator has been converted to a tryptophan codon with concommitant conversion of the second FGF-5 codon from serine to glycine. A longer ORF-1 in mutant G143 ends 40 nucleotides 3' to the FGF-5 initiation codon. In the δ-122 mutant, deletion of the first 122 nucleotides of the cDNA has abolished ORF-1.

FGF-5 RNA was synthesized *in vitro* using wild-type and mutant cDNA templates, and 100 nanograms of each RNA was translated in reticulocyte lysates. The translation products were immunoprecipitated with FGF-5 antibody and analyzed by gel electrophoresis. As shown in FIGURE 3, polypeptides of the same size were

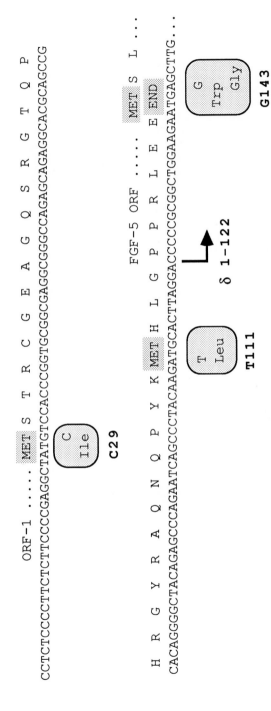

FIGURE 2. Wild-type and mutant FGF-5 cDNAs. The 5' end of the wild-type human FGF-5 cDNA sequence is shown, along with the predicted amino acid sequences for ORF-1 and FGF-5. The 5' end of the δ1-122 mutant, generated by EcoO109 cleavage of wt cDNA, is indicated by the arrow. The nucleotide substitutions in the site-directed mutants C29, T111, and G143 are circled, along with the altered amino acids encoded by mutant codons. The C29/T111 double mutant was generated by sequential cycles of mutagenesis.

generated with each RNA, the largest and most abundant having the predicted molecular weight of the FGF-5 precursor (29.5 kDa). However, translation efficiency varied substantially among the RNAs. FGF-5 RNA lacking ORF-1 due to 5' truncation (lane f) was translated 10–20 times more effectively than wild-type RNA (lane a). The C29 and T111 mutant RNAs were each translated 3- to 4-fold better than wild-type RNA (lanes c and d), while the C29/T111 double mutant was translated as effectively as the deletion mutant (lane e). These data demonstrate that the two ORF-1 AUG codons fully account for the suppressed FGF-5 translation in the native mRNA. In this experiment, the translation efficiency of G143 RNA was equivalent to wild type (lane b). In a different experiment, G143 RNA directed FGF-5 synthesis at ~50% of wild-type level (data not shown). This result argues against a major contribution to FGF-5 synthesis through translation reinitiation. We presume that FGF-5 translation follows the infrequent bypass of scanning ribosomes past the two ORF-1 AUG codons.

These cDNA mutations manifest comparable effects upon FGF-5 synthesis *in vivo,* as monitored indirectly by the transformation efficiencies of cDNA expression vectors transfected into NIH 3T3 cells. In focus formation assays, expression vectors harboring the δ1-122 and C29/T111 mutant cDNAs were 40 to 50-fold more potent than the wt cDNA vector, while vectors bearing the C29 and T111 single mutant cDNAs transformed to an intermediate degree (data not shown). The different transformation efficiencies of vectors are most likely due to differences in *in vivo*

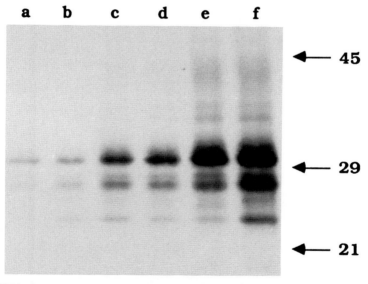

FIGURE 3. *In vitro* translation of wild-type and mutant FGF-5 RNAs. Wild-type and mutant FGF-5 cDNAs cloned into pSP65 were linearized with HindIII to generate templates for *in vitro* transcription with SP6 RNA polymerase. Equivalent amounts of each RNA product were translated *in vitro* using rabbit reticulocyte lysates and ^{35}S-methionine. FGF-5 was immunoprecipitated from reaction products with antipeptide antibodies and detected by SDS-PAGE and fluorography. **Lane a,** wt RNA; **b,** G143; **c,** C29; **d,** T111; **e,** C29/T111; **f,** δ1-122. These proteins were not detected in translation reactions lacking exogenous RNA, and comparable variation in FGF-5 synthesis is seen upon direct electrophoretic analysis without immunoprecipitation (data not shown). The positions of ^{14}C-radiolabeled protein standards are indicated (in kDa).

TABLE 1. Embryonic Sites of FGF-5 Gene Expression

Expression Site	Days of Gestation	Comments
Epiblast and surrounding visceral endoderm	5.3– 7.5	Negative in E3.5 blastocyst; Negative in egg cylinder mesoderm and extra-embryonic tissues
Lateral splanchnic meso-derm adjacent to hepatic bud	9.5–10.5	Negative in all other splanchnic mesoderm
Region of lateral somatic mesoderm	10.5–12.5	Prominent expression near arterial vessels
Myotomal muscle lineages	10.5–12.5 weak at 13.5	Negative in limb muscle lineages
Mastication muscle lineage	11.5–14.5	Negative in other cranial skeletal muscles E11.5 expression in myoblasts or mesenchyme
Mesenchyme near base of limb (hind > fore)	12.5–14.5	Negative in limb muscle and cartilage
Acoustic ganglion	12.5–14.5	Negative in other cranial ganglia

translation efficiency, since all vectors were comparably well transcribed, as monitored by northern blot analysis of transfected cell populations (data not shown).

Detection of FGF-5 RNA in Mouse Embryos by In Situ Hybridization

In order to localize FGF-5 expression during embryogenesis, deparaffinized sections of prefixed mouse embryos at various stages of development were hybridized *in situ,* using radiolabeled antisense and sense FGF-5 RNA probes. Two different antisense probes, transcribed from nonoverlapping segments of FGF-5 cDNA, detected identical profiles of gene expression. Hybridization to deparaffinized sections gives signals 3 to 4-fold weaker than to postfixed, fresh-frozen sections (our unpublished observations); in fact, our *in situ* analysis of FGF-5 expression in adult mouse brain required the use of fresh-frozen sections to detect the very low abundance of FGF-5 RNA in this tissue (see below). However, morphology in sections of fresh-frozen embryos is very poor, mandating our use of paraffin-embedded material in these studies. With this technique, we have detected seven distinct sites of FGF-5 RNA expression in the mouse between embryonic days 3½ to 15½ postconceptus,[b] and these findings are summarized in TABLE 1.

Embryonic expression can be divided temporally into two categories: (1) early expression commencing postimplantation E5¼ and terminating by E8, and (2) later expression commencing at E9½ and extending through E14½. Later expression is detected at six distinct sites restricted in time, tissue, and space.

[b]We use standard nomenclature to define conception as midnight preceding the morning of vaginal plug detection. E3½ = embryonic day 3½ postconceptus.

Expression of FGF-5 RNA in Early Embryogenesis

Embryonic FGF-5 expression first occurs soon after uterine implantation. As can be seen in FIGURE 4, there is no detectable expression in preimplantation blastocysts (panels E and F). FGF-5 RNA was detected in embryos at the next stage examined, E5¼ (panels A and B), while no signal was seen with the negative control (sense-strand) probe (panels C and D). The postimplantation induction of expression is consistent with the observed rapid induction of FGF-5 mRNA levels when embryonal carcinoma cells differentiate as embryoid bodies *in vitro.*[11]

FGF-5 gene expression in E5¾ embryos is restricted to the embryonic ectoderm and adjacent visceral endoderm (FIGURE 4, panels G and H). By contrast, extraembryonic ectoderm, its flanking visceral endoderm, and other extraembryonic tissues are negative. While expression in the embryonic ectoderm of the egg cylinder persists at E7, newly formed mesoderm is negative (panels K and L). A cross-section through the E7 egg cylinder illustrates FGF-5 RNA expression throughout the epiblast, without substantial lateral/medial or anterior/posterior variation (panels I and J). The E7½ embryo shows weaker FGF-5 RNA signals, and expression is undetectable in the E8 embryo (data not shown).

FGF-5 RNA in Lateral Mesoderm

The salient feature of FGF-5 expression in lateral mesoderm and limb mesenchyme (see below) is its spatial restriction. Such expression defines regions of mesoderm that are not morphologically apparent or previously appreciated with other molecular probes.

On embryonic day 9½, the FGF-5 gene is expressed in one small region of splanchnic mesoderm ventral to the portion of the foregut bearing the hepatic bud (FIGURE 5, panels A-C), and expression in this mesenchyme is undetectable by E11½. The onset of expression is approximately coincident with that of rapid liver growth, which proceeds by the invasion of hepatic cords from the hepatic bud into the underlying mesenchyme.

FGF-5 RNA is expressed in a region of somatic mesoderm between E10½ and E12½. On embryonic day 10½, this region extends rostrocaudally from the level of the newly formed sixth aortic arch to approximately the level of the liver primordium. FIGURE 5 shows this expression in a transverse section through an E10½ embryo (panels F and G) and a parasaggital section through an E11½ embryo (panels H and I). FGF-5 expression also occurs in myotomes at these times (see below). The somatic mesoderm signal is readily distinguished from that of skeletal muscle precursors by hybridization of adjacent sections with a probe for α-cardiac actin mRNA,[22] which is expressed in striated muscle cells and their precursors. This is illustrated in FIGURE 5, panel J, where the actin probe detects myotomes and cardiac muscle in the E11½ embryo, but does not hybridize to somatic mesoderm.

FGF-5 Expression in Skeletal Muscle Precursor Cells

FGF-5 RNA is expressed in myotomes starting on E10½ (FIGURE 5, panels D-G). This induction occurs later than the morphological appearance of myotomes (E8½) and their expression of α-cardiac actin mRNA (E8½-E9½). On embryonic day 10½, only myotomes anterior to the middle of the hindlimb bud are FGF-5 positive. More caudal myotomes also express FGF-5 RNA a day later, reflecting their delayed

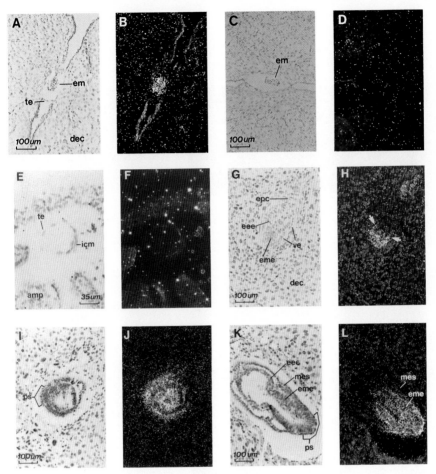

FIGURE 4. FGF-5 RNA in early postimplantation embryos. Antisense and sense FGF-5 ^{35}S-riboprobes were hybridized to egg cylinders in decidua and to blastocysts manually loaded into ampulae of pseudopregnant mice. Dark-field imaging (panels B, D, F, H, J, and K) makes exposed silver grains from NTB2 emulsion luminesce. Panels C and D used sense probe, while all other panels used the antisense probe. Panels **A**, **B**, **C**, and **D**, E5¼ egg cylinder; **E** and **F**, E3½ blastocyst; **G** and **H**, E5¾ egg cylinder, showing expression in embyronic ectoderm and adjacent visceral endoderm (arrows); **I** and **J**, cross-section through E7 egg cylinder; **K** and **L**, transverse section through E7 egg cylinder, showing expression in embyronic ectoderm, but not mesoderm. Embryo (em), deciduum, (dec), blastocyst inner cell mass (icm), trophectoderm (te), ampulla (amp), embryonic ectoderm (eme), extraembryonic ectoderm (eee), visceral endoderm (ve), ectoplacental cone (epc), mesoderm (mes), primitive streak (ps), microns (um).

development, but tail region myotomes never express FGF-5 (data not shown). Myotomal cells continue to express FGF-5 RNA on E11½ (FIGURE 5, panels H-J) and E12½ (data not shown) during their ventral and lateral migration. Expression in trunk muscle precursor cells is undetectable by E14½. By contrast to trunk muscle

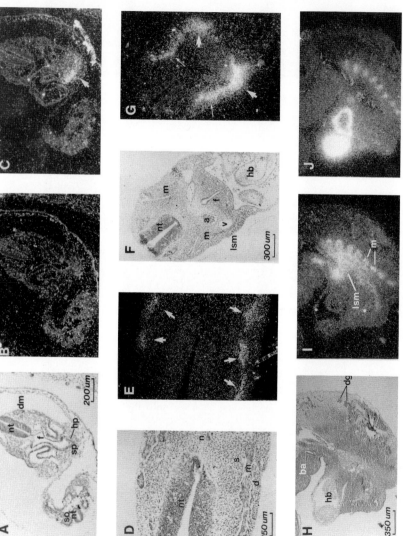

FIGURE 5. FGF-5 RNA in lateral mesoderm and myotomes. Hybridizations were with antisense FGF-5, sense FGF-5, and α-cardiac actin ^{32}P-riboprobes. Panels B, C, E, G, I, and J are dark field images. **Panels A–C**, adjacent E9½ sections hybridized with sense (A and B) or antisense probe (C) with FGF5-positive splanchnic mesoderm (arrow); **panels D and E**, E10½ longitudinal section through rostral somites, antisense probe with FGF5-positive myotomes (arrows); **panels F and G**, E10½ transverse section through rostral tip of heart bulge, antisense probe, with FGF5-positive myotomes (thin arrows) and lateral somatic mesoderm (thick arrows); **panels H–J**, adjacent E11½ parasaggital sections hybridized with antisense FGF-5 (H and I) and α-cardiac actin (J) probes. Hepatic bud (hp), foregut (f), neural tube (nt), notocord (n), dermatome (d), myotome (m), dermamyotome (dm), somite (so), sclerotome (s), dorsal aorta (a), cardinal vein (v), heart bulge (hb), lateral somatic mesoderm (lsm), branchial arch (ba).

differentation, precursors of limb muscles, identifiable by α-cardiac actin RNA expression (see below), do not express FGF-5 RNA at any stage in their development.

A prominent site of FGF-5 RNA expression in the head of the E13½ embryo corresponds to mastication muscle precursor cells (data not shown), as revealed by coexpression of α-cardiac actin RNA. Facial, tongue, and ocular muscle lineages, as recognized by α-cardiac actin hybridization, are also FGF-5 negative (data not shown). The strong FGF-5 mastication muscle signal persists on E14½, but is completely absent 24 hours later (data not shown). This site of FGF-5 RNA expression originates on embryonic day 11½ as bilateral patches, each situated near the juncture of the maxillary process and the mandibular arch. Curiously, α-cardiac actin RNA expression is very weak at this site, and only becomes prominent on E12½ (data not shown).

FGF-5 RNA in Developing Limbs

FGF-5 RNA has been detected in developing limbs from embryonic days 12½ through 14½. Expression is limited to a patch of cells near the base of each limb, and is more prominent in hindlimbs than forelimbs. FIGURE 6 shows the expression patch in a transverse section through the E12½ hindlimb (panels A and B). In the E13½ embryo, parasagittal sections which effectively represent cross-sections through the base of the limb show that the mesenchymal region of FGF-5 expression (panel E) is ventral to the femur (panel D), which is undergoing chondrification. The expression site is distinct from the various developing muscle groups, which are visualized in adjacent section by their expression of α-cardiac actin RNA (panel E). The tracing in panel G illustrates the spatial relationship of FGF-5 positive mesenchyme to developing muscle and cartilage. The weaker FGF-5 RNA signal in forelimbs is positioned similarly to that in hindlimbs (data not shown).

FGF-5 RNA in the Acoustic Ganglion

FGF-5 RNA has been detected in the acoustic branch of the eighth cranial ganglion within the inner ear on embryonic days 12½ through 14½ (FIGURE 6, panels G-J). All other cranial ganglia lack detectable FGF-5 RNA, including the vestibular branch of the eighth ganglion (FIGURE 6, panels G, H) and the trigeminal fifth ganglion (FIGURE 5, panels D and F).

Expression of the FGF-5 Gene in the Adult Brain

Northern blot hybridization has detected FGF-5 RNA in most dissected portions of brain from 8-week-old mice.[9] Expression levels overall were quite low, with somewhat higher expression in the hippocampus and cerebral cortex, and undetectable expression in the hypothalamus. Detection of FGF-5 RNA *in situ* was not possible using deparaffinized sections of brain. Only with postfixed fresh-frozen tissue sections were weak FGF-5 RNA signals in hippocampus, cerebral cortex, and thalamus observable.

Hippocampal FGF-5 RNA was found in the dentate gyrus, presubiculum, and the pyramidal cell layer (data not shown). This lattermost site is composed almost exclusively of pyramidal neuron soma; hence, at least some of FGF-5 RNA synthe-

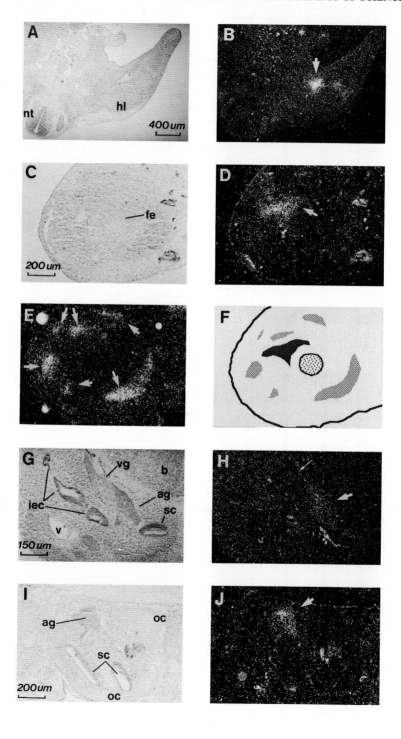

sized in brain is of neural origin. Curiously, only pyramidal neurons in the CA3 field were FGF-5 positive, while adjacent pyramidal neurons in CA2 and CA1 expressed little or no FGF-5 RNA.

DISCUSSION

Human FGF-5 is secreted from cells as glycoprotein molecules bearing N-linked and, most likely, O-linked sugars. Glycosylation is not required to give the protein its mitogenic activity. We have recently expressed and purified recombinant human FGF-5 in *E sclierichia coli,* and as little as one nanogram of this material can maximally stimulate DNA synthesis in Balb/c 3T3 fibroblasts (J.-K. Wang and M. Goldfarb, unpublished data). The biological significance of FGF-5 carbohydrate modification is currently unknown.

The regulation of murine FGF-5 gene expression during development is highly complex. Not only is the gene's expression tissue and temporally specific, but even within a given tissue, expression is spatially restricted. Such regulation has defined subdivisions of splanchnic mesoderm, somatic mesoderm, and limb mesenchyme that are not morphologically evident. The spatial restriction of expression may, in part, be governed by combinations of transcription factors such as homeobox proteins. For example, FGF-5 gene expression in limb mesenchyme may result from the combined actions of homeobox genes that are differentially expressed along anterior-posterior and proximal-distal axes of the limb.[26]

The structurally related INT-2 gene is also expressed in a complex spatiotemporal pattern, although the sites of expression differ from those for the FGF-5 gene.[29,30] At certain stages in development, FGF-5 and INT-2 are expressed in close proximity, such as FGF-5 RNA in E7½ embryonic ectoderm vs. INT-2 RNA in E7½ embryonic mesoderm. Since both of these growth factors are secreted proteins, we suspect that FGF receptors expressed in these regions differentially react with or respond to these two FGFs. While the known spectra of *in vitro* biological activities are similar among FGFs, there are contrasts[6,16,28] which are presumably mediated by differences in ligand-receptor interactions.

FGF-5 is likely to mediate a diverse set of events during embryogenesis. The FGF-5 expression profile along with known biological effects of other FGFs in *in vitro* and *in vivo* assays allows for speculation regarding FGF-5's native functions.

FIGURE 6. FGF-5 RNA in hindlimb and acoustic ganglion. Sections were hybridized with antisense FGF-5 and α-cardiac actin ^{32}P-riboprobes. Panels B, D, E, H and J are dark-field images. **Panels A** and **B,** hybridization of FGF-5 antisense probe to E12½ transverse section through hindlimb, showing positive patch near base of limb (arrow). **Panels C–E,** adjacent E13½ parasagittal sections through base of hindlimb hybridized with FGF-5 antisense (C and D) and α-cardiac actin (E) probes. Arrows in D and E denote FGF5-positive mesenchyme and limb muscle groups, respectively. **Panel F,** E13½ hindlimb schematic derived by tracings from panels C-E, showing femur cartilage (spotted), α-cardiac actin-positive muscle precursors (light shaded), and FGF-5 positive mesenchyme (dark shaded). **Panels G** and **H,** longitudinal section through head of E12½ embryo hybridized with FGF-5 antisense probe, showing expression in acoustical branch of eighth ganglion (bold arrow), but not in more lateral vestibular branch (thin arrow). **Panels I** and **J,** parasagittal section through head of E14½ embryo hybridized with antisense FGF-5 probe. Arrow denotes expression in acoustic ganglion. Hindlimb (h1), neural tube (nt), femur (fe), acoustic ganglion (ag), vestibular ganglion (vg), brain (b), spiral canal (sc), other canals of inner ear (iec), otic capsule (oc), head vein (v).

Uterine implantation of the blastocyst is followed by rapid growth of the egg cylinder and rapid growth of the uterine deciduum. FGF-5 expressed in embryonic ectoderm postimplantation could serve as an autocrine or paracrine factor to mediate these growth events. Egg cylinder expression of the FGF-5 gene might also contribute to mesoderm induction, a phenomenon inducible *in vitro* with amphibian ectodermal explants and bFGF.[12,25] Plausible arguments against a direct role for FGF-5 in mammalian mesoderm formation are (1) the onset of expression ($\leq$ E5¼) far precedes the start of gastrulation (E6½), and (2) FGF-5 RNA is not spatially restricted within the embryonic ectoderm.

FGF-5 might act as an angiogenic factor in lateral mesoderm, which becomes more highly vascularized than dorsal mesoderm.[15,24] In splanchnic mesoderm, local FGF-5 expression might promote vascularization required for the induction of neighboring hepatic cord proliferation.[24] In somatic mesoderm, local FGF-5 expression could contribute to the ongoing remodeling of the arterial vasculature.[33]

Fibroblast growth factors can inhibit the terminal differentiation of myoblasts, as monitored by biochemical markers and by myotube formation.[14,23] FGF-5 expression commences in myotomes on E10½ well after their formation (E8½) and commitment to muscle lineage, as gauged by the onset of α-cardiac actin (E9) and MyoD (E9½) gene expression.[21] FGF-5 expression diminishes by E13½, before the first appearance of myotubes on E15.[33] FGF-5 could inhibit the terminal differentiation of myotomal myoblasts during their migration through the trunk.

In vitro culture of dissociated chicken limb bud cells has revealed an FGF-dependent subcomponent of myoblasts which require treatment with basic FGF in order to ultimately form colonies of terminally differentiated muscle[23] (Seed and Hauschka, 1988). It is possible that E12½-14½ FGF-5 expression in limb mesenchyme acts to induce the development of FGF-dependent limb myoblasts.

The synthesis of FGF-5 is influenced by an upstream open reading frame in the messenger RNA. Each of the two ORF-1 AUG codons are preceded by a purine in the −3 position, making them favorable sites for translation initiation.[32] FGF-5 translation initiation would follow the infrequent scanning of ribosomes past the ORF-1 AUG codons. Other examples of the effects of upstream AUG codons in mRNAs have been documented.[17-19] The proximity of the ORF-1 termination codon and the FGF-5 initiation codon may allow for a low frequency of FGF-5 synthesis by ribosome reinitiation following ORF-1 translation termination, as has been documented for mutated preproinsulin mRNA.[13]

What does ORF-1 do? This region is evolutionarily conserved between mouse and man. However, the hypothetical ORF-1 protein has several nonconservative amino acid substitutions between species, and the ORF-1 region has no silent, third base mutations. Based on evolutionary considerations, we do not believe that ORF-1 protein is biologically active. Rather, we suggest that the ORF-1 nucleotide sequence exists to impart translational regulation upon FGF-5 synthesis in *in vivo* contexts yet to be identified.

The experiments described here have allowed for speculations, but no certainties, regarding FGF-5 function. The identification and localization of FGF-5 receptors is of obvious importance, and such studies are now possible with the availability of purified growth factor. Additionally, the use of genetic methods to disrupt FGF-5 expression in some or all of its natural sites may generate informative phenotypes which, together with our knowledge of the FGF-5 expression profile, can reveal the biological functions of the growth factor during development.

REFERENCES

1. ACLAND, P., M. DIXON, G. PETERS & C. DICKSON. 1990. Subcellular fate of the Int-2 oncoprotein is determined by choice of initiation codon. Nature **343:** 662–665.
2. BATES, B., J. HARDIN, X. ZHAN, K. DRICKAMER & M. GOLDFARB. 1991. Biosynthesis of human fibroblast growth factor-5 Mol. Cell. Biol. **11:** 1840–1845.
3. BURGESS, W. H. & T. MACIAG. 1989. The heparin-binding (fibroblast) growth factor family of proteins. Annu. Rev. Biochem. **58:** 575–606.
4. DELLI-BOVI, P., A. M. CURATOLA, K. M. NEWMAN, Y. SATO, D. MOSCATELLI, R. M. HEWICK, D. B. RIFKIN & C. BASILICO. 1988. Processing, secretion, and biological properties of a novel growth factor of the fibroblast growth factor family with oncogenic potential. Mol. Cell. Biol. **8:** 2933–2941.
5. DIXON, M., R. DEED, P. ACLAND, R. MOORE, A. WHYTE, G. PETERS & C. DICKSON. 1989. Detection and characterization of the fibroblast growth factor-related oncoprotein INT-2. Mol. Cell. Biol. **9:** 4896–4902.
6. FINCH, P. W., J. S. RUBIN, T. MIKI, D. RON & S. A. AARONSON. 1989. Human KGF is FGF-related with properties of a paracrine effector of epithelial cell growth. Science **245:** 752–755.
7. FLORKIEWICZ, R. Z. & A. SOMMER. 1989. Human basic fibroblast growth factor gene encodes four polypeptides: three initiate from non-AUG codons. Proc. Nat. Acad. Sci. USA **86:** 3978–3981.
8. GOLDFARB, M. 1990. The fibroblast growth factor family. Cell Growth Differ. **1:** 439–445.
9. HAUB, O., B. DRUCKER & M. GOLDFARB. 1990. Expression of murine fibroblast growth factor-5 in the adult central nervous system. Proc. Nat. Acad. Sci. USA **87:** 8022–8026.
10. HAUB, O. & M. GOLDFARB. 1991. Expression of the fibroblast growth factor-5 gene in the mouse embryo. Development **112:** 397–406.
11. HEBERT, J. M., C. BASILICO, M. GOLDFARB, O. HAUB & G. R. MARTIN. 1990. Isolation of cDNAs encoding four mouse FGF family members and characterization of their expression during embryogenesis. Dev. Biol. **138:** 454–463.
12. KIMELMAN, D. & M. KIRSCHNER. 1987. Synergistic induction of mesoderm by FGF and TGF-β and the identification of an mRNA coding for FGF in the early *Xenopus* embryo. Cell **51:** 869–877.
13. KOZAK, M. 1987. Effects of intercistronic length on the efficiency of reinitiation by eucaryotic ribosomes. Mol. Cell. Biol. **7:** 3438–3445.
14. LATHROP, B., E. OLSON & L. GLASER. 1985. Control by fibroblast growth factor of differentiation in the BC3H1 muscle cell line. J. Cell Biol. **100:** 1540–1547.
15. LE DOUARIN, N. M. 1975. An experimental analysis of liver development. Med Biol. **53:** 427–455.
16. LIPTON, S. A., J. A. WAGNER, R. D. MADISON & P. A. D'AMORE, 1988. Acidic fibroblast growth factor enhances regeneration of processes by postnatal mammalian retinal ganglion cells in culture. Proc. Nat. Acad. Sci. USA **85:** 2388–2392.
17. MARTH, J. D., R. W. OVERALL, K. E. MEIER, E. G. KREBS & R. M. PERLMUTTER. 1988. Translational activation of the *lck* proto-oncogene. Nature **332:** 171–173.
18. MUELLER, P. P. & A. G. HINNEBUSCH. 1986. Multiple upstream AUG codons mediate translational control of GCN4. Cell **45:** 201–207.
19. OZAWA, K., J. AYUB & N. YOUNG. 1988. Translational regulation of B19 parvovirus capsid protein production by multiple upstream AUG triplets. J. Biol. Chem. **263:** 10922–10926.
20. PRATS, H., M. KAGHAD, A. C. PRATS, M. KLAGSBRUN, J. M. LELIAS, P. LIAUZUN, P. CHALON, J. P. TAUBER, F. AMALRIC, J. A. SMITH & D. CAPUT. 1989. High molecular mass forms of basic fibroblast growth factor are initiated by alternative CUG codons. Proc. Nat. Acad. Sci. USA **86:** 1836–1840.
21. SASSOON, D., G. LYONS, W. E. WRIGHT, V. LIN, A. LASSAR, H. WEINTRAUB, & M. BUCKINGHAM. 1989. Expression of two myogenic regulatory factors myogenin and MyoD1 during mouse embryogenesis. Nature **341:** 303–307.
22. SASSOON, D. A., I. GARNER & M. BUCKINGHAM. 1988. Transcripts of α-cardiac and

α-skeletal actins are early markers for myogenesis in the mouse embryo. Development **104:** 155–164.

23. SEED, J. & S. D. HAUSCHKA. 1988. Clonal analysis of vertebrate myogenesis. VIII. Fibroblast growth factor (FGF)-dependent and FGF-independent muscle colony types during chick wing development. Dev. Biol. **128:** 40–49.

24. SHERER, G. K. 1975. Tissue interaction in chick liver development: a reevaluation. I. Epithelial morphogenesis: the role of vascularity in mesenchymal specificity. Dev. Biol. **46:** 281–295.

25. SLACK, J. M. W., B. G. DARLINGTON, J. K. HEATH & S. F. GODSAVE. 1987. Mesoderm induction in early *Xenopus* embryos by heparin-binding growth factors. Nature **326:** 197–200.

26. SMITH, S. M., K. PANG, O. SUNDIN, S. E. WEDDEN, C. THALLER & G. EICHELE. 1989. Molecular approaches to vertebrate limb morphogenesis. Development (Suppl.) **107:** 121–131.

27. TZAMARIAS, D., D. ALEXANDRAKI & G. THIREOS. 1986. Multiple *cis*-acting elements modulate the translational efficiency of GCN4 mRNA in yeast. Proc. Nat. Acad. USA **83:** 4849–4853.

28. VALLES, A. M., B. BOYER, J. BADET, G. C. TUCKER, D. BARRITAULT & J. P. THIERY. 1990. Acidic fibroblast growth factor is a modulator of epithelial plasticity in a rat bladder carcinoma cell line. Proc. Nat. Acad. Sci. USA **87:** 1124–1128.

29. WILKINSON, D. G., S. BHATT & A. P. MCMAHON. 1989. Expression pattern of the FGF-related proto-oncogene *int*-2 suggests multiple roles in fetal development. Development **105:** 131–136.

30. WILKINSON, D. G., G. PETERS, C. DICKSON & A. P. MCMAHON. 1988. Expression of the FGF-related proto-oncogene *int*-2 during gastrulation and neurulation in the mouse. EMBO J. **7:** 691–695.

31. ZHAN, X., B. BATES, X. HU & M. GOLDFARB. 1988. The human FGF-5 gene encodes a novel protein related to fibroblast growth factors. Mol. Cell. Biol. **8:** 3487–3495.

32. KOZAK, M. 1986. Point mutations define a sequence flanking the AUG initiator codon that modulates translation by eukaryotic ribosomes. Cell **44:** 283–292.

33. RUGH R. 1968. The Mouse: Its Reproduction and Development. Burgess. Minneapolis, Minn.

The *FGF6* Gene within the *FGF* Multigene Family[a]

FRANÇOIS COULIER, VINCENT OLLENDORFF,
IRÈNE MARICS, OLIVIER ROSNET, MICHÈLE BATOZ,
JACQUELINE PLANCHE, SYLVIE MARCHETTO,
MARIE-JOSÈPHE PEBUSQUE, ODILE deLAPEYRIERE,
AND DANIEL BIRNBAUM

U. 119 INSERM
27 Boulevard Leï Roure
13009, Marseille, France

Cell growth and differentiation are controlled at multiple levels through various pathways. External signals and their specific receptors represent important protagonists of this control. Among the signaling molecules are peptide regulatory factors including numerous neuropeptides, cytokines, and growth factors, often grouped in families. Growth factors and their receptors are involved in various processes including cell proliferation, cell differentiation, cell-cell interactions and development.

The FGF (historically meaning fibroblast growth factor) family presently comprises seven related members (hereafter designated FGF1 to FGF7) but the actual size of the family remains unknown. Like most of the signaling molecules, they have various known properties as well as suspected functions in basic processes as diverse as cell proliferation, angiogenesis, tissue differentiation and regeneration, and embryogenesis.[1] Due to these activities, FGFs may also play an important role in carcinogenesis. This involvement in carcinogenesis is strongly suspected but is still, in early 1991, a matter of discussion. The FGFs can behave as oncogene products or potent angiogenic and mitogenic factors in *in vitro* cell cultures and in animal models.[2-4] Moreover, amplifications of some of the *FGF* (*INT2* and *HST/FGFK*) and *FGF* receptor (*FLG* and *BEK*) genes are observed in human carcinomas.[5-8] In cancer, each member of the couple growth factor–growth factor receptor represents a particular potential therapeutical target aimed at blocking the specific interaction. Therefore, it is paramount to characterize every member of the FGF and FGFR families.

The sixth member of the *FGF* family, the *FGF6* gene, was isolated a couple of years ago[9] by virtue of its sequence similarities with *FGF4* (*HST/FGFK*). The human[10] and murine[10] genes have been characterized, mapped on their respective chromosome, and their expression and transforming capacity studied. These results are the object of this report.

[a]This work was supported by grants from INSERM and from the following organisms: Association Française contre la Myopathie, Association pour la Recherche sur le Cancer, Comités Départementaux de Bouches-du-Rhône et du Var de la Ligue Contre le Cancer, FNCLCC, MGEN.

STRUCTURE OF THE *FGF6* GENE AND ENCODED PROTEINS

The *FGF6* gene is localized on chromosome 12, band p12, in the human and on chromosome 6, region F3-G1, in the mouse. It is composed of three coding exons separated by large introns. Human and mouse deduced amino acid sequences are shown in FIGURE 1. There is 93% amino acid sequence identity between the human and mouse gene products. At the nucleotide level (not shown), the calculated rate of nonsynonymous substitutions is 0.17×10^{-9} *per* site *per* year (assuming a time of divergence between human and rodent genes of 80×10^{6} years ago). This is well below average[11] and indicates a strong conservation.

The localization of the splice junctions is similar to that of the other members of the family. There are four stop codons located immediately upstream of the beginning of the coding sequence, followed by three in-frame ATG codons. With the first ATG starts an open reading frame of 624 nucleotides, encoding a putative protein of 208 amino acids. A potential signal peptide, constituted of a hydrophobic region of 25 amino acids, starts shortly after the methionine residue corresponding to the second ATG, extends from position 16 (considering the first ATG as position 1) to position 40. Two potential proteolytic cleavage sites are located towards the end of the signal peptide. *In vitro* translation experiments (not shown) seem to indicate that it is the first site that is used. A potential asparagine-linked glycosylation site is immediately adjacent, at position 45. This site is functional in *in vitro* translation experiments (not shown).

The comparison of the deduced amino acid sequence of the human *FGF6* gene product with those of the other FGFs (FIGURE 2) shows that the closest relative is the *FGF4* gene product, as was expected considering the cross-reactivity of the respective molecular probes. FGFs have sequence similarities in a "core" portion coded by the 3' half of exon 1, exon 2, and exon 3, but differ greatly in the portion coded by the 5' half of exon 1. In addition, some FGF proteins present specific insertions and extensions. Paralogous FGF6 proteins have more similarities than FGF6 and FGF4 proteins of the same species; this is especially true when considering the portion coded by the 5' part of exon 1. It is possible that exon 1 corresponds to a merger between two ancestral exons, created as a result of the loss of an intron. The structural features of the *FGF6* gene product are summarized in FIGURE 3.

```
Human   MALGQKLFITMSRGAGRLQGTLWALVFLGILVGMVVPSPAGTRANNTLLDSRGWGTLLSR   60
Mouse   =====r============v====Q======v============a===G============

Human   SRAGLAGEIAGVNWESGYLVGIKRQRRLYCNVGIGFHLQVLPDGRISGTHEENPYSLLEI   120
Mouse   =========s==================================P===============

Human   STVERGVVSLFGVRSALFVAMNSKGRLYATPSFQEECKFRETLLPNNYNAYESDLYQGTY   180
Mouse   =============k====i==========t====Hd==================R===

Human   IALSKYGRVKRGSKVSPIMTVTHFLPRI*                                208
Mouse   ============================*
```

FIGURE 1. Conservation between human and mouse FGF6 deduced amino acid sequences. The human FGF6 amino acid sequence is shown using the single-letter code. The mouse corresponding sequence is shown beneath using an = symbol for identical amino acids. Different but conserved amino acids are in lowercase letters. Specific features are indicated as follows: signal peptide, black bar; three potential methionine initiation codons, bold and underlined; N-glycosylation site and cysteine residues, bold, underlined with a symbol. The overall amino acid identity is 93.3% (97.6% when considering conservative changes).

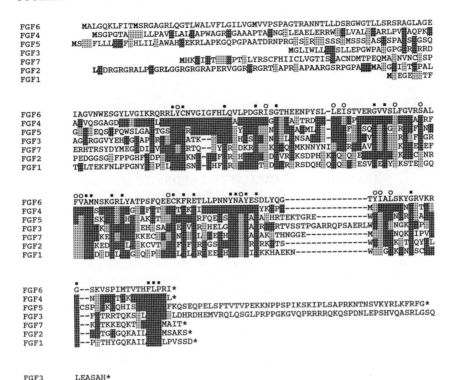

FIGURE 2. Comparison between seven human FGF amino acid sequences. Sequences are aligned, using the single-letter code, to allow comparison with the FGF6 amino acid sequence. Identical amino acids are shown in dark shaded area and conserved amino acids in light shaded area. Initiation methionines are in bold, stops in asterisks. Black squares indicate seven identical residues; and circles, seven identical and conserved residues. FGFs are designated 1 to 7. The correspondence with commonly used names is as follows: FGF1 = aFGF or FGFA or HBGF1; FGF2 = bFGF or FGFB or HBGF2; FGF3 = INT2; FGF4 = HST or HSTF1 or K-FGF or FGFK; FGF7 = KGF. A 5' extension for INT2 has been demonstrated in mouse and probably also exists in human but was not represented. Sequences were taken from References 10 and 27–32.

The structure of the *FGF* genes is very similar in human and mouse. The overall phylogenetic conservation is however difficult to evaluate. A member of the *FGF* gene family has been cloned in frog.[12] We have looked for evidence of such a conservation in lower species using zoo blot hybridizations (FIGURE 4). A murine Fgf6 probe hybridizes well with the human *FGF6* gene as well as with the human *FGF4* (*HST/FGFK*) gene [lighter bands of 6 kilobases (kb) and 2.7 kb]. It lights up clear bands in chicken and frog DNAs. Bands are also present in the *Drosophila* lane. Except for human, it is not known presently whether these cross-reacting bands correspond to homologous genes.

In view of the high degree of sequence similarities existing between them, we can readily surmise that the *FGF* genes are homologues. A phylogenetic tree based on sequence data is shown in FIGURE 5A. *FGF* genes have probably evolved by successive duplications but the evolution may have followed complex schemes at certain stages. Among many possibilities, two such schemes could be hypothesized. In the first, a recent duplication is at the origin of *FGF3* (*INT2*) and *FGF4* (*HST/FGFK*), both located on chromosomal band 11q13.3, separated by less than 40 kb.[13] Then, *FGF3* evolved separately and rapidly diverged from *FGF4,* which remained very similar to *FGF6.* The second scheme hypothesizes a putative gene similar to *FGF3* on human chromosome 12. In FIGURE 5, we have tentatively depicted these hypotheses (note that proteins and not gene sequences were used in this figure).

TRANSFORMING CAPACITY OF CLONED *FGF6* GENE

The normal *FGF6* gene has a transforming capacity comparable to that of *FGF4.*[14-16] This was tested by bioassays involving the transfection of human and murine cloned genes into cultured murine NIH 3T3 fibroblasts followed by either a tumorigenicity assay or a focus assay. Clones of the normal *FGF6* gene were either genomic cosmid clones or cDNA clones inserted into mammalian expression vectors. It is noteworthy that the sequence encoding the signal peptide is necessary for the transforming activity of the transfected *FGF6* gene (Coulier *et al.,* in press). Supernatants from mass culture of the transformed cells and proteins from *in vitro* translated *FGF6* gene are mitogenic for fibroblasts (FIGURE 6). Moreover, *FGF6*-transformed cells are tumorigenic when injected into nude mice.

Despite this intrinsic transforming capacity, the *FGF6* gene has never been detected so far in transfection assays, in contrast to *FGF4* which has been frequently identified in these assays.[17,18] Furthermore, the gene has never been found altered in human tumors. The intrinsic oncogenic capacity of the *FGF* genes must be repressed *in vivo.* However, the mechanisms of this repression are not known.

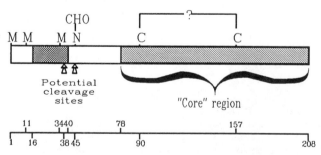

FIGURE 3. Structure of the FGF6 protein. The structure of the FGF6 protein was predicted from cDNA sequence and *in vitro* translation experiments. A long open reading frame starting with three in-frame ATG codons is able to code for a protein of 208 residues. Only the two first ATG codons seem to be used for translation initiation. A stretch of 25 hydrophobic residues, indicated by a hatched box (a bar in FIGURE 1), may act as a signal peptide. Two potential cleavage sites are indicated. A unique N-linked glycosylation site, which appears to be used *in vitro,* is indicated by N. Two cysteine residues, conserved among the seven FGFs described to date, are indicated; they may be involved in a disulfide bridge. A "core" region (hatched) represents a region having sequence similarities in all members. The corresponding amino acid positions are indicated below.

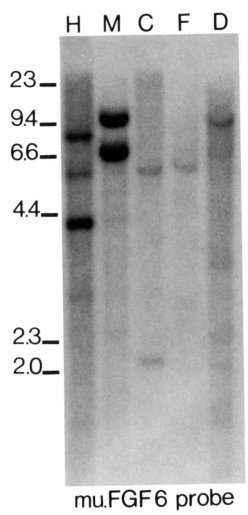

FIGURE 4. Conservation of the *FGF6* gene. Southern "zoo blot" hybridization was carried out at 42°C, in 35% formamide, 5 × SSC, 5 × Denhardt's, on 10 μg of DNA from various species as follows: H, human; M, mouse; C, chicken; F, frog (*Xenopus laevis*); D, fly (*Drosophila melanogaster*). The probe used (probe PN) was a murine *Fgf6* cDNA spanning the three coding exons.

EXPRESSION OF THE *FGF6* GENE IN ADULT AND EMBRYONIC TISSUES

FGF genes are expressed in a wide variety of cells, in the adult as well as in the embryo. *FGF* genes present various patterns of expression in the developing embryo, where they could play an important physiological role as inducers of cell determination and migration.[12,19]

Fgf6 expression was analyzed by northern blot hybridization in the adult and in the embryonic mouse.[10] In adult tissues the *Fgf6* gene is expressed in skeletal muscle and heart, but not in smooth muscle. An *Fgf6* transcript was also detected in adult testis. Use of northern blot hybridization with sense and antisense RNA probes showed that this testis-specific expression corresponds to an antisense transcript from the *Fgf6* locus (FIGURE 7). Presently, nothing is known of its structure and role.

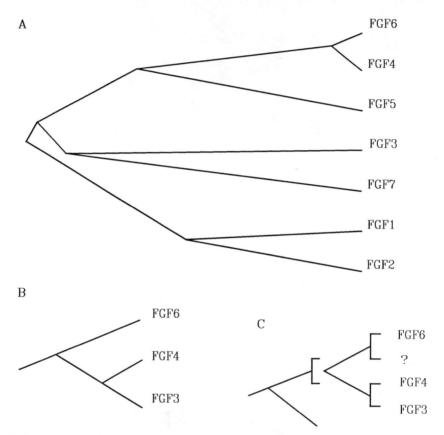

FIGURE 5. Phylogenetic FGF protein trees. **A:** Unrooted phylogenetic tree inferred from the amino acid sequences of the FGFs. The operational units (OU) present at the external nodes are the seven FGF proteins with names used in FIGURE 2. The tree shows one topology only (out of 954 theoretically possible given 7 OU; Reference 11). Branches are scaled, i.e., their lengths are proportional to the number of substitutions. **B** and **C:** portions of unrooted trees inferred from sequence and chromosomal localization data, with focus on three OU; *FGF3* and *FGF4* are clustered on chromosomal band 11q13,[13] while *FGF6* lies on band 12pter.[10] In these particular trees, a specific rapid divergence of FGF3 (B) or a cluster duplication (C) is hypothesized.

The presence of an antisense transcript originating from an *FGF* locus has already been noted in the case of *FGF2* in *Xenopus laevis.*[20]

In mouse embryos, *Fgf6* is expressed during middle and late gestation with a peak at day 15. The transcript has a size of 4.8 kb. As a rule, the expression level is low, necessitating use of poly(A)$^+$ RNA to be detected.

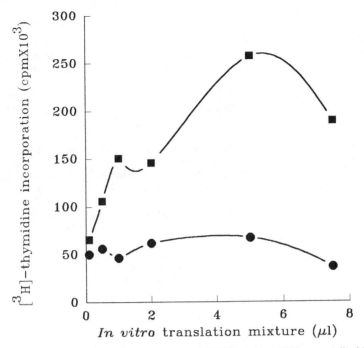

FIGURE 6. Mitogenic activity of *in vitro* translated FGF6 peptides. RNAs transcribed *in vitro* were used in a rabbit reticulocyte lysate system, in the presence of canine pancreatic microsomal membranes, to translate FGF6 peptides. Increasing amounts of *in vitro* translation mixtures were tested for stimulation of [³H]thymidine incorporation of Balb/c 3T3 cells. Samples tested included translation products from antisense (circles) or sense (squares) transcripts.

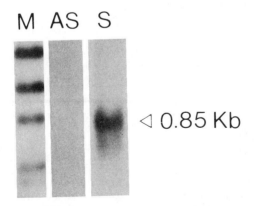

FIGURE 7. Detection of an antisense transcript from the *FGF6* locus by northern blot hybridization. Poly (A⁺) RNA was isolated from 2-month-old mouse testis and hybridized with [³²P]CTP-labeled RNA, transcribed either in the *Fgf6* sense orientation by T7 (lane S) or in the antisense orientation by T3 (lane AS) polymerase from the appropriately linearized cDNA clone p173. This clone represents the mouse *Fgf6* gene first exon. Markers on the left represent Hae III DNA fragments from phage PhiX 174. The sense probe reveals an antisense transcript of 0.85 kilobase pair (kbp) (arrowhead).

CONCLUSION

When the number of the known FGFs suddenly rose from two to seven, it became important to define their specific role and how they assume it. Characterization of the FGF receptors could have helped in resolving this question. However, the analyses of ligand-receptor interactions reveal a peculiar pattern characterized by an ambiguity in the interacting FGF-FGFR couples; for the time being this contributes to darkening the picture a little more. How the cells discriminate between the various FGFs remains unknown. Furthermore, nothing is known about possible interactions between FGFs themselves. Is there such a thing as a coordinate regulation of *FGF* gene expression, at least during certain periods of development, under certain circumstances, or in certain areas? Study of the *FGF6* gene has so far only introduced an element of complexity into these problems. It has shown that some members of the family can be more closely related than the average. It has confirmed that some FGF genes do present an intrinsic transforming activity even if their role in carcinogenesis—especially of those having this capacity—is far from being established. Finally, the restricted pattern of expression of the gene points to a specific activity of FGF6 in the embryo and in cardiac and skeletal muscle cells. Regulating muscular activity or development appears to be an important function of the FGFs, and members of the family seem to exert a control on myoblasts.[21-26] The precise role of FGF6 with respect to muscle function will have to be defined.

ACKNOWLEDGMENTS

We thank C. Mawas for critical reading of the manuscript, and M. Marillet and S. Kerridge for the gift of frog and *Drosophila* DNAs, respectively.

REFERENCES

1. GOLDFARB, M. 1990. Cell Growth Diff. **1:** 439–445.
2. DICKSON, C., R. SMITH, S. BROOKES & G. PETERS. 1984. Cell **37:** 529–536.
3. MULLER, W., F. LEE, C. DICKSON, G. PETERS, P. PATTENGALE & P. LEDER. 1990. EMBO J. **9:** 907–913.
4. HALABAN, R., R. LANGDON, N. BIRCHALL, C. CUONO, A. BAIRD, G. SCOTT, G. MOELLMAN & J. MCGUIRE. 1988. J. Cell. Biol. **107:** 1611–1619.
5. LIDEREAU, R., R. CALLAHAN, C. DICKSON, G. PETERS, C. ESCOT & I. U. ALI. 1988. Oncogene Res. **2:** 285–291.
6. FANTL, V., M. RICHARDS, R. SMITH, G. LAMMIE, G. JOHNSTONE, D. ALLEN, W. GREGORY, G. PETERS, C. DICKSON & D. BARNES. 1990. Eur. J. Cancer **26:** 423–429.
7. TSUDA, T., E. TAHARA, G. KAJIYAMA, H. SAKAMOTO, M. TERADA & T. SUGIMURA. 1989. Cancer Res. **49:** 5505–5508.
8. ADNANE, J., P. GAUDRAY, C. DIONNE, G. CRUMLEY, M. JAYE, J. SCHLESSINGER, P. JEANTEUR, D. BIRNBAUM & C. THEILLET. 1991. Oncogene **6:** 659–663.
9. MARICS, I., J. ADELAIDE, F. RAYBAUD, M.-G. MATTEI, F. COULIER, J. PLANCHE, O. DELAPEYRIERE & D. BIRNBAUM. 1989. Oncogene **4:** 335–340.
10. DELAPEYRIERE, O., O. ROSNET, D. BENHARROCH, F. RAYBAUD, S. MARCHETTO, J. PLANCHE, F. GALLAND, M. G. MATTEI, N. G. COPELAND, N. A. JENKINS, F. COULIER & D. BIRNBAUM. 1990. Oncogene **5:** 823–831.
11. LI, W-H. & D. GRAUR. 1991. Fundamentals in Molecular Evolution. Sinauer Assoc. Sunderland, Mass.
12. KIMELMAN, D., J. ABRAHAM, T. HAAPARANTA, T. PALISI & M. KIRSCHNER. 1988. Science **242:** 1053–1056.

13. NGUYEN, C., D. ROUX, M. G. MATTEI, O. DELAPEYRIERE, M. GOLDFARB, D. BIRNBAUM & B. JORDAN. 1988. Oncogene **3:** 703–708.
14. SAKAMOTO, H., T. YOSHIDA, M. NAKAKUKI, H. ODAGIRI, K. MIYAGAWA, T. SUGIMURA & M. TERADA. 1988. Biochem. Biophys. Res. Commun. **151:** 965–972.
15. MIYAGAWA, K., H. SAKAMOTO, T. YOSHIDA, Y. YAMASHITA, Y. MITSUI, M. FURUSAWA, S. MAEDA, F. TAKAKU, T. SUGIMURA & M. TERADA. 1988. Oncogene **3:** 383–389.
16. DELLI BOVI, P., A. M. CURATOLA, K. NEWMAN, Y. SATO, D. MOSCATELLI, R. HEWICK, D. RIFKIN & C. BASILICO. 1988. Mol. Cell. Biol. **8:** 2933–2941.
17. SAKAMOTO, H., M. MORI, M. TAIRA, T. YOSHIDA, S. MATSUKAWA, K. SHIMIZU, M. SEKIGUCHI, M. TERADA & T. SUGIMURA. 1986. Proc. Nat. Acad. Sci. USA **83:** 3997–4001.
18. DELLI BOVI, P. & C. BASILICO. 1987. Proc. Nat. Acad. Sci. USA **84:** 5660–5664.
19. WILKINSON, D., G. PETERS, C. DICKSON & A. MCMAHON. 1988. EMBO J. **7:** 691–695.
20. VOLK, R., M. KOSTER, A. POTING, L. HARTMANN & W. KNOCHEL. 1989. EMBO J. **8:** 2983–2988.
21. GOSPODAROWICZ, D., J. WESEMAN, J. MORAN & J. LINDSTROM. 1976. J. Cell. Biol. **70:** 395–405.
22. ALLEN, R., M. DODSON & L. LUITEN. 1984. Exp. Cell. Res. **152:** 154–160.
23. LATHROP, B., K. THOMAS & L. GLASER. 1985. J. Cell. Biol. **101:** 2194–2198.
24. SEED, J. & S. HAUSCHKA. 1988. Dev. Biol. **128:** 40–49.
25. VAIDYA, T., S. RHODES, E. TAPAROWSKY & S. KONIECZNY. 1989. Mol. Cell. Biol. **9:** 3576–3579.
26. ALTERIO, J., Y. COURTOIS, J. ROBELIN, D. BECHET & I. MARTELLY. 1990. Biochem. Biophys. Res. Commun. **166:** 1205–1212.
27. JAYE, M., R. HOWK, W. BURGESS, G. RICCA, I. M. CHIU, M. RAVERA, S. O'BRIEN, W. MODI, T. MACIAG & W. DROHAN. 1986. Science **233:** 541–544.
28. PRATS, H., M. KAGHAD, A. C. PRATS, M. KLAGSBRUN, J. M. LELIAS, P. LIAUZUN, P. CHALON, J. P. TAUBER, F. AMALRIC, J. SMITH & D. CAPUT. 1989. Proc. Nat. Acad. Sci. USA. **86:** 1836–1840.
29. BROOKES, S., R. SMITH, G. CASEY, C. DICKSON & G. PETERS. 1989. Oncogene **4:** 429–436.
30. TAIRA, M., T. YOSHIDA, K. MIYAGAWA, H. SAKAMOTO, M. TERADA & T. SUGIMURA. 1987. Proc. Nat. Acad. Sci. USA. **84:** 2980–2984.
31. ZHAN, X., B. BATES, X. HU & M. GOLDFARB. 1988. Mol. Cell. Biol. **8:** 3487–3495.
32. FINCH, P., J. RUBIN, T. MIKI, D. RON & S. AARONSON. 1989. Science **245:** 752–755.

Keratinocyte Growth Factor

A Fibroblast Growth Factor Family Member with Unusual Target Cell Specificity

STUART A. AARONSON,[a] DONALD P. BOTTARO,[a]
TORU MIKI,[a] DINA RON,[b] PAUL W. FINCH,[c]
TIMOTHY P. FLEMING,[a] JAMES AHN,[a]
WILLIAM G. TAYLOR,[a] AND JEFFREY S. RUBIN[a]

[a]Laboratory of Cellular and Molecular Biology
National Cancer Institute
Building 37, Room IE24
9000 Rockville Pike
Bethesda, Maryland 20892

[b]Biology Faculty
Technion-Israel Institute of Technology
Technion City
Haifa-32000, Israel

[c]Department of Neurosurgery
Rhode Island Hospital
593 Eddy Street
Providence, Rhode Island 02903

INTRODUCTION

Growth factors are important mediators of intercellular communication. These potent molecules are released by cells and act to influence proliferation of the same or other cell types.[1] Interest in growth factors has been heightened by evidence of their potential involvement in neoplasia. The v-*sis* transforming gene of simian sarcoma virus encodes a protein that is homologous to the B chain of platelet-derived growth factor.[2,3] Moreover, a number of oncogenes are homologues of genes encoding growth factor receptors.[4] Thus, increased understanding of growth factors and their receptor-mediated signal-transduction pathways is likely to provide insights into mechanisms of both normal and malignant cell growth.

Recognizing that most human malignancies arise in epithelial tissues[5] where cell populations are continuously turning over, we sought to identify growth factors specific for these cell types. To screen for epithelial-specific mitogens, we employed the mouse keratinocyte line BALB/MK[6] as a prototypical epithelial cell and the NIH/3T3 fibroblast[7] as its nonepithelial counterpart. Preliminary analysis of conditioned medium from a variety of sources revealed that fibroblast cell lines produced factors capable of inducing DNA synthesis in both cell types. Whereas boiling or acid treatment eliminated the activity for BALB/MK, the activity for NIH/3T3 cells remained intact. We reasoned that the fibroblast lines were secreting heat- and acid-labile mitogen(s) with an apparent epithelial cell specificity. This interpretation also was consistent with increasing evidence that mesenchymal interactions presumably mediated by diffusible substances had a major impact on epithelial cell proliferation.[8–10] Efforts to purify and characterize the putative agent(s) responsible for the

62

activity on BALB/MK resulted in the identification of keratinocyte growth factor (KGF),[11] a mitogen structurally related to the fibroblast growth factors (FGFs), but with the distinctive properties of a paracrine mediator of epithelial cell growth.[12]

Further studies with recombinant KGF defined the biochemical characteristics of its cell surface receptors, establishing the basis for its target cell specificity.[13] Using a novel expression cloning strategy, we isolated a cDNA encoding the high-affinity KGF receptor.[14] Analysis of this receptor cDNA has provided insights into its relationship to other members of the newly emerging FGF receptor family.

PURIFICATION OF A GROWTH FACTOR SPECIFIC FOR EPITHELIAL CELLS

M426, a human embryonic lung fibroblast line,[15] was selected as the best source for purification of the putative growth factor(s) responsible for BALB/MK mitogenic activity. Ultrafiltration provided a convenient way of reducing the volume of conditioned medium to a suitable level for subsequent chromatography. Heparin-Sepharose affinity chromatography, which has been used in the purification of other growth factors,[16-21] was the most efficient purification step. While estimates of recovered activity were uncertain at this stage because of the likely presence of multiple factors, the apparent yield was 50–70% with a corresponding enrichment of ~ 1000-fold. More than 90% of the BALB/MK mitogenic activity was eluted with 0.6 M NaCl and was not associated with any activity on NIH/3T3 cells.[11] Prompt concentration of 10- to 20-fold was essential for stability, which then could be maintained at $-70°C$ for several months.

Final purification was achieved by reverse-phase high-performance liquid chromatography (RP-HPLC) (Vydac C_4 column), a preparative method suitable for amino acid sequence analysis. While the yield of activity from this step was only a few percent, the loss could be attributed to the solvents used. In other experiments, exposure to 0.1% trifluoroacetic acid/50% (vol/vol) acetronitrile for 1 hour at room temperature reduced the mitogenic activity of the preparation by 98%.[11] Nonetheless, a single peak of BALB/MK stimulatory activity was obtained (FIGURE 1A), coinciding with a distinct peak in the absorption profile. Peak fractions contained a single band with a molecular mass of 28 kDa as estimated on a silver-stained, sodium dodecyl sulfate-polyacrylamide gel (FIGURE 1B), and mitogenic activity (FIGURE 1C) correlated with the intensity of this band across the chromatographic profile. This molecular mass was in good agreement with the elution position of mitogenic activity on two different sizing columns run in solvents expected to maintain native conformation.[11] From these data, we concluded the mitogen was a single polypeptide chain with a mass of 25–30 kDa.

Its distinctive target-cell specificity was demonstrated by comparing it on a variety of cell types with other growth factors known to possess epithelial cell mitogenic activity. The factor exhibited a strong mitogenic effect on BALB/MK cells and stimulated thymidine incorporation in other epithelial cells tested (TABLE 1). In contrast, the factor had no detectable effects on fibroblasts or human saphenous vein endothelial cells,[11] or on melanocytes or PC-12 cells (unpublished observations). By comparison, TGF-α and EGF showed good activity on fibroblasts, while acidic FGF (aFGF) and basic FGF (bFGF) were mitogenic for endothelial cells as well (TABLE 1). Because of its specificity for epithelial cells and the sensitivity of keratinocytes in particular, the mitogen was designated "keratinocyte growth factor."

To establish that KGF not only would stimulate DNA synthesis but also would support sustained cell growth, we attempted to grow BALB/MK cells in a fully

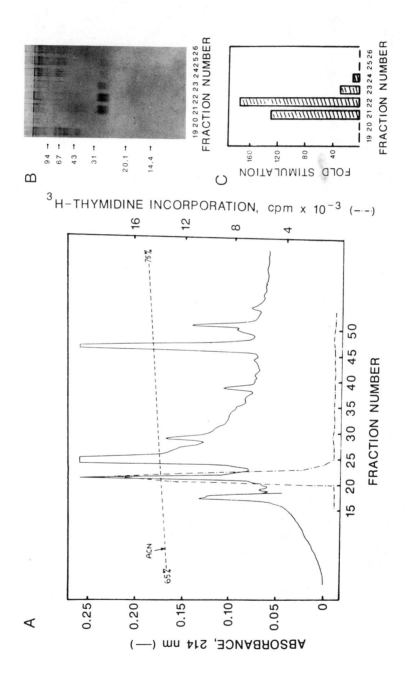

TABLE 1. Target-Cell Specificity of Growth Factors[a]

Growth Factor	Fold Stimulation of Thymidine Incorporation				
	Epithelial Cell Line			Fibroblast	Endothelial
	BALB/MK	B5/589	CCL208	NIH/3T3S	Cell Line[b]
KGF	500–1000	2–3	5–10	<1	<1
EGF	100–200	20–40	10–30	10–20	ND
TGFα	150–300	ND	ND	10–20	ND
aFGF[c]	300–500	2–3	5–10	50–70	5
bFGF	100–200	2–3	2–5	50–70	5

[a]Comparison of maximal thymidine incorporation stimulated by KGF and other growth factors in a variety of cell lines, expressed as fold stimulation over background. These data represent a summary of four different experiments. ND, not determined.

[b]Human saphenous vein cells.

[c]Maximal stimulation by aFGF required the presence of heparin (Sigma) at 20 μg/ml. aFGF and bFGF were recombinant preparations.

defined, serum-free medium supplemented with this growth factor. KGF served as an excellent substitute for EGF but not for insulin (or insulinlike growth factor I) in this chemically defined medium.[11] Thus, KGF acts through the major signaling pathway shared by EGF, aFGF, and bFGF for proliferation of BALB/MK cells.[22]

HUMAN KGF IS FGF RELATED WITH PROPERTIES OF A PARACRINE EFFECTOR OF EPITHELIAL CELL GROWTH

Amino acid sequence analysis of C_4-purified KGF ($\sim$150 pmol) yielded a single sequence with unambiguous assignment for cycles 2–13 as follows: Xaa-Asn-Asp-Met-Thr-Pro-Glu-Gln-Met-Ala-Thr-Asn-Val.[11] Oligonucleotide probes were generated on the basis of this experimentally determined amino acid sequence and then used to screen an oligo (dT)-primed cDNA library prepared from M426 human embryonic lung fibroblasts, the initial source of the factor. Of 10 plaque-purified clones analyzed, 1 (designated clone 49) had an insert of 3.5 kilobases (kb), whereas the rest had inserts ranging from 1.8 to 2.1 kb. Analysis of the smaller clones revealed several common restriction sites. Nucleotide sequencing of a representative clone (desig-

---

FIGURE 1. A: C_4 reverse phase high performance liquid chromatography (HPLC) of BALB/MK mitogenic activity. Active fractions eluted from the heparin-Sepharose column with 0.6 M NaCl were concentrated and loaded directly onto a Vydac C_4 column (4.6 × 250 mm) that had been equilibrated in 0.1% trifluoroacetic acid/20% acetonitrile. After the column was washed with 4 ml of equilibration buffer, the sample was eluted with a modified linear gradient of increasing percentage of acetonitrile. Fraction size was 0.2 ml, and flow rate was 0.5 ml/min. Aliquots for the assay of [^{3}H]thymidine incorporation in BALB/MK cells were promptly diluted 1:10 with 50 μg of bovine serum albumin per ml/20 mM Tris-HCl, pH 7.5, and tested at a final dilution of 1:200. **B:** Sodium dodecyl sulfate-polyacrylamide gel electrophoresis (SDS/PAGE) analysis of selected fractions from the C_4 chromatography shown in A. Half of each fraction was dried, redissolved in SDS/2-mercaptoethanol, heat denatured, and electrophoresed in a 14% polyacrylamide gel which was subsequently silver-stained. The position of each molecular mass marker (in kDa) is indicated by an arrow. **C:** DNA synthesis in BALB/MK cells triggered by the fractions analyzed in B. Activity is expressed as the fold stimulation over background, which was 100 cpm.

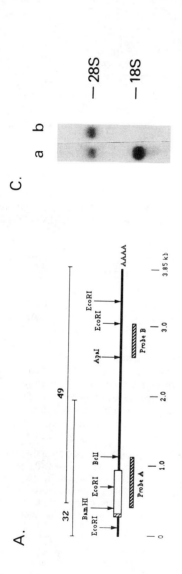

nated clone 32) along with clone 49 demonstrated that they were overlapping cDNAs (FIGURE 2A), which when aligned established a continuous 3.85-kb sequence that contained the complete KGF coding domain[12] (FIGURE 2B).

A likely ATG initiation codon was located at nucleotide position 446, establishing a 582-base-pair (bp) open reading frame, which ended at a TAA termination codon at nucleotide position 1030. This open reading frame encodes a putative 194-amino-acid polypeptide with a calculated molecular size of 22,512 daltons. A 19-amino-acid sequence, which was consistent with the experimentally determined NH$_2$-terminal sequence of purified human KGF, began 32 amino acids downstream of the proposed initiation codon. The predicted KGF amino acid sequence contained one potential N-linked glycosylation site (Asn-X-Ser) from residues 45 through 47.

To search for homology between KGF and any known protein, we analyzed the National Biomedical Research Foundation data base with the FASTP program of Lipman and Pearson.[23] The predicted primary structure of KGF was related to those of aFGF and bFGF, as well as int-2-, hst/KGFG, FGF-5, and FGF-6-encoded proteins. The FGFs are heparin-binding mitogens with broad target cell specificities.[24] FGF-5[25] and hst/KGFG[26,27] are transforming genes, originally detected by DNA-mediated gene transfer, whereas int-2 was identified as an oncogene by proviral integration of mouse mammary tumor virus.[28,29] FGF-6 also is a transforming gene which was initially identified on the basis of homology to a human hst/KFGF probe.[30] Alignment of the seven proteins revealed two major regions of homology, spanning amino acids 65 to 156 and 162 to 189 in the predicted KGF sequence, which were separated by a short nonhomologous series of amino acids. In the aligned regions, KGF was 30 to 45% identical to the other six members of the FGF family.

The primary KGF translation product, like those of hst/KFGF, FGF-5, and FGF-6, contains a hydrophobic NH$_2$-terminal region. Evidence that this NH$_2$-terminal domain is not present in the mature KGF molecule (FIGURE 2B) indicates that it represents a signal peptide sequence.[31] Acidic and basic FGF are synthesized apparently without signal peptides.[32,33] The int-2-encoded protein contains an atypically short region of NH$_2$-terminal hydrophobic residues,[34] which apparently functions as a signal sequence.[35] The int-2- and FGF-5-encoded proteins also contain long COOH-terminal extensions compared to the other family members.

A probe spanning most of the KGF coding sequence (FIGURE 2A, probe A) detected a predominant 2.4-kb transcript as well as a less abundant, ~5-kb transcript

---

FIGURE 2. Nucleotide sequence and deduced amino acid sequence of KGF cDNA. **A:** Representation of human KGF cDNA clones. Overlapping clones 32 and 49, used in sequence determination, are shown above a diagram of the complete coding sequence as well as adjacent 5' and 3' untranslated regions. Untranslated regions are represented by a line; the coding sequence is boxed. The hatched region represents sequences that encode the putative signal peptide. Selected restriction sites are indicated. The derivation of two cDNA probes used for RNA blot analysis is indicated. **B:** Complementary DNA nucleotide sequence encoding the predicted KGF amino acid sequence. Nucleotides are numbered from the left; amino acids are numbered throughout. The NH$_2$-terminal peptide sequence derived from purified KGF is underlined. The hydrophobic NH$_2$-terminal domain is shown in italics. The potential asparagine-linked glycosylation site is overlined. **C:** Identification of KGF mRNAs by RNA blot analysis. An RNA blot of poly(A)$^+$-selected M426 RNA was hybridized with a ^{32}P-labeled 695-bp Bam HI-Bcl I fragment from clone 32 (probe A in **A**), lane a, or a 872-bp fragment from the 3' untranslated region of clone 49 (probe B in **A**) generated by the polymerase chain reaction technique.

by RNA blot analysis of polyadenylated [poly(A)$^+$] M426 RNA (FIGURE 2C).[12] A probe derived from the 3' untranslated region of clone 49, distal to the end of clone 32 (FIGURE 2A, probe B), only hybridized to the larger message (FIGURE 2C). Thus, it appears that the KGF gene is transcribed as two alternative mRNAs. Two other members of the FGF gene family, bFGF[32,33] and int-2,[36] also express multiple RNAs. The 3' untranslated region of the 5-kb KGF cDNA contained many ATTTA sequences, which have been proposed to be markers for the selective degradation of transiently expressed, unstable RNAs[37] and might in part account for the low abundance of the larger KGF transcript.

To investigate the functional role of KGF, we examined the expression of its transcript in a variety of human cell lines and tissues. The predominant 2.4-kb KGF transcript was detected in each of several stromal fibroblast lines derived from epithelial tissues of embryonic, neonatal, and adult sources. In contrast, the transcript was not detected in normal glial cells, or in a variety of epithelial cell lines. The transcript was also evident in RNA extracted from normal adult kidney and organs of the gastrointestinal (GI) tract, but not from lung or brain.[12]

To further explore the stromal pattern of KGF expression, whole skin tissue was dissected from newborn mice and separated into dermal and epidermal layers by mild tryptic digestion.[6] Total cellular RNA was extracted from each layer, as well as from whole skin, and screened for KGF expression. The KGF transcript was observed in whole skin and was specifically detected in the dermis but not in the epidermal layer (FIGURE 3).[12] As controls for the enrichment for each tissue layer, we used DNA probes for vimentin[38,39] and keratin 1,[40,41] which are specific for mesenchymal and epithelial cells, respectively. The striking specificity of KGF RNA expression in stromal cells from epithelial tissues supports the concept that this factor is important in the normal mesenchymal stimulation of epithelial cell growth.

BIOCHEMICAL CHARACTERIZATION OF A HIGH-AFFINITY RECEPTOR FOR KGF: EVIDENCE FOR MULTIPLE FGF RECEPTORS

Consistent with its target cell specificity in mitogenesis bioassays, we detected saturable, specific high-affinity binding of ^{125}I-KGF to the surface of BALB/MK but not NIH/3T3 cells (FIGURE 4A and B). ^{125}I-KGF binding on BALB/MK was competed efficiently by aFGF but with 20-fold lower efficiency by bFGF (FIGURE 4A), in agreement with their relative potency in assays of DNA synthesis (FIGURE 5). The contrast in binding properties of BALB/MK and NIH/3T3 cells was reinforced by the pattern of ^{125}I-aFGF-receptor interactions on the two cell types. ^{125}I-aFGF exhibited specific high-affinity binding to both the keratinocytes and fibroblasts, but KGF competed only for the binding to BALB/MK. On the other hand, bFGF was a significantly better competitor of aFGF binding on NIH/3T3 compared to BALB/MK cells (FIGURE 4C and D).

Scatchard analysis of ^{125}I-KGF binding suggested major and minor high-affinity receptor components (dissociation constant = 400 and 25 pM, respectively) as well as a third high-capacity/low-affinity heparinlike component. The latter was also present on NIH/3T3 cells, and therefore must be insufficient for KGF-induced mitogenic signal transduction. Covalent affinity cross-linking of ^{125}I-KGF revealed two species of 115 and 140 kDa on BALB/MK cells which were absent from NIH/3T3 cells, and presumably corresponded to the high-affinity receptors required for mitogenic signaling. KGF also stimulated the rapid tyrosine phosphorylation of a 90-kDa protein in BALB/MK cells but not in NIH/3T3 fibroblasts. Tyrosine phosphorylation following addition of growth factor and, in particular, the prompt labeling of a 90-kDa substrate has also been observed in response to aFGF and

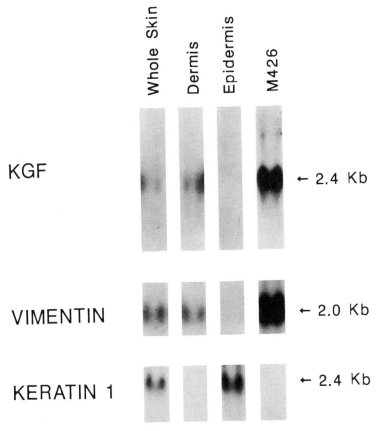

FIGURE 3. Expression of KGF in the skin of newborn mice. Whole skin was removed from 1-day-old mice and incubated overnight at 4°C in 0.25% trypsin solution. On the following day the dermal and epidermal layers were separated and RNA was extracted from these two layers as well as from intact mouse skin. RNA (20 μg) from each specimen, including the human fibroblast line M426, was screened for KGF transcript by RNA blot analysis with a ^{32}P-labeled Pvu II-Ssp I fragment of the human KGF cDNA (nucleotides 162 to 1380). Detection of vimentin and keratin I (K1) transcripts was done with human vimentin and mouse K1 cDNA derived probes. The arrows indicate the location of the transcript detected by each probe.

bFGF.[42–44] Together these results indicated that BALB/MK keratinocytes possess high-affinity KGF receptors to which the FGFs also bind, with aFGF having a greater affinity than bFGF. However, these receptors are distinct from the receptor(s) for aFGF and bFGF on NIH/3T3 fibroblasts, which fail to interact with KGF.

EXPRESSION cDNA CLONING OF THE KGF RECEPTOR BY CREATION OF A TRANSFORMING AUTOCRINE LOOP

Earlier studies, either with FGF family members such as *hst*/K-FGF[26,27] or FGF-5[25] that possess signal peptides, or with bFGF constructs to which a signal

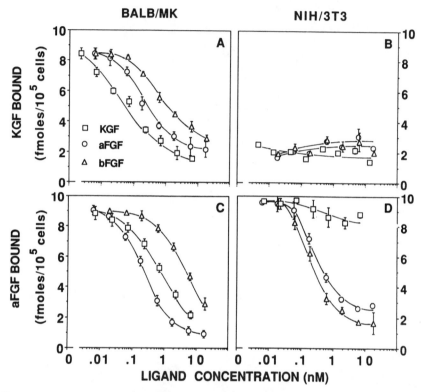

FIGURE 4. A: Specific binding of ^{125}I-KGF (1 ng/ml) to Balb/MK cells, expressed as femtomoles bound per 10^5 cells, competed by increasing concentrations (nM) of unlabeled KGF (squares), aFGF (circles), or bFGF (triangles). Values shown are the mean of triplicate samples ± standard deviation (SD). Where no error bars are shown, the error is less than the symbol size. Similar results were obtained using either low concentrations of heparin (1–3 μg/ml) or brief salt extraction to block low-affinity ligand binding in all competition studies shown. **B:** Specific ^{125}I-KGF binding on NIH/3T3 cells, competed by unlabeled KGF, aFGF, or bFGF. **C:** Specific binding of ^{125}I-aFGF (1 ng/ml) to Balb/MK cells, competed by unlabeled KGF, aFGF, or bFGF. **D:** Specific ^{125}I-aFGF binding on NIH/3T3 cells, competed by unlabeled KGF, aFGF, or bFGF.

peptide sequence had been added,[45,46] demonstrated that transfection of NIH/3T3 cells with vectors encoding these secreted factors would result in transformation. The presumed mechanism for this effect was the creation of a functional transforming autocrine loop. As indicated above, NIH/3T3 cells lack the high-affinity KGF receptors required to mediate mitogenic signal transduction. However, these cells,

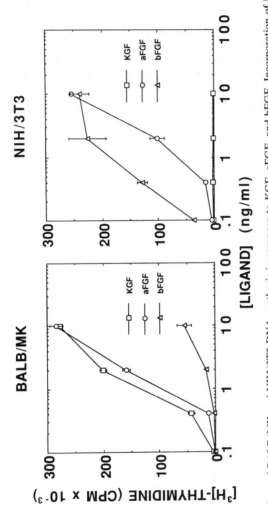

FIGURE 5. Comparison of BALB/MK and NIH/3T3 DNA synthesis in response to KGF, aFGF, and bFGF. Incorporation of [^{3}H]thymidine into trichloroacetic acid–insoluble DNA was measured as a function of the concentration of the indicated growth factors. The assays were performed in the absence of heparin, using bacterially expressed human KGF and bovine-brain-derived aFGF and bFGF. The results are mean values ± standard deviation of triplicate measurements.

like other fibroblast lines, synthesize and secrete KGF.[14] Thus, we reasoned that the introduction of KGF receptors via transfection might cause transformation by creation of a similar, KGF-dependent autocrine mechanism. Screening foci of transformed cells for evidence of high-affinity KGF binding sites would enable us to identify those cells that had become transformed as a consequence of acquiring such an autocrine loop.

This approach was successfully carried out with a newly designed expression cDNA cloning vector that incorporated the attributes of stable transfection of long cDNA inserts with a high cloning efficiency and the capacity for plasmid rescue.[47] When a cDNA library prepared with mRNA from BALB/MK cells was introduced into NIH/3T3 using this vector, 15 transformed foci were detected which were associated with three distinct morphological phenotypes. A single cDNA clone rescued from each transformant was found to possess high-titered transforming activity ranging from 10^3 to 10^4 focus-forming units per nanomole of DNA. Transfectants induced by the individual plasmids containing these epithelial-cell-transforming cDNAs (designated *ect1*, *ect2*, and *ect3*) were used in subsequent analyses.

To investigate the possibility that any of the three genes might encode the KGF receptor, we performed binding studies with recombinant ^{125}I-labeled KGF. BALB/MK cells showed specific high-affinity binding of ^{125}I-labeled KGF, which was not observed when NIH/3T3 cells were used. Expression of the *ect1* gene by NIH/3T3 cells resulted in the acquisition of 3.5-fold more ^{125}I-labeled KGF binding sites than BALB/MK cells (FIGURE 6). Under the same conditions, control NIH/3T3 as well as transfectants containing either *ect2* or *ect3* did not bind the labeled growth factor. These results suggested that *ect1* encoded the KGF receptor (KGFR), whose introduction into NIH/3T3 cells had completed a transforming autocrine loop.

A single *ect1* transcript of around 4.2 kb was observed in BALB/MK cells. Thus, our cDNA clone of 4.2 kb represented essentially the complete *ect1* transcript.[14] In NIH/3T3 cells, a transcript of comparable size was only faintly detectable under stringent hybridization conditions. Therefore, if this transcript were to represent *ect1* rather than a related gene, its expression was markedly lower in fibroblasts as compared to epithelial cells.

Nucleotide sequence analysis of the 4.2-kb *ect1* cDNA revealed a long open reading frame of 2235 nucleotides (nucleotide position 562 to 2796). Two methionine codons were found at nucleotide positions 619 and 676, respectively. The second methionine codon matched the Kozak's consensus for a translational initiator sequence (A/GC-CATGG).[48] Moreover, it was followed by a characteristic signal sequence of 21 residues, 10 of which were identical to those of the putative signal peptide of the mouse basic FGF receptor.[49–51] Thus, it seems likely that the second ATG is the authentic initiation codon. If so, the receptor polypeptide would comprise 707 amino acids with a predicted size of 82.5 kD (FIGURE 7A).

The amino acid sequence predicted a transmembrane tyrosine kinase structurally related to but distinct from the mouse bFGF receptor (bFGFR) encoded by the *flg* gene. The percent similarity between both proteins is shown in FIGURE 7B. The putative KGFR extracellular portion contained two immunoglobulin (Ig) like domains, exhibiting 77% and 60% similarity with the Ig-like domains 2 and 3, respectively, of the mouse bFGFR. Studies have revealed a variant form of the bFGFR, in which the extracellular domain also contains only these two corresponding Ig-like domains.[50,51] The sequence NH$_2$ terminal to the first Ig-like domain of the KGFR was 63 residues long, compared to 88 residues found in the shorter form of the mouse bFGFR. Both chicken and mouse bFGFRs contain a stretch of eight consecutive acidic residues between the first and second Ig-like domains[16-18] not found in the KGFR (FIGURE 7B).

The kinase domain of the KGFR was 90% related to the bFGFR tyrosine kinase (FIGURE 7B). The central core of the catalytic domain was flanked by a relatively long juxtamembrane sequence, and the tyrosine kinase domain was split by a short insert of 14 residues, similar to that observed in mouse, chicken, and human bFGF receptors.[49-55] Hanafusa and coworkers isolated a partial cDNA for a tyrosine kinase gene, designated *bek,* by bacterial expression cloning with phosphotyrosine antibodies.[56] The reported sequence of *bek* was identical to the KGFR in the tyrosine kinase domain (FIGURE 7B).

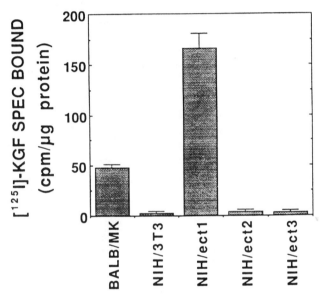

FIGURE 6. Comparison of [^{125}I]KGF specific binding to BALB/MK, NIH/3T3, and NIH/3T3-transfected cells expressing either of the epithelial-cell-transforming cDNAs designated *ect1, ect2,* and *ect3.* Bound cpm were normalized according to protein content of SDS extracts. Specific binding was determined by subtracting normalized cpm of samples incubated with a 100-fold excess of unlabeled KGF from the normalized cpm bound in the presence of [^{125}I]KGF alone. The results are mean values ± SD of triplicate measurements.

Scatchard analysis of ^{125}I-labeled KGF binding to the NIH/*ect1* transfectant revealed expression of high-affinity receptors comparable to the high-affinity KGF binding sites displayed by BALB/MK cells.[13,14] The pattern of KGF and FGF competition for ^{125}I-labeled KGF binding to NIH/*ect1* cells was also very similar to that observed with BALB/MK cells. When ^{125}I-labeled KGF cross-linking was performed with NIH/*ect1* cells, we observed a single species corresponding in size to the smaller 137-kDa complex in BALB/MK cells. Moreover, detection of this band was specifically and efficiently blocked by unlabeled KGF. When glycosylation is considered, the size of the KGFR predicted by sequence analysis corresponds reasonably well with the corrected size (115 kDa, taking into account the size of KGF) of the cross-linked KGFR in the *ect1* transfectant.[14]

ANNALS NEW YORK ACADEMY OF SCIENCES

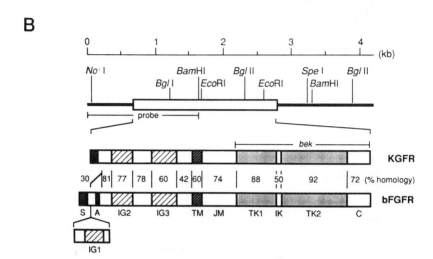

A

```
  1 MVSWGRFICLVLVTMATLSLARPSFSLVEDTTLEPEGAPYWTNTEKMEKRLHAVPAANTVKFRCPAGGNPTPTMRWLKNG
 81 KEFKQEHRIGGYKVRNQHWSLIMESVVPSDKGNYTCLVENEYGSINHTYHLDVVERSPHRPILQAGLPANASTVVGGDVE
161 FVCKVYSDAQPHIQWIKHVEKNGSKYGPDGLPYLKVLKHSGINSSNAEVLALFNVTEMDAGEYICKVSNYIGQANQSAWL
241 TVLPKQQAPVREKEITASPDYLEIAIYCIGVFLIACMVVTVIFCRMKTTTKKPDFSSQPAVHKLTKRIPLRRQVTVSAES
321 SSSMNSNTPLVRITTRLSSTADTPMLAGVSEYELPEDPKWEFPRDKLTLGKPLGEGCFGQVVMAEAVGIDKDKPKEAVTV
401 AVKMLKDDATEKDLSDLVSEMEMMKMIGKHKNIINLLGACTQDGPLYVIVEYASKGNLREYLRARRPPGMEYSYDINRVP
481 EEQMTFKDLVSCTYQLARGMEYLASQKCIHRDLAARNVLVTENNVMKIADFGLARDINNIDYYKKTTNGRLPVKWMAPEA
561 LFDRVYTHQSDVWSFGVLMWEIFTLGGSPYPGIPVEELFKLLKEGHRMDKPTNCTNELYMMMRDCWHAVPSQRPTFKQLV
641 EDLDRILTLTTNEEYLDLTQPLEQYSPSYPDTRSSCSSGDDSVFSPDPMPYEPCLPQYPHINGSVKT
```

B

FIGURE 7. Primary structure of the KGF receptor. **A:** Amino acid sequence deduced from the coding region of the KGF receptor cDNA. Amino acids are numbered from the putative initiation site of translation. Potential sites of N-linked glycosylation are underlined. The potential signal peptide and transmembrane domains are boxed. The interkinase domain is shown by underlined italic letters. Glycine residues considered to be involved in ATP (adenosine triphosphate) binding are indicated by asterisks. Cysteine residues delimit two Ig-like domains in the extracellular portion of the molecule shown by boldface. Nucleotide sequence was determined by the chain termination method. **B:** Structural comparison of the predicted KGF and bFGF receptors. The region used as a probe for Southern and Northern analysis (Figure 1B and C) is indicated. The region homologous to the published *bek* sequence[56] is also shown. The schematic structure of KGF receptor is shown below the restriction map of the cDNA clone. Amino acid sequence similarities with the smaller and larger bFGF receptor variants are indicated. S, signal peptide; A, acidic region; IGI, IG2, and IG3, Ig-like domains; TM, transmembrane domain; JM, juxtamembrane domain; TK1 and TK2, tyrosine kinase domains; IK, interkinase domain; C, COOH-terminus domain.

Our expression cloning of the KGFR was based on its transforming activity for NIH/3T3 cells that synthesize KGF. Thus, its detection could reflect activation of an autocrine loop involving KGF and the normal receptor. Alternatively, the cDNA might have been detected fortuitously as a constitutively activated KGFR mutant. Suramin, which interferes with ligand-receptor interactions,[57,58] inhibited DNA synthesis of KGFR transfectants. We also observed specific inhibition of proliferation of such cells in response to a KGF monoclonal antibody, which neutralizes KGF mitogenic activity.[14] Together these findings argue that induction of the transformed phenotype resulted from autocrine KGF stimulation of an ectopically expressed normal KGFR cDNA.

There have been reports concerning human or avian cDNAs closely related to the KGFR.[59-62] The external portions of *bek* (human) and *cek3* (chicken) proteins contain three Ig-like domains.[59-61] These molecules also differ from the KGFR in that each contains an acidic region and is completely divergent in the COOH-terminal half of its third Ig-like domain from the KGFR. Binding studies with the three Ig-like domain human *bek* variant have indicated similar high affinities for aFGF and bFGF.[59] Since the affinity of the KGFR for aFGF was substantially higher than for bFGF, differences in FGF binding by these receptor molecules must relate to these regions of divergence. In BALB/MK cells, we detected a higher molecular weight KGF-cross-linked species, corresponding in size to the three Ig-like domain *bek* variant.[13] Whether it represents this variant or the product of a distinct gene remains to be determined.

A gene, designated K-*sam*, was recently identified as an amplified sequence in a human stomach carcinoma.[62] A cDNA clone corresponding to one of the overexpressed K-*sam* transcripts predicts a two Ig-like domain *bek* variant, whose Ig-like domains correspond to those of the KGFR. However, it differs in that it contains an acidic region and may be truncated at its COOH terminus as well.[62] These molecules likely reflect alternative transcripts of the same gene, as has also been suggested for two and three Ig-like domain forms of the bFGFR.[63]

SUMMARY AND FUTURE DIRECTIONS

KGF is a fibroblast-derived member of the FGF family, with potent mitogenic activity on epithelial cells but no corresponding activity on fibroblasts, endothelial cells, melanocytes, or other nonepithelial targets of FGF action. Biochemical analysis established that KGF receptors bound aFGF with a high affinity but bFGF with at least an order of magnitude lower affinity. Expression cDNA cloning of a KGF receptor was accomplished by creation of a transforming autocrine loop. The full-length cDNA encoded a transmembrane, tyrosine kinase molecule which resembled the bFGF receptor encoded by *flg*, and was even more similar to the *bek* gene product. Future study will be aimed at determining differences responsible for the binding specificities that distinguish the KGF receptor from the *bek* and *flg* gene products. Using molecular probes to both KGF and its receptor to study their expression during development and in the adult should help define their role in normal growth and repair processes as well as possible pathologic roles in disease. This information, along with experiments testing the effects of KGF *in vivo*, could serve to identify situations in which KGF or antagonists to its actions would be of therapeutic benefit.

REFERENCES

1. JAMES, R. & R. A. BRADSHAW. 1984. Annu. Rev. Biochem. **53:** 259–292.
2. DOOLITTLE, R. F., M. W. HUNKAPILLER, L. E. HOOD, S. G. DEVARE, K. C. ROBBINS, S. A. AARONSON & M. N. ANTONIADES. 1983. Science **221:** 275–277.
3. WATERFIELD, M. D., G. J. SCRACE, N. WHITTLE, P. STROOBAND, A. JOHNSON, A. WASTETON, B. WESTERMARK, C.-H. HELDIN, J. S. HUANG & T. F. DEUEL. 1983. Nature London **304:** 35–39.
4. HUNTER, T. & J. A. COOPER. 1985. Annu. Rev. Biochem. **54:** 897–930.

5. WRIGHT, N. & M. ALLISON. 1984. The Biology of Epithelial Cell Populations. Oxford University Press. New York, N.Y.
6. WEISSMAN, B. E. & S. A. AARONSON. 1983. Cell 32: 599–606.
7. JAINCHILL, J. L., S. A. AARONSON & G. J. TODARO. 1969. J. Virol. 4: 549–553.
8. CUNHA, G. R., L. W. K. CHUNG, J. M. SHANNON, O. TAGUCHI & H. FUJII. 1983. Recent Prog. Horm. Res. 39: 559–598.
9. SAWYER, R. H. & J. F. FALLOWS, Eds. 1983. Epithelial-Mesenchymal Interactions during Development. Praeger. New York, N.Y.
10. SCHOR, S. L., A. M. SCHOR, A. HOWELL & D. CROWTHER. 1987. Exp. Cell Biol. 55: 11–17.
11. RUBIN, J. S., H. OSADA, P. W. FINCH, W. G. TAYLOR, S. RUDIKOFF & S. A. AARONSON. 1989. Proc. Nat. Acad. Sci. USA 86: 802–806.
12. FINCH, P. W., J. S. RUBIN, T. MIKI, D. RON & S. A. AARONSON. 1989. Science 245: 752–755.
13. BOTTARO, D. P., J. S. RUBIN, D. RON, P. W. FINCH, C. FLORIO & S. A. AARONSON. 1990. J. Biol. Chem. 265: 12,767–12,770.
14. MIKI, T., T. P. FLEMING, D. P. BOTTARO, J. S. RUBIN, D. RON & S. A. AARONSON. 1991. Science 251: 72–75.
15. AARONSON, S. A. & G. J. TODARO. 1968. Virology 36: 254–261.
16. RAINES, E. W. & R. ROSS. 1982. J. Biol. Chem. 257: 5154–5160.
17. SHING, Y., J. FOLKMAN, R. SULLIVAN, C. BUTTERFIELD, J. MURRAY & M. KLAGSBRUN. 1984. Science 223: 1296–1299.
18. GOSPODAROWICZ, D., J. CHENG, G.-M. LUI, A. BAIRD & P. BOHLEN. 1984. Proc. Nat. Acad. Sci. USA 81: 6963–6967.
19. MACIAG, T., T. MEHLMAN, R. FRIESEL & A. B. SCHREIBER. 1984. Science 225: 932–935.
20. CONN, G. & V. B. HATCHER. 1984. Biochem. Biophys. Res. Commun. 124: 262–268.
21. LOBB, R. R. & J. W. FETT. 1984. Biochemistry 23: 6295–6299.
22. FALCO, J. P., W. G. TAYLOR, P. P. DI FIORE, B. E. WEISSMAN & S. A. AARONSON. 1988. Oncogene 2: 573–578.
23. LIPMAN, D. J. & R. W. PEARSON. 1985. Science 227: 1435–1441.
24. THOMAS, K. 1987. FASEB J. 1: 434–440.
25. ZHAN, X., B. BATES, X. HU & M. GOLDFARB. 1988. Mol. Cell. Biol. 8: 3487–3495.
26. TAIRA, M., T. YOSHIDA, K. MIYAGAWA, H. SAKAMOTO, M. TERADA & T. SUGIMURA. 1987. Proc. Nat. Acad. Sci. USA 84: 2980–2984.
27. DELLI-BOVI, P. & C. BASILICO. 1987. Proc. Nat. Acad. Sci. USA 84: 5660–5664.
28. JAKOBOVITS, A., G. M. SHACKLEFORD, H. E. VARMUS & G. R. MARTIN. 1986. Proc. Nat. Acad. Sci. USA 84: 7806–7810.
29. PETERS, G., S. BROOKES, R. SMITH & C. DICKSON. 1983. Cell 33: 369–377.
30. MARICS, I., J. ADELAIDE, F. RAYBAUD, M.-G. MATTEI, F. COULIER, J. PLANCHE, O. DE LAPEYRIERE & D. BIRNBAUM. 1989. Oncogene 4: 335–340.
31. VON HEIJNE, G. 1986. Nucleic Acids Res. 14: 4683–4690.
32. JAYE, M., R. HOWK, W. BURGESS, G. A. RICCA, I. CHIER, M. W. RAVERA, S. J. O'BRIEN, W. S. MODI, T. MACIAG & W. DROHAN. 1986. Science 233: 541–545.
33. ABRAHAM, J. A., A. MERGIA, J. L. WHANG, A. TUMOLO, J. FRIEDMAN, K. A. HJERRILD, D. GOSPODAROWICZ & J. C. FIDDES. 1986. Science 233: 545–548.
34. MOORE, R., G. CASEY, S. BROOKES, M. DIXON, G. PETERS & C. DICKSON. 1986. EMBO J. 5: 919–924.
35. ACLAND, P., M. DIXON, G. PETERS & C. DICKSON. 1990. Nature 343: 662–665.
36. MANSOUR, S. L. & G. R. MARTIN. 1988. EMBO J. 7: 2035–2041.
37. SHAW, G. & R. KAMEN. 1986. Cell 46: 659–667.
38. BLOEMENDAL, M., W. QUAX, Y. QUAX-JEUKEN, R. VAN DEN HEUVEL, W. VREE EGBERTS & L. VAN DEN BROEK. 1985. Ann. N.Y. Acad. Sci. 455: 95–105.
39. FERRARI, S., R. BATTINI, L. KACZMAREK, S. RITTLING, B. CALABRETTA, J. KIM DE RIEL, V. PHILIPONIS, J.-F. WEI & R. BASERGA. 1986. Mol. Cell. Biol. 6: 3614–3620.
40. ROOP, D. R., P. HAWLEY-NELSON, C. K. CHENG & S. H. YUSPA. 1983. J. Invest. Dermatol. 81: 144s–149s.
41. ROOP, D. R., T. M. KRIEG, T. MEHREL, C. K. CHENG & S. H. YUSPA. 1988. Cancer Res. 48: 3245–3252.

42. HUANG, S. S. & J. S. HUANG. 1986. J. Biol. Chem. **261:** 9568–9571.
43. COUGHLIN, S. R., P. J. BARR, L. S. COUSSENS, L. J. FRETTO & L. T. WILLIAMS. 1988. J. Biol. Chem. **263:** 988–993.
44. FRIESEL, R., W. H. BURGESS & T. MACIAG. 1989. Mol. Cell. Biol. **9:** 1857–1865.
45. ROGELJ, S., R. A. WEINBERG, P. FANNING & M. KLAGSBRUN. 1988. Nature **331:** 173–175.
46. BLAM, S. B., R. MITCHELL, E. TISCHER, J. S. RUBIN, M. SILVA, S. SILVER, J. C. FIDDES, J. A. ABRAHAM & S. A. AARONSON. 1988. Oncogene **3:** 129–136.
47. MIKI, T., T. MATSUI, M. A. HEIDARAN & S. A. AARONSON. 1990. *In* Recombinant Systems in Protein Expression. K. K. Alitato, M.-L. Huhtala, J. Knowles & A. Vaheri, Eds.: 125–136. Elsevier Science Publishers. Amsterdam, the Netherlands.
48. KOZAK, M. 1987. Nucleic Acids Res. **15:** 8125–8148.
49. SAFRAN, A., A. AVIVI, A. ORR-URTEREGER, G. NEUFELD, P. LONAI, D. GIVOL & Y. YARDEN. 1990. Oncogene **5:** 635–643.
50. REID, H. M., A. F. WILKS & O. BERNARD. 1990. Proc. Nat. Acad. Sci. USA **87:** 1596–1600.
51. MANSUKUHANI, A., D. MOSCATELLI, D. TALARICO, V. LEVYTSKA & C. BASILICO. 1990. Proc. Nat. Acad. Sci. USA **87:** 4378–4382.
52. LEE, P. L., D. E. JOHNSON, L. S. COUSENS, V. A. FRIED & L. T. WILLIAMS. 1989. Science **245:** 57–60.
53. PASQUALE, E. G. & S. J. SINGER. 1989. Proc. Nat. Acad. Sci. USA **86:** 5449–5453.
54. RUTA, M., R. HOWK, G. RICCA, W. DROHAN, M. ZABELSHANSKY, G. LAUREYS, D. E. BARTON, U. FRANCKE, J. SCHLESSINGER & D. GIVOL. 1988. Oncogene **3:** 9–15.
55. RUTA, M., W. BURGESS, D. GIVOL, J. EPSTEIN, N. NEIGER, J. KAPLOW, G. CRUMLEY, C. DIONNE, M. JAYE & J. SCHLESSINGER. 1989. Proc. Nat. Acad. Sci. USA **86:** 8722–8726.
56. KORNBLUTH, S., K. E. PAULSON & H. HANAFUSA. 1988. Mol. Cell. Biol. **8:** 5541–5544.
57. BETSHOLTZ, C., A. JOHNSSON, C.-H. HELDIN & B. WESTERMARK. 1986. Proc. Nat. Acad. Sci. USA **84:** 6440–6444.
58. FLEMING, T., T. MATSUI, C. J. MOLLOY, K. C. ROBBINS & S. A. AARONSON. 1989. Proc. Nat. Acad. Sci. USA **86:** 8063–8067.
59. DIONNE, C. A., G. CRUMLEY, F. BELLOT, J. M. KAPLOW, G. SEARFOSS, M. RUTA, W. H. BURGESS, M. JAYE & J. SCHLESSINGER. 1990. EMBO J. **9:** 2685–2692.
60. PASQUALE, E. G. 1990. Proc. Nat. Acad. Sci. USA **87:** 5812–5816.
61. HOUSSAINT, E., P. R. BLANQUET, P. CHAMPION-ARNAUD, M. C. GESNEL, A. TORRIGLIA, Y. COURTOIS & R. BREATHNACH. 1990. Proc. Nat. Acad. Sci. USA **87:** 8180–8184.
62. HATTORI, Y., H. ODAGIRI, H. NAKATANI, K. MIYAGAWA, K. NAITO, H. SAKAMOTO, O. KATOH, T. YOSHIDA, T. SUGIMURA & M. TERADA. 1990. Proc. Nat. Acad. Sci. USA **87:** 5983–5987.
63. JOHNSON, D. E., P. L. LEE, J. LU & L. T. WILLIAMS. 1990. Mol. Cell. Biol. **10:** 4728–4736.

Chemical Characterization of the Cysteines of Basic Fibroblast Growth Factor

STEWART A. THOMPSON AND JOHN C. FIDDES

California Biotechnology, Inc.
2450 Bayshore Parkway
Mountain View, California 94043

Recombinant expression systems enable the preparation of large amounts of proteins that are normally present in natural sources in small quantities. Unfortunately, recombinant proteins do not always experience the same posttranslational events as do the natural proteins. This may be particularly true when a mammalian protein is expressed intracellularly in bacteria. For example, if a protein naturally has an intramolecular disulfide bond, it would not be expected for this bond to form in the reducing environment of *Escherichia coli* cytoplasm. As a result, the final form of a recombinant protein may be different from that of the natural protein and this can affect biological activity.[1] Therefore, the disulfide structure must be addressed when characterizing a recombinant protein.

The 155 amino acid form of basic fibroblast growth factor (bFGF) has four cysteines at positions 34, 78, 96, and 101.[2] Of these, the first and fourth cysteines are perfectly conserved within the seven members of the FGF family,[3-9] suggesting the presence of an intramolecular disulfide bond between the conserved residues at positions 34 and 101. We have expressed human bFGF as a soluble protein in *E. coli* and have purified the protein to homogeneity.[10] To ensure that the recombinant bFGF has the correct disulfide structure, we have attempted to compare the cysteine chemistry of both recombinant and natural bFGF.

HETEROGENEITY OF RECOMBINANT BASIC FGF

Basic FGF purified from our *E. coli* expression system is in the fully reduced form, and when chromatographed on heparin-TSK elutes as a single peak at 1.4 M NaCl. If the purified protein is stored under oxidizing conditions it converts from the homogeneous, fully reduced species to a form that appears heterogeneous by heparin-TSK chromatography (FIGURE 1). Treatment of the oxidized protein with reducing reagents reverts the protein to the homogeneous form indicating that the heterogeneity is due to the formation of disulfide bonds. To characterize the nature of the heterogeneity, the six heparin-TSK bFGF fractions indicated in FIGURE 1 were purified and analyzed by sodium dodecyl sulfate-polyacrylamide gel electrophoresis (SDS-PAGE). Under reducing conditions, the protein present in all six fractions migrated identically (FIGURE 2A). However, if the samples were not reduced prior to electrophoresis, the protein in each fraction still migrated predominantly as monomer with small amounts of higher molecular weight forms, but the six fractions were not significantly different from one another (FIGURE 2B).

This behavior in polyacrylamide gels is in contrast to size exclusion chromatography experiments which indicated that the fractions identified as 1, 2, and 3 in FIGURE

1 represented monomeric, dimeric, and trimeric forms of the protein respectively (data not shown). This apparent discrepancy between the experimental results obtained by the two techniques could be explained by the occurrence of rapid thiol-disulfide exchange when the multimeric bFGF was mixed with the denaturing sample buffer prior to SDS-PAGE analysis. To test this theory, the six fractions of bFGF isolated in FIGURE 1 were treated with the thiol-alkylating reagent iodoaceta-mide prior to addition of sample buffer. The result of electrophoresis of these samples is shown in FIGURE 2C. Following iodoacetamide treatment, the first three fractions migrated as monomer, dimer, and trimer respectively, consistent with the result observed by size exclusion chromatography, and fractions 4, 5, and 6 migrated

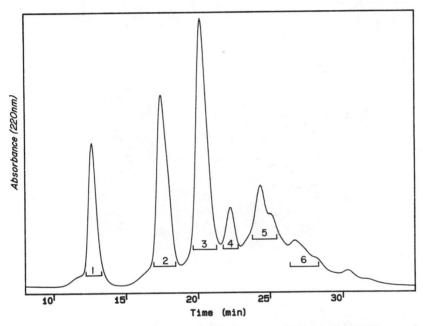

FIGURE 1. Heparin-TSK chromatography of heterogeneous recombinant bFGF. The protein was eluted with 20 mM Tris pH 7.5 and a multilinear NaCl gradient of 0.72 M to 1.2 M over 1 minute, then to 3 M over 24 minutes and continued at 3 M for 5 minutes. The six fractions indicated were collected, and EDTA was added to a final concentration of 5 mM.

as trimeric and higher molecular weight species. It should be noted that this result is only achieved at very high iodoacetamide concentrations (200 mM or greater); we routinely use 500 mM iodoacetamide. At concentrations of 50 mM there is only partial trapping of the high molecular weight forms, and at 10 mM no high molecular weight forms are observed.

To determine which cysteines were responsible for intermolecular disulfide bond formation, the four cysteines of bFGF were individually mutated to serine residues. The mutant proteins were purified and stored in the absence of antioxidants so that multimerization of the protein would occur. The results are shown in FIGURE 3. Mutation of either cysteine 34 or 101 (C34S and C101S) resulted in proteins that

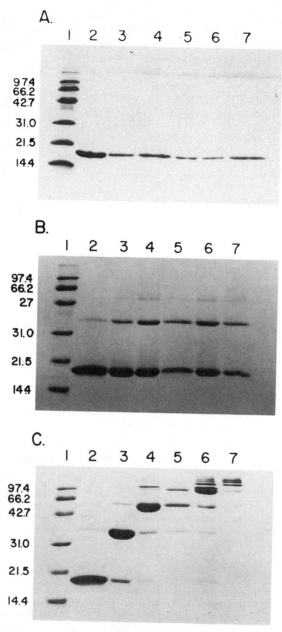

FIGURE 2. 15% SDS-PAGE of the bFGF fractions isolated in Figure 1. **Lane 1:** Molecular weight markers. **Lanes 2–7:** Fractions 1 through 6, respectively. Samples shown in FIGURE 2A and 2B were electrophoresed under reducing and nonreducing conditions respectively. Samples in FIGURE 2C were treated with an equal volume of 1 M iodoacetamide, 8 M urea, 50 mM sodium phosphate for 5 minutes and then precipitated with trichloroacetic acid. The protein pellets were resuspended and electrophoresed under nonreducing conditions.

could form multimeric species, whereas mutation of either cysteine 78 or 96 (C78S and C96S) gave proteins that only appeared as the monomeric and dimeric forms. This result suggests that both residues 78 and 96 are involved in the multimerization process. This conclusion was confirmed by the observation that mutation of both of these residues to serine resulted in a form of bFGF (C78/96S) that was exclusively monomeric, and would not multimerize even in the presence of oxidants such as CuSO$_4$ which will cause rapid multimerization of native sequence recombinant bFGF. This was in agreement with the results observed by Seno *et al.*[11]

We conclude, therefore, that the heterogeneity observed on heparin-TSK is caused by intermolecular disulfide bonds involving cysteines 78 and 96 resulting in multimerization of the protein. After denaturation such as that caused by the sample buffer for SDS-PAGE gels, the disulfides rapidly undergo thiol-disulfide rearrangements resulting in a disulfide pattern different from that of the starting material. A

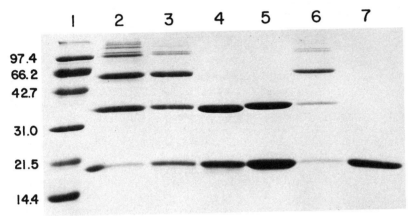

FIGURE 3. 15% SDS-PAGE of oxidized forms of bFGF cysteine mutants. Purified protein was stored at 4°C in the absence of antioxidants to generate the heterogeneity. The protein was then treated was iodoacetamide as in FIGURE 2C. Lane 1, molecular weight markers; 2, bFGF; 3, C34S; 4, C78S; 5, C96S, 6, C101S; 7, C78/96S.

high concentration of thiol-alkylating reagent is required to compete with the thiol exchange reaction in order to preserve the original disulfide structure.

COMPARISON OF BOVINE PITUITARY AND HUMAN RECOMBINANT BASIC FGFs

To determine the cysteine structure of natural bFGF we compared our recombinant human protein to bFGF purified from a natural source, bovine pituitaries.[12,13] (Only two amino acid differences exist between the human and bovine proteins.) The natural protein differed from the recombinant by three criteria. First, bovine pituitary bFGF failed to form the intermolecular disulfide bonds that had been observed in the recombinant protein even when the natural protein was treated with CuSO$_4$. Second, when the recombinant protein and reduced natural protein were cochromatographed on reversed-phase high-performance liquid chromatography

(HPLC), they comigrated perfectly. But, when the natural protein in its nonreduced state was chromatographed on reversed-phase HPLC it eluted approximately seven minutes earlier than the fully reduced protein (FIGURE 4), indicating that a disulfide bond exists in the bovine pituitary protein. This disulfide did not seem to generate multimeric forms of the protein since natural bovine pituitary bFGF was monomeric on nonreducing SDS-PAGE even after iodoacetamide treatment, comigrated with the reduced form of the protein on heparin-TSK chromatography, and failed to coelute with the multimeric forms of the recombinant protein on reversed-phase HPLC (data not shown). Third, thiol titration of the cysteine residues with Ellmans reagent[14] indicated that the bovine pituitary bFGF contained only two free thiols while the recombinant protein had four free thiols. These observations are consistent with the presence of an intramolecular disulfide in the bovine pituitary bFGF.

GLUTATHIONE MODIFICATION OF BASIC FGF

Since cysteines 34 and 101 are perfectly conserved in the FGF family it was considered likely that a disulfide bond existed between those residues in natural bFGF. Surprisingly, we were unable to generate an intramolecular disulfide in recombinant bFGF by established techniques.[15] Therefore, we considered the possibility that instead of there being one intramolecular disulfide bond in natural bFGF, there were two intermolecular disulfides between the protein and a small thiol-containing molecule. Glutathione is a very abundant thiol compound, and several groups have already reported the S thiolation of proteins by gluthathione.[16-18] Therefore, glutathione was thought to be the most likely candidate for a compound forming a disulfide linkage to bovine pituitary bFGF.

To test this theory we treated recombinant bFGF with glutathione disulfide to generate a glutathione-conjugated form of bFGF, and compared its elution by reversed-phase HPLC to that of bovine pituitary bFGF. In addition, we also treated recombinant bFGF with the disulfide compounds cystine, cystamine, and coenzyme A disulfide. It was found that the product of the glutathione disulfide treated protein comigrated on reversed-phase HPLC with the natural protein, whereas the product of bFGF conjugated with the other disulfide compounds did not comigrate with the natural protein, although all of their retention times were very similar (FIGURE 5).

Repeating this experiment with tritiated glutathione disulfide resulted in the incorporation of two equivalents of tritium into the recombinant protein. Thiol titration of this protein with Ellmans reagent[14] indicated two equivalents of remaining free thiol. Therefore, the product of this reaction which comigrated on reversed phase HPLC with bovine pituitary bFGF had two intermolecular disulfide bonds to glutathione moieties and no intramolecular disulfides. (We refer to this form of the protein as GS$_2$-bFGF.)

MAPPING OF GLUTATHIONE ADDUCTS TO RECOMBINANT BASIC FGF

The locations of the radiolabeled glutathione adducts were determined by peptide mapping. The tritium-labeled modified protein was first carboxamidomethylated with 0.5 M iodoacetamide to prevent thiol-disulfide exchange reactions. If this was not done, much of the radiolabel was lost from the protein and eluted from reversed-phase HPLC at the position of glutathione. This result was not unexpected

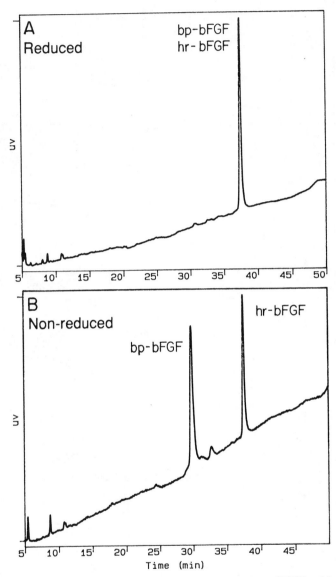

FIGURE 4. Reversed-phase chromatography of human recombinant bFGF and bovine pituitary bFGF. A Vydac C18 column was eluted with a 20% to 40% acetonitrile gradient containing 0.1% trifluoroacetic acid over 40 minutes, then maintained at 40% acetonitrile for 5 minutes. In FIGURE 4A, bovine and recombinant bFGF pretreated with 10 mM DTT were coinjected. The samples in FIGURE 4B were as in FIGURE 4A, except they were not DTT treated.

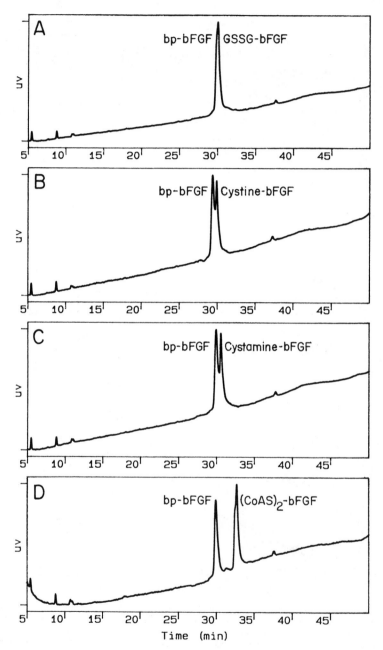

FIGURE 5. Reversed-phase chromatography of bovine pituitary bFGF and S-thiolated forms of recombinant bFGF. HPLC conditions were the same as those in FIGURE 4. The recombinant protein was treated with (a) glutathione disulfide; (b) cystine; (c) cystamine; and (d) coenzyme A disulfide.

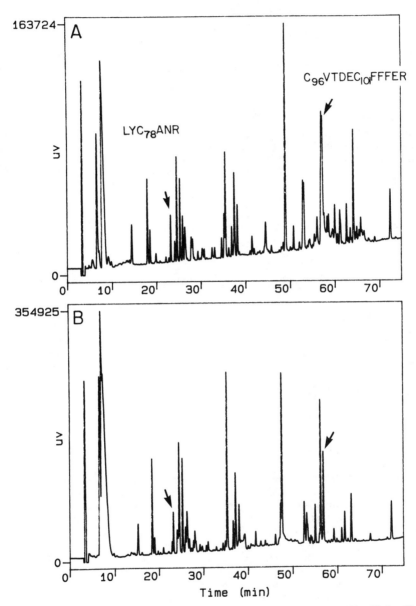

FIGURE 6. Tryptic digests of bFGF on reversed-phase chromatography. The Vydac C18 column was eluted with a 0% to 40% acetonitrile gradient containing 0.1% trifluoroacetic acid over 80 minutes. GS₂-bFGF and bovine pituitary bFGF are shown in A and B respectively.

since we had earlier shown that intermolecular disulfides involving cysteines 78 and/or 96 are unstable in the unfolded protein.

Radiolabeled GS_2-bFGF was digested with trypsin and fractionated by reversed-phase chromatography. The resulting peptide map contained only two radiolabeled fragments (FIGURE 6A). Treatment of the trypsin-digested protein with dithiothreitol (DTT) resulted in the loss of both of these fragments, and all of the radiolabel eluted in the position of glutathione. Sequencing of the radiolabeled fragments demonstrated that the first fragment corresponded to amino acid residues 76–81, and the second to amino acid residues 96–106. Since the second fragment contained two cysteine residues, 96 and 101, we subdigested this peptide with protease V8, which cleaved the peptide between residues 100 and 101, and repurified the two resulting fragments, 96–100 and 101–106. In that case, all of the radiolabel was associated with the 96–100 fragment. Therefore, the two residues in recombinant bFGF that were S thiolated with glutathione are cysteines 78 and 96, the same two residues that are responsible for the multimerization of bFGF.

TABLE 1. Titration of Cysteine Thiols in bFGF, S-Thiolated bFGF, and S-Alkylated bFGF[a]

Recombinant	Conditions	
	Denaturing	Nondenaturing
bFGF	4	2
GS_2-bFGF	2	0
NEM-treated bFGF	2	0
NEM-treated GS_2-bFGF	2	0
Bovine pituitary bFGF (NEM purified)	2	0

[a]Denaturing conditions were used to determine the total number of free thiols in the proteins. The denaturing buffer was 100 mM sodium phosphate pH 7.5, 1 mM EDTA, 6 M guanidine HCl. The nondenaturing buffer was as above except lacking guanidine HCl.

MAPPING OF GLUTATHIONE MOIETIES IN PITUITARY BASIC FGF

To determine the location of the disulfides in bovine pituitary bFGF, we carboxamidomethylated and digested the protein with trypsin exactly as was done with the recombinant protein. Comparison of the tryptic maps demonstrated that they were almost identical (FIGURE 6B). Treatment of the digest with DTT indicated that only two peptides were sensitive to reduction, and both of these eluted at the same positions as the two recombinant GS_2-bFGF fragments containing glutathione. Sequencing of these fragments proved them to be peptides 76–81 and 96–106. Subdigesting the 96–106 fragment with protease V8 produced two peaks as with the recombinant protein, and only the 96–100 fragment was DTT sensitive. Coinjection on reversed-phase HPLC of these peptide fragments with synthetic peptides demonstrated that fragments 76–81 and 96–100 comigrated with the glutathione-conjugated synthetic peptide, while fragment 101–106 comigrated with the carboxamidomethylated synthetic peptide. Therefore, all of the data are consistent with the occurrence of glutathione S thiolation of cysteines 78 and 96 in bovine pituitary bFGF.

PURIFICATION OF PITUITARY BASIC FGF LACKING GLUTATHIONE MODIFICATION

Although S-thiolated proteins have been reported frequently, it is apparent that many of those cases represent artifacts of the purification procedures.[19] To determine if this is the situation with bovine pituitary bFGF, we modified the purification procedure to include homogenization of the pituitaries in the presence of 75 mM N-ethylmaleimide (NEM) to trap free thiol groups. Basic FGF purified in this manner no longer comigrated on reversed-phase HPLC with GS_2-bFGF, was DTT insensitive, and instead comigrated on reversed-phase HPLC with NEM-treated, reduced recombinant bFGF. Therefore, we concluded that the glutathione modification is an artifact of the purification procedure.

THIOL TITRATION OF PITUITARY AND RECOMBINANT BASIC FGF

Finally, thiol titrations of various forms of bFGF were conducted in both the native and unfolded states. As shown in TABLE 1 the recombinant protein has four free thiol equivalents in the unfolded state and two in the native state. The GS_2-bFGF, NEM-treated-bFGF, and NEM-treated GS_2-bFGF (all recombinant) have no free thiols in the native state and two free thiols in the unfolded form. Together, these data indicate that in native bFGF, only cysteines 78 and 96 are reactive with NEM and that cysteines 34 and 101 are sequestered. Thiol titration of bovine pituitary bFGF purified in the presence of NEM also indicated no free cysteines in the native form and two free cysteines in the unfolded state. Since we have shown that NEM is only able to alkylate cysteines 78 and 96 and not cysteines 34 and 101, we concluded that the two remaining free thiols must be cysteines 34 and 101. These data further support the conclusion that there is no intramolecular disulfide in bovine pituitary bFGF as isolated by our purification procedure.

REFERENCES

1. CREIGHTON, T. E. 1984. Methods Enzymol. **107:** 305–329.
2. ABRAHAM, J. A., A. MERGIA, J. L. WHANG, A. TUMOLO, J. FRIEDMAN, K. A. HJERRILD, D. GOSPODAROWICZ & J. C. FIDDES. 1986. Science **233:** 545–548.
3. BURGESS, W. H. & T. MACIAG. 1989. Annu. Rev. Biochem. **58:** 575–606.
4. DICKSON, C. & G. PETERS. 1987. Nature **326:** 833.
5. YOSHIDA, T., K. MIYAGAWA, H. ODAGIRI, H. SAKAMOTO, P. F. R., LITTLE, M. TERADA & T. SUGIMURA. 1987. Proc. Nat. Acad. Sci. USA **84:** 7305–7309.
6. DELLI BOVI, P., A. M. CURATOLA, F. G. KERN, A. GRECO, M. ITTMANN & C. BASILICO. 1987. Cell **50:** 729–737.
7. ZHAN, X., B. BATES, X. HU & M. GOLDFARB. 1988. Mol. Cell. Biol. **8:** 3487–3495.
8. MARICS, I., J. ADELAIDE, F. RAYBAUD, M-G. MATTEI, F. COULIER, J. PLANCHE, O. DELAPEYRIERE & D. BIRNBAUM. 1989. Oncogene **4:** 335–340.
9. FINCH, P. W., J. S. RUBIN, T. MIKI, D. RON & S. AARONSON. 1989. Science **245:** 752–755.
10. THOMPSON, S. A., A. A. PROTTER, L. BITTING, J. C. FIDDES & J. A. ABRAHAM. 1990. Methods Enzymol. **198:** 96–116.
11. SENO, M., R. SASADA, M. IWANE, K. SUDO, T. KUROKAWA, K. ITO & K. IGARASHI. 1988. Biochem. Biophys. Res. Commun. **151:** 701–708.
12. GOSPODAROWICZ, D., S. MASSOGLIA, J. CHENG, G-M. LUI & P. BOHLEN. 1987. J. Cell. Physiol. **122:** 323–332.

13. ESCH, F., A. BAIRD, N. LING, N. UENO, F. HILL, L. DENOROY, R. KEPPLER, D. GOSPO-
     DAROWICZ, P. BOHLEN & R. GUILLEMIN. 1985. Proc. Nat. Acad. Sci. USA **82:** 6507–
     6511.
14. ELLMAN, G. L. 1959. Arch. Biochem. Biophys. **82:** 70–77.
15. SAXENA, V. P. & D. B. WETLAUFER. 1975. Biochemistry **25:** 5015–5023.
16. GILBERT, H. F. 1984. Methods Enzymol. **107:** 330–351.
17. HARRAP, K. R., R. C. JACKSON, P. G. RICHES, C. A. SMITH & B. T. HILL. 1973. Biochim.
     Biophys. Acta **310:** 104–110.
18. SIES, H., R. BRIGELIUS & P. GRAF. 1987. Adv. Enzyme Regul. **26:** 175–189.
19. ZIEGLER, D. M. 1985. Annu. Rev. Biochem. **54:** 305–329.

Structure-Function Studies of Acidic Fibroblast Growth Factor[a]

WILSON H. BURGESS

Laboratory of Molecular Biology
Jerome H. Holland Laboratory for the Biomedical Sciences
American Red Cross
15601 Crabbs Branch Way
Rockville, Maryland 20855

The fibroblast (or heparin-binding) growth factor family consists of at least seven structurally related polypeptides.[1,2] Two of the proteins, acidic fibroblast growth factor (aFGF) and basic fibroblast growth factor (bFGF) have been characterized under many different names and are often referred to as heparin-binding growth factors 1 and 2, respectively. In this report the structural studies of aFGF will be described as will the various functions that have been shown to be associated with the growth factor. Three areas of structure-function analysis will be summarized: (1) those that result from functional analysis of different truncated forms of the protein, (2) those that relate to chemical or enzymatic modification and synthetic peptide work, and (3) recent studies related to site-directed mutagenesis and expression of recombinant forms of aFGF.

STRUCTURES

The complete structure of bovine aFGF was first reported as a 140-residue protein.[3] Subsequently, the sequences of full-length human and bovine aFGF were shown to consist of 154 amino acids.[4,5] In addition, it was established using fast atom bombardment mass spectrometry that the full-length forms of both human and bovine aFGF contained blocked amino termini resulting from acetylation of alanine.[4,5] Recently, we have established that chicken aFGF also consists of 154 amino acids with acetylation of the amino terminal alanine residue.[6] To date the amino terminal truncated forms of aFGF that have been characterized are limited to the residue 15–154 form described above and a form consisting of residues 21–154.[3,5] Both of these truncated forms are probable artifacts of the purification procedures used in their isolation. It should be noted that the biological activities of these truncated forms are very similar to those of the intact protein. Further amino terminal deletions reduce significantly the mitogenic activities of the protein (see below). It should also be noted that the list of potential biological activities associated with aFGF continues to grow and significant differences in the activities of the full-length and truncated forms of the protein may yet be established. For example, differences in the subcellular localization of different amino-terminally extended forms of the closely related protein bFGF have been established.[7,8]

[a]This work was supported in part by National Institutes of Health grant HL35762 and a grant-in-aid from the American Heart Association (891047) with funds contributed in part by the American Heart Association Maryland Affiliate, Inc.

In addition to the protein sequence data described above, the cDNA sequences of aFGF from human, rat, bovine, and hamster sources are available.[9-12] The derived protein sequences from the cDNAs are in agreement with the amino acid sequences established for bovine and human aFGF. The sequences are shown in FIGURE 1. It is clear from this figure that the primary structure of aFGF is highly conserved (90% or more) among mammalian and nonmammalian vertebrates. The highly conserved sequences imply that multiple functional domains exist within the primary structure of aFGF. A brief summary of the diverse functions associated with the protein is provided below.

FUNCTIONS AND MECHANISMS OF ACTION

The observation by Shing *et al.* that an endothelial cell growth factor (later identified as bFGF) could be purified to homogeneity using heparin-Sepharose affinity based chromatography[13] was perhaps the single most important development to our current understanding of the fibroblast growth factor family of polypeptides. This finding provided the basis for the development of rapid and efficient purification schemes for both acidic and basic FGF,[13-15] which in turn provided the quantities of material necessary for extensive characterization of these polypeptides. The observation that other structurally related proteins such as the products of the hst/K-fgf and FGF-5 oncogenes also bind with high affinity to immobilized heparin provided the basis for the designation of the heparin-binding or fibroblast growth factor family of polypeptide mitogens.[1,2]

The relatively high affinity of acidic and basic FGF for immobilized heparin may have significant functional implications. Heparin has been shown to potentiate the biological activities of aFGF but not bFGF (reviewed in References 1 and 2). It is known that heparin can increase the apparent affinity of aFGF for cell surface receptors.[16] In addition, heparin protects aFGF and bFGF from heat and acid inactivation[17,18] and protects them from proteolytic degradation.[18-20] Finally, the affinity of acidic and basic FGF for heparin is thought to provide the basis for their binding to so-called low-affinity cell-surface receptors and for their sequestration in the extracellular matrix. Recently, the cDNA of a cell-surface-associated heparin sulfate proteoglycan has been cloned and its product shown to mediate binding of bFGF to the cell surface.[21] The potential significance of aFGF-heparin interactions to the mitogenic activity of the protein is discussed in more detail below.

The potential biological significance of the FGFs *in vivo* has been reviewed.[1,2,22,23] Briefly, FGFs have been shown to be potent mitogens for a variety of cell types including fibroblasts, endothelial cells, smooth muscle cells, myoblasts, chondrocytes, granulosa cells, epithelial cells, and glial cells.[1,2] Both acidic and basic FGF have been shown to stimulate directed migration or chemotaxis of endothelial cells and fibroblasts.[24,25] More recently, evidence has accumulated that acidic and basic FGF may play important roles in various stages of development. First, direct effects of FGFs in promoting neuronal survival and neurite extension in cultured cells have been observed.[26-30] Second, bFGF has been shown to promote neuronal survival *in vivo.*[31] Third, FGFs have been shown to inhibit the terminal differentiation of myoblasts,[32] and fourth, FGFs have been shown to be potent inducers of mesoderm formation in the *Xenopus* embryo.[33,34] These observations, taken together with the observation that expression of aFGF mRNA in the developing chick tends to correlate more with stages of differentiation rather than proliferation,[35] indicate that the FGFs function as both growth and differentiation factors. All of these isolated activities probably relate to the fact that the FGFs are capable of stimulating

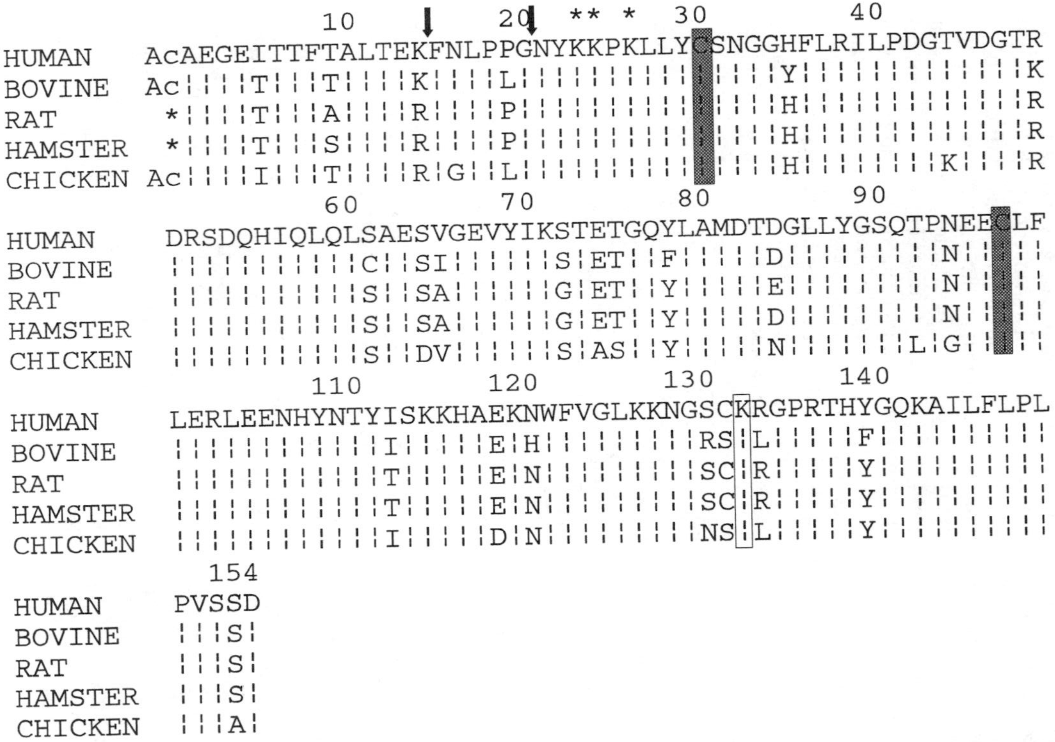

```
                    10            20  ** *    30                    40
HUMAN    ACAEGEITTFTALTEKFNLPPGNYKKPKLLYCSNGGHFLRILPDGTVDGTR
BOVINE   AC!-------!T!----T!----K!----L!--------------------!K
RAT      *----------!T!----A!----R!----P!-------------------!R
HAMSTER  *!----------!T!----!S!----R!----P!------------------!R
CHICKEN  AC!-------!!!----!T!----R!G!----L!-------------K!----!R

         DRSDQHIQLQLSAESVGEVYIKSTETGQYLAMDTDGLLYGSQTPNEECLF
         60                  70              80              90
HUMAN
BOVINE   --------!C!-SI!-----!S!ET!-F!----D!------N!-------
RAT      --------!S!-SA!-----!G!ET!-Y!-----E!------N!-------
HAMSTER  --------!S!-SA!-----!G!ET!-Y!-----D!------N!-------
CHICKEN  --------!S!-DV!-----!S!AS!-Y!-----N!------!L!G!----

         LERLEENHYNTYISKKHAEKNWFVGLKKNGSCKRGPRTHYGQKAILFLPL
         110              120             130             140
HUMAN
BOVINE   -------!H!------!H!------!E!H!-------!RS!L!-F!------
RAT      -------!T!------!T!------!E!N!-------!SC!R!-Y!------
HAMSTER  -------!T!------!T!------!E!N!-------!SC!R!-Y!------
CHICKEN  -------!I!------!I!------!D!N!-------!NS!L!-Y!------

         154
HUMAN    PVSSD
BOVINE   --!S!
RAT      --!S!
HAMSTER  --!S!
CHICKEN  --!A!
```

FIGURE 1. Amino acid sequences of aFGF from different species. The complete sequence of human aFGF is shown. Those residues in aFGF from other species that differ from the human sequence are also shown. The AC at the amino terminus of the human, bovine, and chicken sequences represents the acetylation of the amino terminus that has been characterized directly. The vertical arrows indicate the positions of amino-terminal cleavages that occur during purification that maintain biological activity. The stars indicate the positions of three lysine residues that have been changed by site-directed mutagenesis (see text). The lysine at position 132 changed by site-directed mutagenesis is in a clear box, and the two conserved cysteines are enclosed in a shaded box.

complex responses *in vivo* that require migration, proliferation, and differentiation. These include angiogenesis and wound repair.

Relatively little is known regarding the mechanisms by which the FGFs regulate such diverse cellular responses. A family of FGF receptors with protein tyrosine kinase activities have been identified[36-38] as have receptors or binding proteins that lack tyrosine kinase activity.[39-41] Addition of acidic FGF has been shown to stimulate the phosphorylation on tyrosine residues of phospholipase C-γ.[42] However, the role of stimulated phosphoinositide turnover in the signal transduction pathways affected by acidic or basic FGF remains unclear.[43-45] Both acidic and basic FGF have been shown to induce rapid and transient expression of the mRNAs encoding the protooncogene proteins c-myc, c-fos, and c-jun.[46-49] In addition, increases in actin mRNA levels,[50] ribosomal gene transcription,[51] and synthesis of a number of unidentified cellular proteins[50] have been reported to occur following the addition of FGF to various cells. The role of these observed changes in gene expression to the mechanisms of FGF action remain unknown. However, they provide additional biochemical markers for structure-function studies of the FGFs. To date the evidence from our work and that of others[52,57] indicates that the majority of the FGF-stimulated events described above can be considered necessary but not sufficient to complete the mitogenic signal initiated by aFGF (see below).

TABLE 1. Cell Number ($\times 10^{-5}$)

	Growth Factor Concentration (ng/ml)					
	0	0.1	0.5	1	5	10
aFGF 21-154	2.0	2.5	2.5	3.1	13	18
aFGF 1-154	2.1	1.8	3.0	5.2	25	38

STRUCTURE-FUNCTION RELATIONSHIPS: MULTIPLE FORMS OF aFGF ISOLATED FROM TISSUE SOURCES

As described above, three forms of aFGF have been isolated and characterized. These include the full-length 1–154 residue form, and the amino-terminally truncated forms consisting of residues 15–154 and 21–154. The residue 15–154 form is the one most commonly used in aFGF studies. It has been demonstrated that the heparin-binding, receptor-binding, and mitogenic activities of the three forms are very similar. The amino-terminal truncations arise as artifacts of purification as judged by western blot analysis of cells and tissues treated directly with sodium dodecyl sulfate (SDS) sample buffer.[1] The direct comparisons of the activities of the full-length and truncated species have been limited for the most part to analysis of ³H-thymidine incorporation into DNA. Recently we have observed differences in the abilities of recombinant full-length and the residue 21–154 form of aFGF to support the growth of human umbilical vein endothelial cells in culture (TABLE 1). It is not yet clear whether the observed differences are unique to bacterially expressed proteins. These results do indicate that complete structure-function studies of aFGF should include studies of the different amino-terminal forms of the growth factor as well as site-directed mutagenesis or chemical modification studies of selected residues.

STRUCTURE-FUNCTION RELATIONSHIPS: SYNTHETIC PEPTIDES AND CHEMICAL MODIFICATIONS

Our initial efforts towards understanding the structural basis for the various activities associated with aFGF focused on the use of synthetic peptides and chemical modification of the growth factor to identify functional domains of the protein. In contrast to studies of bFGF structure function, we were unable to identify synthetic peptides or peptides derived from proteolytic cleavage of native protein that were able to compete with intact aFGF for receptor binding or peptides that could function as agonists or antagonists of aFGF's mitogenic activity. These observations are consistent with the fact that following quantitative reduction and alkylation of the cysteine residues of human recombinant aFGF we are unable to detect receptor activity associated with the modified protein.[53] Whether the results of these experiments are related directly to the cysteine chemistry of aFGF or relate to folding issues is a current topic of investigation.

Two synthetic peptides based on the primary structure of bFGF that retain heparin-binding activity have been characterized.[54] We used a gel overlay procedure to study the interaction of heparin-binding proteins with ^{125}I-labeled fluorescence heparin.[55] With this assay one synthetic peptide corresponding to residues 49 to 71 of aFGF was shown to compete for binding of derivatized heparin to intact aFGF. This region of aFGF is homologous to one of the two regions of heparin-binding activity found in bFGF. To date this is the only functional domain that has been defined for aFGF using synthetic peptides.

Chemical modification studies indicate that lysine 132 of aFGF is important to several of the functions of the growth factor. Harper and Lobb showed that limited reductive methylation of bovine aFGF with formaldehyde and cyanoborohydride resulted in stoichiometric methylation only of lysine 132[56] (using the numbering system for full-length, 154-residue aFGF). They reported 90% modification of this residue with 60% dimethyllysine. The modified protein showed a reduced apparent affinity for immobilized heparin, a fourfold reduction in its ability to stimulate DNA synthesis in NIH 3T3 fibroblasts, and a similar reduction in its ability to compete with labeled ligand in a radioreceptor assay. A lysine residue is conserved at this position in both aFGF and bFGF from all species characterized to date. Together, these data indicated a critical role for lysine 132 in aFGF function. Additional evidence for the importance of this residue or secondary consequences of changes at this position is provided below.

STRUCTURE-FUNCTION RELATIONSHIPS: SITE-DIRECTED MUTAGENESIS

Despite the fact that cDNA clones for both acidic and basic FGF have been available since 1986, relatively little has been reported regarding site-directed mutagenesis and expression of altered forms of the growth factors. The published studies have focused primarily on the cysteines of acidic and basic FGF. These studies are the subject of other papers in this volume and will not be reviewed here. Another report of site-directed mutagenesis of aFGF involved amino terminal deletions beyond the amino terminus of the des 1-20 form of aFGF which is known to be mitogenic.[57] The results of this study suggested that the deletion of an additional 7 residues reduces the mitogenic activities of the protein without affecting its apparent affinity for immobilized heparin. Further, it was reported that mitogenic activity was not observed at concentrations of the recombinant growth factor that were sufficient to stimulate intracellular kinase activity and induce c-fos expression. It was noted that deletion of the additional 7 residues removed a sequence (N-Y-K-K) that fits a consensus sequence for nuclear translocation.[58] Finally, it was reported that construction of a chimera protein containing the nuclear translocation sequence of yeast histone 2B adjacent to the des 1-27 form of aFGF resulted in restoration of full mitogenic activity. The results presented below summarize recent work from this laboratory designed to address the implications of the chemical modification studies of Harper and Lobb[56] regarding the importance of lysine 132 to aFGF function and the importance of nuclear translocation sequences to the mitogenic activities of aFGF using site-directed mutagenesis.

Our initial mutagenesis experiments of aFGF involved replacement of lysine 132 with a glutamic acid residue by site-directed mutagenesis of the human cDNA and expression of the mutant protein in *Escherichia coli* to obtain sufficient quantities for detailed functional studies.[52] Purification of the recombinant protein revealed that its apparent affinity for immobilized heparin was reduced significantly relative to wild-type protein (elutes at 0.45 M NaCl vs. 1.1 M NaCl for wild type). Mitogenic assays designed to measure [³H]thymidine incorporation into the DNA of NIH 3T3 cells or to measure the proliferation of human umbilical vein endothelial cells established that the change of lysine 132 to glutamic acid drastically reduces the specific mitogenic activity of aFGF. These results are in good agreement with the results of the chemical modification studies of Harper and Lobb.[56]

The mechanisms responsible for the poor mitogenic activity of the mutant protein are not clear. It is known that human aFGF and human recombinant aFGF in particular are highly dependent on the presence of heparin for optimal mitogenic activity.[52] Although the glutamic acid mutant shows a reduced apparent affinity for heparin, it will bind heparin at physiologic ionic strength. It is known that the presence of added heparin is not necessary for near optimal high-affinity receptor binding of wild-type aFGF.[52] Further, we have shown that added heparin is not required to observe optimal induction of expression of the protooncogenes c-fos, c-jun, or c-myc.[52,59] Similarly, we demonstrated that in the presence of added heparin, the glutamic acid mutant is able to compete fully with wild-type protein for binding to high-affinity cell-surface receptors and is capable of inducing the same pattern of protooncogene expression with the same dose response and time course as the wild-type protein.[52] In addition, the glutamic acid mutant is capable of activating the tyrosine kinase activity of the high-affinity FGF receptors and leads to tyrosine phosphorylation of phospholipase C-γ.[52] Together these observations indicate that the mitogenic deficiencies of the glutamic acid mutant are not related to the early events associated with aFGF signal transduction and lead to the conclusion that activation of tyrosine kinase activity and induction of protooncogene expression are events that may be necessary but not sufficient to initiate a mitogenic response to the growth factor.

The results described above raise the question as to whether additional events to high-affinity receptor binding are required of aFGF to initiate mitogenesis. This second event could include binding to a distinct cell surface receptor or binding protein or to interaction with intracellular targets following receptor-mediated internalization of the growth factor. A result consistent with the second possibility is the observation that in contrast to wild-type aFGF, overexpression of the glutamic acid mutant in transfected NIH 3T3 cells does not result in a transformed phenotype or loss of contact inhibition of cell growth.[52,59] In addition, we have shown that the gross subcellular distribution of wild-type or mutant aFGF in these transfected cells is the same; the majority of both proteins are found associated with the nuclear fraction.[59] We have not rigorously determined whether this mutant aFGF has an altered affinity for any nuclear macromolecule relative to the wild-type protein.

As described above the shortest amino-terminal truncated form of aFGF that retains mitogenic activity corresponds to residues 21–154. The sequence of residues 23–26 (K-K-P-K) is in good agreement with a consensus sequence for nuclear translocation.[58] Imamura et al. demonstrated that amino-terminal deletion mutants of aFGF that extend into this region exhibited reduced mitogenic activity that could not be accounted for by reduced receptor-binding activity.[57] They suggested that nuclear translocation of aFGF may be important for mitogenic activity based on the observation that mitogenic activity of the deletion mutant was restored when an unrelated nuclear translocation sequence was added to the amino terminus of the mutant. We used site-directed mutagenesis to generate three point mutations that change each of the lysine residues (residues 23, 24, and 26) individually to a glycine residue. Each of these mutations would disrupt this particular consensus sequence for nuclear translocation found in aFGF. The subcellular distribution of these mutants has not yet been established; however, their heparin-binding, receptor-binding, mitogenic, and neurite-promoting activities are similar to those of the wild-type protein.[59]

In summary, the results described here provide a basis for future studies of the structural basis for aFGF function. First, it may be important to consider functional differences among the amino-terminal truncations of aFGF and the full-length protein. The majority of studies related to aFGF function have utilized the equivalent of a 14-residue deletion mutant. Second, it appears unlikely that a simple, linear peptide-based agonist or antagonist of aFGF function will be identified. The possible exception is that synthetic peptides corresponding to heparin-binding domains of the growth factor may be identified. Third, the functional analysis of aFGF containing a glutamic acid residue in place of lysine at residue 132 demonstrates that the various functions of aFGF can be dissociated at the structural level and indicate that it may be possible to generate mutants of aFGF that retain certain (i.e., chemotactic, mitogenic, or mesoderm-inducing activities) but not other biological functions characteristic of the wild-type protein. Finally, the role of postreceptor or intracellular activities of aFGF remains unclear. However, the fact that the full-length translation product lacks classical sequences for secretion indicates that intracellular pathways may be important to the normal function of this growth factor.

REFERENCES

1. BURGESS, W. H. & T. MACIAG. 1989. Annu. Rev. Biochem. **58**: 575–606.
2. KLAGSBRUN, M. 1989. Prog. Growth Factor Res. **1**: 207–235.
3. GIMENEZ-GALLEGO, G., J. RODKEY, C. BENNET, M. RIOS-CANDELORE, J. DISALVO & K. THOMAS. 1985. Science **230**: 1385–1388.
4. CRABB, J. W., L. G. ARMES, S. A. CARR, C. M. JOHNSON, G. D. ROBERT, R. S. BORDOLI & W. L. MCKEEHAN. 1986. Biochemistry **25**: 4988–4993.
5. BURGESS, W. H., T. MEHLMAN, D. R. MARSHAK, B. A. FRASER & T. MACIAG. 1986. Proc. Nat. Acad. Sci. USA **83**: 7216–7220.
6. MEHLMAN, T. & W. H. BURGESS. (Manuscript in preparation.)
7. RENKO, M., N. QUARTO, T. MORIMOTO & D. B. RIFKIN. 1990. J. Cell. Phys. **144**: 108–114.
8. BUGLER, B., F. AMALRIC & H. PRATS. 1991. Mol. Cell. Biol. **11**: 573–577.
9. JAYE, M., R. HOWK, W. H. BURGESS, G. A. RICCA, I.-M. CHIU, M. W. RAVERA, S. J. O'BRIEN, W. S. MODI, T. MACIAG & W. N. DROHAN. 1986. Science **233**: 541–545.
10. GOODRICH, S. P., G.-C. YAN, K. BAHRENBURG & P. E. MANSSON. 1989. Nucleic Acids Res. **17**: 2867.

11. HALLEY, C., Y. COURTOIS & M. LAURENT. 1988. Nucleic Acids Res. **16:** 10913.
12. HALL, J. A., M. A. HARRIS, M. MALARK, P.-E. MANSSON, H. ZHOU & S. E. HARRIS. 1990. J. Cell. Biochem. **43:** 17–26.
13. SHING, Y., J. FOLKMAN, R. SULLIVAN, C. BUTTERFIELD, J. MURRAY & M. KLAGSBRUN. 1984. Science **223:** 1296–1299.
14. LOBB, R. R. & J. W. FETT. 1984. Biochemistry **23:** 6295–6299.
15. BURGESS, W. H., T. MEHLMAN, R. FRIESEL, W. V. JOHNSON & T. MACIAG. 1985. J. Biol. Chem. **260:** 11389–11392.
16. SCHREIBER, A. B., J. KENNEY, W. J. KOWALSKI, R. FRIESEL, T. MEHLMAN & T. MACIAG. 1985. Proc. Nat. Acad. Sci. USA **82:** 6138–6142.
17. GOSPODAROWICZ, D. & J. CHENG. 1986. J. Cell Physiol. **128:** 475–484.
18. ROSENGART, T. K., W. V. JOHNSON, R. FRIESEL, T. MEHLMAN & T. MACIAG. 1988. Biochem. Biophys. Res. Commun. **152:** 432–440.
19. SAKSELA, O., D. MOSCATELLI, A. SOMMER & D. B. RIFKIN. 1988. J. Cell Biol. **107:** 743–751.
20. LOBB, R. R. 1988. Biochemistry **27:** 2572–2578.
21. KIEFER, M., J. C. STEPHENS, K. CRAWFORD, K. OKINO & P. J. BARR. 1990. Proc. Nat. Acad. Sci. USA **87:** 6985–6989.
22. RIFKIN, D. B. & D. MOSCATELLI. 1989. J. Cell Biol. **109:** 1–6.
23. LOBB, R. R. 1988. Eur. J. Clin. Invest. **18:** 321–326.
24. PRESTA, M., D. MOSCATELLI, J. J. SILVERSTEIN & D. B. RIFKIN. 1986. Mol. Cell. Biol. **6:** 4060–4066.
25. TERRANOVA, V. P., R. DIFLORIO, R. M. LYALL, S. HIC, R. FRIESEL & T. MACIAG. 1985. J. Cell Biol. **101:** 2330–2334.
26. WALICKE, P., M. COWAN, N. UENO, A. BAIRD & G. GUILLEMIN. 1986. Proc. Nat. Acad. Sci. USA **83:** 3012–3016.
27. MORRISON, R. S., A. SHARMA, J. DEVELLIS & R. A. BRADSHAW. 1986. Proc. Nat. Acad. Sci. USA **83:** 7537–7541.
28. LIPTON, S. A., J. A. WAGNER, R. D. MADISON & P. A. D'AMORE. 1988. Proc. Nat. Acad. Sci. USA **85:** 2388–2392.
29. NEUFELD, G., D. GOSPODAROWICZ, L. DODGE & D. K. FUJII. 1987. J. Cell. Physiol. **131:** 131–140.
30. SCHUBERT, D., N. LING & A. BAIRD. 1987. J. Cell Biol. **104:** 635–643.
31. ANDERSON, K. J., D. DAM, S. LEE & C. W. COTMAN. 1988. Nature **332:** 360–361.
32. CLEGG, C. H., T. A. LINKHART, B. B. OLWIN & S. D. HAUSCHKA. 1987. J. Cell Biol. **105:** 949–956.
33. KIMMELMAN, D. & M. KIRSCHNER. 1987. Cell **51:** 869–877.
34. SLACK, J. M. W., B. G. DARLINGTON, J. K. HEATH & S. F. GODSAVE. 1987. Nature **326:** 197–200.
35. SCHNURCH, H. & W. RISAU. 1991. Development **111:** 1143–1154.
36. DIONNE, C. H., G. CRUMLEY, F. BELLOT, J. M. KAPLOW, G. SEARFOSS, M. RUTA, W. H. BURGESS, M. JAYE & J. SCHLESSINGER. 1990. EMBO J. **9:** 2685–2692.
37. MANSUKHARI, A., D. MOSCATELLI, D. TALARICO, V. LEVYTSKA & C. BASILICO. 1990. Proc. Nat. Acad. Sci. USA **87:** 4378–4382.
38. JOHNSON, D. E., P. L. LEE, J. LU & L. T. WILLIAMS. 1990. Mol. Cell. Biol. **10:** 4728–4736.
39. MOSCATELLI, D. 1987. J. Cell Physiol. **131:** 123–130.
40. KAN, M., D. DISORBO, J. HON, H. HOSHI, P. MANSSON & W. L. MCKEEHAN. 1988. J. Biol. Chem. **263:** 11306–11313.
41. BURRUS, L. W. & B. B. OLWIN. 1989. J. Biol. Chem. **264:** 18647–18653.
42. BURGESS, W. H., C. A. DIONNE, J. KAPLOW, R. MUDD, R. FRIESEL, A. ZILBERSTEIN, J. SCHLESSINGER & M. JAYE. 1990. Mol. Cell. Biol. **10:** 4770–4777.
43. MAGNALDO, I., G. L'ALLEMAIN, J. C. CHAMBARD, M. MOENNER, D. BARRITAULT & J. POUYSSEGUR. 1986. J. Biol. Chem. **261:** 16916–16922.
44. KAIBUCHI, K., T. TSUDA, A. KIKUCHI, T. TANIMOTO, T. YAMASHITA & Y. TAKAI. 1986. J. Biol. Chem. **261:** 1187–1192.
45. TSUDA, T., K. KAIBUCHI, Y. KAWAHARA, H. FUKUZAKI & Y. TAKAI. 1985. FEBS Lett. **191:** 205–210.
46. KELLY, K., B. H. COCHRAN, C. D. STILES & P. LEDER. 1983. Cell **35:** 603–610.

47. MÜLLER, R., R. BRAVO, J. BURCKHARDT & T. CURRAN. 1984. Nature **312:** 716–720.
48. RYDER, K. & D. NATHANS. 1988. Proc. Nat. Acad. Sci. USA **85:** 8464–8467.
49. GAY, C. G. & J. A. WINKLES. 1990. J. Biol. Chem. **265:** 3284–3292.
50. RYBAK, S. M., R. R. LOBB & J. W. FETT. 1988. J. Cell. Physiol. **136:** 312–318.
51. BOUCHE, G., N. GAS, H. PRATS, V. BALDIN, J.-P. TAUBER, J. TEISSIÉ & F. AMALRIC. 1987. Proc. Nat. Acad. Sci. USA **84:** 6770–6774.
52. BURGESS, W. H., A. M. SHAHEEN, M. RAVERA, M. JAYE, P. J. DONOHUE & J. A. WINKLES. 1990. J. Cell Biol. **111:** 2129–2138.
53. JAYE, M., W. H. BURGESS, A. B. SHAW & W. N. DROHAN. 1987. J. Biol. Chem. **262:** 16612–16617.
54. BAIRD, A., D. SCHUBERT, N. LING & R. GUILLEMIN. 1988. Proc. Nat. Acad. Sci. USA **85:** 2324–2328.
55. MEHLMAN, T. & W. H. BURGESS. 1990. Anal. Biochem. **188:** 159–163.
56. HARPER, J. W. & R. R. LOBB. 1988. Biochemistry **27:** 671–678.
57. IMAMURA, T., K. ENGLEKA, X. ZHAN, Y. TOKITA, R. FOROUGH, D. ROEDER, A. JACKSON, J. A. M. MAIER, T. HLA & T. MACIAG. 1990. Science **249:** 1567–1570.
58. LANFORD, R. E., R. G. WHITE, R. G. DUNHAM & P. KAMDA. 1988. Mol. Cell. Biol. **8:** 2733–2739.
59. BURGESS, W. H., A. M. SHAHEEN, B. HAMPTON, P. J. DONOHUE & J. A. WINKLES. 1991. J. Cell. Biochem. **45:** 131–138.

Structure/Activity Relationships in Basic FGF

ANDREW SEDDON, MILDRED DECKER,
THOMAS MÜLLER, DOUGLAS ARMELLINO,
IMRE KOVESDI, YAKOV GLUZMAN,
AND PETER BÖHLEN

Medical Research Division
American Cyanamid
Pearl River, New York 10965

INTRODUCTION: BRIEF REVIEW OF STRUCTURE/ACTIVITY RELATIONSHIPS IN BASIC FGF[a]

Redox State and Function of Cysteine Residues

Two of four cysteine residues in basic fibroblast growth factor (bFGF) (Cys-34 and Cys-101) are conserved in all members of the FGF family. This has widely led to speculation that these cysteines are linked by a disulfide bond. Fox *et al.* presented sulfhydryl titration and peptide mapping data suggesting that Cys-34 and Cys-101 form a disulfide structure whereas Cys-78 and Cys-96 exist in the reduced form.[2] They also stated that Cys-78 and Cys-96 of bFGF can be carboxymethylated without loss of mitogenic activity or change in protein conformation. Furthermore, mutation of one or both of these two cysteines to serine resulted in FGF analogues that were fully biologically active[2,3] and exhibited greater stability to inactivation at low pH.[3] Such mutations also reduced the commonly observed heterogeneity of recombinant bFGF which is attributed to intermolecular disulfide bond formation involving the free sulfhydryls of Cys-78 and/or Cys-96 on the protein surface.[3] In contrast, replacement of Cys-34 with serine is associated with a considerable loss of activity.[3] A mutant in which all cysteines are replaced by serines was reported to possess only 10% of the mitogenic activity of wild-type bFGF by one group,[3] but was found to be fully active and conformationally indistinguishable from wild-type bFGF by another.[4] Although contradictory in some aspects, most of these data are compatible with the possibility of a disulfide linkage between Cys-34 and Cys-101 of bFGF. Essentially similar conclusions regarding the presence of a free sulfhydryl and a disulfide bond were drawn from early studies on aFGF.[5,6] However, more recent biochemical[7] and x-ray crystallographic[8–10] data show that recombinant bFGF and aFGF do not contain a disulfide structure.

Heparin-Binding Domain

FGFs, and in particular bFGF, bind with high affinity to heparinlike structures, which confers protection from proteolytic degradation and denaturation. A function-

---

[a]The numbering system used to identify specific amino acids in bFGF is based on the 155-residue bFGF protein as defined by the cDNA sequence of Abraham *et al.*[1] The methionine corresponding to the translation initiation codon is residue no. 1.

ally relevant association between the FGFs and heparinlike substances located on cell surfaces and in extracellular matrices is suggested from several lines of experimental evidence (see elsewhere in this volume). Furthermore, the importance of the heparin-binding site of bFGF for the proper functioning of the growth factor has recently been highlighted by the finding that cell-surface-bound heparin sulfate proteoglycans (a cellular low-affinity bFGF receptor) and free heparinlike substances may function by inducing a change in the conformation of bFGF enabling the growth factor to bind to the high-affinity bFGF receptor (induced fit).[11] This hypothesis is supported by the finding that heparin increases the binding affinity of aFGF to the FGF receptor by two- to threefold.[12]

Studies aimed at elucidating the heparin-binding site(s) of FGFs have been attempted by several groups of investigators. Baird *et al.* synthesized FGF peptide fragments and found that peptides related to the sequences of bFGF(33–77) and bFGF(102–129) bound to heparin with appreciable affinity.[13] These data suggested that binding of heparin to bFGF may involve two sequence domains. In contrast, Seno *et al.*, using recombinant N-terminal and C-terminal deletion mutants of bFGF, concluded that the C-terminal structure of bFGF contributes mostly to heparin binding.[14] In support of the latter, the finding that chemical modification of Lys-133 in aFGF (which is homologous to Lys-134 in bFGF) by reductive alkylation[15] or mutation of this residue to glutamic acid[16] leads to loss of high-affinity heparin binding. Interestingly, crystal structure data implicate Lys-134, together with Arg-129, as a sulfate ion binding site.[9,10] Sulfate binding in this location also involves Asn-36.[9] A second sulfate binding site is formed by Lys-128 and Lys-138.[10] The combined evidence strongly suggests that this cluster of four C-terminal basic residues may be part of the heparin-binding site of bFGF, possibly in conjunction with an N-terminal sequence strand containing Asn-36.

Receptor-Binding Domain

The identification of bFGF residues participating in the binding to the FGF receptor has yielded somewhat conflicting results. Baird *et al.* reported that synthetic peptides related to the sequences bFGF(33–77) and bFGF(102–129) act as weak partial agonists and antagonists in a mitogenic assay.[13] Interestingly, these peptides correspond to the same sequence domains identified by the authors as heparin-binding domains, suggesting that binding to the receptor and heparin may involve the same two surface domains of bFGF. In a different approach, Seno *et al.* demonstrated that recombinant truncated forms of bFGF lacking up to 49 and 33 residues at the N and C termini, respectively, retained significant mitogenic activity.[14] The results imply that there may be a single receptor-binding domain located within the sequence 50–122 of bFGF.

X-Ray crystallographic data show that the sequence bFGF(115–124), which was previously implicated in receptor binding,[13] lies on the surface of the protein molecule.[9,10] This sequence also contains Thr-121, the phosphorylation of which with protein kinase A was shown to enhance the affinity of bFGF for its receptor.[17] This change in affinity might be caused by a phosphorylation-induced change of the conformation of this sequence domain.[10] Although these data support the notion that the bFGF sequence 115–124 is involved in receptor binding, they do not define the complete receptor-binding domain. Since bFGF(115–124) possesses only minimal receptor affinity, additional factors must be important. Finally, it will be important to determine whether heparin- and receptor-binding sites of bFGF are functionally interdependent. Although there is biochemical/biological[14,18] and crystallographic[8–10] evidence suggesting that the putative heparin- and receptor-binding sites in bFGF are associated with different molecular domains, there are other data in support of a functional linkage between heparin and receptor binding. Heparin binding may induce a conformational change in bFGF necessary for proper receptor binding (induced fit),[11] and in aFGF, heparin binding was reported to increase the binding affinity of the protein to its receptor.[12] Recent evidence suggests further that ligand-induced signal transduction may depend on additional sequences in bFGF. To this end, deletion of residues 27–32 in bFGF abolishes activation of plasminogen activator gene expression but has no effect on receptor affinity, tyrosine phosphorylation, or mitogenic activity.[19]

EXPERIMENTAL

Construction of Expression Plasmid

A synthetic gene encoding the 155–residue form of human bFGF[1] cloned into pUC18 was purchased from British Bio-technology (Oxford, United Kingdom) and cloned into the Nde1/Bam H1 site of the expression vector pT7 Kan 5, a derivative of pET-3a containing the T7 promoter for RNA polymerase.[20] Using the T7 expression system the yield of recombinant bFGF was low (0.4 mg/l). It was found (data to be reported elsewhere) that protein yield was significantly increased (4 mg/l) using a mutated bFGF gene in which codons for Ala-3 and Ser-5 were replaced with glutamic acid codons to give a chimeric FGF gene in which the first 8 codons are those of acidic FGF. Recombinant Glu^{3,5}-bFGF was found to be indistinguishable from wild-type bFGF with respect to mitogenic activity and affinity to heparin (data not shown). The expression plasmid pT7 Glu^{3,5}-bFGF was used as a template for site-directed mutagenesis using appropriate oligonucleotides and the polymerase chain reaction (PCR) to change Cys-78 and Cys-96 of bFGF to serine codons. The DNA encoding Glu^{3,5}Ser^{78,96}-bFGF was then cloned into the T7 expression vector pET-3a(M13), a derivative of pET-3a (details to be reported elsewhere).

Expression and Purification of Glu^{3,5}Ser^{78,96}-bFGF

Following sequence verification,[21] the gene encoding the bFGF mutant was transformed into competent BL21 plys S cells. *Escherichia coli* harboring the plasmid were grown in Luria broth containing ampicillin (100 μg/ml) and chloramphenicol (34 μl/ml) at 37°C to about 0.6 absorbance units at 600 nm, and bFGF synthesis induced by addition of isopropyl-β-D-thiogalactopyranoside (1 mM). The cells were then harvested 2 hours postinduction by centrifugation at 4°C. Cell pellets from 1-liter cultures were resuspended in 50 mM Tris, 0.1 mM EDTA buffer (pH 7.6; 30 ml) and lysed by 3 rapid freeze/thaw cycles. The lysate was then treated with DNase I (20 μg/ml) in the presence of 5 mM MgCl$_2$ for 20 minutes at 4°C and centrifuged to remove cell debris (10,000 × g; 20 minutes). bFGF was purified from the supernatant solution by heparin Sepharose affinity chromatography essentially as described[22] using a linear salt gradient from 0.6–3.0 M NaCl. Fractions containing growth factor were pooled, diluted with Tris buffer (10 mM; pH 7.6) to 0.6 M NaCl, and loaded onto a TSK Heparin-5PW column (0.75 × 7.5 cm; TosoHaas, Philadelphia, Pa.) equilibrated with 10 mM Tris, 0.6 M NaCl (pH 7.6). Elution of bound material was monitored at 280 nm and was accomplished using a linear salt gradient (0.6–3.0 M

NaCl in 60 minutes) at a flow rate of 0.7 ml/minute. Growth factor purified in this manner was judged homogeneous by reverse-phase high-performance liquid chromatography (HPLC), monitoring elution at 210 nm (C_4, Vydac; the Separations Group, Hesperia, Calif.), using a 0.1% trifluoroacetic acid/acetonitrile gradient (0–28% CH_3CN in 15 minutes; 28–60% CH_3CN in 90 minutes) at a flow rate of 0.7 ml/minute, N-terminal sequence analysis and sodium dodecyl sulfate-polyacrylamide gel electrophoresis (SDS-PAGE) on 10–15% gradient and 20% homogeneous gels using a silver-stain detection system (Phastgel System, Pharmacia/LKB).

Proteolytic Digestion of Glu³,⁵Ser⁷⁸,⁹⁶-bFGF Bound to Heparin Sepharose

A solution of $Glu^{3,5}Ser^{78,96}$-bFGF (710 µg in 0.475 ml 50 mM Tris-HCl, pH 7.6, 250 mM NaCl) was added to a drained slurry (0.2 ml) of heparin Sepharose (Pharmacia/LKB, Uppsala, Sweden), mixed and incubated at 4°C. After 60 minutes the supernatant solution was removed and the gel washed with 50 mM Tris-HCl pH 7.6 (0.4 ml). The gel slurry was resuspended in Tris-HCl buffer (0.2 ml) containing pronase (Sigma Chemical Co.; Type XXV) to give a ratio of bFGF to pronase of 0.75:1.0 (w/w). The mixture was incubated at 37°C with agitation. After 24 hours the gel mixture was centrifuged, the supernatant solution removed, and the drained gel washed 3 times with 0.4 ml of 10 mM Tris buffer (pH 7.6). The gel was further washed with Tris buffer containing 0.6 M NaCl (2 × 0.4 ml) to remove nonspecifically bound proteolytic products. Elution of more strongly bound material was effected by washing the gel with Tris buffer containing 3 M NaCl (2 × 0.2 ml).

Analysis of Proteolytic Fragments of Glu³,⁵Ser⁷⁸,⁹⁶-bFGF

Heparin-binding proteolytic fragments of the bFGF mutant in the 3 M NaCl eluent from heparin Sepharose were analysed under reducing conditions by SDS-PAGE using a silver-stain detection system. Peptide fragments were resolved by C_4 reverse-phase HPLC using a 0.1% trifluoroacetic acid/acetonitrile gradient at a flow rate of 0.7 ml/minute. N-terminal sequence analysis of reverse-phase HPLC-purified peptides was performed on a Model 477A pulsed-liquid phase sequencer (Applied Biosystems, Foster City, Calif.) equipped with an on-line PTH-derivative analyzer (Model 120A, Applied Biosystems). Amino acid compositions were determined after HCl gas phase hydrolysis (5.7 M HCl/0.1% phenol; 24 hours at 110°C) using a model 420A phenylisothiocyanate derivatizer equipped with an on-line model 130A separation system (Applied Biosystems) according to the manufacturer's protocol.

Mitogenic Assays

Reverse-phase HPLC-purified samples of bFGF and bFGF peptide fragments were diluted (1:5) in Dulbecco's modified Eagle media (DMEM) 1% bovine serum albumin (BSA) and tested for mitogenic activity using bovine vascular endothelial cells derived from the adult aortic arch as described.[22] After 5 days in culture, duplicate plates were trypsinized and cell densities determined by cell counting in a Coulter counter or by a colorimetric method.[23]

RESULTS

Analysis by SDS-PAGE of the protein material in the 3 M NaCl eluent from heparin Sepharose chromatography, after pronase digestion of the heparin/ Glu[3,5]Ser[78,96]-bFGF complex, showed complete digestion of the 18-kDa band corresponding to Glu[3,5]Ser[78,96]-bFGF and the generation of 2 lower molecular weight peptides that migrated as 11- and 9-kDa species (FIGURE 1). Reverse-phase HPLC of this material revealed 2 major peaks, termed heparin-binding fragment 1 and 2 (HFB-1 and HFB-2) (FIGURE 2). The earlier eluting fragment, HFB-1, was identi-

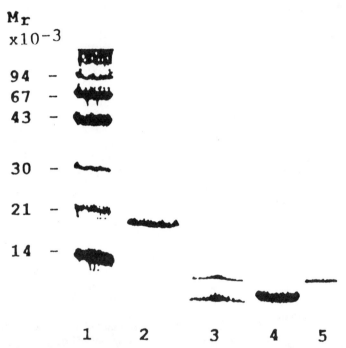

FIGURE 1. Analysis by SDS-PAGE of bFGF peptides protected from pronase digestion by association with heparin Sepharose. Molecular weight markers (lane 1), Glu[3,5]Ser[78,96]-bFGF (lane 2), bFGF peptides eluted with 3M NaCl from heparin Sepharose after pronase digestion (lane 3), HBF-1 (lane 4), and HBF-2 (lane 5) from reverse-phase HPLC (see FIGURE 2).

fied by SDS-PAGE as the 9-kDa species, whereas the more retarded fragment (HBF-2) migrated in a position identical to that of the 11-kDa fragment (FIGURE 1). N-terminal sequence analysis of the 2 reverse-phase HPLC-purified fragments gave single sequences consistent with Glu[3,5]Ser[78,96]-bFGF peptide regions that begin at Lys-27 and Gly-70 for HBF-1 and HFB-2, respectively. The amino acid compositions of HBF-1 and HBF-2 correspond, within experimental error, to the Glu[3,5]Ser[78,96]-bFGF sequences 27–69 and 70–153, respectively. HBF-1 and HBF-2 each contain a single cysteine residue corresponding respectively to the positions 34 and 101 in bFGF, which some data indicate to be disulfide linked in bFGF.[2] Since HBF-1 and

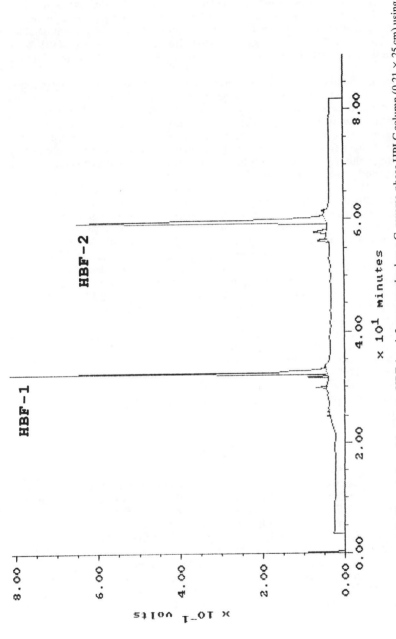

FIGURE 2. Reverse-phase HPLC of heparin-bound fragments. HBF-1 and -2 were resolved on a C$_4$ reverse-phase HPLC column (0.21 × 25 cm) using a 0.1% trifluoroacetic acid/acetonitrile gradient; 0–26% acetonitrile in 15 minutes, 26–38% acetonitrile in 60 minutes at a flow rate of 0.075 ml/minute. Detection was at 210 nm.

HBF-2 were separated by reverse phase HPLC in the absence of a thiol reducing agent, and N-terminal sequence analysis of HPLC-purified HBF peptides indicated no other sequences, it must be inferred that cysteines 34 and 101 are not disulfide linked. The possibility that the single cysteine residues in HBF-1 or HBF-2 are disulfide linked as HBF homodimers was ruled out by mass spectrometry (data not shown).

Heparin-bound mutant bFGF is rather well protected from digestion by pronase. Pronase treatment resulted in cleavage of the growth factor at two restricted locations, i.e., between Phe-26 and Lys-27, and between Arg-69 and Gly-70 resulting in the generation of HBF-1 and HBF-2. No peptide was found that corresponds to the N-terminal fragment (residues 2–26), presumably because this fragment is either completely digested or does not interact with heparin. It is noteworthy that HBF-1 and HBF-2, although not covalently linked (see above), were not separated by cation exchange and heparin-affinity HPLC (data not shown).

The 3 M NaCl eluate from heparin-Sepharose containing an approximately equimolar mixture of HFB-1 and HFB-2 was examined for mitogenic activity. The mixture induced a dose-dependent proliferation of bovine endothelial cells, with the dose for half-maximal growth stimulation being about 10-fold lower (ED$_{50}$ about 3 ng/ml)[b] when compared to intact Glu^{3,5}Ser^{78,96}-bFGF (FIGURE 3, upper panel). A comparison of the mitogenic activities of HBF-1, HBF-2, and Glu^{3,5}Ser^{78,96}-bFGF as determined with peptides obtained under denaturing reverse-phase HPLC conditions is shown in FIGURE 3 (lower panel). Consistent with earlier data,[24] reverse-phase HPLC-purified Glu^{3,5}Ser^{78,96}-bFGF is approximately 10-fold less active than the bFGF mutant not exposed to HPLC solvent conditions. Presumably, this loss of activity is the result of solvent-induced protein denaturation. Reverse-phase HPLC-purified HBF-1 was not mitogenic at the doses tested whereas HBF-2 exhibited a dose-dependent stimulatory response with a potency (ED$_{50}$ ~ 10 nM or 100 ng/ml) that was approximately 50-fold lower than that of reverse-phase HPLC-purified Glu^{3,5}Ser^{78,96}-bFGF. It should be noted that it is possible that HBF-2, like bFGF, may be denatured by reverse-phase HPLC conditions, and that the observed potency of HPLC-purified HBF-2 may therefore be lower than that of its nondenatured counterpart.

DISCUSSION

In the present work advantage was taken of the interaction between heparin and bFGF to more closely define structural domains that interact with heparin and the receptor for bFGF. Three major conclusions regarding structure/activity principles in bFGF are drawn from the results.

First, analysis of the present data shows unambiguously that in Glu^{3,5}Ser^{78,96}-bFGF a disulfide bond between Cys-34 and Cys-101 is not formed. Our biochemical results are in agreement with the interpretations of x-ray crystallographic data.[8–10]

Second, our study also appears to further define the heparin-binding domain of bFGF. It is intriguing that the HBF fragments generated by pronase digestion cochromatograph on high-resolution ion-exchange and heparin-affinity HPLC and

[b]The ED$_{50}$ determined in this study differ somewhat from those reported elsewhere. This discrepancy derives from the manner in which the bioassays are performed. In this study bFGF was added to the cells at the beginning of a 5-day growing period, whereas in other studies[22] growth factor was added at day 0 and day 2 of a 5- to 6-day culture period.

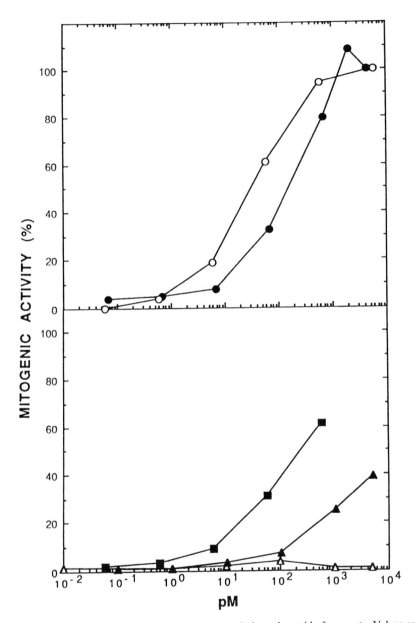

FIGURE 3. Mitogenic activities of bFGF and heparin-bound peptide fragments. Values are expressed as percent of maximum stimulation for $Glu^{3,5}Ser^{78,96}$-bFGF. **Upper panel:** $Glu^{3,5}Ser^{78,96}$-bFGF (open circles), 3 M NaCl eluent from heparin Sepharose containing HBF-1 and -2 (filled circles). **Lower panel:** reverse-phase HPLC-purified $Glu^{3,5}Ser^{78,96}$-bFGF (squares), HBF-1 (open triangles), and HBF-2 (filled triangles).

are indistinguishable from the parent bFGF with respect to chromatographic behavior and heparin affinity (data not shown) but are readily separable under reverse-phase HPLC conditions. The data indicate that pronase cleavage of the heparin-bound bFGF analogue at peptide bonds 26–27 and 69–70 (the only major apparent cleavage sites) does not significantly disrupt interactions that normally stabilize the native FGF structure. The fact that the pronase-digested material retains high biological activity and heparin affinity strongly suggests that HBF-1 and -2 remain associated. In contrast, reverse-phase HPLC conditions cause sufficient denaturation of the complex, leading to dissociation of HBF-1 and HBF-2 and reduced biological activity. In this context, it should be noted that reverse-phase HPLC-purified bFGF, HBF-1, and HBF-2 do not retain their affinity for heparin, even after removal of HPLC solvents (data not shown). In conclusion, we propose that heparin binding of bFGF depends on an intact 3-dimensional structure of bFGF that is maintained and stabilized by noncovalent association of N-terminal as well as C-terminal sequence domains. This model is consistent with x-ray crystallographic data which suggest the involvement of 2 separate bFGF domains [(Asn-36, bFGF(128–138)] in heparin binding,[9] but is not in agreement with earlier data. In our hands, bFGF(27–69) (HBF-1) and Ser[78,96]-bFGF(70–153) (HBF-2), after "denaturation" by HPLC solvent, do not bind to heparin although these fragments are remarkably similar to the synthetic fragments bFGF(33–77) and bFGF(112–155) reported to be heparin binding.[13] Furthermore, our data do not seem to support a conclusion by Seno *et al.,*[14] who provided evidence for a C-terminal location of the heparin-binding domain of bFGF. The C-terminal fragment HBF-2, which encompasses the putative heparin-binding site proposed by Seno *et al.,* lacks high affinity for heparin. However, since there is uncertainty as to whether denaturation of HBF-2 is responsible for loss of heparin affinity, it remains to be seen, whether recombinant HBF-2, like recombinant C-terminally truncated bFGF analogues,[14] can assume a conformation enabling high-affinity heparin-binding.

Finally, this study presents additional evidence that only relatively large fragments of bFGF [bFGF(50–155),[14] bFGF(10–122),[14] and bFGF(70–153) (this study)] exhibit relatively high mitogenic activity. Short or intermediate length synthetic peptides [e.g., bFGF(34–46), bFGF(33–77), bFGF(115–124), bFGF(112–155)][13] appear to be much less active. The increased activity of the larger fragments may be explained if association of two distant sequence domains are required for strong binding as suggested by the finding of Baird *et al.*[13] Alternatively, higher mitogenic activity of larger bFGF fragments may derive from a requirement for 3-dimensional structural integrity of a single receptor-binding sequence domain which may only be provided by relatively large fragments. Our results and those of Seno *et al.*[14] favor the latter possibility. The fragments bFGF(50–155), bFGF(10–122), and bFGF(70–153) all possess at least 100- to 1000-fold higher mitogenic activity than the most potent synthetic peptides. The sequence 70–122, common to all three fragments, might be considered a minimally required sequence for maintenance of a stable, receptor-binding conformation. Interestingly, this sequence almost completely encompasses that of the weak partial agonist/antagonist peptide bFGF(115–124)[13] which, as determined by x-ray crystallography, forms a surface loop on the bFGF molecule. It remains to be seen whether this sequence alone, if held in a stable conformation by the remainder of the protein, conveys the necessary elements for full receptor binding or whether additional surface sequence(s) participate in binding. In summary, present knowledge is still rudimentary with regard to defining the receptor-binding domain(s) of bFGF. However, this knowledge, in conjunction with that derived from the FGF receptor genes, now permits us to formulate much more refined hypotheses the testing of which will lead to a greatly increased understanding of FGF-receptor interaction.

SUMMARY

Although the FGFs have been subject to extensive biological studies, only limited progress has been made so far in determining the critical elements of structure-activity relationships in the FGFs. Among the recognized structural elements with potential to affect the biological activity of FGFs are the cysteine residues, and the heparin- and receptor-binding domains. These features have been studied using a variety of experimental approaches, but the available data are inconclusive. For example, ambiguity regarding the presence of a disulfide structure in FGFs was not resolved until the availability of x-ray crystal structure data. Furthermore, the functionally important heparin- and receptor-binding domains have been poorly characterized, with some interpretations being controversial. In this report, we describe a novel fragment of basic FGF (bFGF) with high biological activity [$Ser^{78,96}$-bFGF(70–153)]. This fragment was generated by pronase treatment of heparin-bound recombinant $Glu^{3,5}Ser^{78,96}$-bFGF mutant and is active *in vitro* at an ED_{50} of about 100 ng/ml. The structure of the fragment and the manner by which it was generated provide additional insight into important aspects of structure-activity relationships in FGFs. Specifically, we conclude that (a) the cysteines in our bFGF mutant do not form a disulfide bond, (b) the high-affinity heparin binding of bFGF critically depends on an intact 3-dimensional structure of the growth factor rather than on specific heparin-binding sequence domains, and (c) the bFGF sequence between residues 70 and 122 is important for high biological activity.

ACKNOWLEDGMENTS

We wish to express our gratitude to Andrew Baird for many occasions of sharing unpublished data, and for providing synthetic peptides and antibodies. We also thank Z. Misulovin for excellent technical assistance.

REFERENCES

1. ABRAHAM, J. A., A. MERGIA, J. L. WHANG, J. F. TUMOLO, J. FRIEDMAN, K. A. HJERRIELD, D. GOSPODAROWICZ & J. FIDDES. 1986. Science **233**: 545–548.
2. FOX, G. M., S. G. SCHIFFER, M. F. ROHDE, L. B. TSAI, A. R. BANKS & T. ARAKAWA. 1988. J. Biol. Chem. **263**: 18452–18458.
3. SENO, M., R. SASADA, M. IWANE, K. SUDO, T. KUROKAWA, K. ITO & K. IGARASHI. 1988. Biochem. Biophys. Res. Commun. **151**: 701–708.
4. ARAKAWA, T., Y. R. HSU, S. G. SCHIFFER, L. B. TSAI, C. CURLESS & G. M. FOX. 1989. Biochem. Biophys. Res. Commun. **161**: 335–341.
5. STRYDOM, D. J., J. W. HARPER & R. R. LOBB. 1986. Biochemistry **25**: 945–951.
6. CRABB, J. W., L. G. ARMES, S. A. CARR, C. M. JOHNSON, G. D. ROBERTS, R. S. BORDOLI & W. L. MCKEEHAN. 1986. Biochemistry **25**: 4988–4993.
7. LINEMEYER, D. L., J. G. MENKE, L. J. KELLY, J. DISALVO, D. SODERMAN, M.-T. SCHAEFFER, G. ORTEGA, G. GIMENEZ-GALLEGO & K. THOMAS. 1990. Growth Factors **3**: 287–298.
8. ZHU, X., H. KOMIYA, A. CHIRINO, S. FAHAM, G. M. FOX, T. ARAKAWA, B. T. HSU & D. C. REES. 1991. Science **251**: 90–93.

9. ERIKSSON, A. E., L. S. COUSSENS, L. H. WEAVER & B. W. MATTHEWS. 1991. Proc. Nat. Acad. Sci. USA **88:** 3441–3445.
10. ZHANG, J., L. S. COUSSENS, P. J. BARR & S. R. SPRANG. 1991. Proc. Nat. Acad. Sci. USA **88:** 3446–3450.
11. YAYON, A., M. KLAGSBRUN, J. D. ESKO, P. LEDER & D. M. ORNITZ. 1991. Cell **64:** 841–846.
12. KAPLOW, J. M., F. BELLOT, G. CRUMLEY, C. A. DIONNE & M. JAYE. 1990. Biochem. Biophys. Res. Commun. **172:** 107–112.
13. BAIRD, A., D. SCHUBERT, N. LING & R. GUILLEMIN. 1988. Proc. Nat. Acad. Sci. USA **85:** 2324–2328.
14. SENO, M., R. SASADA, T. KUROKAWA & K. IGARASHI. 1990. Eur. J. Biochem. **188:** 239–245.
15. HARPER, J. W. & R. R. LOBB. 1988. Biochemistry **27:** 671–678.
16. BURGESS, W. H., A. M. SHAHEEN, M. RAVERA, M. JAYE, P. J. DONOHUE & J. A. WINKLES. 1990. J. Cell Biol. **111:** 2129–2138.
17. FEIGE, J. J. & A. BAIRD. 1989. Proc. Nat. Acad. Sci. USA **86:** 3174–3178.
18. KUROKAWA, M., S. R. DOCTOROW & M. KLAGSBRUN. 1989. J. Biol. Chem. **264:** 7686–7691.
19. ISACCHI, A., M. STATUTO, R. CHIESA, L. BERGONZONI, M. RUSNATI, P. SARMIENTOS, G. RAGNOTTI & M. PRESTA. 1991. Proc. Nat. Acad. Sci. USA **88:** 2628–2632.
20. ROSENBERG, A. H., B. N. LADE, D. CHUI, J. J. DUNN & F. W. STUDIER. 1987. Gene **56:** 125–135.
21. SANGER, F., S. NICKLEN & A. R. COULSON. 1977. Proc. Nat. Acad. Sci. USA **74:** 5463–5467.
22. GOSPODAROWICZ, D., J. CHENG, G. M. LUI, A. BAIRD & P. BÖHLEN. 1984. Proc. Nat. Acad. Sci. USA **81:** 6963–6967.
23. CONNOLLY, D. T., M. B. KNIGHT, N. K. HARAKAS, A. J. WITTWER & J. FEDER. 1986. Anal. Biochem. **152:** 136–140.
24. GOSPODAROWICZ, D., S. MASSOGLIA, J. CHENG, G. M. LUI & P. BÖHLEN. 1985. J. Cell. Physiol. **122:** 323–332.

Basic Fibroblast Growth Factor Gene Expression

R. Z. FLORKIEWICZ, F. SHIBATA, T. BARANKIEWICZ,
A. BAIRD,[a] A-M. GONZALEZ,[a] E. FLORKIEWICZ,
AND N. SHAH

Department of Biochemistry
[a]Department of Molecular and Cellular Growth Biology
The Whittier Institute for Diabetes and Endocrinology
9894 Genesee Avenue
La Jolla, California 92037

INTRODUCTION

Basic fibroblast growth factor (bFGF) is a multifunctional molecule having activities that have been associated with mitogenesis, proliferation, and differentiation.[1-6] In addition, bFGF has been purified from many species and has a wide tissue-specific distribution as well as target cell specificity. The nearly ubiquitous presence of bFGF suggests that there may be important functions for this factor during normal cell growth and development. It seems likely that differences in both the qualitative and quantitative profiles of bFGF gene expression could influence these different bFGF activities. Consequently, the spatial as well as temporal synthesis of bFGF may also be regulated and tightly controlled. We present our data beginning with a functional analysis of the bFGF promoter, then present the rather novel bFGF translation strategy functionally diversifying this single copy gene, and finally we show that when synthesized *de novo*, the multiple isoforms of bFGF have different intracellular sites of localization.

RESULTS

To understand, in part, the regulation bFGF gene expression, we have partially characterized the organization of the human gene, including its regulatory elements. Our organizational data are consistent with previous publications, showing that intron 1 is at least 16 kilobases (kb) long while intron 2 is 16 kb long and that, including its three exons, the entire bFGF gene is at least 36 kb long.[7,8] Within genomic sequences 5' of exon 1 there are five GC boxes, which may represent SP-1 binding sites, and one potential AP-1 binding site. At this time we have, at least partially, defined the bFGF core promoter region structurally and functionally by primer extension analysis of bFGF mRNA and/or by the ability of noncoding 5'-flanking sequences to control the synthesis of the heterologous reporter protein, bacterial chloramphenicol acetlytransferase (CAT). Two human cell lines were used in these studies, the human medulloblastoma cell line TE671 and the human hepatoma cell line SK-HEP-1. Both cell lines were first characterized to determine their endogenous profile of bFGF gene expression. To do this, the number and sizes of bFGF mRNA being transcribed *de novo* as well as the profile of bFGF proteins being synthesized were determined. Total RNA was prepared, fractionated through

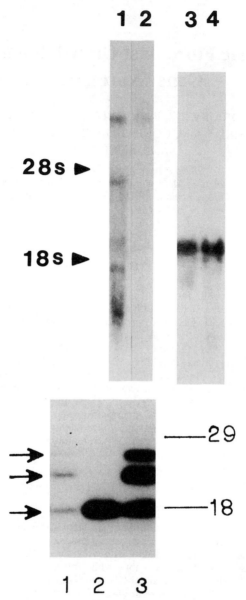

FIGURE 1. Comparative northern and immunoblot analysis of TE671 and SK-Hep-1 cells. **Top panel:** 20 μg of total RNA from both cell lines was fractionated on formaldehyde-agarose gels, transferred to nitrocellulose, and hybridized with ^{32}P-labeled human bFGF cDNA probe, lanes 1 and 2; or beta-actin probe, lanes 3 and 4. SK-Hep-1 and TE671 RNA are shown in lanes 1 and 3, and lanes 2 and 4, respectively. **Bottom panel:** Immunoblot analysis of heparin-Sepharose purified proteins from TE671 (lane 1) and SK-Hep-1 (lane 3) cells. Blots were probed with guinea pig anti-bFGF (18-kDa) antibody and ^{125}I-protein A. Signal is visualized by autoradiography. Lane 2, human recombinant bFGF (5 ng) size marker. Arrows mark the location of three bFGF specific immunoreactive bands as has been previously described.[9] Molecular weight standards are shown in kilodaltons.

denaturing agarose gels, and transferred to nitrocellulose. These RNA transfers were probed by hybridization with ^{32}P-labeled bFGF cDNA. As shown in FIGURE 1 (top), two major transcripts, 7 and 3.7 kb in size, are detected along with several minor transcripts in RNA isolated from both cell lines, lanes 1 and 2. Although quantitatively different, qualitatively the profiles in both cell lines are identical. When parallel transfers were hybridized with ^{32}P-labeled beta-actin probe, equal levels of beta-actin RNA were detected from both preparations, lanes 3 and 4. To determine if the level of bFGF proteins being synthesized corresponded with the levels of mRNA detected, immunoblots were prepared using heparin-Sepharose

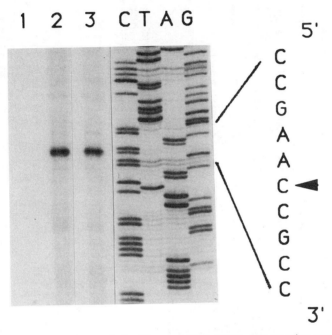

FIGURE 2. Primer extension analysis of human bFGF mRNA. Twenty micrograms of total RNA prepared from TE6714 and SK-Hep-1 cells was hybridized to ^{32}P-labeled primer (32-mer) at 30°C overnight. Hybrids were ethanol precipitated and primer extended with reverse transcriptase carried out at 42°C for 90 minutes. The primer-extended products were separated on a sequencing gel as described in Methods. Sequencing reactions with same primer (unlabeled) were run at the same time. The arrow indicates the transcription start site.

purified fractions from TE671 and SK-HEP-1 cell lysates (FIGURE 1, bottom). Both TE671 and SK-Hep-1 cell lysates, lanes 1 and 3 respectively, show the identical three-band pattern (18, 22/23, and 24 kDa) of immunoreactive bFGF at levels consistent with the levels of mRNA detected in FIGURE 1 top.

In order to identify the bFGF mRNA transcription start site, primer extension analysis was performed. A synthetic 32-base oligonucleotide complementary to nucleotide positions +101 to +132 (see FIGURE 3) was hybridized to total RNA isolated from SK-HEP-1 and TE671 cells. After addition of reverse transcriptase the primer-extended reaction products were separated on a 6% sequencing gel. A

```
-1001 AAGCTTCCCCA AATCTCCTGC CTCCCCACGC TGAGTTATCC GATGTCTGAA ATGTCACAGC ACTTAGTCTT ACTCTTCTAT GGCCTACTTT CTACTGCTAT
      Δ F1.2
 -900 TTGTGTTACT CATGCTACCC ATCTTATCTC CCTCAGTGTG TGAGACGCTG GCATCAGATT TGGCATCTCA CACACACTCA ACATTATGTG TTGCACACAG
 -800 TAGGTACTCA ATACATGCAA GTTTTCTGAA TAGATATTTT CCTAGTCATC TGTGGCACCT GCTATATCCT ACTGAAAATT ACCAAAATGC AATTAACTTC
                                                                      Δ 12/5
 -700 AATTTTACAT TTGGGATTTA CAGAAAATAA CTCTCTCTCC AAGAAAATGCA TAACAATTTA GCTAGGGCAA ATGCCAGTC CGAGTTAAGA CATTAATGCG
 -600 CTTCGATCGC GATAAGGATT TATCCTTATC CCCATCCTCA TCTTTCTCGG TCGTCTAATT CAAGTTAGT CAGTAAAGGA AACCTTTTCG TTTTAGCAAC
 -500 CCAATCTGCT CCCTTCTCT GGCCTCTTTG TCTTCCTTTG TTGGTAGACG ACTTCAGCCT CTGTCCTTTA ATTTTAAGT TTATGCCCCA CTTGTACCCC
               Δ 16/8
 -400 TCGTCTTTTG GTGATTTAGA GATTTTCAAA GCCTGCTCTG ACACAGACTC TTCCTTGAT TGCAACTTCT CTACTTTGG GTGGAAACGG CTTCTCCGTT
 -300 TTGAAACGCT AGCGGGGAAA AAATGGGGA GAAAGTTGAG TTTAAACTT TAAAAGTTGA GTCACGGCTG GTTGCCGAGC AAAAGCCCCG CAGTGTGGAG
 -200 AAAGCCTAAA CGTGGTTTCG GTGGTGCGGG GGTTGGGCGG GGGTGACTTT TGGGGGATAA GGGGCCGTGG AGCCCAGGGA ATGCCAAAGC CCTGCCGCGG
                                                      Δ 22/2
 -100 CCTCCGAGCG GCGCCCCCGG CCCCTGCCT CTCCCCGGCC CCGACTCGAG GCCGGGCTCC CCGCCGGACT GATGTCGCGC GCTTGCGTGT TGTGGCCGAA
                                                                                                      Δ 6N
   +1 CCGCCGAACT CAGAGGCCGG CCCCAGAAAA CCCGAGCCAG TAGGGGGCGG CGCGCAGGAG GGAGAGAAAC TGGGGGCCCG GGAGGGTTGGT GGGTGTGGGG
      *
 +101 GGTGGAGATG TAGAAGATGT GACGCCGCGG CCCGGCGGGT GCCAGATTAG CCCAGATTAG CGGACGGTGC AACGGGATCC CGGGCCGCTGC AGCTTGGGAG
 +201 GCGGCTCTCC CCAGGCGGCG TCCGCGGAGA CACCCATCCG TGAACCCCAG GTCCCGGGCC GCCGGCTCGC CGCGGCCACCAG GGGCCGGCGG ACAGAAGAGC
 +301 GGCCGAGGCTG GGGGACCGCG GGCCGCGCCG CGCGCTCCGG TGGGGGGCCG TGGGGGGCCG GGGGGGGCGG CGTGCCCCGG AGCGGGTCGG
      L      G  D  R  G   R  G  R  G   R  L   P   G  G  R    G  G  R    A  P  E    R  V  G
 +401 AGGCCGGGGC CGGGGCCGGG GGACGGCCGGC TCCCCGCGGG GCTCCAGCGG CTCGGGGATC CCGGCCAGGGA CCATGGCAGC CGGGAGCATC
      G  R  G    R  G  R  G   T  A  A   P  R  A   A  R  P    G   A  G    T   M  A    A   G   S   I
 +501 ACCACGCTGCC CCGCCTTGCC CGAGGATGGC GGAGCGGGA CCTTCCCGCC CGGCCACTTC AAGGACCCCA AGGCGGCTGTA CTGCAAAAAC GGGGGCTTCT
      T  T  L   P  A  L   P  E  D  G   G  S  G   A  F  P   P  G  H  F   K  D  P  K   R  L  Y   C  K  N   G  G  F  F
 +601 TCCTGGCGAT CCACCCCGAC GGCCGAGTTG ACGGGGTCCC ACGGGGTCCC GGAGAAGAGC GACCCTCACA gtgagtgccg accggctctc tccgcctcat ttccattcg
      L  R  I   H  P  D   G  R  V  D   G  V  R   E  K  S    D  P  H  (I)
```

dideoxy sequencing ladder, generated using the same synthetic 32-base oligonucleotide, was also included to accurately determine reaction product length and position. One band of primer-extended product, 132 nucleotides in length, was detected from both TE671 and SK-HEP-1 RNA (FIGURE 2). These data imply that one start site is used to transcribe the multiple size classes of bFGF mRNAs. The data presented in FIGURE 2 identify the 5' end of bFGF mRNA as starting 483 nucleotides 5' of the AUG-methionine codon used to initiate translation of the 18-kDa bFGF polypeptide. Relative to the first of three previously identified alternative CUG translation initiation codons,[9] this would represent 317 nucleotides of 5' untranslated sequence. To position the bFGF RNA start site within the genomic clones containing exon 1 and 5'-flanking sequences, their relevant nucleotide sequence was determined. The sequence of two independent genomic clones (λHBF11 and 17) containing the 5' portion of exon 1 was determined. Both sequences were identical and included the proposed mRNA start site with additional upstream (5'-flanking) sequences as shown in FIGURE 3. The region upstream (5') of the proposed transcription initiation site (RNA start site, designated +1) is G+C rich (65%) and as previously shown does not contain consensus CAAT or consensus TATA box motifs.[7,8] However, there are five potential binding sites for the transcription factor Sp1 (GGGCGG or CCGCCC), located at nucleotide positions −166, −139, −83, and −65. In addition, there is one potential AP-1 transcription factor binding site (TGA(C/G)TCA) located at nucleotide position −243.[10] This consensus core DNA sequence TGA(C/G)TCA, named TRE (TPA-responsive promoter element), is the binding site of nuclear transcription factor AP-1, and is located at position −243 in human bFGF promoter region (FIGURE 3).[11,12] This sequence has not been found within the promoter regions of other FGF family members. Although we need to determine the functional relevance of this observation, it is consistent with published data describing an increased level of bFGF mRNA synthesis in response to the phorbol ester TPA in the human astrocytoma cell line (U87-MG).[13]

Since TE671 cells actively express the bFGF gene as shown in FIGURE 1, they are well suited to the further analysis of genomic clones with predicted promoter activity. Hybrid expression vectors were constructed that contain fragments of bFGF genomic DNA, including the mRNA start site, fused with the coding sequences for bacterial chloramphenicol acetyltransferase (CAT). The level of CAT enzyme activity was used as a measure of promoter activity following transient transfection.[14,15] The starting plasmid pBLCAT2, in which the CAT gene is regulated by the herpes simplex virus thymidine kinase (TK) gene promoter, was used as a positive control.[16] The promoter minus pBLCAT4 plasmid was used as a negative control. The relative level of promoter function was determined by measuring CAT activity in TE671 cells transfected with these plasmids. Extracts of TE671 cells transfected with

---

FIGURE 3. DNA sequence of the 5'-flanking region of exon 1 of the human bFGF gene. All numbering of deletions and sequence elements is relative to the transcriptional initiation site determined in FIGURE 2. The transcriptional initiation site is shown as +1 (*). The multiple translation initiation sites (three CTG codons and one ATG codon) within exon 1 are single underlined. Amino acid sequence is shown in the single letter code. The positions of relevant 5' deletions are indicated by triangles below the first nucleotide of that particular deletion and labeled (abbreviated) to correspond with data presented in FIGURE 4. The position of the GC box motifs is double underlined, and the potential AP-1 binding site is boxed. The sequence with dyad symmetry is underlined and arrowed; mismatches within this sequence are shown by small circles under the particular nucleotide. The locations of the two restriction sites (Bam H1 and Xho 1) used to define the 3' ends on constructs described in FIGURE 4 are shown underscored with single and double dash lines respectively.

pF2.1CAT produced minimal (background) levels of CAT activity, as did the negative (promoter minus) pBLCAT4 control (FIGURE 4). However, extracts of TE671 cells transfected with pF2.0CAT (a 3' deletion mutant) had levels of CAT activity fourfold greater than background. This suggests that the genomic DNA fragment in the pF2.0CAT construct contains, as predicted by sequence, a functional bFGF gene promoter. To more accurately identify the core promoter and potential regulatory sequences, a nested series of 5'-deletion mutants were prepared from pF2.0CAT. Deletions spanning the sequence from the 5' end down to nucleotide position −1001 (named pF1.2CAT) or to nucleotide position −747 (named pFd-CAT12/5) have the same relative levels of CAT activity as pF2.0CAT. However, deletions extended to nucleotide position −480 (named pFdCAT16/8) stimulated the highest highest level of CAT activity, an additional fourfold increase as compared to the pF2.0CAT construct. Further deletions to nucleotide position −160 (named pFdCAT22/2) or to position −21 (named pFdCAT6N) have reduced levels of CAT activity, 88% and 31% respectively, compared to the maximal activity detected for pFd CAT16/8. The data presented in FIGURE 4 indicate that the genomic sequences within pF2.1CAT and pF2.0CAT contain a functional human bFGF promoter, as well as sequences regulating its activity in TE671 cells.

Taken together, these data suggest that the human bFGF gene promoter contains two regions representing negative regulatory domains. The first domain is positioned 3' to the GC box promoter core. Based on primer extension analysis, this sequence is present in the 5' untranslated region of all known bFGF mRNAs that we have detected. This first domain is positioned between the BamHI site (+175) and an XhoI site (+311), includes 136 base pairs (bp) of sequence, and has no obvious structure or homology with published sequences that have been assigned a negative (or positive) regulatory function.[17–21] However, this sequence has a rather high G+C content (78%) which may reflect a nonspecific attenuation of mRNA translation, mimicking a negative regulatory function on transcription. Further experiments are required to differentiate between specific or nonspecific regulatory effects on bFGF gene expression. The second region, with negative regulatory function, resides between deletion 12/5 (−854) and deletion 16/8 (−521). Although this 333-bp sequence has an average G+C content of 31%, it contains a near perfect (13 nucleotides out of 15) inverted repeat sequence between nucleotide positions −596 to −565. Since there is no direct homology to published regulatory sequences, it is possible that this sequence with dyad symmetry is functionally specific in the regulation of bFGF gene expression. However, there is some homology with a negative regulatory sequence motif found in the cMyc gene promoter(s).[22] The similarities, between Myc and bFGF in this region, suggest the possibility that Fos or Fos-like proteins may be a part of the protein(s) complex that mediates regulatory activity. Whether the bFGF DNA sequence domains having negative regulatory properties act through factor-dependent or factor-independent mechanisms remains to be determined.

Following transcription, bFGF mRNA is translated into its polypeptide gene product. Recent reports have shown that the single-copy human bFGF gene encodes not one protein, but a complex set of four coexpressed isoforms, with apparent molecular masses of 24, 23, 22, and 18 kDa.[9,23] Analysis of human bFGF cDNAs by *in vitro* transcription/translation, transient COS cell expression, or from continuous cell lines has demonstrated that the multiple isoforms are cotranslated from a single mRNA transcript. In addition, these reports have suggested that the translation of the larger molecular forms (24, 23, and 22 kDa), but not the 18-kDa form, initiates from novel CUG (leucine) codons as opposed to the classical AUG (methionine) codon. Alternative translation initiation could serve as a posttranscriptional mecha-

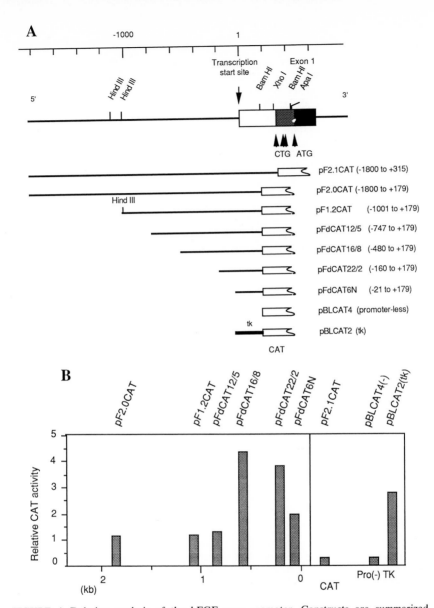

FIGURE 4. Deletion analysis of the bFGF gene promoter. Constructs are summarized schematically in panel A. The series of deletion mutants were constructed by Xho 1 or Bam H1 restriction endonuclease digestion (defining the 3′ termini) and Bal 31 exonuclease digestion (defining the 5′ termini) of clone HBF 17. The transcription start site for exon 1 is named and labeled as 1, and is indicated by the long arrow. Relevant restriction endonuclease sites are named. The CTG translation initiation codons within the hatched box and the single ATG translation initiation codon are shown by short arrows, and define the beginning of the coding sequences for the multiple bFGF isoforms. Constructs joining the bFGF genomic (promoter) sequences (solid line) with bacterial CAT coding sequences (open box) are named at the right along with the bFGF sequence boundaries as indicated. The negative control (promoterless) and positive control tk-promoter constructs are also shown and named as such. All constructs were transfected into TE671 cells, with the levels of ensuing CAT activity are shown in panel B. CAT activities were calculated relative to the activity from cells transfected with pF2.0CAT. For convenience each bar is named (sequentially) above. Variation in transfection efficiency was corrected for by using CH110 (β-gal) as an internal control. The activity values presented here are the mean of four parallel transfected plates.

nism for regulating bFGF gene expression and intracellular localization. Although bioactivities of the 18-kDa protein are currently used to define bFGF gene function, it is not yet known if the three larger proteins have these same bioactivities or whether they will serve to define new bFGF gene functions. It is difficult to design experimental model systems that would address issues concerning their individual

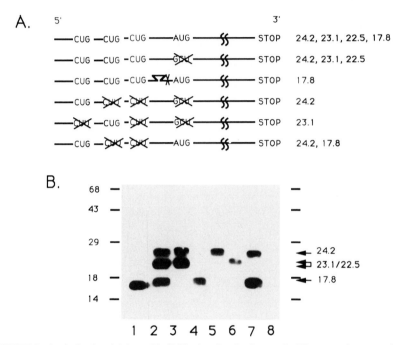

FIGURE 5. Analysis of multiple and individual molecular forms of wild-type and mutagenized human bFGF by cDNAs expressed in COS cells. **Panel A:** Schematic representation of the original (wild-type) human-bFGF cDNA and each mutant derived from it. Positions of CUG and AUG codon(s) are lettered with mutations ablating specific codons lettered and crossed over. The sizes of protein(s) predicted are shown at the right in kilodaltons. Diagrams are ordered to correspond with consecutive lanes (2 through 7). **Panel B:** The pattern of heparin-Sepharose purified anti-bFGF immunoreactive proteins from transfected COS cells. Blots were probed with guinea pig anti-bFGF (18 kDa) antibody and ^{125}I-protein A. Signal is visualized by autoradiography. Lane 1, marker recombinant 18-kDa bFGF; lane 2, the wild-type cDNA encoded isoforms; lane 3, mutated AUG codon and consequent 18-kDa protein ablated; lane 4, all three CUG codons silenced by frame-shifted translation, only the 18 kDa signal remains; lane 5, mutated second and third CUG codons and AUG codon, 24-kDa isoform alone; lane 6, mutated first and third CUG codons and AUG codon, 23-kDa isoform alone; lane 7, mutated second and third CUG codons, the 18 plus 24-kDa isoforms remain; lane 8, control COS cells. Molecular weight standards are shown in kilodaltons.

physiological significance. Our approach has been to express these proteins individually and compare aspects of their *de novo* biosynthesis that might suggest functional differences.[9] To do this, expression vectors containing either the wild-type or mutagenized bFGF cDNAs were prepared. The specific cDNAs are presented schematically in FIGURE 5A to show the location of each mutation, followed by the calculated molecular weight (kDa) of the protein(s) predicted to be synthesized. The anti-bFGF immunoreactive profile of proteins synthesized after COS cell transfection are shown in FIGURE 5B. As shown, mutation of a specific CUG codon to CUU ablates the synthesis of the corresponding higher molecular weight protein, while mutation of all but one CUG codon plus mutation of the single AUG codon results in the synthesis of individual higher molecular weight bFGF isoforms. As previously suggested, the middle band is actually a doublet of proteins approximately 23 and 22 kDa.[9] Only when both middle CUG codons are simultaneously changed (lane 7) is the middle protein band ablated. The similar intensities observed for all bands shown in FIGURE 5B suggests that blocking CUG-mediated translation initiation at one position or in different combinations apparently does not have a significant effect on the level of protein synthesis initiating from nonmutagenized CUG codons, or from the single AUG codon initiating synthesis of the 18-kDa isoform. Mock transfected controls have no detectable signal (lane 8). These data support the notion that in-frame CUG codons initiate translation of each individual high molecular weight protein species of human bFGF and represent a system for further characterization of individual bFGF isoforms. Furthermore, there is a predicted stem-loop structure ($\Delta G - 65$ kcal) positioned between the three CUG codons and the single AUG codon. We would speculate that, under certain conditions, this structure might be recognized by specific cytosolic factors and be used to posttranscriptionally regulate or modify bFGF gene expression. The simultaneous synthesis of multiple colinear extended forms of bFGF is not peculiar to human cell cultures because a rat cDNA expressed transiently in COS-1 cells also encodes a similar set of multiple bFGF isoforms (FIGURE 6, panel A). This same set of bFGF isoforms are also detected in heparin-Sepharose purified primary rat brain extracts (FIGURE 6, panel B). Besides the presence of potential CUG translation initiation codons to initiate their synthesis, there is a conserved amino acid sequence contained within the amino terminal extensions in the rat and human proteins (FIGURE 7).

The presence of multiple isoforms of bFGF might imply that there are either as yet unknown functions of bFGF or that some of the activities previously assigned to the 18-kDa protein are actually functions of specific higher molecular weight bFGF isoforms. A detailed analysis of the *de novo* synthesis of these isoforms would be a first step towards understanding and functionally distinguishing their individual physiological significance. We have previously suggested that the different bFGF isoforms might be differentially stored or that they may individually localize to different subcellular compartments.[9] Therefore, we have characterized the biosynthesis and the intracellular site of localization of these proteins. Transient COS cell expression experiments were conducted to characterize the *de novo* biosynthesis of individual and multiple bFGF isoforms in more detail. Cell and media fractions were prepared at various times postlabeling with [35]S(methionine and cysteine) and analyzed by immunoprecipitation. In FIGURE 8, cDNAs encoding all four isoforms (panel a), the 18-kDa isoform alone (panel b), the 24-, 23-, 22-kDa protein set (panel c), or the 24-kDa isoform alone (panel d) was tested. None of the bFGF isoforms were immunoprecipitated from media fractions. Furthermore, the higher molecular weight isoforms do not appear to constitute precursors of a lower molecular weight isoform and all isoforms appear equally stable during the course of these experiments. Similar results were obtained using the cDNA encoding the 23-kDa isoform alone (data not shown). The results presented here are consistent with previously published cDNA sequence data, indicating that bFGF does not contain a hydrophobic peptide signal sequence and is not released into the surrounding culture medium.[7-9,24-26]

However, the 18-kDa isoform, but not the three higher molecular weight isoforms, is exported onto the cell surface of transfected COS cells. Expression of the wild-type and mutagenized human bFGF cDNAs combined with double-label indirect immunofluorescence allows a comparative analysis of individual isoforms based upon intracellular localization. The data show that all three higher molecular weight isoforms localize exclusively to the nucleus, while the 18-kDa isoform is the

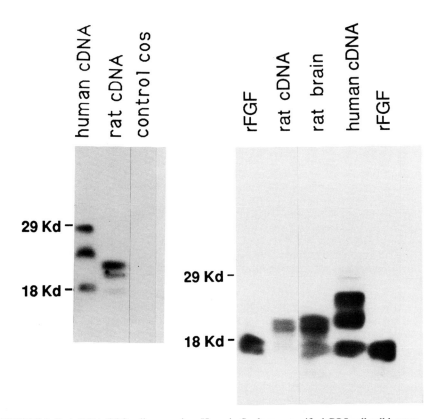

FIGURE 6. Rat cDNA COS cell expression. Heparin-Sepharose purified COS cell cell lysates were prepared as already described herein. Lane 1, the human cDNA insert; lane 2, the rat cDNA insert; lane 3, COS cell control. **Right panel:** Rat brain synthesizes multiple isoforms of bFGF. HS purified samples were fractionated by 12% SDS-PAGE, Western transferred, and probed with protein-A-purified rabbit anti-bFGF antibodies and secondly with ^{125}I protein-A. Lanes 1 and 5, recombinant FGF marker; lane 2, COS cells transfected with the expression vector containing rat cDNA; lane 3, rat brain; lane 4, COS cells transfected with the expression vector containing human cDNA.

only isoform exported onto the cell surface, but it also colocalizes in the nucleus (FIGURE 9). Transfected COS cells expressing the wild-type cDNA synthesizing all four isoforms (panels a and b) shows both surface and nuclear signal. The mutant encoding the 18-kDa isoform alone (panels c and d) shows both nuclear and surface signal. The mutant cDNAs encoding the 24-, 23-, and 22-kDa isoform set (panels e and f) or the 24- and 23-kDa isoforms individually (panels g and h; panels i and j,

MULTIPLE FORMS OF HUMAN BASIC FGF

```
LGDRGRGRALPGGHLGGRGRGRAPGRVGGRGRGRCTAAPRALPAAFGSRPCPAGTMAAGSITTLPAL.  .  .END
           LPGGHLGGRGRGRAPGRVGGRGRGRCTAAPRALPAAFGSRPCPAGTMAAGSITTLPAL.  .  .END
                LGGRGRGRAPGRVGGRGRGRCTAAPRALPAAFGSRPCPAGTMAAGSITTLPAL.  .  .END
```

D1   D2

MULTIPLE FORMS OF RAT BASIC FGF

```
LAARGRAALGGRGRGRGRGAPRAAAAGSRGRGGAMAAGSITSLPAL.  .  .END
         LGGRGRGRGRGAPRAAAAGSRGRGGAMAAGSITSLPAL.  .  .END
                                   MAAGSITSLPAL.  .  .END
```

D1   D2

FIGURE 7. Comparison of predicted amino-terminal extensions. Shaded, boxed sequence domains of the human proteins indicate homologous with boxed domains of rat protein. Domain 1 (D1) is 8 out of 8. Domain 2 (D2) is 13 out of 18.

respectively) show only nuclear fluorescence. The observed patterns of cell surface and nuclear signals can be ascribed to the specific bFGF proteins being synthesized because (1) the majority of cells are nontransfected and therefore serve as internal negative controls, (2) only cells that show intracellular signal have a corresponding cell surface signal, and (3) preimmune serum controls are also negative (data not

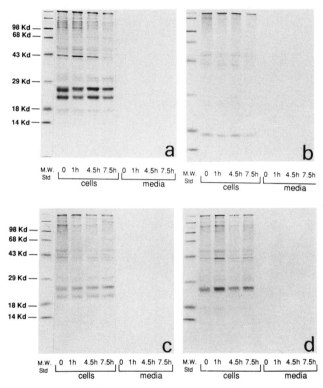

FIGURE 8. *De novo* biosynthesis of individual bFGF isoforms in transfected COS cells. Forty to 48 hours posttransfection COS cells were ^{35}S pulse labeled. Cell lysates and corresponding media fractions were prepared at various times postlabeling as indicated. Immune complexes were precipitated with protein A sepharose and fractionated by 12% SDS-PAGE. Labeled proteins are visualized by fluorography. **Panel a,** the original (wild-type) cDNA, encodes four bFGF isoforms. **Panel b,** the 18-kDa isoform alone. **Panel c,** the three larger isoforms, 24, 23, and 22 kDa. **Panel d,** the 24-kDa isoform alone. Molecular weight markers are shown in kilodaltons.

shown). Surface immunoreactive signal (panels a and d) indicates that the 18-kDa protein is selectively exported from the cytosol onto the cell surface despite the absence of a classical peptide signal sequence. However, as previously shown (FIGURE 8), the 18-kDa protein is not released into the surrounding cell culture medium. The selective cell surface localization of 18 kDa bFGF was confirmed by surface-specific labeling and immunoprecipitation (data not shown).

In conclusion, bFGF gene transcription, as defined in this presentation, appears to be under strong negative regulatory control, but the multiple mRNAs that are transcribed contain the same start site. Although there is no evidence for regulating bFGF gene expression posttranscription, it is likely that the presence and utilization of CUG codons (three in the human, two in the rat) plus one AUG codon to cosynthesize multiple bFGF isoforms functionally diversifies the single copy gene. We have shown that the different human bFGF isoforms localize to different subcellular compartments. As a consequence, this observation suggests that the different isoforms might have different functions dependent upon localization. In these experiments, the three higher molecular weight isoforms are exclusively localized in the nucleus. However, in the absence of the conventional transport signal sequence(s), the 18 kDa isoform is selectively exported onto the cell surface. Whether bFGF utilizes the classical secretory pathway or an alternative pathway is not yet known; but, regardless of which pathway is utilized, it is probable that a new set of criteria will emerge characterizing a distinct mechanism mediating and/or regulating export of bFGF.

METHODS

Cell Culture and Transient Expression

COS-1 cells were grown in Dulbecco-Voigt modified Eagle medium (MEM) supplemented with 10% fetal bovine serum, 2 mM L-glutamine, 12 mM pyruvic acid, nonessential amino acids, and antibiotics. For immunoblotting or immunoprecipitation, 60-mm plates of COS cells (60% confluent) were transfected with 10 μg of the appropriate DNA as previously described.[27] Briefly, CsCl-purified DNA was mixed with 1 ml of Tris/saline and then DEAE-dextran was added to 500 μM. This mixture was added to COS cell monolayers and incubated at 37°C for 30 minutes, then removed and replaced with complete medium supplemented with 100 μM chloroquine for an additional 90 minutes incubation. Transient expression was examined 40–48 hours after DNA transfection.

Mutagenesis

Mutagenized cDNAs were constructed as previously described for those encoding the 24-, 23- and 22-kDa isoform set and for the 18-kDa isoform alone.[9] Briefly, oligonucleotide site-directed mutagenesis changing the first ATG to GCT resulted in the synthesis of the three larger bFGF isoforms (no 18 kDa); and an oligonucleotide inserted at a unique Apa1 site (frame shift) resulted in the synthesis of only 18-kDa bFGF. In this report, additional mutants were constructed using the following strategies and according to the schematic presented below.

```
5' EcoR1 Xho1    Not1         Apa1                    EcoR1 3'
    ▼     ▼       ▼            ▼                         ▼
_____ CUG __ CUG _ CUG __ AUG __ ʃʃ __ STOP ___
```

The cDNA mutant encoding the 24- plus 18-kDa proteins was constructed by replacement of the Xho1/Apa1 fragment in wild-type cDNA with four overlapping synthetic oligonucleotides incorporating both middle CTG to CTT alterations. cDNAs that encode the 24- or 23-kDa isoforms alone were constructed by subcloning

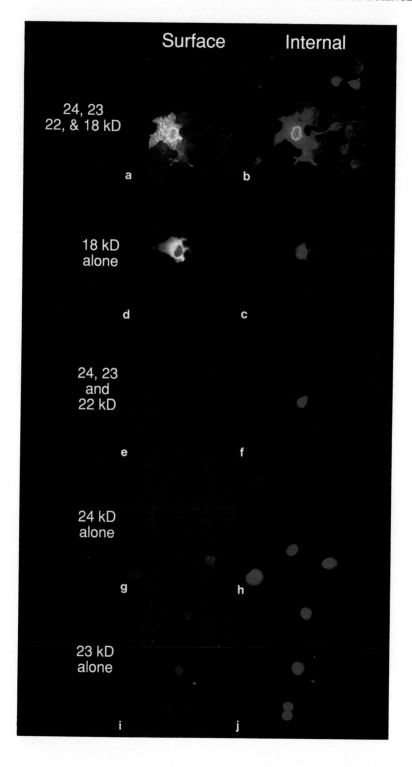

as follows. For the mutant encoding the 24-kDa isoform alone, we utilized the EcoR1/Apa1 fragment from the double middle CTG to CTT mutant, plus the Apa1/EcoR1 fragment from the ATG to GCT mutant (previously described)[9]; for the mutant encoding the 23-kDa isoform we used the EcoR1/Not1 fragment from the mutant in which the first CTG was changed to CTT, plus the Not1/Apa1 fragment from the mutant with the third CTG changed to CTT plus the Apa1/EcoR1 fragment containing the ATG to GCT change. All mutants were confirmed by standard DNA sequence analysis and/or restriction endonuclease mapping.

Immunoblot Analysis

COS cell extracts were prepared for immunoblot analysis according to the following procedures. Forty to 48 hours posttransfection, culture fluid was removed and monolayers washed with phosphate-buffered saline (PBS). Cell extracts were prepared by adding 1 ml of lysis buffer (ice cold) containing 1% NP40, 0.5% deoxycholate, 20 mM Tris (pH 7.5), 5 mM EDTA, 2 mM EGTA, 150 mM NaCl, 0.01 mM PMSF, 10 ng/ml Aprotinin (Sigma, St. Louis, Mo), 10 ng/ml leupeptin (ICN Biomedicals, Costa Mesa, Calif.), and 10 ng/ml pepstatin (ICN Biomedicals, Costa Mesa, Calif.). Lysates were clarified by microfuge centrifugation (16,000 × g) for 10 minutes at 4°C. Supernatants were removed and incubated with pre-washed heparin-Sepharose (HS, Pharmacia, Piscataway, N.J.) for 2 hours at 4°C. HS (and bound protein) was pelleted and washed twice with lysis buffer; three times with buffer containing 20 mM Tris (pH 7.5), 0.5 M NaCl, 5 mM EDTA, 2 mM EGTA plus protease inhibitors; and finally, three times in buffer containing 1.0 M NaCl. Proteins were eluted directly into sodium dodecyl sulfate (SDS) sample buffer for 12% SDS-polyacrilamide gel electrophoresis (SDS-PAGE),[28] and transferred to nitrocellulose in buffer containing 25 mM AMPSO (Sigma, St. Louis, Mo.) pH 9.5 in 20% methanol as described previously.[9,29] Transfers were blocked in buffer containing 5% powdered milk, incubated with guinea pig anti–18-kDa bFGF antisera diluted 1:500 and then ^{125}I-protein A.[9,30] Signal was visualized by autoradiography.

[35]S Labeling and Immunoprecipitation

Pulse labeling and immunoprecipitation were carried out as previously described.[27,31] Briefly, transfected COS cell cultures were incubated with 100 μCi of ^{35}S-methionine and ^{35}S-cysteine (Trans[35]S-label, ICN Biomedicals, Inc., Costa Mesa, Calif.) for 20 minutes (pulse label) and then replaced with medium supplemented

---

FIGURE 9. Intracellular localization of different bFGF isoforms by double-label indirect immunofluorescence microscopy. Transfected COS cell were fixed with 3% paraformaldehyde and surface stained with rabbit bFGF anti-peptide antibody and fluorescein conjugated goat anti–rabbit immunoglobulin G (IgG) second antibody. For internal staining, the same cells were then permeabilized with 0.3% triton, incubated with guinea pig anti–18-kDa bFGF antibody, and then rhodamine-conjugated goat anti–guinea pig IgG second antibody. Each set of panels represents surface (green) and internal (red) fluorescence, respectively. The cDNAs used are listed according to their encoded bFGF isoforms as follows: **panels a and b,** the wild-type cDNA (all four isoforms encoded); **panels c and d,** 18-kDa isoform alone; **panels e and f,** multiple larger isoforms (no 18-kDa protein); **panels g and h,** 24-kDa isoform alone; **panels i and j,** 23-kDa isoform alone.

with excess (10 mM) unlabeled methionine and cysteine (chase). At various times postlabeling, cell extracts were prepared using lysis buffer as described above. Corresponding media fractions were also prepared, clarified, and immunoprecipitated. Radiolabeled cell and media fractions were incubated with guinea pig anti–18-kDa bFGF antisera (1:200). Immune complexes were collected as protein A sepharose precipitates (Spectrum Medical Industries, Inc., Los Angeles, Calif.) and washed five times with RIPA buffer (ice cold). Immunoprecipitated proteins were eluted directly into SDS-sample buffer and fractionated by 12% SDS-PAGE, fluorographed, and exposed to x-ray film.

Double-Label Indirect Immunofluorescence

COS cells were transfected (using 2 μg of the appropriate DNA) on tissue culture chamber slides (Tissue-Tek, Miles Scientific, Naperville, Ill.). Forty to 48 hours posttransfection, the cells were rinsed with PBS and fixed for 15 minutes with 3% paraformaldehyde in PBS (final pH 7.4), at room temperature. Slides were rinsed and incubated for 30 minutes with 1.5% normal goat serum in PBS. For detection of cell surface signal, cells were incubated with polyclonal guinea pig anti–18-kDa bFGF antibody (raised against purified recombinant 18-kDa protein), in PBS containing 5% bovine serum albumin (BSA). For internal signal, the same cells were then incubated with polyclonal rabbit bFGF antipeptide (no. 773, amino acids 9–33) antibody (generously provided by Dr. Andrew Baird) in PBS containing 0.3% Triton plus 5% BSA. After 1 hour, cells were rinsed and incubated with fluorescein conjugated goat antirabbit (surface) and rhodamine conjugated goat anti–guinea pig (internal) labeled second antibody (Tago, Inc., Burlingame, Calif.) in PBS. Finally, slides were rinsed, mounted, and analyzed by fluorescence microscopy. All incubation steps were done at room temperature.

Analysis of the 5′ End of the Human Basic FGF mRNA

A 32-base oligonucleotide primer, corresponding to nucleotides +101 to +132 (FIGURE 4), was end labeled with $[\gamma\text{-}^{32}P]ATP$ using T4 polynucleotide kinase. Twenty micrograms of total RNA and 1×10^5 cpm of labeled primer were coprecipitated with ethanol. The pellet was dissolved in 30 μl of 3 M NaCl, 0.5 M HEPES pH 7.5, and 1 mM EDTA and allowed to hybridize at 30°C overnight. Hybrids were ethanol precipitated and resuspended in buffer containing 50 mM Tris-HCl pH 8.0, 50 mM KCl, 5 mM $MgCl_2$, 5 mM dithiothreitol (DTT), and 0.5 mM of all four dNTPs. The primer was extended with the addition of reverse transcriptase at 42°C for 90 minutes. After ribonuclease A digestion and ethanol precipitation, the primer-extended products were fractionated on a sequencing gel.

Plasmid Construction for Chloramphenicol Acetyltransferase
Assay and Deletion Analysis

Plasmid pBLCAT2 was used as a positive control.[16] As a negative control plasmid pBLCAT4 (i.e., promoterless) was constructed by removing the thymidine kinase promoter region from pBLCAT2. Plasmid pF2.0CAT was constructed by ligation of pBLCAT4 with a 2.0-kb bFGF genomic DNA fragment bounded 3′ at the first Bam HI site in exon 1 (nucleotide positions −1800 to +177, see FIGURE 1). Similarly, pF2.1CAT was constructed by using a 2.1 kb bFGF genomic DNA fragment bounded

3' at the unique Xho I site (nucleotide position +311) in exon 1. In this context pF2.0CAT represents a 3' deletion mutant of pF2.1CAT. The nested set of 5'-deletion mutants were constructed using pF2.0CAT and Bal 31 nuclease digestion.

Chloramphenicol Acetyltransferase Assay

Sample plasmid (15 µg) and standardized internal control plasmid pCH110 (Pharmacia, Piscataway, N.J.) (5 µg) were transfected into TE671 cells (1 × 10⁶ cells/10-cm/dish) by calcium phosphate coprecipitation.[14] Plasmid pUC18 was used as carrier, if necessary. The precipitate was removed after 6 hours, and transfected cells were cultured for 42 hours. Cell extracts were prepared for CAT assay after four cycles of freeze thawing.[15] To monitor transfection efficiency, β-galactosidase activity was measured as previously described.[32]

REFERENCES

1. BAIRD, A., F. ESCH, P. MORMEDE, N. UENO, N. LING, P. BOHLEN, S-Y. YING, W. B. WEHRENBERG & R. GUILLEMIN. 1986. Molecular characterization of fibroblast growth factor: distribution and biological activities in various tissues. Rec. Prog. Horm. Res. **42:** 143–205.
2. GOSPODAROWICZ, D., G. NEUFEILD & L. SCHWEIGERER. 1987. Fibroblast growth factor: structural and biological properties. J. Cell. Physiol. Suppl. **5:** 15–26.
3. KIMELMAN, D., J. A. ABRAHAM, T. HAAPARANTA, T. M. PALISI & M. W. KIRSCHNER. 1988. The presence of fibroblast growth factor in the frog egg: its role as a natural mesoderm inducer. Science **242:** 1053–1056.
4. BURGESS, W. H. & T. MACIAG. 1989. The heparin-binding (fibroblast) growth factor family of proteins. Annu. Rev. Biochem. **58:** 575–606.
5. RIFKIN, D. B. & D. J. MOSCATELLI. 1989. Recent developments in the cell biology of basic fibroblast growth factor. J. Cell Biol. **109:** 1–6.
6. BAIRD, A. & P. BOHLEN. 1990. *In* Handbook of Experimental Pharmocology, eds. M. B. Sporn & A. B. Roberts, Eds. **95/1:** 369–418. Springer-Verlag. Berlin & Heidelberg.
7. ABRAHAM, J. A., J. L. WHANG, A. TUMOLO, A. MERGIA, J. FRIEDMAN, D. GOSPODAROWICZ & J. C. FIDDES. 1986. Human basic fibroblast growth factor: nucleotide sequence and genomic organization. EMBO J. **5:** 2523–2528.
8. ABRAHAM, J. A., A. MERGIA, J. L. WHANG, A. TUMOLO, J. FRIEDMAN, K. A. HJERRIELD, D. GOSPODAROWICZ & J. C. FIDDES. 1986. Nucleotide sequence of a bovine clone encoding the angiogenic protein, basic fibroblast growth factor. Science **233:** 545–548.
9. FLORKIEWICZ, R. Z. & A. SOMMER. 1989. Human basic fibroblast growth factor gene encodes four polypeptides: three initiate translation from non-AUG codons. Proc. Nat. Acad. Sci. USA **86:** 3978–3981.
10. BOHMANN, D., T. J. BOS, A. ADMON, T. NISHIMURA, P. K. VOGT & R. TJIAN. 1987. Human proto-oncogene *c-jun* encodes a DNA binding protein with structural and functional properties of transcription factor AP-1. Science **238:** 1386–1392.
11. LEE, W., P. MITCHELL & R. TJIAN. 1987. Purified transcription factor AP-1 interacts with TPA-inducible enhancer elements. Cell **49:** 741–752.
12. ANGEL, P., M. IMAGAWA, R. CHIU, B. STEIN, R. J. IMBRA, H. J. RHAMSDORF, C. JONAT, P. HERRLICH & M. KARIN. 1987. Phorbol ester–inducible genes contain a common *cis* element recognized by a TPA-modulated *trans*-acting factor. Cell **49:** 729–739.
13. MURPHY, P. R., Y. SATO, R. SATO & H. G. FRIESEN. 1988. Regulation of multiple basic fibroblast growth factor messenger ribonucleic acid transcripts by protein kinase C activators. Mol. Endocrinol. **2:** 1196–1201.
14. WIGLER, M., A. PELLICER, S. SILVERSTEIN & R. AXEL. 1978. Biochemical transfer of single-copy eucaryotic genes using total cellular DNA as donor. Cell **14:** 725–731.
15. GORMAN, C. M., L. F. MOFFAT & B. H. HOWARD. 1982. Recombinant genomes which

express chloramphenicol acetyltransferase in mammalian cells. Mol. Cell. Biol. **2:** 1044–1051.

16. LUCKOW, B. & G. SCHUTZ. 1987. CAT constructions with multiple unique restriction sites for the functional analysis of eukaryotic promoters and regulatory elements. Nucleic Acids Res. **15:** 5490.

17. BANIAHMAD, A., M. MULLER, C. STEINER & R. RENKAWITZ. 1987. Activity of two different silencer elements of the chicken lysozyme gene can be compensated by enhancer elements. EMBO J. **6:** 2297–2303.

18. HAY, N., J. M. BISHOP & D. LEVENS. 1987. Regulatory elements that modulate expression of human *c-myc*. Genes Dev. **1:** 659–671.

19. HERBST, R. S., N. FRIEDMAN, J. E. DARNELL JR. & L. E. BABISS. 1989. Positive and negative regulatory elements in the mouse albumin enhancer. Proc. Nat. Acad. Sci. USA **86:** 1553–1557.

20. PECH, M., C. D. RAO, K. C. ROBBINS & S. A. AARONSON. 1989. Functional identification of regulatory elements within the promoter region of platelet-derived growth factor 2. Mol. Cell. Biol. **9:** 396–405.

21. WINOTO, A. & D. BALTIMORE. 1989. $\alpha\beta$ lineage-specific expression of the α T cell receptor gene by nearby silencers. Cell **59:** 649–655.

22. HAY, N., M. TAKIMOTO & J. M. BISHOP. 1989. A FOS protein is present in a complex that binds a negative regulator of MYC. Genes Dev. **3:** 293–303.

23. PRATS, H., M. KAGHAD, A. C. PRATS, M. KLAGSBRUN, J. M. LELIAS, P. LIAUZUN, P. CHALON, J. P. TAUBER, F. AMALRIC, J. A. SMITH & D. CAPUT. 1989. High molecular mass froms of basic fibroblast growth factor are initiated by alternative CUG codons. Proc. Nat. Acad. Sci. USA **86:** 1836–1840.

24. ROGELJ, S., R. P. WEINBERG, P. FANNING & M. KLAGSBRUN. 1988. Basic fibroblast growth factor fused to a signal peptide transforms cells. Nature **331:** 173–175.

25. SASADA, R., T. KUROKAWA, M. IWANE & K. IGARASHI. 1988. Transformation of mouse BALB/c 3T3 cells with human basic fiboblast growth factor cDNA. Mol. Cell. Biol. **8:** 588–594.

26. ACLAND, P., M. DIXON, G. PETERS & C. DICKSON. 1990. Sub-cellular fate of the int-2 oncoprotein is determined by choice of initiation codon. Nature **343:** 662–665.

27. MACHAMER, C. E., R. Z. FLORKIEWICZ & J. K. ROSE. 1985. A single N-linked oligosaccharide at either of the two normal sites is sufficient for transport of vesicular stomatitis virus G protein to the cell surface. Mol. Cell. Biol. **5:** 3074–3083.

28. LAEMMLI, U. K. 1970. Clevage of structural proteins during the assembly of Bacteriophage T4. Nature London **263:** 680–685.

29. SZEWCZYK, B. & L. M. KOZLOFF. 1985. A method for the efficient blotting of strongly basic proteins from sodium dodecyl sulfate-polyacrylamide gels to nitrocellulose. Anal. Biochem. **150:** 403–407.

30. GLENNY, J. R. 1986. Antibody probing of western blots which have been stained with india ink. Anal. Biochem. **156:** 315.

31. FLORKIEWICZ, R. Z., A. SMITH, J. E. BERGMANN & J. K. ROSE. 1983. Isolation of stable mouse cell lines that express cell surface and secreted forms of the vesicular stomatitis virus glycoprotein. J. Cell Biol. **97:** 1381–1388.

32. HERBOMEL, P., B. BOURACHOT & M. YANIV. 1984. Two distinct enhancers with different cell specificities coexist in the regulatory region of polyoma. Cell **39:** 653–662.

Nuclear Translocation of Basic Fibroblast Growth Factor[a]

F. AMALRIC, V. BALDIN, I. BOSC-BIERNE, B. BUGLER,

B. COUDERC, M. GUYADER, V. PATRY, H. PRATS,

A. M. ROMAN, AND G. BOUCHE

Center for Research in Biochemistry and Cellular Genetics of the CNRS
118, Route de Narbonne
31062 Toulouse Cédex, France

INTRODUCTION

Basic fibroblast growth factor (bFGF) belongs to the family of heparin-binding growth factors that includes oncogenes and the two prototypic nononcogenic members, acidic and basic FGF.[1-6] bFGF is a potent mitogen for a wide variety of cell types of mesodermal and neuroectodermal origin that express FGF receptors on their surface.[7] Some of these cell lines also produce this factor; in particular, bovine aortic endothelial cells synthesize their own bFGF but they still require exogenous bFGF to grow.[8] In these cells, the endogenous growth factor which does not contain signal sequence for secretion is not recovered in the culture medium. An autocrine model of growth could predict that, as long as bFGF has access to its receptor, activation of the receptor would lead to proliferation. The dependence of exogenous bFGF suggested a different compartmentalization of the synthesized factor versus the receptor in the cell. This hypothesis was confirmed by data carried out on NIH 3T3 cells in which bFGF fused to signal peptide transformed cells even though no bFGF was detected in the medium.[9] The growth factor may bind its receptor at an intracellular location and induce mitogenic response. In endothelial cells, three forms of bFGF are synthesized differing in their N terminus, initiated at either an AUG (156 aa) or one of two CUG codons (195 and 210 aa).[10] The AUG-initiated proteins were cytoplasmic, while the CUG-initiated forms were nuclear.[11] Similarly, the mouse int-2 gene transfected in COS cells was shown to give rise to an AUG-initiated product that followed the secretory pathway and a CUG-initiated fom with an additional N-terminal peptide that was localized in the nucleus.[12] This process of alternative translation could be a general process that regulates the subcellular localization of growth factors.

On the other hand, recent evidence demonstrated that exogenous AUG-initiated bFGF is translocated to the nucleus of ABAE cells.[13] This process is cell- cycle-dependent and occurs at the transition G_1-S of the cell cycle.[14] Similarly, exogenous aFGF added to murine lung capillary endothelial cells was shown to be translocated to the nucleus and this process conferred mitogenic activity to the factor.[15]

This paper reviews selected aspects of the uptake of exogenous bFGF and the cellular localization of cell-synthesized bFGF in endothelial cells and in ABAE-derived cell lines that stably express bFGF without transcriptional control.

[a]This work was supported by grants from the Centre National de la Recherche Scientifique, Fondation pour la Recherche Médicale, and Association pour le Développement de la Recherche sur le Cancer.

RESULTS

Fate of Exogenous bFGF

bFGF and Cell Growth

Quiescent sparse ABAE or quiescent confluent Swiss 3T3 cells were obtained after 48 hours of culture in the absence of serum and bFGF. Addition of bFGF alone induced transition G1-S of the cell cycle after 2 hours and 8 hours in ABAE and 3T3 cells, respectively, as shown by incorporation of labeled thymidine (FIGURE 1).

To follow the uptake of exogenous bFGF by cells, iodinated bFGF was added to the culture medium. The factor was detected in the cytoplasm and in the nucleus of

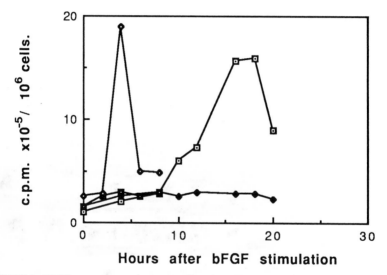

FIGURE 1. bFGF promotes synchronous cell cycle progression of G1 arrested adult bovine aortic endothelial cells (ABAE; open diamonds) and G0 arrested Swiss 3T3 cells (squares). ABAE and 3T3 arrested cells were stimulated by bFGF alone (5 ng/ml and 20 ng/ml respectively). Cells were pulse labeled for 15 minutes with 10 mCi/ml of methyl-3H thymidine at different times after the addition of bFGF. (Filled diamonds) no bFGF added.

exponentially growing cells both in ABAE and NIH 3T3 cells but no longer in the nucleus of confluent cells. A strong accumulation is routinely observed in the nucleolus by indirect immunofluorescence using affinity purified anti-bFGF immunoglobulin G (IgG). The amount of ^{125}I bFGF per 10^6 nuclei was estimated to be 55 pg in ABAE and NIH 3T3 cells corresponding to 7% and 8% of the amount present in the cytoplasm, respectively (TABLE 1). In the cytoplasm, the 18.4-kDa form is rapidly processed to a 16.5-kDa form (half-life, 1 hour), while it is more stable in the nucleus (half-life, 3 hours).

Thus in two cell lines that both possess bFGF receptors on their cell surface and that express high levels (ABAE cells) or very low levels of endogenous bFGF (NIH 3T3 cells), the uptake of exogenous bFGF and the intracellular distribution are identical.

TABLE 1. Intracellular Distribution of ^{125}I bFGF in Exponential Growing ABAE and Swiss 3T3 Cells

Cells	Nucleus			Cytoplasm	
	pg[a]	Molecules[b]	Percent[c]	pg	Molecules
ABAE	55 ± 10	1800	8.1	610 ± 63	20400
Swiss 3T3	54 ± 12	1750	6.6	820 ± 123	26500

[a]Values (pg per 10^6 cells) calculated from specific activity of radiolabeled bFGF are the mean ± standard error of the mean from three experiments.
[b]The number of bFGF molecules per subcellular fraction corresponding to one cell was calculated using Avogadro's number and the specific activity of ^{125}bFGF.
[c]Percentage of bFGF in the nucleus.

The same experiments were carried out using the two longer native forms initiated at one of the two CUGs (22.5 kDa and 21 kDa) and an NH2-truncated form (131 residues; TABLE 2). These three factors presented almost the same mitogenic activity as the 18.4-kDa factor (FIGURE 2). They are similarly recovered in the cytoplasm and in the nucleus. The 131-residue form is stable in the cytoplasm and in the nucleus, suggesting that the 16.5-kDa protein that accumulated in cells grown in the presence of the 18.4-kDa bFGF resulted from an NH2 cleavage of the molecule (FIGURE 3). The two CUG-initiated forms behave like the 18.4-kDa form.

Nuclear Translocation and Fate of bFGF during the Cell Cycle

Quiescent G1 arrested ABAE cells were stimulated by addition of unlabeled bFGF and the kinetics of bFGF uptake was followed by coaddition of iodinated bFGF at different times during the cell cycle. The uptake and the nuclear accumulation of bFGF were estimated after cell fractionation. As shown in FIGURE 4A, bFGF entered the cytoplasm continuously with a maximum in G2 between 8 and 10 hours after stimulation, then after mitosis the bFGF amount decreased sharply. On the other hand, nuclear translocation occurred only at transition G1-S.

The fate of the exogenous factor is different in the nucleus and in the cytoplasm. As shown in FIGURE 4B the concentration of bFGF inside the nucleus is maximal at the begining of the S phase then went down regularly to become almost undetectable

TABLE 2. Structure of bFGFs[a]

		Residues	kDa
1 100	a	210	22.5
15	b	196	21
56	c	155	18.4
66	d	146	18
81	e	131	16.5
	f	?	16.5

[a]Structure of the NH2-terminal domain: a and b, CUG1 and CUG2 initiated proteins; c, AUG initiated protein; d and e, natural maturation products of bFGF; f, mutant with 81 residues of the N-terminal domain deleted.

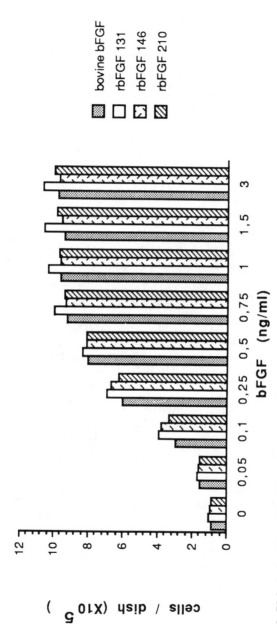

FIGURE 2. Biological activity of the different forms of bFGF in ABAE cell growth. 10^4 cells were seeded per dish. The different forms of bFGF were then added to the indicated final concentration. Media were changed every two days. On day 5, the cells were detached and viable cell numbers were quantitated with a hemacytometer. All values were the average of five samples.

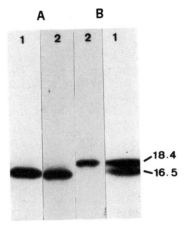

FIGURE 3. Detection of ^{125}I bFGF in the cytoplasm and the nucleus of ABAE cells. Two hours after the addition of ^{125}I bFGF (131 residues, A; 146 residues, B), 10^6 cells were harvested and fractionated in cytoplasm (1) and nuclei (2). Proteins of each fraction were analyzed by sodium dodecyl sulfate-polyacrylamide gel electrophoresis (SDS-PAGE), and the gel was submitted to autoradiography for 48 hours.

at the beginning of the G2 phase. These data clearly demonstrate that the intracellular localization of the exogenous growth factor is strictly controlled during the cell cycle and that the nuclear translocation is not related to the cytoplasmic concentration of the factor.

In additional experiments, a decoupling of cytoplasmic and nuclear uptakes was achieved by addition of chloroquine to the culture medium. The nuclear transport was increased by a factor two to three, while the uptake of the 18.4-kDa form to the cytoplasm was significantly decreased, as was the processing of the 18.4-kDa to the

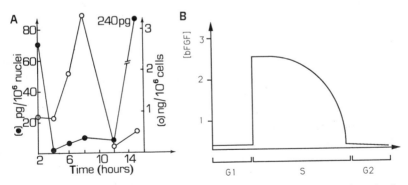

FIGURE 4. bFGF uptake along the cell cycle. G1 arrested ABAE cells were first stimulated with unlabeled bFGF (5 ng/ml). At different times in the cell cycle cells were pulse labeled with ^{125}I bFGF during 2 hours and cells were harvested. After cellular fractionation, cytoplasmic and nuclear proteins were analyzed by SDS-PAGE and quantification of bFGF uptake was carried out by counting. A: Filled circles, nuclei; open circles, cytoplasm. B: Quantification of bFGF amount of exogenous origin in nuclei along the cell cycle.

16.5-kDa form. This result suggests that the cytoplasmic maturation of bFGF occurs in lysosomes (FIGURE 5). The process of bFGF internalization was also followed in the presence of cycloheximide. The drug was added to the culture medium of quiescent cells together with bFGF. A 50% drop in the the cytoplasmic uptake was observed after two hours, while the nuclear uptake was unaffected (FIGURE 6). Furthermore, the maturation of the 18.4-kDa form was completely inhibited after 6 hours of treatment. These results suggest that the process of nuclear transport of FGF is independent of the classical pathway through receptor internalization and endocytosis used for cytoplasmic uptake.

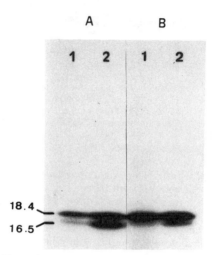

FIGURE 5. Effects of chloroquine on uptake and turnover of cytoplasmic and nuclear bFGF. Chloroquine (50 μM) was added to G1 arrested cells that were stimulated with ^{125}I bFGF (5 ng/ml); After 2 hours, the cells were harvested and fractionated in nuclei (1) and cytoplasm (2). Analysis was carried out as described in the legend of FIGURE 3. **A:** Control cells. **B:** Chloroquine-treated cells.

Nuclear Targets of bFGF

As shown in FIGURE 4, a close relationship exists between the appearance of bFGF in the nucleus and the start of DNA replication on one hand, the replication level and the concentration of nuclear bFGF on the other hand. These relationships argue in favor of a direct involvement of bFGF in replication. Furthermore, preliminary results suggested that bFGF could bind DNA efficiently with nucleotide sequence specificity (data not shown).

In previous experiments, we have also shown that an early effect induced by the addition of bFGF to quiescent sparse cells was the stimulation of rDNA transcription. Cells were harvested at different times after the addition of bFGF, and nuclei were prepared to carry out "run-on" experiments. The amount of synthesized rDNA was determined by hybrid selection. As shown in FIGURE 7, the transcription of rDNA, which is 10% in nuclei of resting cells compared to growing cells, increased rapidly after the addition of bFGF to cell culture, to reach a maximum after 2 hours.

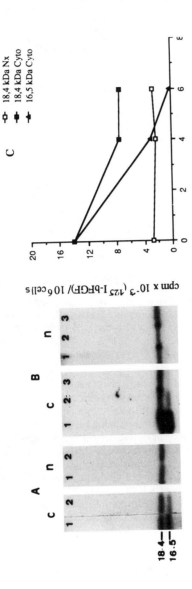

FIGURE 6. Effects of cycloheximide on the time course of ^{125}I bFGF accumulation and turnover in cytoplasm (c) and nuclei (n). **A:** Control cells. **B:** Cycloheximide-treated cells. ^{125}I bFGF and cycloheximide (20 μg/ml) were added simultaneously for 2 hours (1), 4 hours. (2), 6 hours (3). **C:** Relative amount of the different bFGFs.

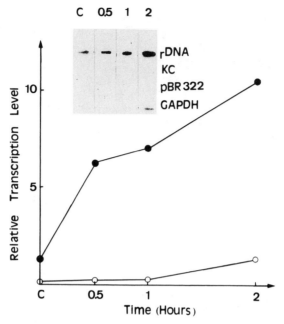

FIGURE 7. rDNA transcription in ABAE cells nuclei. Nuclei (3×10^5) were prepared from quiescent sparse cells (C) or from cells stimulated different times by bFGF (5 ng/ml). Run on assays were carried out as previously described.[13] Newly synthesized RNA was hybridized with a constant amount of rDNA, KC, GAPDH, and pBR322 DNA. The relative transcriptional level was obtained after normalization of the densitometer scanning of autoradiograms using GAPDH as reference. rDNA (filled circles); KC (open circles).

In a second set of experiments, bFGF was added in run-on assays. A 70% increase in rDNA transcription was observed only in quiescent cell nuclei. These results suggested that bFGF acted on rDNA transcription through a regulatory factor that should modulate the level of transcription. Such a factor could be the protein kinase CKII which uses several factors implicated in rDNA transcription as substrates. CKII is found in limited amounts in the nuclei of confluent cells. Addition of bFGF to these nuclei induced a strong increase in the phosphorylation of a subset of protein and among them the substrates of CKII. A similar result was observed by addition to the assay of exogenous CKII. Thus, one of the nuclear targets of bFGF could be the protein kinase CKII which controls not only the activity of factors involved in rDNA transcription but also several transactivators of RNA polymerase II.

Localization of Endogenous bFGF

ABAE cells synthesized a high level of bFGF, but its fate in the cells is difficult to determine since they require exogenous bFGF to proliferate. In cells cultivated with the 131-residue bFGF as exogenous growth factor, the three forms of endogenous bFGF that are initiated at the AUG (90%) or at one of the two CUGs (10%), were detected. In quiescent cells kept in the absence of serum and exogenous bFGF, endogenous bFGF is still present in small amounts.

To determine whether the endogenous bFGF is translocated to the nucleus, a series of experiments was carried out in which COS cells were transfected by plasmids expressing chimeric genes constructed by fusion of the bFGF and chloramphenicol acetyl transferase (CAT) open reading frames. Translation of this fused gene started at either one of the CUGs or at the AUG. The chimeric proteins were both nuclear and cytoplasmic, while the native CAT protein was strictly cytoplasmic (FIGURE 8). In a second set of experiments, the CAT gene was fused either to the bFGF gene initiated at the AUG or to the 55 residues of the N terminus located between the CUG and the AUG. The chimeric protein initiated at the AUG was strictly cytoplasmic while the N-terminal peptide induced a nuclear localization. This peptide contains an RGR(X)5R motif, repeated three times. This motif is also present in an additional N-terminally extended int-2 protein initiated at an in-frame CUG codon. This domain is responsible for the protein location in the nucleus.

In an additional experiment a chimeric protein initiated at residue 26 was produced. This protein which does not contain the repeated motif is partially localized in the nucleus. Furthermore, direct mutagenesis experiments suggested

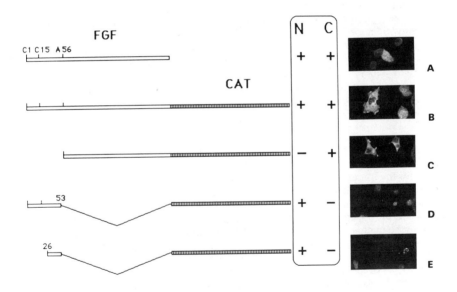

FIGURE 8. Subcellular location of bFGF and bFGF-CAT protein. Shown are constructs inserted into the pSVL expression vector (Pharmacia) and fluorescent staining micrographs of COS-7 cells transfected with the corresponding DNA.[11] A part of the N-terminus nuclearization signal sequence is presented below the micrographs.

that arginines 47 and 50 are required to have a complete translocation of bFGF to the nucleus (Bugler *et al.*, manuscript in preparation). This N-terminal domain, although rich in arginine residues, shows no sequence similarity to nuclear signals identified so far.

Thus, the subcellular localization of bFGF is controlled by a process of alternative initiation of translation at two CUGs or at an AUG. The relative amount of each form could modulate the autocrine cell growth and the dependence on exogenous growth factors.

In addition, it was routinely observed that the overexpressed endogenous bFGF large forms were distributed homogeneously in the nucleus while in ABAE cells growing in the presence of exogenous bFGF the growth factor was concentrated in the nucleolus (FIGURE 9). All these data suggest the existence of two distinct translocation pathways for bFGF to the nucleus from outside the cell or from the cytoplasm.

To get an insight into the function of the three bFGF forms, ABAE cell lines that expressed constitutively one of the three forms were established by transfection with a retroviral vector. Different phenotypes were observed according to the bFGF gene introduced into the cell. The cell lines in which CUG-initiated proteins were produced did not require exogenous bFGF to grow, while AUG-initiated protein did not confer this property (Couderc, manuscript in preparation). It must be noted that this new property (bFGF-independent growth) was not due to an overexpression of the growth factor ($\pm 50\%$ compared to the amount of bFGF in control ABAE cells) but may be due to a deregulation of bFGF synthesis during the cell cycle (data not shown). The cells in which the nuclear large forms were synthesized without control lost their cell cycle dependence on exogenous bFGF.

CONCLUSION

Adult bovine aortic endothelial cells synthesize three forms of intracellular bFGF which are localized in the cytoplasm or in the nucleus according to the codon used for the initiation of translation. These cells required exogenous growth factor to proliferate. This factor is translocated to the nucleus at the transition G1-S of the cell cycle, and this process could be the key step for the mitogenic activity. Since these cells also produced a small amount of large forms of bFGF susceptible to translocation to the nucleus, we must postulate that the amount of nuclear endogenous bFGF is insufficient for proliferation. This model could explain whether different endothelial cell lines are dependent or not on exogenous bFGF for proliferation, depending on the level of large forms synthesized. A strict compartimentalization of the growth factor could control the autocrine model of growth and maybe that of transformation.[16] Thus the CUG-initiated bFGFs would be involved in the mitogenic activity, while the AUG-initiated form could be implicated in differentiation processes.

The intracellular route of exogenous bFGF to the nucleus remains to be established. It seems clear that, while cytoplasmic uptake involves lysosomes, nuclear transport follows another unusual pathway. A puzzling find is the similarity between the nuclear uptake of exogenous bFGF and of heparan sulfate, which are both independent of lysosomes.[17] Since bFGF has a strong affinity for heparin and since heparan sulfate inhibited the binding of bFGF to its receptor, we express the hypothesis that a cotransport to the nucleus of the two molecules occurs in endothelial cells. This would explain how the exogenous 18.4-kDa bFGF, which does not contain a nuclearization signal sequence, is recovered in the nucleus.

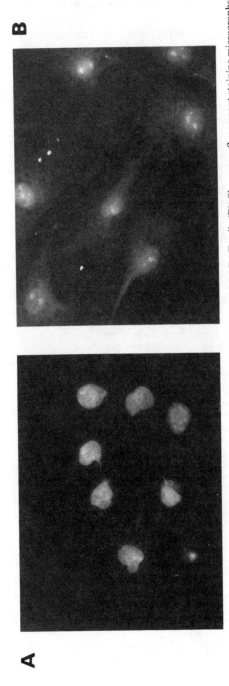

FIGURE 9. Subcellular location of bFGF in bFGF gene transfected COS-7 cells (A) and ABAE cells (B). Shown are fluorescent staining micrographs.

ACKNOWLEDGMENTS

We are grateful to J. Maurel for the typing of the manuscript.

REFERENCES

1. ABRAHAM, J. A., J. L. WHANG, A. TUMOLO, A. MERCIA, J. FRIEDMAN, D. GOSPODAROWICZ & J. C. FIDDES. 1986. Human basic fibroblast growth factor: nucleotide sequence and genomic organization. EMBO J. **5:** 2523–2518.

2. JAYE, M., R. HOWK, W. BURGESS, G. A. RICCA, I. M. CHIU, M. W. RAVERA, S. J. O'BRIEN, W. S. MODI, T. MACIAG & W. N. DROHAN. 1986. Human endothelial cell growth factor: nucleotide sequence, and chromosome localization. Science **233:** 541–545.

3. YOSHIDA, T., K. MIYAGAWA, H. ODAGIRI, H. SAKAMOTO, P. F. R. LITTLE, M. TERADA & T. SUGIMARA. 1987. Genomic sequence of hst, a transforming gene encoding a protein homologous to fibroblast growth factors and the int-2 encoded protein. Proc. Nat. Acad. Sci. USA **84:** 7305–7309.

4. DELLIBOVI, P., A. M. CURATOLA, F. KERN, A. GRECO, M. ITTMAN & C. BASILICO. 1987. An oncogene isolated by transfection of Kaposi's sarcoma DNA encodes a growth factor that is a member of the FGF family. Cell **50:** 729–737.

5. ZHAN, X., B. BATES, X. HU & M. GOLDFARB. 1988. The human FGF 5 oncogene encodes a novel protein related to fibroblast growth factors. Moll. Cell. Biol. **8:** 3487–3495.

6. MARICS, I., J. ADELAIDE, F. RAYBAUD, M. G. MATTEI, F. COULIER, J. PLANCHE, O. DE LAPEYRIERE & D. BIRNBAUM. 1989. Characterization of the hst related FGF 6 gene, a new member of the fibroblast growth factor gene family. Oncogene **4:** 335–340.

7. GOSPODAROWICZ, D., G. NEUFELD & L. SCHWEIGERER. 1986. Fibroblast growth factor. Mol. Cell. Endocrinol. **46:** 187–204.

8. RIFKIN, D. B. & D. MOSCATELLI. 1989. Recent developments in the biology of basic fibroblast growth factor. J. Cell Biol. **109:** 1–6.

9. ROGELJ, S., R. A. WEINBERG, P. FANNING & M. KLAGSBRUN. 1988. Basic fibroblast growth factor fused to a signal peptide transforms cells. Nature **331:** 173–175.

10. PRATS, H., M. KAGHAD, A. C. PRATS, M. KLAGSBRUN, J. M. LÉLIAS, P. LIAUZUN, P. CHALON, J. P. TAUBER, F. AMALRIC, J. SMITH & D. CAPUT. 1989. High molecular mass forms of basic fibroblast growth factor are initiated by alternative CUG codons. Proc. Nat. Acad. Sci. USA **86:** 1836–1840.

11. BUGLER, B., F. AMALRIC & H. PRATS. 1991. Alternative initiation of translation determines cytoplasmic or nuclear localization of basic fibroblast growth factor. Mol. Cell. Biol. **11.**

12. ACLAND, P., M. DIXON, G. PETERS & C. DICKSON. 1990. Subcellular fate of the Int-2 oncoprotein is determined by choice of initiation codon. Nature **343:** 662–665.

13. BOUCHE, G., N. GAS, H. PRATS, V. BALDIN, J. P. TAUBER, J. TEISSIE & F. AMALRIC. 1987. Basic fibroblast growth factor enters the nucleolus and stimulates the transcription of ribosomal genes in ABAE cells undergoing G0 → G1 transition. Proc. Nat. Acad. Sci. USA **84:** 6770–6774.

14. BALDIN, V., A. M. ROMAN, I. BOSC-BIERNE, F. AMALRIC & G. BOUCHE. 1990. Translocation of bFGF to the nucleus is G1 phase specific in bovine aortic endothelial cells. EMBO J. **9:** 1511–1517.

15. IMAMURA, T., K. ENGLEKA, X. ZHAN, Y. TOKITA, R. FOROUGH, D. ROEDER, A. JACKSON, J. A. M. MAIER, T. HLA & T. MACIAG. 1990. Recovery of mitogenic activity of a growth mutant with a nuclear translocation sequence. Science **249:** 1567–1574.

16. ROGELJ, S., R. A. WEINBERG, P. FANNING & M. KLAGSBRUN. 1989. Characterization of tumors produced by signal peptide–bFGF-transformed cells. J. Cell. Biochem. **39:** 13–23.

17. ISHIHARA M., N. S. FEDARKO & E. CONRAD. 1986. Transport of heparan sulfate into the nuclei of hepatocytes. J. Biol. Chem. **261:** 13575–13580.

Mitogenic Effects of Fibroblast Growth Factors in Cultured Fibroblasts[a]

Interaction with the G-Protein-Mediated Signaling Pathways

SONIA PARIS[b] AND JACQUES POUYSSÉGUR

Biochemistry Center, CNRS
University of Nice
Parc Valrose
06034 Nice, France

INTRODUCTION

A first step toward understanding the mechanism of action of growth factors is the identification of the early cellular responses that immediately follow the interaction of the growth factors with their receptors. Cultured cells can serve as useful models for such studies, provided that they can be efficiently and reversibly arrested in the G0 resting phase of the cell cycle. The Chinese hamster lung fibroblast line CCL39 fulfills this condition since these cells become quiescent upon removal of serum from the culture medium and can be stimulated to reinitiate DNA synthesis by the addition of serum or purified growth factors. A limited set of pure hormones or growth factors are mitogenically active on CCL39 cells, including α-thrombin, fibroblast growth factor (FGF), epidermal growth factor (EGF), serotonin, and insulin (substituting for insulinlike growth factor I, IGF-I). But among these mitogens, only two of them, namely, α-thrombin and FGF, are potent enough to trigger a significant response alone, in the absence of any other growth factor. The other ones are active only when used in synergistic combinations.

The present report will focus on the mitogenic effects of FGF in CCL39 cells. Both late and early responses are described with a special emphasis given to the interactions with the G-protein-mediated pathways.

EFFECTS OF FGF ON DNA SYNTHESIS

As illustrated in FIGURE 1 (left), basic FGF alone (either purified from bovine brain or recombinant bovine FGF, both preparations giving identical results) promotes DNA synthesis in G0-arrested CCL39 cells, with half-maximal stimulation around 10 ng/ml and maximal effect at 30 ng/ml. At optimal concentrations, bFGF triggers DNA synthesis in 20–30% of the cells, as measured by labeled nuclei. The acidic form of FGF is also mitogenic for CCL39 cells, but higher concentrations

[a] This work was supported by grants from the Centre National de la Recherche Scientifique (UPR 7300), the Institut National de la Santé et de la Recherche Médicale, the Fondation pour la Recherche Médicale, and the Association pour la Recherche contre le Cancer.

[b] Present affiliation: CNRS–Institute of Pharmacology, 660, route des Lucioles Sophia Antipolis, F-06560 Valbonne, France.

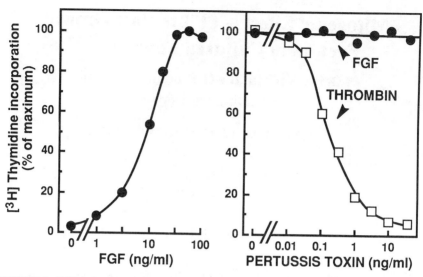

FIGURE 1. FGF-induced reinitiation of DNA synthesis is insensitive to pertussis toxin. Left: G0-arrested CCL39 cells were incubated for 24 hours in serum-free medium with [^{3}H]thymidine (1 μCi/ml, 4.5 μM) and various concentrations of recombinant basic FGF. right: G0-arrested CCL39 cells were preincubated for 4 hours with [^{3}H]thymidine and the indicated concentrations of pertussis toxin. Incubation was continued for 24 hours in the presence of either 1 nM α-thrombin (squares) or 100 ng/ml basic FGF (circles). The concentration of thrombin was chosen to give a mitogenic response equivalent to the maximal effect of FGF.

(50–300 ng/ml, depending on the preparation) are required for a maximal response. Although basic FGF has been used in most of our studies, we will refer to it simply as FGF since similar results would be obtained with the acidic form. It is likely that both forms bind to the same receptor(s).[1]

Insensitivity to Pertussis Toxin

Treatment of CCL39 cells with pertussis toxin, a bacterial toxin known to ADP ribosylate certain G proteins,[2] does not affect at all FGF-induced DNA synthesis, whereas it inhibits up to 95% of thrombin-induced mitogenesis (FIGURE 1, right).[3] Similarly, pertussis toxin has been reported in other cell systems to selectively inhibit the mitogenic actions of bombesin,[4,5] serotonin,[6] and lysophosphatidate,[7] while it does not affect the stimulation of DNA synthesis by platelet-derived growth factor (PDGF)[4-6] or EGF.[7] Thus mitogens capable of triggering the G0 → S phase transition alone can be classified into two groups according to their sensitivity to pertussis toxin: the first group, highly sensitive to pertussis toxin, is composed of factors known (or supposed) to bind to G-protein-coupled receptors, whereas the second group, insensitive to the toxin, is composed of growth factors known to activate receptor tyrosine kinases.[8-11]

Synergism with G-Protein-Activating Factors

FGF strongly potentiates the mitogenic effects of α-thrombin in CCL39 cells. As shown in FIGURE 2, FGF both increases the maximal response elicited by thrombin and decreases the thrombin concentration required to produce a half-maximal response (by three orders of magnitude). It is noteworthy that a combination of FGF and α-thrombin at optimal concentrations is often even more mitogenic than 10% fetal calf serum ($\geq 80\%$ labeled nuclei).

FGF-induced mitogenesis is also potentiated by G-protein-activating agents which are not mitogenic *per se*. Thus serotonin, which inhibits adenylate cyclase and activates a phosphoinositide-specific phospholipase C through G proteins in CCL39 cells, markedly enhances the mitogenicity of FGF.[12,13] A similar effect can be obtained through the direct activation of G proteins by GTPγS. Indeed, we have shown that GTPγS can activate G proteins in intact CCL39 cells when the cells are depolarized in high-KCl medium (135 mM KCl, 0 Na$^+$). Although the mechanism underlying this activation is unknown, this finding provides a simple method to examine long-term effects of G-protein activation, since cell viability is not altered by this treatment.[14] Thus, when quiescent CCL39 cells are exposed to GTPγS for 30 minutes in high-KCl medium and then returned to a normal serum-free culture medium supplemented with FGF, a strong synergistic interaction is observed between GTPγS and FGF (FIGURE 3). Whereas the pretreatment with GTPγS is not sufficient by itself to promote a significant mitogenic response, it induces a striking dose-dependent potentiation of FGF's mitogenicity.

Altogether these observations suggest that activation of G-protein-mediated pathways provides mitogenic signals that FGF is not able to elicit by itself. However, as far as the pertussis-toxin-sensitive signal is concerned, it should be noted that FGF must be able either to induce this signal by a different pathway (not involving a pertussis-toxin-sensitive G protein) or to replace this signal by an equivalent one, because at high concentration of thrombin the sensitivity to the toxin is nearly

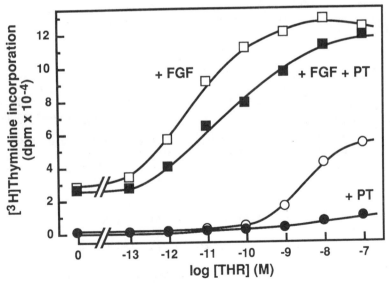

FIGURE 2. FGF potentiates thrombin-induced DNA synthesis and suppresses the inhibitory effects of pertussis toxin. G0-arrested CCL39 cells were preincubated for 5 hours with [³H]thymidine and with (filled symbols) or without (open symbols) 50 ng/ml pertussis toxin. Incubation was continued for 24 hours in the presence of the indicated concentrations of thrombin either alone (circles) or combined with 20 ng/ml FGF (squares).

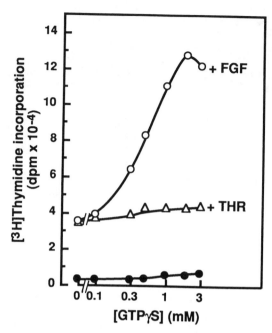

FIGURE 3. GTPγS potentiates FGF-induced DNA synthesis. G0-arrested cultures were exposed for 30 minutes to the indicated concentrations of GTPγS in an isotonic salt solution containing 135 mM KCl and no Na⁺. GTPγS was then washed away, and incubation was continued for 18 hours in regular serum-free culture medium with [³H]thymidine and the indicated growth factor: 25 ng/ml FGF (open circles), 10 nM thrombin (triangles), or no growth factor (filled circles).

abolished by FGF (FIGURE 2). Along this line, we have also observed that in the presence of FGF the dose-dependence curve for pertussis toxin inhibition of the synergistic effects of low concentrations of thrombin or GTPγS is markedly shifted to the right,[14] as compared to that obtained in cells stimulated with thrombin alone (FIGURE 1, right).

Therefore it can be conclued from these observations that tyrosine kinase– and G-protein-activated signaling pathways, while complementary, must also share some common targets. In an attempt to determine these points of signal integration, we have next examined the early responses of CCL39 cells to FGF.

EARLY EVENTS ELICITED BY FGF

Phosphoinositide Hydrolysis and Ca²⁺ Mobilization

The breakdown of phosphatidylinositol 4,5-bisphosphate (PIP$_2$) by a phospholipase C (PLC) generates at least two important intracellular messengers: inositol 1,4,5-trisphosphate, which releases Ca²⁺ from intracellular stores, and diacylglycerol, which activates protein kinase C (PKC).[15,16] It is now accepted that stimulation of PIP$_2$-PLC can be achieved by two distinct mechanisms, either through receptor-mediated activation of G proteins[17] or by tyrosine phosphorylation of the γ isozyme of PLC.[18] It was therefore of interest to examine the effects of FGF on this signaling pathway.

When used alone, FGF fails to induce any significant release of inositol phosphates in quiescent CCL39 cells, even after prolonged incubations in the presence of Li[+] which allows accumulation of inositol monophosphate.[19] Accordingly, the intracellular Ca^{2+} concentration is not increased by FGF when external Ca^{2+} is chelated with EGTA.[19] These results are in agreement with the findings of Rozengurt's group in Swiss 3T3 cells,[20] but in conflict with those of Brown et al., who reported that both acidic and basic recombinant FGFs stimulate inositol phosphate production and Ca^{2+} mobilization in the same Swiss 3T3 cells.[21] The reasons for these discrepancies may be differences between various strains either in the number of FGF receptors or in the level of the PLCγ isozyme.

Although unable to activate on its own the turnover of inositol lipids in CCL39 cells, FGF can potentiate the stimulation of PLC induced by G-protein-activating agents. This is illustrated in FIGURE 4 for α-thrombin. The synergistic effect is best revealed when PLC is weakly stimulated. Thus the potentiation is immediately detectable at low thrombin concentrations (FIGURE 4) whereas at high thrombin concentrations it becomes pronounced only after desensitization of PLC to thrombin.[22] Moreover, FGF also potentiates the stimulation of PLC by AlF_4^-, which directly activates the G protein(s) coupled to PLC.[22] To explain this potentiating effect we have proposed that FGF, like other tyrosine kinase–activating growth factors (EGF and PDGF), enhances the coupling between G protein(s) and PLC, presumably through the phosphorylation of one of these proteins. We believe

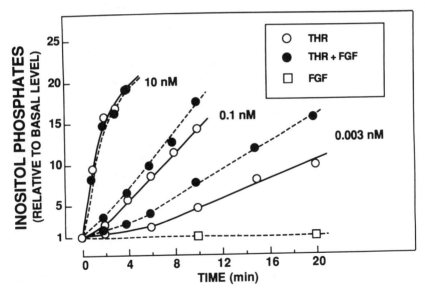

FIGURE 4. FGF potentiates the formation of inositol phosphates induced by low thrombin concentrations. G0-arrested CCL39 cells, labeled to equilibrium with [³H]inositol, were incubated with 20 mM LiCl and the indicated growth factors: thrombin (THR) at 0.03, 0.1 or 10 nM either alone (open circles) or with 100 ng/ml FGF (filled circles), or FGF alone at 100 ng/ml (squares). Total inositol phosphates were determined at the indicated times.[22]

however that phosphorylation of PLCγ is not responsible for this effect, since G-protein-coupled phosphoinositide responses seem to be mediated by other PLC isozymes.[18] It is worth noting that EGF and FGF have been similarly reported to reinforce bradykinin-induced Ca^{2+} rise in various cell types.[23,24]

Diacylglycerol Production and Protein Kinase C Activation

Despite its lack of effect on the hydrolysis of phosphoinositides, FGF induces a marked increase in the diacylglycerol content of CCL39 cells.[19] It is now recognized that phosphatidylinositol turnover is not the only source of diacylglycerol. Phosphatidylcholine breakdown by phospholipases of the C and D types has been reported to occur upon stimulation by a variety of agonists including growth factors such as EGF or PDGF.[25–27] It is therefore very likely that FGF also stimulates phosphatidylcholine hydrolysis in CCL39 cells, but this remains to be demonstrated.

Activation of PKC by FGF has not been detected in CCL39 cells when it is measured indirectly by down modulation of EGF binding sites.[19] In contrast, FGF has been clearly demonstrated to activate PKC in Swiss 3T3 cells.[20,28] Whether these contradictions are due to different experimental conditions used in PKC assays or to real differences in PKC activation between cell types is not known.

Adenylate Cyclase

FGF alone does not alter the basal level of cAMP in CCL39 cells, or the cAMP accumulation induced by the phosphodiesterase inhibitor IBMX. However FGF potentiates (by up to 50%) the cAMP accumulation induced by prostaglandin E_1, forskolin, or cholera toxin (FIGURE 5).[29] The insensitivity of this potentiation to pertussis toxin indicates that FGF does not modulate cAMP synthesis by inhibition of G_i (inhibitory G protein) activity. We have also excluded that this effect is mediated via PKC activation as it persists in PKC down-regulated CCL39 cells. We have therefore proposed that FGF enhances the activation of adenylate cyclase by G_s, the stimulatory G protein.[29]

Similar effects of FGF on stimulated adenylate cyclase have been reported in Swiss 3T3 cells but these effects have been mostly attributed to the activation of PKC by FGF in these cells.[20] It is interesting to note however that a significant potentiating effect of FGF on adenylate cyclase persists in Swiss 3T3 cells after down regulation of PKC[20] and the residual stimulation could correspond to the PKC-independent effect that we have described. Such differences in the contribution of PKC between the two cell types strongly suggest that FGF-induced activation of PKC is indeed more important in Swiss 3T3 cells than in CCL39 cells (see above). It should be noted also that PKC-mediated potentiation of cAMP accumulation is pertussis-toxin-sensitive in Swiss 3T3 cells[30] whereas it is not in CCL39 cells,[31] which points to definite differences between the signaling networks of the two cell types.

Most interestingly, EGF also can potentiate the stimulation of adenylate cyclase in CCL39 cells,[29] which is in good agreement with recent findings in A-431 cells.[32] It can therefore be proposed that tyrosine phosphorylation might be responsible, directly or indirectly, for the enhanced coupling between adenylate cyclase and G_s. Whether a common mechanism underlies the potentiations of PLC (see above) and adenylate cyclase is not known.

Early Responses Common to FGF and G-Protein-Activating Mitogens

Both FGF and α-thrombin activate the amiloride-sensitive Na^+/H^+ antiporter present in the plasma membrane of CCL39 cells, leading to a cytoplasmic alkalinization.[19] Activation of Na^+/H^+ exchange has been shown to be associated with the phosphorylation of the antiporter.[33] The observation that the phosphorylation occurs at serine residues upon stimulation by both thrombin and EGF[33] suggests that the antiporter is not a direct substrate for tyrosine kinase receptors but rather the final target of a phosphorylation cascade, which could involve at some step a "switch

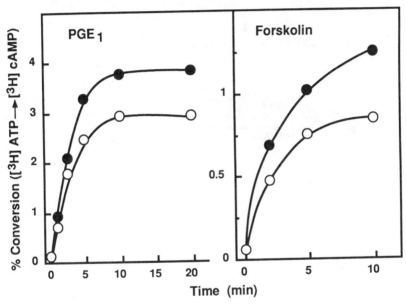

FIGURE 5. FGF potentiates receptor- and non-receptor-mediated activation of adenylate cyclase. CCL39 cells were labeled with [³H]adenine during the 30-hour incubation in serum-free medium. Adenylate cyclase activity was measured by the conversion of [³H]ATP into [³H]cAMP. The cultures were treated for 10 min with 1 mM IBMX in the presence (filled circles) or absence (open circles) of 20 ng/ml FGF before being stimulated (at time 0) with 10 μM PGE_1 (left) or 50 μM forskolin (right). Formation of [³H]cAMP was determined at the times indicated as described.[29]

kinase," capable of integrating signals coming from tyrosine kinase receptors as well as from G-protein-coupled receptors (see FIGURE 6).

Potential candidates for this role of "switch kinase" are *raf* and MAP (mitogen-activated protein) kinases.[34] Indeed both FGF and α-thrombin activate these serine/threonine protein kinases. In the case of MAP kinases at least, activation correlates with an increased phosphorylation of the kinases on tyrosine residues for both mitogens. However, the finding that pertussis toxin specifically inhibits thrombin-induced MAP-kinase activation and its phosphorylation on tyrosine residues whereas the toxin does not affect the FGF response clearly demonstrates that distinct signaling pathways are activated by thrombin and FGF which apparently converge on MAP kinases.[35]

Similarly, the phosphorylation of the ribosomal protein S6 is stimulated by both FGF and thrombin,[19] but also as the result of a complex phosphorylation cascade,[36] which might even include the MAP kinase.[37]

Activation of early genes such as *c-myc* is another common response to FGF and thrombin in CCL39 cells,[19] but the transducing pathways that regulate transcriptional factors are still poorly understood.

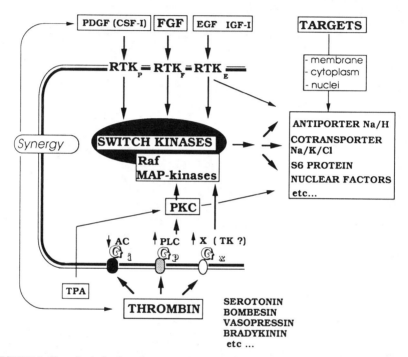

FIGURE 6. Hypothetical scheme representing how the common pleiotypic response could be generated by integration of extracellular signals. Switch kinases (Raf and MAP kinases) integrate signals from receptor tyrosine kinases (RTK) and G-protein-coupled receptors. Switch kinases will in turn propagate the signals by a cascade of ser/thr kinases activating specific targets.

SUMMARY AND CONCLUSIONS

We have shown that FGF (basic or acidic) is mitogenic for quiescent hamster lung fibroblasts (CCL39 line). It is active alone but is much more efficient in synergistic combinations with G-protein-activating agents.

When used alone, FGF appears to exert its mitogenic effects without involving any of the major G-protein-mediated signaling pathways. It causes no significant hydrolysis of phosphoinositides, it does not alter the activity of adenylate cyclase, and its mitogenicity is insensitive to pertussis toxin. It therefore seems likely that all pleiotropic actions of FGF are primarily mediated by the intrinsic protein tyrosine kinase of its receptors.

However, FGF, acting through its receptor tyrosine kinase, and thrombin, acting through G-protein-coupled receptors, induce a common set of early responses detected within seconds or minutes at the level of membranes, cytoplasm, and nuclei. Typical examples of early responses are activation of Na/H antiporter and Na/K/Cl cotransporter, phosphorylation of ribosomal protein S6, and increased transcription of early-immediate genes (c-fos, c-jun, and c-myc). Not only various classes of growth factors acting via distinct transducing mechanisms activate common targets, but also their synergistic effects on reinitiation of DNA synthesis is reflected on the early responses. How does the coordination of these signaling events take place? A partial answer to this question is illustrated in FIGURE 6 in which "switch kinases" play the role of integrators of multiple extracellular signals. Raf and, perhaps more convincingly, MAP kinases that are activated by dual phosphorylation on tyrosine and threonine residues[38] are potential good candidates for this integration. This hypothetical scheme could therefore explain, in part, the coordination and the synergy commonly observed in the mitogenic response. The synergy could be generated at the level of MAP kinases simply by dual activating phosphorylations. With the recent cloning of MAP kinases,[39,40] these questions will be more easily addressed. Another important gap that will have to be filled in future studies is the identification of all the members of the kinase cascade.

When used in synergistic combinations with G-protein-activating agents, FGF does exert in contrast some effects on the G-protein-mediated pathways. It potentiates the G-protein-mediated activations of both PIP_2-PLC and adenylate cyclase. Although it is generally admitted that the existence of synergistic interactions between two factors implies that they produce different and complementary signals, our findings raise the possibility that the synergism might be due also, at least in part, to an increased production of second messengers leading to an amplification of the early signals.

ACKNOWLEDGMENTS

We would like to acknowledge the valuable contribution of Drs. Isabelle Magnaldo, Gilles L'Allemain, and Jean-Claude Chambard. We are grateful to Gabrielle Imbs and Martine Valetti for secretarial assistance and to Franck Aguila for the drawings.

REFERENCES

1. NEUFELD, G. & D. GOSPODAROWICZ. 1986. J. Biol. Chem. **261:** 5631–5637.
2. UI, M. 1984. Trends Pharmacol. Sci. **5:** 277–279.
3. CHAMBARD, J. C., S. PARIS, G. L'ALLEMAIN & J. POUYSSÉGUR. 1987. Nature **326:** 800–803.
4. LETTERIO, J. J., S. R. COUGHLIN & L. T. WILLIAMS. 1986. Science **234:** 1117–1119.
5. ZACHARY, I., J. MILLAR, E. NANBERG, T. HIGGINS & E. ROZENGURT. 1987. Biochem. Biophys. Res. Commun. **146:** 456–463.
6. KAVANAUGH, W. M., L. T. WILLIAMS, H. E. IVES & S. R. COUGHLIN. 1988. Mol. Endocrinol. **2:** 599–605.
7. VAN CORVEN, E. J., A. GROENINK, K. JALINK, T. EICHHOLTZ & W. H. MOOLENAAR. 1989. Cell **59:** 45–54.
8. HUNTER, T. & J. A. COOPER. 1985. Annu. Rev. Biochem. **54:** 897–930.
9. LEE, P. L., D. E. JOHNSON, L. S. COUSENS, V. A. FRIED & L. T. WILLIAMS. 1989. Science **245:** 57–60.

10. RUTA, M., W. BURGESS, D. GIVOL, J. EPSTEIN, N. NEIGER, J. KAPLOW, G. CRUMLEY, C. DIONNE, M. JAYE & J. SCHLESSINGER. 1989. Proc. Nat. Acad. Sci. USA **86:** 8722–8726.
11. KUO, M.-D., S. S. HUANG & J. S. HUANG. 1990. J. Biol. Chem. **265:** 16455–16463.
12. SEUWEN, K., I. MAGNALDO & J. POUYSSEGUR. 1988. Nature **335:** 254–256.
13. VAN OBBERGHEN-SCHILLING, E., V. VOURET-CRAVIARI, R. J. HASLAM, J. C. CHAMBARD & J. POUYSSEGUR. Mol. Endocrinol. **5.** (In press.)
14. PARIS, S. & J. POUYSSEGUR. 1990. J. Biol. Chem. **265:** 11567–11575.
15. BERRIDGE, M. J. & R. F. IRVINE. 1989. Nature **341:** 197–205.
16. NISHIZUKA, Y. 1988. Nature **334:** 661–665.
17. COCKCROFT, S. 1987. Trends Biochem. Sci. **12:** 75–78.
18. MARGOLIS, B., A. ZILBERSTEIN, C. FRANKS, S. FELDER, S. KREMER, A. ULLRICH, S. G. RHEE, K. SKORECKI & J. SCHLESSINGER. 1990. Science **248:** 607–6110.
19. MAGNALDO, I., G. L'ALLEMAIN, J. C. CHAMBARD, M. MOENNER, D. BARRITAULT & J. POUYSSEGUR. 1986. J. Biol. Chem. **261:** 16916–16922.
20. NANBERG, E., C. MORRIS, T. HIGGINS, F. VARA & E. ROZENGURT. 1990. J. Cell. Physiol. **143:** 232–242.
21. BROWN, K. D., D. M. BLAKELEY & D. R. BRIGSTOCK. 1989. FEBS Lett. **247:** 227–231.
22. PARIS, S., J. C. CHAMBARD & J. POUYSSEGUR. 1988. J. Biol. Chem. **263:** 12893–12900.
23. OLSEN, R., K. SANTONE, D. MELDER, S. G. OAKES, R. ABRAHAM & G. POWIS. 1988. J. Biol. Chem. **263:** 18030–18035.
24. PANDIELLA, A. & J. MELDOLESI. 1989. J. Biol. Chem. **264:** 3122–3130.
25. EXTON, J. H. 1990. J. Biol. Chem. **265:** 1–4.
26. LARRODERA, P., M. E. CORNET, M. T. DIAZ-MECO, M. LOPEZ-BARAHONA, I. DIAZ-LAVIADA, P. H. GUDDAL, T. JOHANSEN & J. MOSCAT. 1990. Cell **61:** 1113–1120.
27. PESSIN, M. S., J. J. BALDASSARE & D. M. RABEN. 1990. J. Biol. Chem. **265:** 7959–7966.
28. TSUDA, T., K. KAIBUCHI, Y. KAWAHARA, H. FUKUZAKI & Y. TAKAI. 1985. FEBS **191:** 205–210.
29. MAGNALDO, I., J. POUYSSEGUR & S. PARIS. 1989. Cell. Signal. **1:** 507–517.
30. ROZENGURT, E., M. MURRARY, I. ZACHARY & M. COLLINS. 1987. Proc. Nat. Acad. Sci. USA **84:** 2282–2286.
31. MAGNALDO, I., J. POUYSSEGUR & S. PARIS. 1988. Biochem. J. **253:** 711–719.
32. BALL, R. L., K. D. TANNER & G. CARPENTER. 1990. J. Biol. Chem. **265:** 12836–12845.
33. SARDET, C., L. COUNILLON, A. FRANCHI & J. POUYSSEGUR. 1990. Science **247:** 723–726.
34. ROSSOMANDO, A. J., D. M. PAYNE, M. J. WEBER & T. W. STURGILL. 1989. Proc. Nat. Acad. Sci. USA **86:** 6940–6943.
35. L'ALLEMAIN, G., J. POUYSSEGUR & M. J. WEBER. Cell Regul. (In press.)
36. KOZMA, S. C., S. FERRARI & G. THOMAS. 1989. Cell. Signal. **1:** 219–225.
37. STURGILL, T. W., L. B. RAY, E. ERIKSON & J. MALLER. 1988. Nature **334:** 715–718.
38. ANDERSON, N., J. MALLER, N. TONKS & T. STURGILL. 1990. Nature **343:** 651–652.
39. BOULTON, T., G. YANCOPOULOS, J. GREGORY, C. SLAUGHTER, C. MOOMAW, J. HSU & M. COBB. 1990. Science **249:** 64–66.
40. MELOCHE, S., G. PAGES & J. POUYSSEGUR. EMBO J. (Submitted.)

Expression of Modified bFGF cDNAs in Mammalian Cells

REIKO SASADA, MASAHARU SENO,
TATSUYA WATANABE, AND KOICHI IGARASHI

Biotechnology Research Laboratories
Takeda Chemical Industries, Ltd.
Yodogawa-ku
Osaka 532, Japan

INTRODUCTION

High-level expression of basic fibroblast growth factor (bFGF) yields a transformed phenotype in mammalian cells.[1-7] The absence of a typical amino-terminal signal peptide sequence agrees with the result that very little bFGF is released from the bFGF-producing cells.[8,9] Addition of foreign signal sequence to bFGF failed in the efficient secretion of active bFGF, while the transformation potential of bFGF was raised.[3,4]

Basic FGF contains four cysteine residues, and two of them are well conserved among all members of the FGF family.[10,11] We previously reported the modification of human bFGF by the substitution of cysteine residues by serine using site-directed mutagenesis.[12] The results indicated that Cys^{69} and Cys^{87} of bFGF are not essential for its biological activities and mainly participate in the formation of intermolecular disulfide bonds. The modified bFGF, CS23, was revealed to conserve the same biological activity as that of original bFGF and to be more stable, especially under acidic conditions.

Here we report the effect of the modification, bFGF to CS23, on cellular transformation and on secretion of the molecule using mouse BALB/3T3 cells transfected with bFGF/CS23 cDNA. We also describe the expression of the glycosylated bFGF, CN3, in which an N-glycosylation site is introduced by the substitution of Cys^{87} to Asn.

RESULTS AND DISCUSSION

Expression of CS23 cDNA

To compare the expression of modified bFGF, CS23, cDNA in mammalian cells with that of original bFGF cDNA, eight plasmids were constructed. In these expression plasmids, cDNAs encoding bFGF or CS23 were under the control of MuLV LTR and SV40 early region promoter.[13] As shown in FIGURE 1, plasmid A (pTB1008) or A/CS23 (pTB1081) contained a cDNA encoding the mature form of bFGF or CS23 (146 residues), which was designed to be expressed directly with the initiation codon. Plasmid B (pTB999) or B/CS23 (pTB1000) had a chimeric gene created by linking the human IL-2 leader sequence to the cDNA. We exploited the IL-2 signal peptide for the construction of expression plasmids because it was useful to secrete human EGF outside of the fibroblast cells.[13] Plasmid C (pTB1085) or

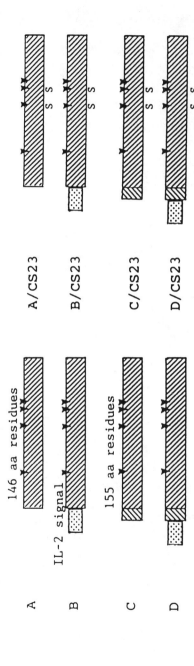

FIGURE 1. Constructs for bFGF/CS23 expression in mammalian cells. All the constructs were under the control of SV40 early promoter and MuLV LTR as an enhancer.[13] Expression plasmids A to D/CS23 were constructed by ligation with appropriate DNA fragments excised from pTB732,[7] pTB503, pTB506,[13] pTB669,[23] and pTB762,[12] using standard procedure.[24] Plasmid A means pTB1008; B, pTB999; C, pTB1085; D, pTB1001; A/CS23, pTB1080; B/CS23, pTB1000; C/CS23, pTB1167; and D/CS23, pTB1079. ▨, IL-2 signal peptide[13]; ▨, 146 amino acid residues of mature bFGF (Pro[1] to Ser[146]) with Met for initiation; ▨, 9 amino acid residues of pro-sequence of bFGF (Met[−9] to Ala[−1]); arrowheads indicate the positions of cysteine residue in bFGF, and the letter S represents the Ser substituted for Cys.

C/CS23 (pTB1167) contained cDNA encoding the 155-residue form of bFGF or CS23 (Met^{-9} to Ser^{146}), and the same cDNA was fused to the IL-2 leader sequence to construct plasmid D (pTB1001) or D/CS23 (pTB1079).

The plasmids carrying the eight forms of bFGF/CS23 were introduced into mouse BALB/3T3 cells. As shown in FIGURE 2, plasmid A or A/CS23 without signal sequence induced no detectable foci. Transfection with the plasmid B for bFGF (146 residues) fused to signal peptide also showed little focus formation. In contrast, morphologically transformed cells appeared in 3 days after transfection with plasmid B/CS23, and many foci were observed at an efficiency of over 100 foci/10 μg DNA per 10^5 cells after 3 weeks. Morphological change of BALB/3T3 cells was obviously induced with plasmid C or C/CS23 in 2 to 3 days after transfection. These "transformed" cells, however, disappeared in prolonged culture and only a few foci were observed on day 20. Transfection with plasmid D induced foci with distinctive morphology at an efficiency of about 30 foci/10 μg DNA per 10^5 cells. Transfection with plasmid D/CS23 induced foci at the highest frequency in this focus formation assay. Almost all the cells in the plate were morphologically transformed.

From the foci in each culture, transformed cells were cloned. In the case of plasmids A, B, and C, G418 resistant cells were also selected after cotransfection with neomycine resistant gene because these plasmids induced no or only a few foci.

The amount of bFGF synthesized by the transformed cells was assayed by the stimulation of [^{3}H]thymidine incorporation in quiescent BALB/3T3 cells (TABLE 1). Mitogenic activity was detected in the culture medium of K1000 (B/CS23) and K1079 (D/CS23) cells. Especially in the culture medium of K1079 cells more than 10 ng/ml per day of FGF activity was detected, while no significant activity was detected in the culture medium of transformed cells with bFGF cDNA with or without signal peptide sequence [K999 (B), K1001 (D), K1008 (A), and K1085 (C)]. No significant difference was observed on the levels of FGF activity detected in the cell lysate comparing K999 (B) with K1000 (B/CS23), 2 to 10 ng/10^6 cells, or K1001 (D) with K1079 (D/CS23), 25 to 50 ng/10^6 cells. The levels of FGF activity detected in the cell lysate of K1085 (C), K1001 (D), and K1079 (D/CS23) were 5- to 10-fold higher than those of K1008 (A), K999 (B), and K1000 (B/CS23), respectively.

The amount of bFGF estimated by two-site, enzyme immunoassay (EIA) was 3- to 5-fold higher than that estimated by mitogenic activity in every culture (culture medium or cell lysate) of transformed cells (TABLE 2). This EIA system was specific for detecting biologically active bFGF/CS23, and molecules closely related to bFGF such as aFGF[14,15] or HST-1/KS3[16,17] had no effect in this assay system. The results of both mitogenic assay and EIA can be summarized in two points: (1) The transformed cells expressing CS23 fused to signal peptide [K1000 (B/CS23) and K1079 (D/ CS23)] efficiently released biologically active molecules into the culture medium, suggesting that the modification of bFGF to CS23 increased the secretion of the biologically active molecules synthesized. (2) The transformed cells expressing 155 amino acid residues of bFGF/CS23 [K1085 (C), K1001 (D), and K1079 (D/CS23)] expressed high levels of FGF activity in the cells, suggesting that the sequence of 9 amino acid residues preceding the 146-residue bFGF plays some functional role in synthesis or storage of the bFGF molecule.

To characterize bFGF/CS23 synthesized in various transformed cells, western blot analysis was performed on the cell lysates and the culture medium (FIGURE 3). In cell lysate (FIGURE 3, i), bFGF was not detected from parental BALB/3T3 cells. In the lysate of TCNO523 cells[7] established by transfecting whole bFGF cDNA, four species of molecules related to bFGF were detected as reported by Florkiewicz *et al.*[18] Cell lines K999-N1 (B) and K1000-F2 (B/CS23) produced a protein of 17 KDa indistinguishable from the mature form of bFGF (146 residues) in size. The cell lines

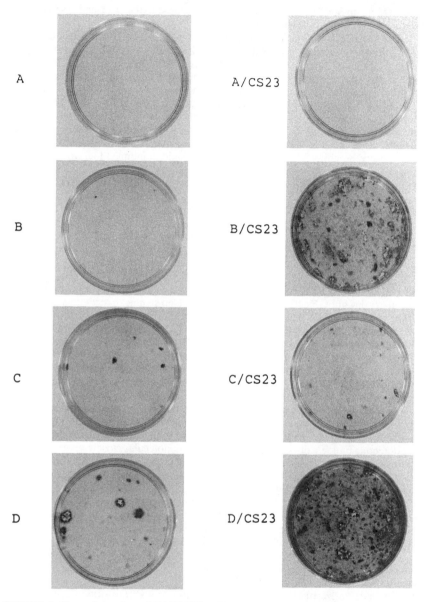

FIGURE 2. Focus formation of BALB/3T3 cells transfected with expression plasmids for bFGF/CS23. Ten micrograms of plasmid DNA was introduced into mouse BALB/3T3 clone A31-1-1 cells[25] (10^5 cells per 60-mm dish) by the calcium phosphate coprecipitation procedure.[26] The transfected cells were cultured in DMEM supplemented with 5% calf serum for 3 weeks and then photographed after crystal violet staining.

TABLE 1. FGF Activity of Transformed BALB/3T3 Cell Cultures

Plasmid	Clone	Activity[c]	
		Medium (ng/ml)	Cell Lysate (ng/10^6 cells)
A	K1008-N2[b]	<0.05	1.1
	K1008-N5	<0.05	1.0
B	K999-N1	<0.05	5.9
	K999-F1[a]	<0.05	12.2
B/CS23	K1000-F2	1.4	8.2
	K1000-F4	0.22	1.9
C	K1085-N2	<0.05	77.4
	K1085-N3	<0.05	16.8
D	K1001-F1	<0.05	29.3
	K1001-F2	<0.05	24.6
D/CS23	K1079-F2	17.0	38.2
	K1079-F3	10.8	40.9

[a]To establish cell lines, foci of transfected cultures were picked up, transferred into new cultures, and clonal cells with distinctive morphology were obtained by limiting dilution procedure, designated F.

[b]G418 resistant cells were selected with 500 μg/ml of Geneticin (G418, Gibco) in DMEM containing 10% calf serum after cotransfection with 10 μg of expression plasmid and 0.5 μg of the plasmid pTB6[7] having neomycine resistant gene, designated N.

[c]Transformed cells cultured to confluence in 60-mm dish were incubated for 24 hour in 5 ml of DMEM supplemented with 1% calf serum. After collecting the medium, cells were washed twice with phosphate-buffered saline, scraped, and suspended in 20 mM Tris-HCl (pH7.6)/1 M NaCl (10^7 cells/ml). The cells were lysed by sonication and kept on ice for 1 hour. Then cell lysates were clarified by centrifugation at 15,000 rpm for 15 minutes. Mitogenic activity was assayed by the incorporation of [^{3}H]thymidine in resting BALB/3T3 cells as described previously.[7]

TABLE 2. Immunoreactive bFGF Produced by Transformed BALB/3T3 Cells

Plasmid	Clone	Immunoreactive bFGF[a]		Ratio[b] (medium/cell)
		Medium (ng/ml)	Cell Lysate (ng/10^6 cells)	
A	K1008-N2	<0.1	3	—
B	K999-N1	<0.1	17	—
B/CS23	K1000-F2	7.8	18	0.27
C	K1085-N2	0.36	320	0.001
D	K1001-F2	0.32	115	0.001
D/CS23	K1079-F3	33.0	205	0.12
E[c]	TCNO523	0.50	400	0.001
—	BALB/3T3	<0.1	<1	—

[a]Enzyme immunoassay was performed as described[21] with some modification. Briefly, 96-well immunoplates were coated with a mixture of two distinct monoclonal antibodies to bFGF, followed by the addition of samples. After appropriate incubation, MAb3H3 labeled with horseradish peroxidase was added and the peroxidase reaction was measured at 415 nm. The samples were the same as those assayed in TABLE 1.

[b]The "ratio" was calculated from the values of FGF activity obtained per each dish.

[c]Plasmid E (pTB732) contains whole bFGF cDNA under the control of MT-I promoter. TCNO523 cells and pTB732 were described previously.[7]

K1001-F2 (D) and K1079-F3 (D/CS23) exhibited a protein of 18 KDa corresponding to the 155-residue form of bFGF. These results suggested that the IL-2 signal peptide sequence was processed out as expected. The relative intensities of the signals on the nitrocellulose filter were in good correlation with the levels of bFGF/CS23 measured by mitogenicity or EIA. In the culture medium (FIGURE 3, ii), no clear immunoreactive band of predicted molecular size was observed from various transformed cells including K1079-F3 cells (D/CS23). Another broad signal of a high molecular weight species (30 to 40 KDa) was detected in the culture medium of transformed cells, K1000-F2 (B/CS23), K1001-F2 (D), and K1079-F3 (D/CS23). These high molecular weight species were not observed in the culture medium of TCNO523 cells (FIGURE 3, lane h) and K1085-N3 (C) cells (data not shown). These results suggest that bFGF/CS23 fused to IL-2 signal sequence may be processed and modified posttranslationally during secretion through the ordinary pathway. This high molecular weight bFGF seems to be the same species found in the culture medium of NIH3T3 cells transfected with an expression plasmid for chimeric bFGF fused to hGH signal peptide.[5]

FGF activity in the culture medium of K1079-F3 (D/CS23) was about 100-fold higher than that of K1001-F2 (D), TCNO523, or K1085-N3 (C) cells (TABLES 2 and 3). However, no difference between K1001-F2 and K1079-F3 was observed in the pattern of signals obtained by western blot in the culture medium. These results indicated that the intensity of the signals of this high molecular weight species on western blot did not reflect the levels of bFGF/CS23 determined by the mitogenic

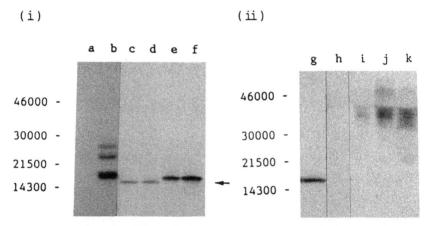

FIGURE 3. Western blot analysis of bFGF produced by transformed BALB/3T3 cells. Confluent cells in a 60-mm dish were cultured for 24 hours in DMEM supplemented with 1% calf serum. After the culture medium was collected, the cells were harvested and suspended in 20 mM Tris-HCl (pH7.6)/1 M NaCl (10^7 cells/ml). Then the cell lysate was prepared by brief sonication and centrifugation at 15000 rpm for 15 minutes. The cell lysate (100 μl) and the culture medium (1 ml) were incubated with anti-bFGF rabbit antiserum (1:50 dilution) for 20 hours at 4°C. The immunoprecipitates were recovered with protein A–Sepharose (Pharmacia), electrophoresed in a 17.5% SDS-PAGE[27] and transferred to nitrocellulose filters. Proteins blotted onto the filter were detected with the monoclonal antibody MAb3H3 using the Proto Blot Western Blot AP System (Promega, USA). **i:** Cell lysate. **ii:** Culture medium. **Lanes a,** BALB/3T3; **b** and **h,** TCN0523; **c,** K999-N1; **d** and **i,** K1000-F2; **e** and **j,** K1001-F2; and **f** and **k,** K1079-F3. The arrow shows the position of recombinant bFGF (146 residues).

TABLE 3. Expression of Modified bFGF CN3 cDNA in COS7 Cells[a]

| | FGF Activity[c] (ng/dish) | | Ratio |
Plasmid[b]	Medium	Cell Lysate	M/C
A	8	100	0.08
B	6	46	0.13
A/CN3	46	246	0.18
B/CN3	64	110	0.65

[a]COS7 cells[22] (3×10^5 cells) were transfected with 10 μg of each expression plasmid DNA per 60-mm dish by calcium phosphate precipitation procedure and the medium was changed to DMEM containing 0.5% fetal calf serum on the next day. The culture medium and cells were harvested after cultivation for 2 days, and subjected for assay.

[b]See FIGURE 1 and FIGURE 5 for plasmid A, B, A/CN3, and B/CN3.

[c]See the legends to TABLE 1 for the assay on mitogenicity.

assay, which showed good correlation with the amount of immunoreactive bFGF estimated by EIA. CS23 and bFGF were suggested to be modified through the secretion pathway in the biologically active and inactive forms respectively. Two cysteine residues, Cys[69] and Cys[87], on bFGF may be coupled with other cellular components by disulfide bond. Without these cysteine residues, CS23 molecules may be released freely, which enhances the secretion of the molecule as a biologically active form.

All the BALB/3T3 cell lines expressing bFGF/CS23 cDNA were morphologically transformed. Degree of the morphological alterations correlated to the amount of bFGF/CS23 synthesized in each cell line. These cells grew well in serum-free medium and formed colonies in soft agar (data not shown). The significant difference in the growth properties *in vitro* was not observed among the transformed cells examined. K1000-F2 (B/CS23), K1001-F2 (D), and K1079-F3 (D/CS23) cells induced tumors with progressive growth in all the nude mice inoculated with 10^6 cells subcutaneously, while K999-N1 (B), K1085-N3 (C), and TCNO523 cells did not. The tumorigenicity correlated well with the frequency of focus formation. These results suggest that the transformed cells acquired tumorigenicity by expressing bFGF/CS23 fused to IL-2 signal peptide. In addition, the tumorigenicity was probably enhanced by the substitution of Cys to Ser.

MAb3H3, monoclonal antibody against bFGF, is revealed to neutralize the mitogenic activity of bFGF/CS23 on human umbilical vein endothelial (HUVE) cells. MAb3H3 rather stimulated the actions of exogenous bFGF on BALB/3T3 cells. This antibody, however, repressed these actions of bFGF in the presence of heparin (unpublished observation). As shown in FIGURE 4, the morphology of the transformed K1000-N2 cells (B/CS23), which released a low level of FGF activity as much as 0.1 ng/ml (assayed by EIA) and showed tumorigenicity in nude mice, appeared to be reversed to the normal flat shape by MAb3H3 with heparin. Without heparin, clear reversion was not observed by this antibody. The role of heparin on the neutralizing effect of MAb3H3 is not obvious. These results suggest that the bFGF/CS23 expressed in transformed cells acts extracellularly. The transformed phenotype of K1000-F2 (B/CS23) or K1079-F3 (D/CS23) cells was hardly reversed by MAb3H3 even in the presence of heparin. In these transformed cells CS23 is produced in greater quantities and the antibody could not neutralize the activity. K1001-F2 cells (D) were also not reversed to the normal morphology effectively while the transformed phenotype of TCNO523 cells and K1085-N3 cells (C) was

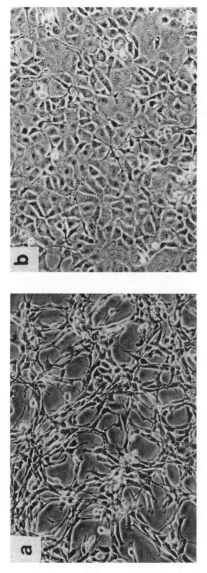

FIGURE 4. Effect of anti-bFGF antibody on the morphology of transformed BALB/3T3 cells. K1000-N2 (B/CS23) cells in a 24-well plate were cultured in DMEM containing 5% calf serum (a), supplemented with MAb3H3 (10 µg/ml) in the presence of heparin (20 µg/ml) (b). On day 4, the cultures were replaced by the same medium, and the cells were photographed on day 7.

repressed. In these cell lines, however, similar levels of FGF activity were detected in the culture medium (TABLE 2), suggesting that the molecules released from K1001 (D) may receive some modifications that inhibit the interaction with bFGF antibody to bFGF.

(I)

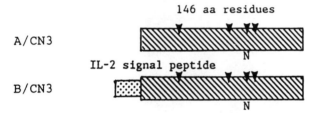

(II)

FIGURE 5. **i:** Modification of bFGF to CN3. **ii:** Structure of expression constructs for CN3. For constructing CN3 expression plasmid, a DNA fragment from pTB1001 containing bFGF cDNA encoding 155 residues with IL-2 leader sequence was cloned into pUC119. Using synthetic oligonucleotide and site-directed mutagenesis techniques, the sequence coding 9 residues (Met^{-9} to Leu^{-1}) of the pro-region of bFGF were precisely deleted and then the codon for Cys^{87} was changed for Asn. To construct an expression plasmid B/CN3 (pTB1163), the modified cDNA encoding CN3 fused to IL-2 signal was cloned into the expression vector as described in the legend to FIGURE 1. Plasmid A/CN3 (pTB1140) was constructed by deleting the IL-2 leader sequence of pTB1163.

Expression of CN3 cDNA

Several biologically active polypeptides were reported to be stabilized by glycosylation.[19,20] bFGF has no N-glycosylation site, and native molecule is not glycosylated.[10,11] To create a glycosylated molecule of bFGF, an N-glycosylation site was introduced into the bFGF by substituting the Cys^{87} to Asn, which is speculated to be exposed outside of the molecule.[12] Two expression plasmids for CN3 (A/CN3; pTB1140) and CN3 fused to IL-2 signal peptide (B/CN3; pTB1163) were constructed (FIGURE 5). These plasmids were expressed transiently in COS7 cells and the mitogenic activity of the transfected cultures was assayed by [³H]thymidine incorporation in BALB/3T3 cells. As shown in TABLE 3, about 10 ng/ml of FGF activity was detected in the culture medium of COS7 transfected with the plasmids. The level of FGF activity was 5- to 10-fold higher than that released from COS7 cells transfected with plasmid A and B having bFGF cDNA. COS7 cells transfected with plasmid B/CN3 having cDNA of signal-peptide-fused CN3 seemed to release modified bFGF efficiently. The products in the conditioned medium were analyzed by sodium dodecyl sulfate-polyacrylamide gel electrophoresis (SDS-PAGE) and western blotting using MAb3H3 (FIGURE 6). Several immunoreactive proteins of about 20 to 23 kDa were observed mainly in the medium of COS7 transfected with plasmid B/CN3. Simultaneously, a series of molecules with 30 to 50 kDa, which seem to be the same species found in the culture medium of transformed BALB/3T3 cells transfected with the plasmids for bFGF/CS23 fused to the signal peptide, were observed (FIGURE 6, lane 3). In the presence of tunicamycin the proteins around 20 kDa disappeared and only an immunoreactive protein of 17 KDa, the same size as

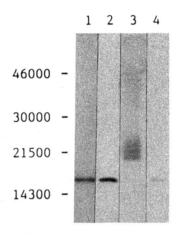

FIGURE 6. Western blot analysis of glycosylated bFGF CN3 in the culture medium of COS7 cells transfected with CN3 expression plasmid. Molecules in the conditioned medium of COS7 transfected with plasmid A/CN3 (pTB1140) or B/CN3 (pTB1163) were precipitated with 5% trichloroacetic acid, washed twice with cold ethanol, and subjected to SDS-PAGE. Western blot was performed using MAb3H3. **Lanes 1,** recombinant human bFGF[23]; **2,** plasmid A/CN3; **3,** plasmid B/CN3; and **4,** plasmid B/CN3 in the presence of 0.5 μg/ml of tunicamycin.

recombinant bFGF, was recognized (FIGURE 6, lane 4). The 17-kDa protein was also detected in the culture medium of COS7 cells transfected with plasmid A/CN3 (FIGURE 6, lane 2). These results indicate that the addition of signal sequence is necessary for glycosylation of CN3 and that the molecules of about 20 kDa were really glycosylated. This CN3 of 20 kDa was likely to be biologically active (TABLE 3). The glycosylated bFGF, CN3, seems to be released efficiently from the cells. Transfection of BALB/3T3 cells with plasmid B/CN3 induced foci at a high frequency, while no foci were observed with plasmid A/CN3 (data not shown). Several transformed cell lines with typical transformed phenotypes were established from the foci of BALB/3T3 cells transfected with the plasmid B/CN3. These cells secreted biologically active CN3 (0.5 to 1.5 ng/ml) in the culture medium. These results suggest that the CN3 modification increased the secretion of biologically active molecule and resulted in enhancement of the transformation potential as in the case of CS23 modification. Further characterization of glycosylated CN3 is now in progress.

CONCLUSIONS

CS23 and CN3 modification drastically increased the secretion of the FGF molecules from mammalian cells when both were linked to IL-2 signal sequence and the secreted molecules were biologically active, which resulted in the apparent enhancement of transforming potential of the molecules. In contrast, expression of original bFGF cDNA in the same manner also caused increased secretion but the molecules did not have full activity. These results suggest that Cys^{69} and Cys^{87} of bFGF may be involved in interaction with other cellular components by disulfide bonding, which may regulate bFGF activity in physiological conditions.

ACKNOWLEDGMENTS

We thank Drs. A. Kakinuma for his encouragement, A. Hori for providing MAb3H3 and tumorigenicity assay, H. Watanabe for EIA, and Ms. K. Ishimaru for her excellent technical assistance.

REFERENCES

1. THOMAS, K. A. 1988. Trends Biochem. Sci. **13:** 327–328.
2. RIFKIN, D. B. & D. MOSCATELLI. 1989. J. Cell Biol. **109:** 1–6.
3. NEUFELD, G., R. MITCHELL, P. PONTE & D. GOSPODAROWICZ. 1988. J. Cell Biol. **106:** 1385–1394.
4. ROGELJ, S., R. A. WEINBERG, P. FANNING & M. KLAGSBRUN. 1988. Nature **331:** 173–175.
5. BLAM, S. B., R. MITCHELL, E. TISCHER, J. S. RUBIN, M. SILVA, S. SILVER, J. C. FIDDES, J. A. ABRAHAM & S. A. AARONSON. 1988. Oncogene **3:** 129–136.
6. MOSCATELLI, D. & N. QUARTO. 1989. J. Cell Biol. **109:** 2519–2527.
7. SASADA, R., T. KUROKAWA, M. IWANE & K. IGARASHI. 1988. Mol. Cell. Biol. **8:** 588–594.
8. ABRAHAM, J. A., A. MERGIA, J. L. WHANG, A. TUMOLO, J. FRIEDMAN, K. A. HJERRILD, D. GOSPODAROWICZ & J. C. FIDDES. 1986. Science **233:** 545–548.
9. KUROKAWA, T., R. SASADA, M. IWANE & K. IGARASHI. 1987. FEBS Lett. **213:** 189–194.
10. BURGESS, W. H. & T. MACIAG. 1989. Annu. Rev. Biochem. **58:** 575–606.
11. GOSPODAROWICZ, D., N. FERRARA, L. SCHWEIGERER & G. NEUFELD. 1987. Endocr. Rev. **8:** 95–114.
12. SENO, M., R. SASADA, M. IWANE, K. SUDO, T. KUROKAWA, K. ITO & K. IGARASHI. 1988. Biochem. Biophys. Res. Commun. **151:** 701–708.
13. SASADA, R., R. MARUMOTO & K. IGARASHI. 1988. Cell Struct. Funct. **13:** 129–141.
14. MACIAG, T., T. MEHLMAN, R. FRIESEL & A. B. SCHREIBER. 1984. Science **225:** 932–935.
15. WATANABE, T., M. SENO, R. SASADA & K. IGARASHI. 1990. Mol. Endocrinol. **4:** 869–879.
16. TAIRA, M., T. YOSHIDA, K. MIYAGAWA, H. SAKAMOTO, M. TERADA & T. SUGIMURA. 1987. Proc. Nat. Acad. Sci. USA **84:** 2980–2984.
17. DELLI-BOVI, P., A. M. CURATOLA, F. G. KERN, A. GRECO, M. ITTMANN & C. BASILICO. 1987. Cell **50:** 729–737.
18. FLORKIEWICZ, R. Z. & A. SOMMER. 1989. Proc. Nat. Acad. Sci. USA **86:** 3978–3981.
19. GROSS, V., K. STEUBE, T.-A. TRAN-THI, D. HAUSSINGER, G. LEGLER, K. DECKER, P. C. HEINRICH & W. GEROK. 1987. Eur. J. Biochem. **162:** 83–88.
20. SAIRAM, M. R. & G. N. BHARGAVI. 1985. Science **229:** 65–67.

21. SENO, M., M. IWANE, R. SASADA, N. MORIYA, T. KUROKAWA & K. IGARASHI. 1989. Hybridoma **8:** 209–221.
22. GLUZMAN, Y. 1981. Cell **23:** 175–182.
23. IWANE, M., T. KUROKAWA, R. SASADA, M. SENO, S. NAKAGAWA & K. IGARASHI. 1987. Biochem. Biophys. Res. Commun. **146:** 470–477.
24. MANIATIS, T., E. F. FRITSCH & J. SAMBROOK. 1982. Molecular Cloning: a Laboratory Manual. Cold Spring Harbor Laboratory. Cold Spring Harbor, N.Y.
25. KAKUNAGA, T. & J. D. CROW. 1980. Science **209:** 505–507.
26. GRAHAM, F. L. & A. J. VAN DER EB. 1973. Virology **52:** 456–467.
27. LAEMMLI, U. K. 1971. Nature **227:** 680–685.

Structural Diversity and Binding
of FGF Receptors

CRAIG A. DIONNE, MICHAEL JAYE, AND
JOSEPH SCHLESSINGER[a]

Molecular Biology Division
Rhône-Poulenc Rorer Central Research
680 Allendale Road
King of Prussia, Pennsylvania 19406

[a]*Department of Pharmacology*
New York University Medical Center
550 First Avenue
New York, New York 10016

Fibroblast growth factors (FGFs) constitute a family of seven closely related polypeptide mitogens with profound biological activities.[1] They are relatively small (15–29 kDa) heparin binding proteins which are mitogenic for a wide variety of cell types from mesenchymal and neuroectodermal origin. Acidic FGF (aFGF) and basic FGF (bFGF), the two best characterized members of the FGF family, were the first to be cloned[2,3] and have been shown to possess neurotrophic and angiogenic activities in addition to their mitogenic properties. Keratinocyte growth factor (KGF) may be a specific mitogen for cells of epithelial origin since it promotes growth of a limited spectrum of cell types.[4] The other FGFs (FGF-5,[5] FGF-6,[6] int-2,[7] and hst/KGF[8,9]) are expressed in specific temporal and spatial patterns in the developing embryo and in certain types of cancers.[10–13] They are not generally found in adult tissues but they appear to have a spectrum of activities similar to aFGF and bFGF *in vitro*.

The large number of FGF ligands and the diversity of their biological effects have prompted much effort toward isolation and characterization of FGF receptors. Three different tyrosine kinase linked FGF receptors have been identified (flg, bek, and CEK-2) and the presence of others has been indicated.[14] Interestingly, all three gene products were initially isolated on the basis of their tyrosine kinase domains. A partial cDNA for human flg (fms-like gene) was first isolated by low stringency hybridization with a cDNA probe corresponding to the tyrosine kinase domain of the CSF-1 receptor.[15] Flg was described as a bFGF receptor when the chicken protein was isolated by bFGF affinity chromatography[16] and demonstrated to be a receptor for both aFGF and bFGF when antibodies to the human flg protein became available.[17]

A partial cDNA for bek (bacterially expressed kinase) was first isolated by phosphotyrosine antibody screening of a mouse liver expression library.[18] Full length cDNAs for human and chicken bek[19–21] were later described, and quantitative binding analyses on recombinant bek overexpressing cell lines demonstrated its activity as a high-affinity receptor for both aFGF and bFGF.[19,21]

Full-length clones for CEK-2 (chicken embryo kinase) as well as chicken flg and bek were also isolated by phosphotyrosine antibody screening of a chicken embryo expression library.[20,22] Although high-affinity binding of CEK-2 to FGFs has not yet been reported, its extremely high homology to flg and bek suggests similar function.

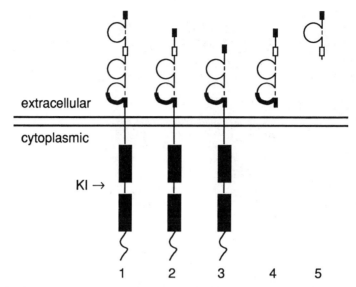

FIGURE 1. Diversity of FGF receptor forms. The diversity of FGF receptor forms probably arises in part from transcription of four different receptor genes, each of which may produce alternatively spliced transcripts to yield alternate protein forms. Some of the isolated forms are indicated. The extracellular regions consist of one to three Ig-like domains and may contain a hydrophilic "acid box" (open box) in the region between the first and second Ig-like domains. All forms begin with the same short sequence at the N terminus (small shaded box) which was probably contained in the first coding exon. The cytoplasmic region, when present, consists of a kinase domain (large shaded box) with a kinase insert (KI). There is alternative usage of exons corresponding to the second half of the third Ig-like domains (thick line). References for the various forms are: Form 1: flg,[16,19,21,22,25-27] bek,[19-21] CEK-2.[20] Form 2: flg,[25,29-31] bek.[32] Form 3: bek.[33,34] Form 4: flg.[30] Form 5: bek.[34]

The cloning and sequencing of the FGF receptor cDNAs has revealed consider-able information concerning the overall structure of the FGF receptor proteins (FIGURE 1). The longest receptor forms predicted by the cDNAs contain 3 immuno-globulin (Ig) like domains in the extracellular ligand binding region of the receptor (FIGURE 1, form 1). The Ig-like domains, which indicate that the FGF receptors belong to the Ig superfamily,[23,24] are inferred from the presence and location of cysteines and other specific conserved amino acids.[16,19-22,25-28] Between the first and second Ig domains is a region of high concentration of acidic residues. This "acid box" is especially noticeable in flg but is also found in bek and CEK-2. The receptors contain 8-9 N-linked glycosylation sites, at least some of which are utilized, since tunicamycin treatment of cells results in lower apparent molecular weight forms of the receptors.[19] A single transmembrane segment joins the extracellular region to the cytoplasmic region which is notable in that it contains a relatively long juxtamem-brane sequence (~ 70 amino acids) and a 14 amino acid insertion in the conserved tyrosine kinase domain.

Each FGF receptor exhibits different forms, especially in the extracellular region, which are apparently the result of alternative splicing events. A predominant alternative form, described for both flg[25,29-31] and bek,[32] is deleted of the first Ig domain but is functional in binding (FIGURE 1, form 2). Another form missing both the first Ig-like domain and "acid box" linker region has so far only been described for products of the BEK gene (FIGURE 1,[33,34] form 3). This form is the smallest extracellular domain yet reported to be functional in high-affinity binding to the FGFs. Other cDNAs that predict soluble, secreted forms of the receptors have also been described. Form 4 which contains the "acid box" and Ig domains 2 and 3 would be expected to bind FGFs.[30] The biological role of this form is unclear, but it may bind FGFs and down regulate FGF signaling or it may function as a carrier molecule to transport FGFs to their proper site of action. The biological function(s) of form 5, which contains only the first Ig-like domain and acid box,[34] is not yet known, but cDNAs which code for this form are relatively abundant in a human neonatal brain-stem library, suggestive of high levels of expression in that issue.

In addition to splicing events that include or exclude entire Ig domains, there is also apparent alternative exon usage in comprising the second half of the third Ig-like domain. Two alternative exons in this region have been noted for bek tyrosine kinase linked forms.[19–21,32–34] Usage of one exon apparently confers high-affinity binding to aFGF and bFGF,[19,21,34] with no information yet reported for KGF binding. Usage of a different exon confers high-affinity binding to aFGF and KGF but not bFGF.[33] In a similar way, an alternative sequence encoding the second half of the third Ig-like domain of flg leads to a secreted form rather than a tyrosine kinase linked receptor.[30] If the BEK and FLG genes are completely analogous, then it would be expected that this event may generate a similar secreted form of the bek extracellular domain. These alternate forms of each receptor probably arise from alternative splicing of a single gene product rather than expression of different genes, since FLG and BEK have been identified as single-copy genes located at human chromosomes 8p12 and 10q25.3-q26, respectively.[15,35]

One major question concerning the FGF/FGF receptor interactions is that of ligand/receptor specificity. There are at least 7 FGFs and 4 tyrosine kinase linked FGF receptors which, due to alternative splicing, generate a highly diverse system. We have shown that the flg and bek forms that contain 3 Ig-like domains exhibit very high affinity for aFGF, bFGF, and hst/kFGF[19,36] (TABLE 1). Similar results were obtained by others for aFGF binding to bek[21] and bFGF and hst/kFGF binding to a flg form that is deleted of the first Ig-like domain.[29] Taken together, these results imply a very high level of redundancy in FGF receptor/ligand interactions in that multiple ligands can bind multiple receptors with essentially equal affinity. However, some level of binding specificity can be achieved by alternative splicing of single exons in the extracellular domain of the receptors. Ultimately, the regulation of binding will be achieved by the spatial and temporal specific expression of the ligand and receptor genes in the intact organism.

The elucidation of overall FGF receptor structure allows comparison to other

TABLE 1. Binding of Overexpressed FGF Receptors[a]

	Apparent Kds (pM)			
Cell Line	Receptors/Cell	aFGF	bFGF	hst/KFGF
NIH 3T3	5,000	60	ND	ND
NNeo4	5,000	60	ND	ND
NBek8	100,000	50	80	80
NF1g26	125,000	25	50	320

[a]The flg and bek transfected 3T3 cells, NF1g26 and NBek8, and the control neomycin-resistant NNeo4 cells were previously described.[19] The apparent dissociation constants were obtained by Scatchard analysis of equilibrium binding data.[19,36]

receptor tyrosine kinase linked receptors.[37] Receptor-linked tyrosine kinases can be characterized into four broad families according to their structure (TABLE 2). The first receptor tyrosine kinase (RTK) family, represented by the EGF receptor, consists of an extracellular domain containing two cysteine-rich regions, a single transmembrane region, and a cytoplasmic tyrosine kinase domain. The insulin receptor family (RTK II) are heterotetrameric structures consisting of two identical heterodimers. Each heterodimer consists of an extracellular chain containing two cysteine-rich regions which is disulfide bonded to another chain which has a single transmembrane and a cytoplasmic tyrosine kinase. The RKT III family, represented by the PDGF receptor, contains five Ig-like domains in its extracellular region, along with a single transmembrane sequence, and a cytoplasmic tyrosine kinase containing an insert of 66–104 amino acids in the middle of the kinase domain. The ligands for this family are dimeric molecules. FGF receptors comprise a fourth RTK family which, although similar to the RTK III family, has the distinct features described earlier.

The structural similarity of FGF receptors to other receptor-linked tyrosine kinases allows certain predictions to be made concerning their modes of signal transduction.[37] Ligand-induced dimerization of the receptors will be a necessary prerequisite for effective signal transduction. Consequently, signal transduction

TABLE 2. Receptor Tyrosine Kinase Families

Family	Receptor Tyrosine Kinases
RTK I	EGF-R, HER2/neu HER3.c-erbB-3, Xmrk
RTK II	Insulin-R, IGF-1-R, IRR
RTK III	PDGF-R-A, PDGF-R-B, CSF-1, c-kit, flt
RTK IV	flg, bek, CEK2

[a]The classification is adapted from Ullrich and Schlessinger[37] and is extended by the addition of CEK-2[20] and flt.[42]

might be inhibited in a "dominant negative fashion" by soluble forms of the extracellular domain of the receptor or by truncated forms that contain only enough of the cytoplasmic domain to anchor the extracellular region at the membrane surface. Also, the kinase activity should be necessary for signal transduction by the receptor. Preliminary evidence in our laboratories indicates that these predictions will indeed turn out to be valid.

One area of similarity between FGF receptors and other receptor-linked tyrosine kinases is in the substrates for tyrosine phosphorylation. The FGF receptors have been shown to activate tyrosine phosphorylation of several substrates in addition to phospholipase C-γ (PLC-γ).[38,39] The EGF and PDGF receptors specifically phosphorylate several proteins including PLC-γ and GAP, and the interaction of receptor with these substrates has been shown to occur between the C terminal region of the receptor and the SH2 domains of the substrate.[40,41] Preliminary evidence in our laboratories suggests that this general reaction mechanism is intact in the FGF receptor system, although affinities for the various substrates may be different than those for the EGF receptor.

In summary the FGF receptors exhibit characteristics that are similar to those of other receptor-linked tyrosine kinases. This is a significant advantage to researchers, because the previous work on the EGF and PDGF receptors has set up many paradigms that are easily extended into the study of the FGF receptors. However, elaboration of these paradigms may uncover signaling pathways that are unique to the FGF receptors. Finally, the extraordinary variation in the number of expressed forms of FGF receptors leads to a high level of redundancy in ligand/receptor interactions which implies exquisite regulation of FGF signaling.

REFERENCES

1. BURGESS, W. H. & T. MACIAG. 1989. Annu. Rev. Biochem. **58:** 575–606.
2. JAYE, M., R. HOWK, W. BURGESS, G. A. RICCA, I-M. CHIU, M. W. RAVERA, S. J. O'BRIEN, W. S. MODI, T. MACIAG & W. N. DROHAN. 1986. Science **233:** 541–545.
3. ABRAHAM, J. A., J. L. WHANG, A. TUMULO, A. MERGIA, J. FRIEDMAN, D. GOSPODAROWICZ & J. C. FIDDES. 1986. EMBO J. **5:** 2523–2528.
4. FINCH, P. W., J. S. RUBIN, T. MIKI, D. RON & S. A. AARONSON. 1989. Science **245:** 752–755.
5. ZHAN, X., B. BATES, X. HU & M. GOLDFARB. 1988. Mol. Cell. Biol. **8:** 3487–3495.
6. MARICS, I., J. ADELAIDE, F. RAYBAUD, M. G. MATTEI, F. COULIER, J. PLANCHE, O. DE LAPEYRIERE & D. BIRNBAUM. 1989. Oncogene **4:** 335–340.
7. DICKSON, C. & G. PETERS. 1987. Nature **326:** 833.
8. DELLI-BOVI, P., A. M. CURATOLA, F. G. KERN, A. GRECO, M. ITTMAN & C. BASILICO. 1987. Cell **50:** 729–737.
9. YOSHIDA, T., K. MIYAGAWA, H. ODAGIRI, H. SAKAMOTO, P. F. R. LITTLE, M. TERADA & T. SUGIMURA. 1987. Proc. Nat. Acad. Sci. USA **84:** 7305–7309.
10. ZHOU, D. J., G. CASEY & M. J. CLINE. 1988. Oncogene **2:** 279–282.
11. VARLEY, J. M., R. A. WALKER, G. CASEY & W. J. BRAMMAR. 1988. Oncogene **3:** 87–91.
12. LIDEREAU, R., R. CALLAHAN, C. DICKSON, G. PETERS, C. ESCOT & I. U. ALI. 1988. Oncogene Res. **2:** 285–291.
13. TSUDA, T., H. NAKATANI, T. MATSUMURA, K. YOSHIDA, E. TAHARA, T. NISHIHIRA, H. SAKAMOTO, T. YOSHIDA, M. TERADA & T. SUGIMURA. 1988. Jpn. J. Cancer Res. Gann **79:** 584–588.
14. PARTANEN, J., T. P. MAKELA, R. ALITALO, H. LEHVASLAIHO & K. ALITALO. 1990. Proc. Nat. Acad. Sci. USA **87:** 8913–8917.
15. RUTA, M., R. HOWK, G. RICCA, W. DROHAN, M. ZABELSHANSKY, G. LAUREYS, D. E. BARTON, U. FRANCKE, J. SCHLESSINGER & D. GIVOL. 1988. Oncogene **3:** 9–15.
16. LEE, P. L., D. E. JOHNSON, L. S. COUSENS, V. A. FRIED & L. T. WILLIAMS. 1989. Science **245:** 57–60.
17. RUTA, M., W. BURGESS, D. GIVOL, J. EPSTEIN, N. NEIGER, J. KAPLOW, G. CRUMLEY, C. DIONNE, M. JAYE & J. SCHLESSINGER. 1989. Proc. Nat. Acad. Sci. USA **86:** 8722–8726.
18. KORNBLUTH, S., K. E. PAULSON & H. HANAFUSA. 1988. Mol. Cell. Biol. **8:** 5541–5544.
19. DIONNE, C. A., G. CRUMLEY, F. BELLOT, J. M. KAPLOW, G. SEARFOSS, M. RUTA, W. H. BURGESS, M. JAYE & J. SCHLESSINGER. 1990. EMBO J. **9:** 2685–2692.
20. PASQUALE, E. B. 1990. Proc. Nat. Acad. Sci. USA **87:** 5812–5816.
21. HOUSSAINT, E., P. R. BLANQUET, P. CHAMPION-ARNAUD, M. C. GESNEL, A. TORRIGLIA, Y. COURTOIS & R. BREATHNACH. 1990. Proc. Nat. Acad. Sci. USA **87:** 8180–8184.
22. PASQUALE, E. B. & S. J. SINGER. 1989. Proc. Nat. Acad. Sci. USA **86:** 5449–5453.
23. WILLIAMS, A. F. & A. N. BARCLAY. 1988. Annu. Rev. Immunol. **6:** 381–405.
24. HUNKAPILLER, T. & L. HOOD. 1989. Adv. Immunol. **44:** 1–63.
25. REID, H. H., A. F. WILKS & O. BERNARD. 1990. Proc. Nat. Acad. Sci. USA **87:** 1596–1600.
26. SAFRAN, A., A. AVIVI, A. ORR-URTEREGER, G. NEUFELD, P. LONAI, D. GIVOL & Y. YARDEN. 1990. Oncogene **5:** 635–643.
27. MUSCI, T. J., E. AMAYA & M. W. KIRSCHNER. 1990. Proc. Nat. Acad. Sci. USA **87:** 8365–8369.
28. ISACCHI, A., L. BERGONZONI & P. SARMIENTOS. 1990. Nuleic Acids Res. **18:** 1906.
29. MANSUKHANI, A., D. MOSCATELLI, D. TALARICO, V. LAVYTSKA & C. BASILICO. 1990. Proc. Nat. Acad. Sci. USA **87:** 4378–4382.
30. JOHNSON D. E., P. L. LEE, J. LU & L. T. WILLIAMS. 1990. Mol. Cell. Biol. **10:** 4728–4736.

31. ITOH, N., T. TERACHI, M. OHTA & M. K. SEO. 1990. Biochem. Biophys. Res. Commun. **169:** 680–685.
32. HATTORI, Y., H. ODAGIRI, H. NAKATANI, K. MIYAGAWA, K. NAITO, H. SAKAMOTO, O. KATOH, T. YOSHIDA, T. SUGIMURA & M. TERADA. 1990. Proc. Nat. Acad. Sci. USA **87:** 5983–5987.
33. MIKI, T., T. P. FLEMING, D. P. BOTTARO, J. S. RUBIN, D. RON & S. A. AARONSON. 1991. Science **251:** 72–75.
34. CRUMLEY, G., F. BELLOT, J. KAPLOW, J. SCHLESSINGER, M. JAYE & C. A. DIONNE. Submitted.
35. DIONNE, C. A., W. S. MODI, G. CRUMLEY, S. J. O'BRIEN, J. SCHLESSINGER & M. JAYE. Submitted.
36. BELLOT, F., J. M. KAPLOW, G. CRUMLEY, C. BASILICO, M. JAYE, J. SCHLESSINGER & C. DIONNE. Submitted.
37. ULLRICH, A. & J. SCHLESSINGER. 1990. Cell **61:** 203–212.
38. BURGESS, W. H., C. A. DIONNE, J. KAPLOW, R. MUDD, R. FRIESEL, A. ZILBERSTEIN, J. SCHLESSINGER & M. JAYE. 1990. Mol. Cell Biol. **10:** 4770–4777.
39. FRIESEL, R., W. H. BURGESS, T. MEHLMAN & T. MACIAG. 1986. J. Biol. Chem. **261:** 7581–7584.
40. MARGOLIS, B., N. LI, A. KOCH, M. MOHAMMADI, D. R. HURWITZ, A. ZILBERSTEIN, A. ULLRICH, T. PAWSON & J. SCHLESSINGER. 1990. EMBO J. **9:** 4375–4380.
41. ANDERSON, D., C. A. KOCH, L. GREY, C. ELLIS, M. F. MORAN & T. PAWSON. 1990. Science **250:** 979–982.
42. SHIBUYA, M., S. YAMAGUCHI, A. YAMANE, T. IKEDA, A. TOJO, H. MATSUSHIME & M. SATO. 1990. Oncogene **5:** 519–524.

The Molecular Biology of Heparan Sulfate Fibroblast Growth Factor Receptors[a]

MICHAEL C. KIEFER, MASAYUKI ISHIHARA,[b]
STUART J. SWIEDLER,[b] KEVIN CRAWFORD,
JAMES C. STEPHANS, AND PHILIP J. BARR[c]

Chiron Corporation
4560 Horton Street
Emeryville, California 94608

[b]*Glycomed Inc.*
860 Atlantic Avenue
Alameda, California 94501

The ligand-affinity or "panning" method of expression cloning has been used extensively for the isolation of cDNAs that encode mammalian cell surface proteins. Since the initial development of the technique,[1,2] numerous cDNAs have been isolated based on the interactions of their encoded products with previously characterized monoclonal antibodies. Such cloned DNAs have included cDNAs that encode the cell surface antigens CD28,[1] CD2,[2] ELAM-1,[3] and CDw44.[4]

In an attempt to develop a general method for the isolation of growth factor receptor cDNAs using this technique, we targeted the molecular cloning of receptors for the fibroblast growth factors (FGFs). Members of this family of polypeptide growth factors are thought to have significant potential in several medically relevant areas, including soft tissue repair,[5] angiogenesis,[6] and neurogenesis[7] and for certain opthalmic indications.[8] More recently still, the FGF-FGF receptor system has been implicated in the entry of the herpes simplex virus (HSV) into susceptible cells.[9,10]

Currently, the FGF family includes the prototypic acidic and basic FGFs,[11–13] the product of the *int-2* protooncogene,[14,15] a factor isolated from sarcoma tissue DNA (*hst* or KS-FGF),[16,17] FGF-5,[18] FGF-6,[19] and the keratinocyte growth factor.[20] The acidic and basic FGFs are mitogenic for a number of cell types including those of mesenchymal, epithelial, or neural origin.[21,22] Consequently, studies directed toward *in vivo* wound healing and nerve regeneration have been initiated.[5,23]

The multiplicity of characterized FGFs and their diverse spectrum of activities have suggested the possibility that several receptors might exist for this growth factor family. Indeed, for the acidic and basic FGFs themselves, two classes of receptor have been described.[24–27] These receptor classes are distinguished by their differing affinities for FGF. For example, binding of basic FGF to a high-affinity site on baby hamster kidney (BHK) cells occurs with a dissociation constant in the range of 20 pM, whereas basic FGF binds to the low-affinity site with a dissociation constant of around 2 nM, and is released with 2 M sodium chloride.[27,28]

Much progress in the molecular characterization of the high-affinity FGF receptors has been reported. Since the initial discovery that these membrane-bound

[a]This work was supported by Chiron Corporation and Glycomed Inc.
[c]Author to whom correspondence should be addressed.

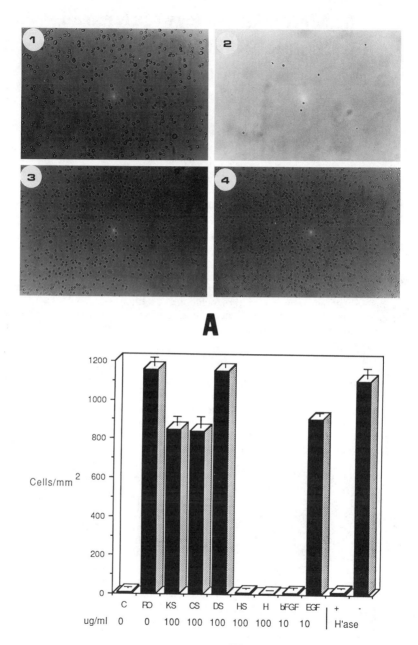

tyrosine kinases were encoded by the *FLG* gene (*fms*-like gene), the molecular biology of this receptor family has been well studied.[29-35] The complexity of this group of proteins is illustrated by the multiplicity of cDNAs that have been isolated using *FLG* probes.[29-35] Additional *FLG*-related cDNAs have also been shown to encode receptors that can bind FGF.[32,36]

In contrast, however, the molecular biology of the low-affinity receptors for the FGFs has not been well defined. Here, we describe the initial isolation of a cDNA encoding a cell surface heparan sulfate proteoglycan that binds basic fibroblast growth factor.[37] This cDNA was isolated from a hamster kidney cell line that had been shown previously to be a rich source of both low- and high-affinity FGF receptors. We also describe the characterization of a cDNA that encodes a human homologue of the hamster proteoglycan, isolated from a human liver cell line.

MOLECULAR CLONING OF A HAMSTER HEPARAN SULFATE PROTEOGLYCAN (HSPG) FGF RECEPTOR

Construction of a cDNA expression library from the BHK-21 cell line in the expression vector EBO-pCD-XN has been described previously.[37] The cDNA library was stably introduced into the human lymphoblastoid cell line WI-L2-729 HF$_2$ (ATCC CRL 8062) by electroporation and selection on hygromycin. The WI-L2-729 HF$_2$ cell line does not bind significantly to basic-FGF-coated culture dishes (FIGURE 1). Repeated panning (three times) of the stably transfected WI-L2-729 HF$_2$ cells on basic-FGF-coated culture dishes allowed for the enrichment of cells that contained two distinct families of recombinant plasmids with similar insert restriction patterns. WI-L2-729 HF$_2$ cells stably transfected with purified plasmids from each family were found to bind efficiently to basic-FGF-coated cultures dishes (FIGURE 1) whereas no binding was observed in areas of the dish not coated with basic FGF (data not shown). The binding was markedly reduced in the presence of free basic FGF, heparin, and heparan sulfate, whereas epidermal growth factor, platelet-derived growth factor, acidic FGF, chondroitin sulfate, dermatan sulfate, and keratan sulfate had little effect. Treatment of the transfected cells with heparinase abolished 98% of the binding (FIGURE 1B).[37]

Inserts from each subgroup (RO-12 and RO-5) were excised and sequenced. The two families were shown to encode identical proteins and differed only in the site of attachment of poly(A) sequences within their 3'-untranslated regions. RO-12 and RO-5 cDNA sequences differed only in that RO-5 was polyadenylated 14 residues beyond the internal polyadenylation signal. The encoded protein sequence is shown

---

FIGURE 1. **A:** Binding of HSPG-FGF-transformed WI-L2-729 HF$_2$ cells to basic FGF. Binding assays were performed essentially as described for panning.[37] Cells were photographed (1–4) or counted after the first panning cycle. Photographs are of BHK-21 cells (1), WI-L2-729 HF$_2$ cells (2), RO-5-transformed WI-L2-729 HF$_2$ cells (3), RO-12-transformed WI-L2-729 HS$_2$ cells (4). **B:** Binding of control (bar C) and RO-12-transformed WI-L2-729 HF$_2$ cells with no additions (bar RO) and in the presence of keratan sulfate (bar KS), chondroitin sulfate (bar CS), dermatan sulfate (bar DS), heparan sulfate (bar HS), heparin (bar H), basic FGF (bar bFGF), or epidermal growth factor (bar EGF) at the indicated concentration (μg/ml). Also shown are transformed cells pretreated with heparinase (H'ase; bar +) or heparinase buffer (bar −), respectively, as described.[27] The values shown represent the mean of three independent experiments. (Reprinted from Reference 37 with permission.)

(FIGURE 2A), and also a schematic representation of the structure of the hamster
HSPG-FGF receptor (FIGURE 3A).

MOLECULAR CLONING OF A HUMAN HOMOLOGUE OF
THE HAMSTER HSPG-FGF RECEPTOR

The hamster HSPG-FGF receptor cDNA (RO-5) was used as a probe to isolate a
human cDNA from a cDNA library that was constructed from RNA isolated from
the human hepatoma cell line Hep G2. Multiple positive signals from the plated
cDNA library and northern blot analysis indicated a moderate abundance of human
HSPG-FGR receptor mRNA. Restriction enzyme analysis of three isolated cDNAs
showed similar patterns for each clone. This suggested that each individual clone
encoded the same human HSPG-FGF receptor. Two hybridizing mRNA species of
approximately 3.2 kilobases (kb) and 2.4 kb were seen on the northern blot. The
same size mRNAs were identified previously in BHK-21 cells.[37] One clone (RO-4)
was analyzed by DNA sequencing, and the structure of the encoded protein is shown
(FIGURE 2B and 3B). We have noted previously[37] the homology of our hamster
HSPG-FGF receptor clone to a cDNA encoding murine syndecan,[38,39] an integral

(A)	MRRAALWLWLCALALRLQPVLPQIVTVNVPPEDQDGSGDDSDNFSGSGTGALPDITLSRQ	60
(B)	MRRAALWLWLCALALSLQPALPQIVATNLPPEDQDGSGDDSDNFSGSGAGALQDITLSQQ	60
(C)	MRRAALWLWLCALALSLQLALPQIVATNLPPEDQDGSGDDSDNFSGSGAGALQDITLSQQ	60
(D)	MRRAALWLWLCALALRLQPALPQIVAVNVPPEDQDGSGDDSDNFSGSGTGALPD.TLSRQ	59
(A)	ASPTLKDVWLLTATPTAPEPTSRDAQATTTSILPAAEKPGEGEPVLTAEVDPGFTARDKE	120
(B)	TPSTWKDTQLLTAIPTSPEPTGLEATAASTSTLPAGEGPKEGEAVVLPEVEPGLTAREQE	120
(C)	TPSTWKDTQLLTAIPTSPEPTGLEATAASTSTLPAGEGPKEGEAVVLPEVEPGLTAREQE	120
(D)	TPSTWKDVWLLTATPTAPEPTSSNTETAFTSVLPAGEKPEEGEPVLHVEAEPGFTARDKE	119
(A)	SEVTTRPRETTQLLITHWVSTARATTAQAPVTSHPHRDVQPGLHETSAPTAPGQPDQQPP	180
(B)	A..TPRPRETTQLPTTHQASTTTATTAQEPATSHPHRDMQPGHHETSTPAGPSQADLHTP	178
(C)	A..TPRPRETTQLPTTHQASTTTATTAQEPATSHPHRDMQPGHHETSTPAGPSQADLHTP	178
(D)	KEVTTRPRETVQLPITQRASTVRVTTAQAAVTSHPHGGMQPGLHETSAPTAPGQPDHQPP	179
(A)	.S..GGTSVIKE.VAEDGATNQLPTGEGSGEQDFTFETSGENTAVAAVEPDQRNQPPVDE	236
(B)	HTEDGGPSAT.ERAAEDGASSQLPAAEGSGEQDFTFETSGENTAVVAVEPDRRNQSPVDQ	237
(C)	HTEDGGPSAT.ERAAEDGASSQLPAAEGSGEQDFTFETSGENTAVVAVEPDRRNQSPVDQ	237
(D)	RVEGGGTSVIKE.VVEDGTANQLPAGEGSGEQDFTFETSGENTAVAAVEPGLRNQPPVDE	238
(A)	GATGASQGLLDRKEVLGGVIAGGLVGLIFAVCLVGFMLYRMKKKDEGSYSLEEPKQANGG	296
(B)	GATGASQGLLDRKEVLGGVIAGGLVGLIFAVCLVGFMLYRMKKKDEGSYSLEEPKQANGG	297
(C)	GATGASQGLLDRKEVLGGVIAGGLVGLIFAVCLVGFMLYRMKKKDEGSYSLEEPKQANGG	297
(D)	GATGASQSLLDRKEVLGGVIAGGLVGLIFAVCLVAFMLYRMKKKDEGSYSLEEPKQANGG	298
(A)	AYQKPTKQEEFYA	309
(B)	AYQKPTKQEEFYA	310
(C)	AYQKPTKQEEFYA	310
(D)	AYQKPTKQEEFYA	311

FIGURE 2. Comparison of the hamster HSPG-FGF receptor amino acid sequence (A) with
the homologous sequences encoded by our liver-derived human cDNA clone (B), human
syndecan (C), and murine syndecan (D). Regions of amino acid sequence that are identical in
all four core proteins are shaded. The predicted transmembrane domains are overlined, and a
predicted signal peptidase cleavage site is arrowed. Since the amino terminus of each mature
protein has not been defined experimentally, numbering is based on the sequence of each
proposed precursor protein.

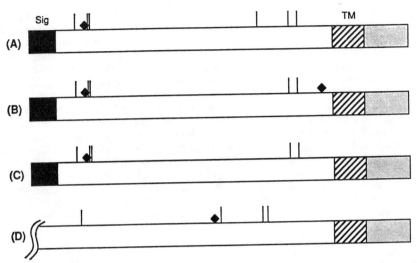

FIGURE 3. Schematic representations of the structures of the hamster HSPG-FGF receptor,[37] the human-liver-derived homologue (present study) (B), murine syndecan[38,39] (C), and the human lung fibroblast cell surface HSPG, 48K5, defined by a partial cDNA clone[42] (D). Proposed signal peptide sequences are noted (Sig), as are transmembrane domains (TM). Also highlighted, are the particularly conserved cytoplasmic domains (shaded) that have been implicated in the binding of actin.[38,40] Asparagine-linked glycosylation sites are marked (◆), and consensus GAG attachment sites[41] is shown as vertical bars. The fourth serine-glycine pair of the hamster core protein (A) is not surrounded by acidic residues, and consequently may not be utilized as a GAG attachment site.

membrane proteoglycan, supporting the proposal that syndecan is the murine equivalent of the HSPG-FGF receptor.[39] In further support of this notion, our encoded human core protein is almost identical to the human syndecan described by Mali *et al.*[40] The only amino acid difference between these two proteins occurs in the proposed signal peptide sequence (Pro19 to Leu) (FIGURE 2B and C), and may represent either a cloning artefact or a polymorphism.

STRUCTURE COMPARISON OF THE HSPG-FGF RECEPTOR AND HOMOLOGUES

An alignment of the amino acid sequences of the hamster HSPG-FGF receptor, the human liver-derived homologue, human syndecan and murine syndecan (FIGURE 2) reveals several highly conserved regions. The cytoplasmic domain shows 100% amino acid sequence identity in all of the molecules. This high level of conservation suggests an important role for the cytoplasmic domain and is probably related to its actin binding function.[38]

The transmembrane (TM) regions of the hamster and human proteins display 100% sequence identity whereas the murine TM differs by only one amino acid (Gly to Ala) resulting in a 96% sequence similarity. The signal peptides also display a high degree of homology ($\geq 91\%$). As was noted above, the single amino acid difference (Pro to Leu) between the human HSPG-FGF and human syndecan was found in this region. Thus, the mature core proteins are identical. The extracellular domains of the core proteins are less well conserved with a range of sequence identity between 68% and 81%. Most of the sequence divergence in this domain is located centrally with the flanking regions that contain the GAG attachment sites being more conserved (see FIGURE 3). The overall sequence identity between the hamster and mouse, the hamster and human, and the mouse and human proteins is 85%, 76%, and 77%, respectively.

The above-described domains of the core proteins—which include a signal peptide, an extracellular domain that contains consensus GAG attachment sites,[41] a transmembrane region, and a cytoplasmic domain homologous to previously reported actin binding domains—have been reported.[38-40] Schematic representations of the core proteins and these various domains are shown in FIGURE 3. Also shown (FIGURE 3D) is a diagram of the known sequence of 48K5.[42] This structure, derived from the sequence of a partial cDNA clone, corresponds to the human lung fibroblast HSPG of M_r 48,000.[42] Mali et al. have demonstrated a 56% identity between 48K5 and human syndecan in the transmembrane and cytoplasmic domains.[40] Also, Kiefer et al. noted 68% identity in the cytoplasmic domains of 48K5 and hamster HSPG-FGF receptor.[37] Particularly evident in these domains is a striking conservation of tyrosine residues.[40]

BIOCHEMICAL CHARACTERIZATION OF EXPRESSED HSPG-FGF RECEPTORS

In order to demonstrate that the expressed proteins serve as cell surface proteoglycan core proteins, we immunoprecipitated[35]SO_4-labeled proteins from trypsinized and nontrypsinized transfected cells with an antibody directed against a carboxyl-terminal peptide (FIGURE 4). Using this method, WI-L2-729 HF$_2$ cells that were transfected with the hamster cDNA were shown to express the HSPG-FGF receptor as judged by the significant increase in $^{35}SO_4$ incorporation above control levels. Furthermore, the trypsin sensitivity of the incorporated counts indicated the extracellular nature of the sulfate incorporation.

To further characterize the glycosaminoglycan (GAG) side chains of the immunoprecipitated molecule, we treated samples with either nitrous acid or chondroitinase AC. The release of 77% of the incorporated counts with nitrous acid indicated the predominance of heparan sulfate incorporation into the expressed protein. Similarly, chontroitinase AC treatment showed that 22% of the incorporated $^{35}SO_4$ was incorporated as chondroitin side chains. This is consistent with previous observations, where the incorporation of both heparan and chondroitin sulfates into murine syndecan was demonstrated.[38]

In similar experiments with the human-liver-derived cDNA, we measured incorporation of $^{35}SO_4$ into immunoprecipitable expressed protein. Surprisingly, incorporation levels were low when compared with the corresponding hamster protein (FIGURE 5) and in addition, trypsin treatment of these transfected cells showed that the majority of incorporated $^{35}SO_4$ remained intracellular (data not shown). Consistent with this finding, we also found that in contrast to the hamster HSPG-FGF receptor transfected cells, these human-cDNA-transfected cells were not capable of binding to basic-FGF-coated plates. In other studies, we showed that trypsin-sensitive $^{35}SO_4$ incorporation into heparan sulfate of cells that were transfected with the hamster cDNA was fivefold over control levels for cells transfected with a control plasmid.

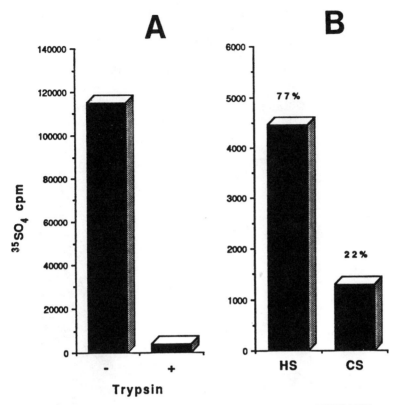

FIGURE 4. Characterization of GAG incorporation into the hamster HSPG-FGF receptor expressed in WI-L2-729 HF_2 cells. **A:** Cells that were stably transfected with the expression vector EBO-pCD-XN containing hamster HSPG-FGF receptor cDNA (RO12) were labeled with 50 μCi $^{35}SO_4$/ml for 18 hours, washed, incubated with (+) and without (−) trypsin, and solubilized. HSPG-FGF receptor was isolated by immunoprecipitation with rabbit polyclonal antiserum directed against a synthetic C-terminal peptide of hamster HSPG-FGF receptor (amino acids 287–305) and protein A Sepharose. **B:** Aliquots from the untreated sample in A (trypsin −) were treated with either buffer, 0.25 M nitrous acid pH 1.5 or chondroitinase AC. Heparan sulfate (HS) and chondroitin sulfate (CS) released from the HSPG-FGF receptor by nitrous acid and chondroitinase AC treatment, respectively, were separated from the proteoglycan by paper chromatography and the counts per minute determined. The percentage of the total cpm found in either HS or CS is shown.

CONCLUSIONS

Our results show that the hamster cDNA that was isolated by panning on FGF-coated plates most likely encodes the low-affinity FGF receptor of BHK-21 cells. Furthermore, the homology of our hamster clone to the previously described murine syndecan indicates that, as proposed by Bernfield and Sanderson,[39] syndecan is most likely the murine low-affinity FGF receptor. Surprisingly, however, the characteristics of the expressed human syndecan, did not parallel those of the expressed hamster HSPG-FGF receptor. The inability of the human molecule to induce binding of the transfected cells to basic FGF-coated plates is most likely a reflection of the low expression levels of human syndecan coupled with low levels of incorporation of GAG side chains into extracellularly located polypeptide. It remains to be established whether or not the expression differences using this cell line result from a species difference or if other, related molecules are responsible for low-affinity FGF binding in humans *in vivo*.

It is also interesting to note that a three-immunoglobulin-domain form of the human high-affinity FGF receptor[35] was also unable to bind basic FGF-coated plastic dishes when expressed using this system. Again, this observation might be a reflec-

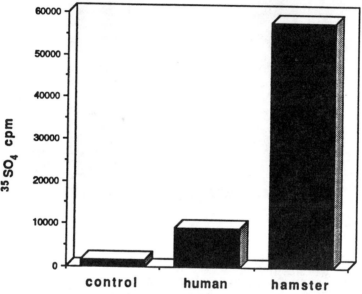

FIGURE 5. Comparison of GAG incorporation into human syndecan and hamster HSPG-FGF receptor expressed in WI-L2-729 HF$_2$ cells. Labeling of cells and immunoprecipitations were performed as in FIGURE 4. Cells were stably transfected with the expression vector EBO-pcD-XN containing no cDNA (control), human syndecan cDNA, isolated from human liver cell line Hep G2 (human), and hamster HSPG-FGF receptor cDNA (hamster). The human syndecan cDNA (clone RO4) was modified by PCR amplification to contain Xbal and Notl restriction enzyme sites which flanked the coding region and facilitated subcloning of the PCR product into EBO-pcD-XN.

tion of low levels of expression and/or transport to the cell surface of this *FLG* protein. Alternatively, it is possible that multivalent interactions with bound FGF, as one might expect from the expressed hamster HSPG-FGF receptor, are required for the adherence of the appropriate transfected cells during panning. It is unlikely that this type of multivalent interaction would occur with the high-affinity FGF receptor. Either of these scenarios can explain the lack of detection of *FLG* or *FLG*-related cDNA clones during our initial planning studies,[37] and serve to demonstrate that this procedure might not represent a general cloning strategy for high-affinity growth factor receptors. In further support of this notion, Harada *et al.* were able to clone the receptor for the lymphokine interleukin 4 (IL-4) by a panning methodology.[43] In this case, however, it was found to be necessary to chemically cross-link biotinylated IL-4 to the expressed receptor prior to panning on immobilized antibiotin antibodies in order to select for the appropriate transfected cells.

SUMMARY

Two distinct classes of cell surface FGF-binding proteins have been identified. These receptors differ in both mode of interaction and in affinity for the FGFs. cDNAs that encode the low-affinity receptor were isolated from a hamster kidney cell line cDNA library by expression cloning. Transfected cells that contained these heparan sulfate proteoglycan FGF receptor cDNAs were enriched for by panning on basic FGF-coated plates. The analogous human cDNA was isolated from a hepatoma cell line cDNA library. The homology of our hamster cDNAs to the previously described murine integral membrane proteoglycan syndecan, together with an exact amino acid sequence match of our human-cDNA-encoded product to human syndecan, clearly indicates the identity of these independently isolated proteoglycans. Further confirmation that the expressed molecule serves as a proteoglycan core protein was achieved by immunoprecipitation of $^{35}SO_4$-labeled material from solubilized transfected cells. Nitrous acid treatment and chondroitinase digestion revealed that 77% of the label was associated with heparan sulfate chains and 22% with chondroitin sulfate chains. These heparan sulfate chains contributed to the fivefold increase in the total heparan sulfate found to be present on the surface of the transfected cells compared with cells transfected with a vector lacking the cDNA insert.

ACKNOWLEDGMENTS

We thank A. Saphire for technical assistance and P. Anderson for preparation of the manuscript.

REFERENCES

1. ARUFFO, A. & B. SEED. 1987. Proc. Nat. Acad. Sci. USA **84:** 8573–8577.
2. SEED, B. & A. ARUFFO. 1987. Proc. Nat. Acad. Sci. USA **84:** 3365–3369.
3. BEVILACQUA, M. P., S. STENGELIN, M. A. GIMBRONE, JR. & B. SEED. 1989. Science **243:** 1160–1165.
4. STAMENKOVIC, I., M. AMIOT, J. M. PESANDO & B. SEED. 1989. Cell **56:** 1057–1062.
5. BUNTROCK, P., K. D. JENTZSCH & G. HEDER. 1982. Exp. Pathol. **21:** 62–67.
6. FOLKMAN, J. & M. KLAGSBURN. 1987. Science **235:** 442–447.
7. SEIVERS, J., B. HAUSMANN, K. UNSICKER & M. BERRY. 1987. Neurosci. Lett **76:** 157–162.
8. FREDJ-REYGROBELET, D., J. PLOUËT, T. DELAYRE, C. BAUDOUIN, F. BOURRET & D. LAPALUS. 1987. Curr. Eye Res. **6:** 1205–1209.
9. KANER, R. J., A. BAIRD, A. MANSUKHANI, C. BASILICO, B. D. SUMMERS, R. A. FLORKIEWICZ & D. P. HAJJAR. 1990. Science **248:** 1410–1413.
10. BAIRD, A., R. Z. FLORKIEWICZ, P. A. MAHER, R. J. KANER & D. P. HAJJAR. 1990. Nature **348:** 344–346.
11. JAYE, M., R. HOWK, W. BURGESS, G. A. RICCA, I.-M. CHUI, M. W. RAVERA, S. J. O'BRIEN, W. S. MODI, T. MACIAG & W. N. DROHAN. 1986. Science **233:** 541–545.
12. ABRAHAM, J. A., A. MERGIA, J. L. WHANG, A. TUMOLO, J. FRIEDMAN, K. A. HJERRID, D. GOSPODAROWICZ & J. C. FIDDES. 1986. Science **233:** 545–548.

13. ABRAHAM, J. A., J. L. WHANG, A. TUMOLO, A. MERGIA, J. FRIEDMAN, D. GOSPODAROWICZ & J. C. FIDDES. 1986. EMBO J. **5:** 2523–2528.
14. MOORE, R., G. CASEY, S. BROOKES, M. DIXON, G. PETERS & C. DICKSON. 1986. EMBO J. **5:** 919–924.
15. JAKOBOVITS, A., G. M. SHACKLEFORD, H. E. VARMUS & G. R. MARTIN. 1986. Proc. Nat. Acad. Sci. USA **83:** 7806–7810.
16. DELLI BOVI, P., A. M. CURATOLA, F. KERN, A. GRECO, M. ITTMANN & C. BASILICO. 1987. Cell **50:** 729–737.
17. TAIRA, M., T. YOSHIDA, K. MIYAGAWA, H. SAKAMOTO, M. TERADA & T. SUGIMURA. 1987. Proc. Nat. Acad. Sci. USA **84:** 2980–2984.
18. ZHAN, X., B. BATES, X. HU & M. GOLDFARB. 1988. Mol. Cell Biol. **8:** 3487–3495.
19. MARICS, I., J. ADELAIDE, F. RAYBAUD, M.-G. MATTEI, F. COULIER, J. PANCHE, O. DE LAPEYRIERE & D. BIRNBAUM. 1989. Oncogene **4:** 335–340.
20. FINCH, P. W., J. S. RUBIN, T. MIKI, D. RON & S. A. AARONSON. 1989. Science **245:** 752–755.
21. THOMAS, K. 1987. FASEB J. **1:** 434–440.
22. GOSPODAROWICZ, D. 1987. Methods Enzymol. **147:** 106–119.
23. CUEVAS, P., J. BURGOS & A. BAIRD. 1988. Biochem. Biophys. Res. Commun. **156:** 611–617.
24. NEUFELD, G. & D. GOSPODAROWICZ. 1985. J. Biol. Chem. **260:** 13860–13868.
25. KAN, M. D., J. DISORBO, H. HOU, H. HOSHI, P.-E. MANSSON & W. L. MCKEEHAN. 1988. J. Biol. Chem. **263:** 11306–11313.
26. WALICKE, P. A., J.-J. FEIGE & A. BAIRD. 1989. J. Biol. Chem. **264:** 4120–4126.
27. MOSCATELLI, D. 1987. J. Cell. Physiol. **131:** 123–130.
28. RIFKIN, D. B. & D. MOSCATELLI. 1989. J. Cell. Biol. **109:** 1–6.
29. RUTA, M., R. HOWK, G. RICCA, W. DROHAN, M. ZABELSHANSKY, G. LAUREYS, D. E. BARTON, U. FRANCKE, J. SCHLESSINGER & D. GIVOL. 1988. Oncogene **3:** 9–15.
30. LEE, P. L., D. E. JOHNSON, L. S. COUSENS, V. A. FRIED & L. T. WILLIAMS. 1989. Science **245:** 57–60.
31. PASQUALE, E. & S. J. SINGER. 1989. Proc. Nat. Acad. Sci. USA **86:** 5449–5453.
32. REID, H. H., A. F. WILKS & O. BERNARD. 1990. Proc. Nat. Acad. Sci. USA **87:** 1596–1600.
33. ISACCHI, A., L. BERGONZONI & P. SARMIENTOS. 1990. Nucleic Acids Res. **18:** 1906.
34. JOHNSON, D. E., P. L. LEE, J. LU & L. T. WILLIAMS. 1990. Mol. Cell. Biol. **10:** 4728–4736.
35. KIEFER, M. C., A. BAIRD, T. NGUYEN, C. GEORGE-NASCIMENTO, O. B. MASON, L. J. BOLEY, P. VALENZUELA & P. J. BARR. Growth factors. (In press.)
36. KORNBLUTH, S., K. E. PAULSON & H. HANAFUSA. 1988. Mol. Cell. Biol. **8:** 5541–5544.
37. KIEFER, M. C., J. C. STEPHANS, K. CRAWFORD, K. OKINO & P. J. BARR. 1990. Proc. Nat. Acad. Sci. USA **87:** 6985–6989.
38. SAUNDERS, S., J. JALKANEN, S. O'FARRELL & M. BERNFIELD. 1989. J. Cell Biol. **108:** 1547–1556.
39. BERNFIELD, M. & R. D. SANDERSON. 1990. Phil. Trans. R. Soc. London Ser. B **327:** 171–186.
40. MALI, M., P. JAAKKOLA, A.-M. ARVILOMMI & M. JALKANEN. 1990. J. Biol. Chem. **265:** 6884–6889.
41. BOURDON, M. A., T. KRUSIUS, S. CAMPBELL, N. B. SCHWARTZ & E. RUOSLAHTI. 1987. Proc. Nat. Acad. Sci. USA **84:** 3194–3198.
42. MARYNEN, P., J. ZHANG, J.-J. CASSIMAN, H. VAN DEN BERGHE & G. DAVID. 1989. J. Biol. Chem. **264:** 7017–7024.
43. HARADA, N., B. E. CASTLE, D. M. GORMAN, N. ITOH, J. SCHREURS, R. L. BARRETT, M. HOWARD & A. MIYAJIMA. 1990. Proc. Nat. Acad. Sci. USA **87:** 857–861.

Interaction of Basic Fibroblast Growth Factor with Extracellular Matrix and Receptors[a]

DAVID MOSCATELLI, ROBERT FLAUMENHAFT,
AND OLLI SAKSELA[b]

*Department of Cell Biology and
the Kaplan Cancer Center
New York University Medical Center
550 First Avenue
New York, New York 10016*

One of the characteristics of fibroblast growth factors (FGFs) is their strong affinity for heparin.[1,2] For basic FGF (bFGF) this affinity for heparin reflects its affinity for heparinlike molecules produced by cultured cells.[3] Thus, when exogenous bFGF is added to cultured cells, it binds to cell-associated heparinlike molecules in addition to high-affinity receptors.[3] The binding to heparinlike molecules is strong, with a dissociation constant of approximately 2×10^{-9} M.[3,4] However, binding to receptors is approximately two orders of magnitude stronger with a dissociation constant of 2×10^{-11} M.[3] Thus, although heparinlike molecules are produced in great abundance compared to receptors by several cell types, bFGF seems to be efficiently transferred from the heparinlike molecules to receptors and internalized by these cells.[5]

bFGF has been shown to have potent angiogenic properties.[6] bFGF induces in cultured endothelial cells increased proliferation, increased motility, and increased production of proteases,[7-9] responses that are thought to be related to the angiogenic response *in vivo*. However, several properties of bFGF may interfere with its effectiveness as an angiogenic agent *in vivo*. First, bFGF has no signal sequence for secretion,[10] and it has been proposed that this molecule may be released primarily from injured or dying cells. If bFGF is released as a bolus upon tissue injury, it is unclear how it can mediate the long-term responses required for angiogenesis. Second, bFGF is very sensitive to proteases, perhaps because it contains a large number of arginines and lysines, and, therefore, the proteases that are generated in areas of neovascularization may inactivate the molecule. Third, when bFGF is released *in vivo*, its strong affinity for heparinlike molecules produced by surrounding cells may prevent it from diffusing away and restrict its availability to only those cells in the immediate vicinity of its point of release. Recent experiments have indicated that the interactions of bFGF with heparinlike molecules are important in its ability to overcome these limits. In the experiments described here, we have identified the heparinlike molecules produced by cultured endothelial cells and have explored the importance of these molecules in the biology of bFGF.

[a]This work was supported by grants CA 42229 and CA 34289 from the National Institutes of Health and BC554C from the American Cancer Society.
[b]Present affiliation: Department of Dermatology, University of Helsinki, Finland.

177

RESULTS AND DISCUSSION

To characterize the cell-associated FGF-binding heparinlike molecules, bovine capillary endothelial (BCE) cells were incubated with $^{35}SO_4$ for 24 hours and then extracted with 4 M guanidine.[11] The $^{35}SO_4$-labeled molecules were partially purified on DEAE-cellulose. The $^{35}SO_4$-containing fractions were incubated with bFGF, and bFGF and molecules forming complexes with bFGF were precipitated with anti-bFGF antibodies. Autoradiography after sodium dodecyl sulfate-polyacrylamide gel electrophoresis (SDS-PAGE) showed that two major $^{35}SO_4$-labeled bFGF-binding molecules were present, with molecular weights of approximately 250,000 and 800,000. Both molecules were sensitive to heparinase but not chondroitinase and could be shifted to lower molecular weights after treatment with proteases, suggesting that they are heparan sulfate proteoglycans (HSPGs). To determine the cellular localization of these HSPGs, $^{35}SO_4$-labeled cells were extracted first with 0.5% Triton X-100 to remove cytoplasmic and plasma membrane structures, and the remaining extracellular matrix was extracted with 4 M guanidine. Purification of the bFGF-binding $^{35}SO_4$-labeled molecules from these fractions demonstrated that the Triton X-100 soluble fraction contained the 250,000 HSPG, while the Triton X-100 insoluble fraction contained the 800,000 HSPG, suggesting that these molecules were components of the plasma membrane and extracellular matrix, respectively.[11]

The role of these cell-associated HSPGs in mediating long-term responses to bFGF was suggested by experiments designed to measure the effects of brief exposures to bFGF.[12] In these experiments, BCE cells were incubated at 37°C with 10 ng/ml bFGF for 30 minutes. The medium was removed, the cells were washed 3 times with phosphate-buffered saline, the medium was replaced with bFGF-free medium, and the cells were incubated at 37°C. At 24 and 48 hours after the medium change, cells were assayed for plasminogen activator (PA) activity. When BCE cells were exposed to bFGF continuously throughout this period, PA activity had increased sevenfold after 24 hours and still remained fivefold higher than controls after 48 hours. BCE cells exposed to medium containing bFGF for only 30 minutes responded similarly with a sevenfold stimulation of PA after 24 hours and a threefold stimulation at 48 hours. These results suggested that during the brief exposure to bFGF, most of the bFGF bound to cell-associated HSPGs and, thus, remained in the culture after the medium change. During the subsequent time period, bFGF was released from the matrix, bound to the high-affinity receptors on the BCE cells, and stimulated PA production.

To test that idea, the experiment was repeated under conditions that disrupt bFGF interactions with cellular HSPGs.[12] Since soluble heparin competes for bFGF binding to the cell-associated HSPGs,[3] the effect of adding heparin during a 10-minute exposure to bFGF was assessed. As observed previously, when bFGF was added alone for 10 minutes at 37°C and then the cells changed to bFGF-free medium, PA production was stimulated to the same extent after 24 hours as in cultures that had been continuously exposed to bFGF. However, if the 10-minute incubation with bFGF was done in the presence of heparin and the cells then washed and changed to medium without bFGF and heparin, the stimulation of PA production was reduced to less than 20% of that observed when the 10-minute incubation was done in the absence of heparin. When cells were incubated with bFGF and heparin continuously, the increase in PA activity at 24 hours was identical to that observed in cells incubated continuously with bFGF alone, demonstrating that there is no direct effect of heparin on PA production. These results show that if bFGF is prevented from binding to cell-associated HSPG by soluble heparin, all or most of the bFGF is removed with the wash, and long-term responses are reduced.

These results were confirmed by several other means. If the cultures were treated with heparinase to remove HSPGs and then exposed to bFGF for a short period of time, no stimulation of PA production was observed after 24 hours, presumably because the storage sites for bFGF were removed.[12] In addition, if the cells were washed after a brief exposure with 2 M NaCl rather than phosphate-buffered saline, no stimulation of PA production was observed.[12] This is due to the fact that 2 M NaCl dissociates bFGF from HSPGs while phosphate-buffered saline does not.[3]

These experiments suggest that the HSPGs can act as a reservoir or a buffer for bFGF. When bFGF is released from cells, it binds to the HSPGs, and then slowly, over a period of time, is released and can interact with the appropriate high-affinity membrane receptor to stimulate the cell. As a result, long-term effects can be obtained from short exposures to high concentrations of bFGF.

A second biological role for the affinity of bFGF for HSPGs is the protection of bFGF from proteolytic inactivation. bFGF will induce the synthesis of two proteases, PA and collagenase, in endothelial cells.[9,13] Since PA can generate the trypsinlike enzyme plasmin from plasminogen, proteolytic activity may be quite high in areas of neovascularization. To explore the effect of HSPGs on the stability of bFGF to proteases that would be present during angiogenesis, bFGF was incubated with plasmin in the presence or absence of heparin or endothelial-cell-derived heparan sulfate.[14] The free molecule is extremely sensitive to enzymatic digestion and was completely degraded after 6 hours, whereas in the presence of heparin or heparan sulfate it was relatively resistant to proteolysis. Others have shown that heparin also protects bFGF from trypsin or chymotrypsin.[15-17] Thus, in addition to serving as a reservoir, the interaction with the HSPGs affords the bFGF protection against proteolysis.

Finally, soluble HSPGs seem to have a role in promoting the dissemination of bFGF from its point of release. The problems confronting bFGF dispersion were illustrated in an experiment in which bFGF was allowed to diffuse under different conditions in agarose, which was used as a model for a heparan sulfate–containing matrix because it is sulfated.[18] When radiolabeled bFGF was added to a 6-mm hole in the agarose, allowed to diffuse, and then the agarose dried down and exposed for autoradiography, the radiolabeled bFGF was found to be restricted to the area immediately surrounding the well. However, if heparin was added to the well along with the bFGF, the radiolabeled bFGF diffused much further and equilibrated throughout the dish.

In a similar experiment, the diffusion of bFGF alone was shown to be restricted over an intact endothelial cell monolayer.[18] In this experiment, a Millicell insert was placed onto a bovine aortic endothelial cell monolayer, and either ^{125}I-bFGF alone or ^{125}I-bFGF and heparin were added to the well. After an overnight incubation, the medium was removed, and the monolayer was dried and exposed for autoradiography. The ^{125}I-bFGF by itself diffused very little, and radioactivity was restricted to the area immediately surrounding the well. However, in the presence of heparin, the radiolabeled bFGF diffused throughout the dish.

To show that the distribution of radiolabeled bFGF reflects biological availability of bFGF, a similar experiment was done in which a 12-mm Millicell insert was placed in the middle of a monolayer of endothelial cells on a 60-mm tissue culture dish, and either bFGF alone or bFGF and heparin were added to the well.[18] After 24 hours incubation, the response of the cells to bFGF was assessed. Since it was known that exposure to bFGF causes BCE cells to assume an elongated morphology, the activity of bFGF was detected by morphological changes in the cells. With bFGF alone, the

BCE cells became elongated up to 5 mm from the center of the dish, whereas at 10 or 25 mm from the center, the cells retained the cobblestone appearance characteristic of endothelial cells. Therefore, it appeared that bFGF alone did not exert a biological effect beyond 5 mm, which is the approximate boundary of the Millicell insert. In contrast, when bFGF was added in the presence of heparin or HSPG, the BCE cells underwent a morphological change throughout the dish. Heparin or HSPG added to the well alone did not affect the morphology of the cells. Thus, in the presence of heparin or HSPG, bFGF was able to diffuse over the whole dish and was biologically available to all of the cells.

Together, this series of experiments suggested that the complex of bFGF with heparin or HSPG partitions preferentially into the aqueous phase, while the free bFGF binds to the matrix and does not diffuse. Thus, the complexed form of bFGF may be biologically relevant in assuring the availability of bFGF to sites distant from the point of release.

What is the source of these soluble forms of HSPG? Soluble HSPGs with molecular characteristics similar to the cell-associated HSPGs have been found in the conditioned medium of BCE cells.[14] These HSPGs seem to be released through the action of plasmin generated by plasminogen activator produced by endothelial cells.[11] Addition of plasmin to cultures of BCE cells results in the release of soluble HSPGs, and addition of aprotinin, an inhibitor of plasmin, blocks their spontaneous release. Furthermore, addition of bFGF which stimulates the production of PA by endothelial cells results in an increase in HSPG release from these cells. In contrast, addition of transforming growth factor-β which stimulates the production of plasminogen activator inhibitor-1 in BCE cells, thereby decreasing PA activity, results in a decrease in the release of HSPGs. To demonstrate that the release of HSPGs was mediated by PA, release of HSPGs could be inhibited by specific antibodies to bovine urokinase PA. Thus, BCE cells have the means of regulating the levels of soluble active bFGF–heparan sulfate complexes through growth factor effects on their PA activity.

These studies suggest that the bFGF-HSPG interaction is critical to the biological activities of bFGF. The cell-associated HSPGs provide a slow release reservoir of bFGF that can mediate long-term responses to brief exposures to bFGF; soluble HSPGs can act as carriers for bFGF, providing a diffusible form of bFGF; and while bFGF is associated with its HSPG reservoir or HSPG carrier, it is protected from proteases.

REFERENCES

1. LOBB, R. R., J. W. HARPER & J. W. FETT. 1986. Purification of heparin-binding growth factors. Anal. Biochem. **154:** 1–14.
2. BURGESS, W. H. & T. MACIAG. 1989. The heparin-binding (fibroblast) growth factor family of proteins. Annu. Rev. Biochem. **58:** 575–606.
3. MOSCATELLI, D. 1987. High and low affinity binding sites for basic fibroblast growth factor on cultured cells: absence of a role for low affinity binding in the stimulation of plasminogen activator production by bovine capillary endothelial cells. J. Cell. Physiol. **131:** 123–130.
4. VIGNY, M., M. P. OLLIER-HARTMANN, M. LAVIGNE, N. FAYEIN, J. C. JEANNY, M. LAURENT & Y. COURTOIS. 1988. Specific binding of basic fibroblast growth factor to basement membrane–like structures and to purified heparan sulfate proteoglycan of the EHS tumor. J. Cell. Physiol. **137:** 321–328.
5. MOSCATELLI, D. 1988. Metabolism of receptor-bound and matrix-bound basic fibroblast growth factor by bovine capillary endothelial cells. J. Cell Biol. **107:** 753–759.

6. SHING, Y., J. FOLKMAN, C. HAUDENSCHILD, D. LUND, R. CRUM & M. KLAGSBRUN. 1985. Angiogenesis is stimulated by a tumor-derived endothelial cell growth factor. J. Cell. Biochem. **29:** 275–287.

7. SHING, Y., J. FOLKMAN, R. SULLIVAN, C. BUTTERFIELD, J. MURRAY & M. KLAGSBRUN. 1984. Heparin affinity: purification of a tumor-derived capillary endothelial cell growth factor. Science **223:** 1296–1299.

8. GOSPODAROWICZ, D., S. MASSOGLIA, J. CHENG & D. K. FUJII. 1986. Effect of fibroblast growth factor and lipoproteins on the proliferation of endothelial cells derived from bovine adrenal cortex, brain cortex, and corpus luteum capillaries. J. Cell. Physiol. **127:** 121–136.

9. MOSCATELLI, D., M. PRESTA & D. B. RIFKIN. 1986. Purification of a factor from human placenta that stimulates capillary endothelial cell protease production, DNA synthesis, and migration. Proc. Nat. Acad. Sci. USA **83:** 2091–2095.

10. ABRAHAM, J. A., J. L. WHANG, A. TUMOLO, A. MERGIA, J. FRIEDMAN, D. GOSPODAROWICZ & J. C. FIDDES. 1986. Human basic fibroblast growth factor: nucleotide sequence and genomic organization. EMBO J. **5:** 2523–2528.

11. SAKSELA, O. & D. B. RIFKIN. 1990. Release of basic fibroblast growth factor–heparan sulfate complexes from endothelial cells by plasminogen activator-mediated proteolytic activity. J. Cell Biol. **110:** 767–775.

12. FLAUMENHAFT, R., D. MOSCATELLI, O. SAKSELA & D. B. RIFKIN. 1989. The role of extracellular matrix in the action of basic fibroblast growth factor: matrix as a source of growth factor for long-term stimulation of plasminogen activator production and DNA synthesis. J. Cell. Physiol. **140:** 75–81.

13. SAKSELA, O., D. MOSCATELLI & D. B. RIFKIN. 1987. The opposing effects of basic fibroblast growth factor and transforming growth factor beta on the regulation of plasminogen activator activity in capillary endothelial cells. J. Cell Biol. **105:** 957–964.

14. SAKSELA, O., D. MOSCATELLI, A. SOMMER & D. B. RIFKIN. 1988. Endothelial cell–derived heparan sulfate binds basic fibroblast growth factor and protects it from proteolytic degradation. J. Cell Biol. **107:** 743–751.

15. GOSPODAROWICZ, D. & J. CHENG. 1986. Heparin protects basic and acidic FGF from inactivation. J. Cell. Physiol. **128:** 475–484.

16. BAIRD, A., D. SCHUBERT, N. LING & R. GUILLEMIN. 1988. Receptor- and heparin-binding domains of basic fibroblast growth factor. Proc. Nat. Acad. Sci. USA **85:** 2324–2328.

17. SOMMER, A. & D. B. RIFKIN. 1989. Interaction of heparin with human basic fibroblast growth factor: protection of the angiogenic protein from proteolytic degradation by a glycosaminoglycan. J. Cell. Physiol. **138:** 215–220.

18. FLAUMENHAFT, R., D. MOSCATELLI & D. B. RIFKIN. 1990. Heparin and heparan sulfate increase the radius of diffusion and action of bFGF. J. Cell Biol. **111:** 1651–1659.

Possible Regulation of FGF Activity by Syndecan, an Integral Membrane Heparan Sulfate Proteoglycan[a]

MERTON BERNFIELD AND KEVIN C. HOOPER

Joint Program in Neonatology
Harvard Medical School
300 Longwood Avenue, Enders 9
Boston, Massachusetts 02115

INTRODUCTION

The fibroblast growth factors (FGFs) have affinity for heparin (for reviews see References 1, 2, and 3). This binding is not likely a physiological interaction because heparin itself is not usually found in the extracellular space (for extensive discussions, see References 4 and 5). Heparin proteoglycans are intracellular, within the secretory granules of mast cells and basophils, and can become extracellular when these cells degranulate at sites of specific immune reactions. Heparin binding by these extracellular growth factors likely represents interactions with the heparinlike molecule, heparan sulfate, that is found within cells, at the cell surface, and within the extracellular matrix. Many cells possess so-called low-affinity receptors for basic FGF (bFGF) (K_d ca. 2 nM, ca. 0.5–2 × 10^6 binding sites per cell) on their surfaces.[6] These "receptors" have all the properties of heparan sulfate proteoglycans; the bound FGF can be displaced by treatment with heparin or heparan sulfate and can be removed by digestion with heparitinase or proteases.[3]

SYNDECAN, A CELL SURFACE HEPARAN SULFATE PROTEOGLYCAN

The cell surface heparan sulfate proteoglycan, syndecan, originally isolated from mouse mammary epithelial cells as a heparan sulfate– and chondroitin sulfate–containing molecule, was the initial cell surface heparan sulfate proteoglycan to be characterized and molecularly cloned[7] (for review see Reference 8). Syndecan represents a family of cell surface heparan sulfate proteoglycans that vary in their extent of glycosaminoglycan (GAG) substitution depending on the type and organization of the cells. Syndecan, found predominantly on epithelia in mature tissues,[9] behaves as matrix receptor for these epithelia; it binds selectively to components of the extracellular matrix, associates intracellularly with the actin cytoskeleton when cross-linked at the cell surface, and its extracellular domain is shed upon cell rounding. Thus, syndecan can reversibly link the cytoskeleton to the extracellular matrix.

Although predominantly on epithelial cell surfaces in mature tissues, syndecan can be expressed by mesenchymal cells. On epithelia, syndecan contains smaller

[a]This work was supported by National Institutes of Health grants HD-06763 and CA-28735, and by the Lucille P. Markey Charitable Trust.

heparan sulfate chains and is predominantly at basolateral cell surfaces of simple epithelia where it colocalizes with bundles of cortical actin. Syndecan on mesenchymal cells contains much larger heparan sulfate chains and is predominantly within the cells. Indeed, NMuMG mammary epithelial cells made deficient in cell surface syndecan by stable transfection with antisense cDNA become mesenchyme like in cell shape and organization and show reduced expression of epithelial adhesion molecules. In neoplastic transformation as well, reduced expression of syndecan at the cell surface correlates with loss of normal epithelial cell behavior.[10]

Syndecan shows changes in amount, location, and structure during development. It appears initially in 4-cell mouse embryos, becomes restricted in preimplantation embryos to the cells that will form the embryo proper, and undergoes marked changes in expression during epithelial-mesenchymal inductive interactions. On embryonic epithelia, it is lost transiently during periods of rapid cell differentiation and permanently upon subsequent terminal differentiation, as in the formation of pulmonary alveolar epithelia.[11] On mesenchymal cells, it is expressed transiently when they are induced to condense and then is lost permanently upon their terminal differentiation, as in the formation of skeletal muscle myotubes.[12] This developmental regulation is consistent with its ability to bind factors that regulate differentiation, viz., components of the extracellular matrix and members of the FGF family.

Syndecan binds bFGF with the affinity and specificity characteristic of the low-affinity "receptors" (FIGURE 1 and FIGURE 2). Indeed the recent expression cloning of hamster syndecan in lymphoblastoid cells provides these cells with such low-affinity binding sites for bFGF.[13] These low-affinity heparan sulfate binding sites appear to be functionally important for the action of bFGF. Cells containing the high-affinity receptor but unable to produce heparan sulfate fail to bind bFGF, but this abnormality can be corrected by adding exogenous heparin[14] (see Klagsbrun *et al.*, this volume). bFGF-responsive cells made deficient in cell surface heparan sulfate, either by enzyme pretreatment or by preventing sulfation, become markedly reduced in their ability to respond to bFGF by either proliferating (Swiss 3T3 cells) or repressing their differentiation (MM14 myoblasts)[15] (see Olwin *et al.*, this volume). Thus, the binding of bFGF to syndecan could be involved in generating a biological response to members of this growth factor family.

While syndecan is the best characterized of these molecules, there are a variety of cell surface heparan sulfate proteoglycans, and they may vary in their properties. The purpose of this review is to summarize the present state of knowledge of cell surface heparan sulfate proteoglycans with regard to their interactions with the FGF family of growth factors and to assess the possible significance of these interactions.

HEPARAN SULFATE IS AT THE CELL SURFACE

The presence of heparan sulfate at cell surfaces has been known for 20 years since Kraemer showed that more than 20 types of adherent cells contained heparan sulfate at their surfaces.[16] The precise nature of this heparan sulfate, however, is only now becoming an object of intense study. However, the structurally similar glycosaminoglycan, heparin, has been extensively used as a chromagraphic affinity material for proteins, implying that a wide variety of proteins can potentially interact with heparan sulfates at cell surfaces (for review see Reference 17). Indeed, of the large number of proteins able to bind to heparin, only proteins with substantial affinity should be considered as possible ligands for heparan sulfate. These include a variety of growth factors, notably the FGFs and the platelet derived growth factor(s) (PDGFs), and a panoply of extracellular matrix molecules, including the fibrillar

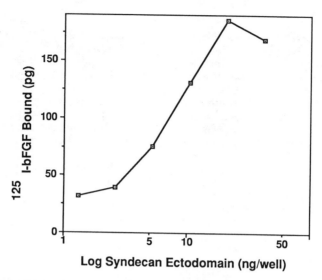

FIGURE 1. bFGF binds to syndecan from mammary epithelial cells. Monoclonal antibody 281-2, directed against the syndecan core protein, or Mel-14, an irrelevant IgG_{2a} antibody, were immobilized onto PVC wells with rabbit anti-rat IgG. ^{125}I-bFGF was incubated overnight at 4°C with a partially purified preparation of proteoglycans from conditioned media of NMuMG cells. The mixture was then incubated with the antibody-containing wells overnight at 4°C, the wells were washed extensively, and the amount of bound bFGF determined.

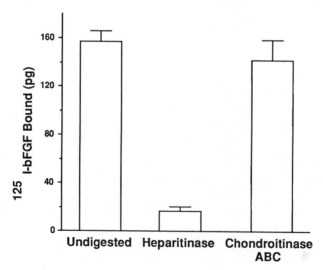

FIGURE 2. bFGF binds to the heparan sulfate chains of syndecan. Monoclonal antibodies were immobilized on the wells as in FIGURE 1. The partially purified proteoglycan preparation from conditioned medium of NMuMG cells was treated with either heparatinase, chondroitinase ABC, or no enzyme for 180 minutes after which the mixture was diluted and added together with ^{125}I-bFGF to the wells. The wells were incubated overnight at 4°C, washed extensively, and the amount of bound bFGF determined.

interstitial collagens (types I, III, and V), fibronectin, laminin, thrombospondin, and tenascin. With regard to other growth factors, there is some suggestion that granulo- cyte macrophage colony stimulating factor (GM-CSF), interleukin-3 (IL-3),[18] and a Schwann cell mitogen[19] can bind heparin or heparan sulfate but the affinities of these growth factors have not been assessed. Additional heparin-binding molecules in- clude the cell adhesion molecule N-CAM, the enzymes acetylcholinesterase, lipopro- tein lipase, and thrombin, and the enzyme inhibitors protease nexin and antithrom- bin III. This is not an exclusive list, and other molecules with substantial affinity for heparin/heparan sulfate will likely be found (see TABLE 1).

Heparan Sulfate Has Unique Structural Properties

Heparan sulfate is a glycosaminoglycan, a polysaccharide containing alternating residues of uronic acid and glucosamine, both potentially substituted with sulfate groups. These molecules are synthesized in the golgi from the requisite uridine nucleotide sugars and, by analogy with heparin, the subsequent modifications of the polymer apparently occur without a template, on a stochastic basis, presumably depending on the availability of the individual modifying enzymes. Much variability occurs because the enzymic reactions do not go to completion.[20] Much of the uronic acid is L-iduronic acid (IdoA) which is produced after polymer formation by C-5 epimerization of D-glucuronic acid. This IdoA is produced in a reaction that requires N-sulfation of the amino group of the adjacent glucosamine. Subsequently, there is 6-O and 3-O sulfation of the glucosamine. Regions rich in IdoA, N-sulfate, and O-sulfate alternate with more glucuronic acid rich, N-acetylated regions and the relative proportions of these highly sulfated regions vary according to the tissue source of heparan sulfate.

This biosynthetic scheme imparts unique structural properties to heparan sul- fate. Unlike other GAGs (e.g., hyaluronan, chondroitin sulfate), heparan sulfate and heparin contain multiple different disaccharides; at least 20 are theoretically possible but only nine are generally found. Heparan sulfate is distinct from heparin because of its clusters of highly sulfated regions separated by regions of unsulfated sugars enriched in N-acetyl groups.[21] Although the N-sulfate regions make heparan sulfate (and heparin) the most highly acidic of the GAGs, their ability to bind proteins is not

TABLE 1. Binding Interactions of Heparin/Heparan Sulfate (Partial Listing)

• Self-association	• Oncogene Products
• Matrix components	*int-1*
Fibronectin	*int-2*
Interstitial collagens	*hst*/K-*fgf*
Laminin	FGF-5
Tenascin	FGF-6
Thrombospondin	• Enzymes
• Cell adhesion molecules	Acetylcholinesterase
N-CAM	Lipoprotein lipase
Purpurin	Thrombin
• Growth factors	• Protease Inhibitors
Acidic FGF	Anti–thrombin III
Basic FGF	Protease nexin
PDGF	• Serum components
	Low-density lipoprotein
	Very-low-density lipoprotein

primarily due to this property. Rather, this binding affinity is markedly enhanced by the presence of the IdoA residues, thought to enhance the conformational flexibility of the GAG chain.[22]

Heparan Sulfate Has Unique Binding Properties

The interaction of heparin with antithrombin III is highly specific and depends on the presence of a unique pentasaccharide sequence.[20] Unique and specific heparan sulfate sequences have not (yet) been found for interactions with other proteins, possibly because the conformational flexibility of this GAG yields distinct three-dimensional shapes.[23] IdoA within the polysaccharide chain is in conformational equilibrium among three conformers, allowing the IdoA-containing regions of the chains to change their shape readily wherever a change occurs in their environment.[24] This extra degree of freedom in the polysaccharide chain can alter the relative orientation of the substituents on the heparan sulfate backbone, enabling the anionic carboxylate and sulfate residues to seek out basic amino acid residues on proteins. The interconversion of internal IdoA conformers would not cause a large rearrangement of the overall conformation of the GAG. On the other hand, the increased interaction of these GAGs with an appropriate region of an interacting protein could alter the local shape of the protein. For example, heparan sulfate apparently induces a conformational change in fibronectin.[25]

With regard to the binding of FGFs, the conformational flexibility of heparan sulfate could both enhance binding and alter the conformation of the growth factor. One might suppose if a unique heparan sulfate or heparin sequence were involved in binding a protein, then only the subpopulation of GAGs that contained this sequence would bind selectively, as is seen with the binding of heparin to antithrombin III.[20] However, if the binding were primarily due to a range of possible polysaccharide sequences that could attain a common shape via conformational flexibility, then most or all of the GAG would bind. Affinity coelectrophoresis of antithrombin III with commercial heparin yields two distinct heparin subpopulations, consistent with the presence of the specific binding sequence on only a portion of the heparin molecules, classically known as high-affinity heparin. An analogous result was seen with fibronectin. However, when acidic FGF (aFGF) and bFGF were analyzed similarly, all the heparin molecules bound, without evidence of subpopulations.[26] Thus, binding to fibronectin may be due to a distinct GAG sequence but binding to the FGFs may be due to the highly charged carboxylate and sulfate residues interacting extensively with these polypeptides because of the conformational flexibility of the GAG chain.

THE STRUCTURE OF CELL SURFACE HEPARAN SULFATE PROTEOGLYCANS DETERMINES THEIR LIGAND BINDING

Heparan sulfate chains are bound to proteins by glycosidic linkage of a distinctive trisaccharide sequence at the reducing end of the polysaccharide chain to a serine residue in the protein. These proteins can be structurally distinct. Indeed, those on the cell surface may vary in their mode of association with the plasma membrane. More than one type of cell surface heparan sulfate proteoglycan may coexist on the same cell type, and there also may be various cell surface chondroitin sulfate proteoglycans at the cell surface (see References 27 and 28). The heparan sulfate proteoglycans may be peripheral, and thus loosely associated with the cell surface, or integral to the plasma membrane, requiring detergent to remove intact. The loosely associated proteoglycans may be bound to various integral membrane proteins either by the core protein, as suggested for the basement membrane proteoglycan perlican,[29] or by their heparan sulfate chains, released by competition with anionic polysaccharides.[30] The integral membrane proteoglycans can be associated with the cell surface either via a hydrophobic transmembrane domain, as in syndecan, or via covalent linkage to phospholipid, as via the phosphatidyl inositol residue of glypican.[31]

The means of association of the proteoglycan with the plasma membrane infers the nature of its metabolism. For example, by analogy with similar proteins, syndecan is likely inserted into the plasma membrane at the basolateral surface of epithelial cells, whereas glypican would likely be inserted at the apical surface.[32] The means of membrane association also determines the biological properties of the proteoglycan. For example, the integral membrane proteoglycans are mobile within the plane of the membrane; both types can be immobilized by interactions with extracellular matrix components, but only the transmembrane proteins are likely immobilized by the intracellular cytoskeleton.

Syndecan represents a family of integral membrane heparan sulfate proteoglycans. Variable extents of glycosylation yield polymorphic forms that differ according to cell type and stage of differentiation.[33] However, the transmembrane and cytoplasmic domain sequences of syndecan are conserved in other integral membrane heparan sulfate proteoglycans. For example, the core protein of fibroglycan, an integral membrane proteoglycan from human lung fibroblasts, has a derived amino acid sequence that is highly homologous to the transmembrane (52%) and cytoplasmic (66%) domains of murine syndecan.[34]

Most of the currently known binding properties of cell surface heparan sulfate proteoglycans is due to their heparan sulfate chains. But there are highly significant exceptions, including a transmembrane proteoglycan that binds via its core protein with high specificity to TGF-β (see References 35 and 36), a property shared by the matrix proteoglycans biglycan and, to a lesser extent, decorin,[37] and CD-44 (also known as Hermes antigen, PgP-1, ECM-RIII, and other names), a transmembrane proteoglycan that binds by its core protein to hyaluronan and to high endothelial venules (see References 38 and 39). Each of these proteoglycans can contain both chondroitin sulfate and heparan sulfate chains, as syndecan, and presumably can also bind to the various ligands that interact with these GAG chains.

Each of the cell surface proteoglycans characterized to date has more than one GAG chain, or at least more than one GAG attachment site, and several can bear both heparan sulfate and chondroitin sulfate chains. Thus these proteoglycans can interact simultaneously with more than one ligand. It appears that the secondary or tertiary structure of the regions surrounding the GAG attachment serine as well as the type of cell determines the type of GAG chain that will be added to the core protein. There are two major canonical GAG attachment sequences, which we will refer to as the "serglycin-type" and the "syndecan-type" sequences from the proteoglycans in which they were initially described. Neither sequence appears to be specific for a type of GAG chain (see TABLE 2). For example, the serglycin-type sequence is linked to chondroitin sulfate chains in serglycin from L2 rat yolk sac tumor cells,[40] but to heparin chains in mouse mastocytoma cells and apparently to both types of chains in rat basophilic leukemia cells.[41] Likewise, the syndecan-type sequence is linked to heparan sulfate chains in syndecan but to chondroitin sulfate chains in versican.

The extent of glycosylation also can vary substantially. The core proteins of the cell surface proteoglycans can be N-glycosylated and O-glycosylated, but a detailed analysis of the extent of these differences has yet to be made. Variations in the GAG chains are more evident. Syndecan, for example, exhibits polymorphic forms that differ in the number and size of GAG chains, especially the heparan sulfate chains. Although proteoglycans are normally heterogeneous because of differences in extent of GAG addition, the polymorphic forms of syndecan differ by localizing at different parts of the cell and by varying in size with the type and organization of the cells.[33]

TABLE 2. Glycosaminoglycan Attachment Sequences

Type of Attachment Site	Amino Acid Sequence	Examples	
		GAG Chains	Proteoglycans
Serglycin	D/EXXSGXG	Chondroitin sulfate	Serglycin
			Syndecan
		Heparan sulfate	Decorin
			Glypican
		Heparin	Serglycin
Syndecan	D/EXSGX$_{(0-2)}$D/E	Heparan sulfate	Syndecan
			Glypican
		Chondroitin sulfate	Biglycan
			Versican

LIGAND BINDING WILL VARY WITH METABOLISM OF PROTEOGLYCANS AT THE CELL SURFACE

Heparan sulfate proteoglycans, similar to other cell surface molecules, can vary in their route to or from the cell surface. The pathways differ, of course, depending on whether the proteoglycan is peripheral or integral to the plasma membrane. These pathways are important to the behavior of any bound FGF because binding to heparan sulfate can prevent degradation of the growth factor while maintaining its activity.[2] Thus, any bound FGF will follow the pathway of the proteoglycan and consequently will vary in its availability to the signal-transducing FGF receptor.

Deposition at the Cell Surface

Proteins that are integral to the plasma membrane undergo a circuitous route to the cell surface.[32] They are first translocated across the endoplasmic reticulum membrane, then, by vesicles, through the golgi apparatus and to the plasma membrane. For syndecan, this pathway appears to function satisfactorily in simple and stratified epithelia where the constitutively expressed proteoglycan is almost entirely on the cell surface where it can interact with the extracellular environment. In mesenchymal cells, however, the syndecan is predominantly intracellular, especially evident in perinuclear vesicles. The derived amino acid sequences of the cytoplasmic and transmembrane domains of hamster fibroblast syndecan[14] are nearly identical to those of murine[7] and human[42] epithelial syndecan, and fibroglycan, a cell surface proteoglycan from human diploid fibroblasts, contains cytoplasmic and transmembrane domains that are highly homologous to those in syndecan.[34] Thus, it is unlikely that mesenchymal cell syndecan differs from epithelial cell syndecan in an amino acid sequence that directs mesenchymal syndecan to intracellular compartments. Rather, these cells may contain an escort or transport protein with a structural motif that influences the intracellular trafficking of the cell surface proteoglycan in mesenchymal cells. For example, this escort role is the function of the invariant chain in targeting class II MHC molecules to the endocytic pathway.[43]

The cellular localization of proteoglycan often provides a clue to its function. Heparan sulfate proteoglycans have been localized to the adhesion plaques of a variety of mesenchymal cells, inferring a role in cell-substratum adhesion, but the molecular nature of these proteoglycans is not yet clear (see, e.g., References 44 and 45). Glypican, as other phospholipid-linked molecules, is most likely localized at the apical surfaces of epithelia.[46] Localization of syndecan varies with the cell type; moreover, a distinct polymorphic form is at each site. Syndecan from stratified epithelia has a modal relative molecular mass of 92 kDa, both *in vivo* and *in vitro,* and localizes over the entire cell surface. Syndecan from simple epithelia has a modal relative mass of 160 kDa and polarizes at basolateral cell surfaces both *in vivo* and *in vitro.* Finally, syndecan from mesenchymal cells (e.g., fibroblasts, endothelial cells) has a modal relative mass of 300 kDa and localizes in small amounts all over the cell surface; the bulk of syndecan in these cells is intracellular. Embryonic mesenchymal cells are exceptional; when these cells condense during an embryonic tissue interaction, syndecan is expressed at their surfaces (see, e.g., Reference 47).

Possibly significant, the heparan sulfate chains of mesenchymally derived syndecan are more than twofold larger than those on simple epithelia and more than threefold larger than those on stratified epithelia. Such large chains could have several IdoA-rich regions and be multivalent, binding several small peptides, such as the members of the FGF family.

Release from the Cell Surface

Mechanisms for the physiological release of peripherally associated proteoglycans are not clear, but integral membrane proteoglycans are lost from the cell surface both by release into the pericellular environment[48] and by internalization via endocytosis.[49] One suggestion places these latter processes in sequence; a phospholipid-linked rat hepatocyte cell surface proteoglycan is proposed to be released by action of an insulin-activated phospholipase, it then binds to a cell surface receptor via a now exposed myoinositol phosphate residue on its core protein, and finally, it is internalized.[50] This process may culminate in a fraction of the proteoglycan being taken to the nucleus where it could regulate cell growth (see References 51 and 52).

When integral membrane proteoglycans are shed from the cell surface, they become nonhydrophobic, as shown for the transmembrane proteoglycan syndecan[48] and the lipid-linked proteoglycan glypican.[31] The mechanisms and presumably the purpose and control of this shedding apparently differ for these two types of linkage. Syndecan appears to be shed by proteolytic cleavage, apparently at a dibasic site immediately adjacent to the plasma membrane that is uniquely susceptible to trypsinlike proteases, including plasminogen activator, plasmin, and thrombin. Shedding is from the apical cell surface of polarized epithelia and can be markedly accelerated by suspending the cells. When polarized NMuMG cells are suspended and placed at 37°C, they change shape and rapidly ($t_{1/2}$ ca. 15 minutes) shed their entire cohort of cell surface syndecan.[48] No syndecan is replaced on these cells as long as they remain suspended. The heparan sulfate proteoglycan(s) that binds bFGF at the surface of endothelial cells is also shed, presumably by plasminogen-activator-mediated proteolysis.[53] Glypican is also shed from cultured cells, yielding a nonhydrophobic, slightly smaller molecule, presumably resulting from action of a phospholipase, such as has been found in a number of tissues and in plasma.[54]

Internalization from the Cell Surface

Integral membrane proteoglycans turn over rapidly,[55] and although a portion of the proteoglycan is shed, a fraction also appears to be internalized.[49] Internalization or endocytosis of these proteoglycans can occur through coated pits, either rapidly as in receptor-mediated endocytosis[56] or at the same rate as the bulk of the membrane,[57] or without involvement of coated pits. Several integral membrane proteins containing short cytoplasmic domains lacking enzymatic activity require specific amino acid sequences in their cytoplasmic domains to localize to coated pits and to internalize.[58] These sequences contain a tyrosine internalization signal that is near the plasma membrane (less than 50 residues from the transmembrane domain). Near the signal tyrosine on the side nearest the transmembrane domain is a cluster of 4–6 amino acids that are frequently found in turns. Thus, the signal appears to be a tyrosine exposed on a small surface loop near the plasma membrane. This tyrosine internalization signal is found in each of the syndecan core protein sequences published to date and in the fibroglycan sequence, making likely their rapid internalization via coated pits. Various growth factor receptors containing large cytoplasmic domains with tyrosine kinase activity apparently do not use this signal although they are also internalized via coated pits.

SIGNIFICANCE OF bFGF BINDING TO HEPARAN SULFATE PROTEOGLYCAN AT THE CELL SURFACE

Members of the FGF family bind to heparan sulfate proteoglycans in the extracellular matrix and at the cell surface. This binding can stabilize the growth factors to degradation, induce changes in conformation, and can potentiate their interaction with the high-affinity signal-transducing receptor. The mechanism of these effects is not clear, but cell surface heparan sulfate appears to be required for the biological activity of bFGF.[15] Cell surface heparan sulfate proteoglycan may act as a proximate reservoir for FGF that transfers the growth factor to the high-affinity receptor.[3]

Based on the known metabolism and interactions of syndecan, other functions are possible. The proteoglycan could act as a carrier molecule, carrying aFGF or bFGF, devoid of N-terminal or internal signal peptides, to the cell surface. The proteoglycan could localize bound FGF to specific sites at the cell surface. Competition with various extracellular molecules known to interact with heparan sulfate (e.g., fibronectin and interstitial collagens) could release the bound FGF, enabling the growth factor to diffuse away or be degraded more readily. Internalization of the proteoglycan, as likely occurs via coated pits, could carry FGF into the cell. Indeed, the nuclear translocation sequence on aFGF could be responsible for the nuclear localization of some internalized heparan sulfate (see References 52, 59, and 60).

The FGFs are nearly ubiquitous in growing and nongrowing and nondifferentiating tissues. Immunolocalization of late fetal tissues has placed bFGF predominantly within basement membranes and not at the cell surface where its effects are exerted.[61] Therefore, the action of FGFs is thought to be highly regulated. Cell surface heparan sulfate proteoglycan may regulate the activity of the FGFs by modifying their conformation and controlling their availability to the signal-transducing receptor. In this regulatory scheme, cell surface heparan sulfate would also interact with the bulk of the FGF that is bound to heparan sulfate in the pericellular matrix. This regulatory process is analogous to the control of a complex and expensive machine (e.g., a coal-burning furnace) by a mobile, rapidly responsive

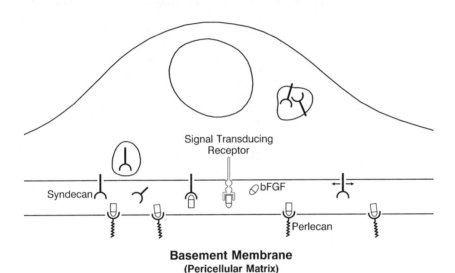

Basement Membrane
(Pericellular Matrix)

FIGURE 3. Possible regulation of FGF activity by heparan sulfate proteoglycans. In this diagram, the basement membrane proteoglycan, perlecan, is used as an example of a proteoglycan in the matrix. For description, see text.

element (e.g., a conveyor belt) that shuttles the coal from its reservoir (e.g., the storage bin). The advantage of this regulatory system is its ability to respond rapidly to changes in the environment.

The high-affinity receptors are specialized machines that convert FGF occupancy into an intracellular signal. The receptors contain two or three extracellular immunoglobulin (Ig) like domains, a transmembrane domain and an intracellular tyrosine kinase domain that cleaves ATP in response to ligand occupancy, initiating a cascade of intracellular events in an on/off manner.[62] Although these transducer machines have higher affinity for FGF compared with cell surface heparan sulfate proteoglycans, they are lower in abundance [0.5–2 × 10^6 per cell compared to 0.5–10 × 10^4 per cell (see Reference 3)], and are apparently not shed from the cell surface. Compared with FGF binding sites in the extracellular matrix, cell surface heparan sulfate proteoglycan turns over much more rapidly. Thus, the cell surface proteoglycan would be able to change abundance quickly in response to environmental changes.

For regulation, we suspect that the cell surface and the pericellular matrix heparan sulfate proteoglycans act in concert (see FIGURE 3). The cell surface proteoglycan, potentially syndecan, can move its bound FGF around the cell, because of its mobility in the plane of the membrane, to present the conformationally appropriate growth factor to high-affinity receptors. This primary regulatory system is supplanted by the heparan sulfate proteoglycan in the matrix which acts as a reservoir of growth factor and as a secondary regulatory system. However, when the cells change shape, as during morphogenesis or in wound repair, they can shed this primary regulator and the matrix proteoglycan would serve as the backup regulatory system. In this way, cellular responsiveness to bFGF could be regulated by controlling the availability of the growth factor to its signal-transducing receptor.

This scheme could be a general mechanism for controlling the availability of otherwise diffusable biological effector molecules to their signal-transducing receptors. Cell surface heparan sulfate proteoglycan acts as the regulator because it can bind a wide variety of polypeptides, is mobile in the plane of the membrane, and can turn over rapidly by shedding or internalization. It is higher in abundance but lower in affinity than the signal-transducing receptor. Because it turns over more rapidly than the heparan sulfate in the extracellular matrix, it may replenish its ligand from this reservoir. This mechanism also fits the situation for TGF-β. This growth factor binds two potential signal-transducing receptors that have a higher affinity but a lower abundance than the TGF-β-binding cell surface proteoglycan, a chondroitin sulfate– and heparan sulfate–containing integral membrane protein that can be shed from the cell surface (see Reference 63 for review). TGF-β binds to biglycan and decorin, proteoglycans of the extracellular matrix which are present in the pericellular space around mesenchymal cells.[37] One difference between the regulatory system described for TGF-β and that for the FGF family is that the TGF-β-binding proteoglycan is not detectable on several cell types that respond to TGF-β, including skeletal muscle myoblasts and vascular endothelial cells. Because TGF-β binds to the proteoglycan core protein, either the GAG-free core protein exists on these cells or they do not have the proposed mode of regulation.

ACKNOWLEDGMENTS

I am grateful to James Svingos for helpful discussions, to Jon Rowland for unpublished data, to Guido David, Michael Klagsbrun and Alan Rapraeger for knowledge of unpublished work, and to Lynda Herrera for preparation of the manuscript.

REFERENCES

1. BURGESS, W. H. & T. MACIAG. 1989. Annu. Rev. Biochem. **58:** 575–606.
2. RIFKIN, D. B. & D. MOSCATELLI. 1989. J. Cell Biol. **109:** 1–6.
3. KLAGSBRUN, M. 1990. Curr. Opinion Cell Biol. **2:** 857–863.
4. OATES, J. A. & A. J. J. WOOD. 1991. N. Engl. J. Med. **324:** 1565–1574.
5. OFOSU, F. A., I. DANISHEFSKY & J. HIRSH, Eds. 1989. Ann. N.Y. Acad. Sci. **556:** 1–498.
6. MOSCATELLI, D. 1987. J. Cell. Physiol. **131:** 123–130.
7. SAUNDERS, S., M. JALKANEN, S. O'FARRELL & M. BERNFIELD. 1989. J. Cell Biol. **108:** 1547–1556.

8. BERNFIELD, M. & R. D. SANDERSON. 1990. Phil. Trans. R. Soc. London **327:** 171–186.
9. HAYASHI, K., M. HAYASHI, M. JALKANEN, J. H. FIRESTONE, R. L. TRELSTAD & M. BERNFIELD. 1987. J. Histochem. Cytochem. **35:** 1079–1088.
10. LEPPÄ S., P. HARKONEN & M. JALKANEN. 1991. Cell Regul. **2:** 1–11.
11. TRAUTMAN, M. S., J. KIMELMAN & M. BERNFIELD. 1991. Development **111:** 213–220.
12. SOLURSH, M., R. S. REITER, K. L. JENSEN, M. KATO & M. BERNFIELD. 1990. Dev. Biol. **140:** 83–92.
13. KIEFER, M. C., J. C. STEPHANS, K. CRAWFORD, K. OKINO & P. J. BARR. 1990. Proc. Nat. Acad. Sci. USA **87:** 6985–6989.
14. YAYON, A., M. KLAGSBRUN, J. D. ESKO, P. LEDER & D. M. ORNITZ. 1990. J. Cell Biol. **111:** 137a.
15. RAPRAEGER, A. C., A. KRUFKA & B. B. OLWIN. 1991. Science **252:** 1705–1708.
16. KRAEMER, P. M. 1971. Biochemistry **10:** 1437–1449.
17. GALLAGHER, J. T. 1989. Curr. Opinion Cell Biol. **1:** 1201–1218.
18. ROBERTS, R., J. T. GALLAGHER, E. SPOONCER, T. D. ALLEN, F. BLOOMFIELD & M. DEXTER. 1988. Nature **332:** 376–378.
19. RATNER, M., D. HONG, M. A. LIEBERMAN, R. P. BURGE & L. GLASER. 1988. Proc. Nat. Acad. Sci. USA **85:** 6992–6996.
20. LINDAHL, U., D. FEINGOLD & L. RODE'N. 1986. Trends Biochem. Sci. **11:** 221–225.
21. GALLAGHER, J. T. 1987. Nature **326:** 136.
22. SCOTT, J. E. 1988. Biochem. J. **252:** 313–323.
23. CASU, B., M. PETITOU, M. PROVASOLI & P. SINAY. 1988. Trends Biochem. Sci. **13:** 221–225.
24. SANDERSON, P. N., T. N. HUCKERBY & A. NIEDUSZYNSKI. 1987. Biochem. J. **243:** 175–181.
25. JOHANSSON, S. & M. HOOK. 1980. Biochem J. **187:** 521–544.
26. LEE, M. K. & A. D. LANDER. 1991. Proc. Nat. Acad. Sci. USA **88:** 2768–2772.
27. LORIES V., J.-J. CASSIMAN, H. VAN DER BERGHE & G. DAVID. 1989. J. Biol. Chem. **264:** 7009–7016.
28. RUOSLAHTI, E. 1989. J. Biol. Chem. **264:** 13369–13372.
29. CLEMENT, B., B. SEGUI-REAL, J. R. HASSELL, G. R. MARTIN & Y. YAMADA. 1989. J. Biol. Chem. **264:** 12467–12471.
30. KJELLEN, L., A. OLDBERG, K. RUBIN & M. HOOK. 1977. Biochem. Biophys. Res. Commun. **74:** 126.
31. DAVID, G., V. LORIES, B. DECOCK, P. MARYNEN, J.-J. CASSIMAN & H. VAN DEN BERGHE. 1990. J. Cell Biol. **111:** 3165–3176.
32. SCHWARTZ, A. L. 1990. Annu. Rev. Immunol. **8:** 195–229.
33. SANDERSON, R. D. & M. BERNFIELD. 1988. Proc. Nat. Acad. Sci. USA **85:** 9562–9566.
34. MARYNEN, P., J. ZHANG, J.-J. CASSIMAN, H. VAN DER BERGHE & G. DAVID. 1989. J. Biol. Chem. **264:** 7017–7024.
35. SEGARINI, P. R. & S. M. SEYEDIN. 1988. J. Biol. Chem. **263:** 8366–8370.
36. CHEIFETZ, S. & J. MASSAGUE. 1989. J. Biol. Chem. **264:** 12025–12028.
37. YAMAGUCHI, Y., D. M. MANN & E. RUOSLAHTI. 1990. Nature **346:** 281–284.
38. JALKANEN, S., M. JALKANEN, R. BARGATZE, M. TAMMI & E. C. BUTCHER. 1988. J. Immunol. **141:** 1615–1623.
39. ARUFFO, A., I. STAMENKOVIC, M. MELNICK, C. B. UNDERHILL & B. SEED. 1990. Cell **61:** 1303–1313.
40. OLDBERG, A., E. G. HAYMAN & E. RUOSLAHTI. 1981. J. Biol. Chem. **256:** 10847–10852.
41. SELDIN, D. C., K. F. AUSTEN & R. L. STEVENS. 1985. J. Biol. Chem. **260:** 11131–11139.
42. MALI, M., P. JAAKOLA, A.-M. ARVILOMMA & M. JALKANEN. 1990. J. Biol. Chem. **205:** 6884–6889.
43. LOTTEAU, V., L. TEYTON, A. PEPERAUX, T. NILSSON, L. KARLSSON, S. L. SCHMID, V. QUARANTA & P. A. C. PETERSON. 1990. Nature **348:** 600–605.
44. IZZARD, C. S., R. RADINSKY & L. A. CULP. 1986. Exp. Cell. Res. **154:** 320–336.
45. MURPHY-ULRICH, J. E. & M. HOOK. 1989. J. Cell Biol. **109:** 1309–1319.
46. CROSS, G. A. M. 1990. Annu. Rev. Cell Biol. **6:** 1–39.
47. THESLEFF, I., M. JALKANEN, S. VAINIO & M. BERNFIELD. 1988. Dev. Biol. **129:** 565–572.
48. JALKANEN, M., A. RAPRAEGER, S. SAUNDERS & M. BERNFIELD. 1987. J. Cell Biol. **105:** 3087–3096.

49. YANAGASHITA, M. & V. C. HASCALL. 1984. J. Biol. Chem. **259:** 10260–10270.
50. ISHIHARA, M., N. S. FEDARKO & H. E. CONRAD. 1987. J. Biol. Chem. **262:** 4708–4716.
51. ISHIHARA, M. & H. E. CONRAD. 1989. J. Cell Physiol. **138:** 467–476.
52. FEDARKO, N. S. & H. E. CONRAD. 1989. J. Cell Biol. **102:** 587–597.
53. SAKSELA, O. & D. B. RIFKIN. 1990. J. Cell Biol. **110:** 767–775.
54. LOW, M. D. & A. R. X. PRESARD. 1988. Proc. Nat. Acad. Sci. USA **85:** 980–984.
55. DAVID, G. & M. BERNFIELD. 1981. J. Cell Biol. **91:** 281–286.
56. GOLDSTEIN, J. L., M. S. BROWN, R. G. W. ANDERSON, D. W. RUSSELL & W. J. SCHNEIDER. 1985. Annu. Rev. Cell Biol. **1:** 1–39.
57. ROTH, M. G., C. DOYLE, J. SAMBROOK & M. J. GETHING. 1986. J. Cell Biol. **102:** 1271–1283.
58. KTISTAKIS, N. T., D. THOMAS & M. G. ROTH. 1990. J. Cell Biol. **111:** 1393–1407.
59. ISHIHARA, M., N. S. FEDARKO & H. E. CONRAD. 1986. J. Biol. Chem. **261:** 13575–13580.
60. IMAMURA, T., K. ENGLEKA, X. ZHAN, Y. TOKITA, R. FOROUGH, D. RAEDER, A. JACKSON, J. A. M. MAIER, T. HLA & T. MACIAG. 1990. Science **249:** 1507–1574.
61. GONZALEZ, A. M., M. BUSCAGLIA, M. ONG & A. BAIRD. 1990. J. Cell Biol. **110:** 753–765.
62. LEE, R. L., A. E. JOHNSON, L. S. COUSENS, V. A. FRIED & L. T. WILLIAMS. 1989. Science **245:** 57–60.
63. MASSAGUE, J. 1990. Annu. Rev. Cell Biol. **6:** 597–641.

Characterization of a Non–Tyrosine Kinase FGF-Binding Protein

BRADLEY B. OLWIN, LAURA W. BURRUS,
MICHAEL E. ZUBER, AND BRUCE LUEDDECKE

Department of Biochemistry
420 Henry Mall
University of Wisconsin
Madison, Wisconsin 53706

INTRODUCTION

Polypeptide growth factors are implicated in both the regulation of vascular development and in the maintenance of the vascular system. A number of specific growth factors that promote wound healing in healing-compromised animals are also potent angiogenic agents.[1–3] Among these factors is basic fibroblast growth factor (bFGF), a member of the FGF family, which is comprised to date of 7 related polypeptides. *In vitro,* bFGF and acidic FGF (aFGF) are potent mitogens for cell types from ectoderm, mesoderm, and endoderm.[3,4] These growth factors are also implicated in the regulation of several developmental processes which include repression of skeletal muscle,[5–7] cardiac muscle,[8] adipocyte,[9] and chondrocyte differentiation,[10] promotion of neuronal outgrowth, and enhancement of neuronal survival.[11–16] Furthermore, aFGF, bFGF, and k-FGF stimulate mesoderm induction in the amphibian embryo,[17–19] mitogenesis in Swiss 3T3 cells,[3] and repression of skeletal muscle differentiation in myoblast cultures (unpublished data).[6,7] These data argue that FGFs are crucial mediators of embryonic development. Although seven distinct polypeptides comprise the FGF family (TABLE 1), differences in the biological activities, sites of action, or metabolism of these related factors have not been established.

Several aspects of FGF biology are unique to this family of growth factors. aFGF and bFGF, the best characterized FGFs, possess no signal peptide sequence yet are ubiquitously distributed and localized in some tissues to the extracellular matrix.[3] Several different forms of bFGF have been identified that arise from alternative translational initiation at CUG codons upstream of the Kozak consensus site for AUG start codons.[20,21] These forms appear to be differentially distributed within the cell; AUG-initiated translational products are found in the cytoplasm and extracellular matrix, while some CUG-initiated bFGF forms appear to be localized to the nucleus.[22] The relationship of cellular localization to the biological activities of FGF has not been established. Int-2, an FGF family member, is similar to aFGF and bFGF as it has a signal peptide sequence that is inefficiently cleaved and is found in both the cytoplasm and nucleus of cells, again as a result of alternative translational initiation at upstream CUG codons.[23] Of the remaining four FGF family members, three have "traditional" signal peptide sequences and appear to be secreted via described mechanisms involving recognition and cleavage of signal peptide sequences.[3]

TABLE 1. Structural Characteristics of the FGF Family

Growth Factor	No. of Amino Acids (Human)	Sequence Identity to bFGF	References
Acidic FGF (FGF-1)	155	53	41, 42
Basic FGF (FGF-2)	155, 196, 210	100	43, 41
Int-2 (FGF-3)	245, 274	42	44, 45
k-FGF/HST (FGF-4)	206	42	46
FGF-5 (FGF-5)	268	43	47
FGF-6 (FGF-6)	198	—	48, 49
KGF (FgF-7)	194	28	50

FGF-binding proteins can be grouped into three distinct classes (TABLE 2) including three membrane-associated tyrosine kinases (cek 1 or flg, cek 2, and cek 3 or bek),[24–29] syndecan, a membrane-associated heparan sulfate proteoglycan (Rapraeger and Olwin, unpublished data),[30–32] and a membrane-associated protein that retains high affinity for aFGF and bFGF.[33] The tyrosine kinase–containing FGF receptors are most similar to ret tyrosine kinase, the platelet-derived growth factor receptors, and the CSF-1 receptor.[24,34,25,29,27] This group of FGF-binding proteins have been proposed to function as "classical" growth factor receptors for several FGFs.[24,34,29] The second distinct class of FGF-binding proteins are the heparan sulfate proteoglycans.[35–38] Although it is likely that several types of HSPG interact with FGF, to date one membrane-bound HSPG, syndecan, and an identified extracellular matrix HSPG have been shown to bind bFGF (Rapraeger and Olwin, unpublished data).[30,38] HSPGs are proposed to function as storage depots for FGF[36,3] since the affinity of these proteins for FGF is several orders of magnitude less than that of the tyrosine kinases (TABLE 2). However, the participation of the HSPGs in FGF signal transduction is essential, as we have recently shown that both high-affinity binding and biological activity of aFGF and bFGF in fibroblasts and myoblasts requires heparan sulfate.[52] An example of the third class of FGF-binding proteins was identified and first characterized by our group.[33] This membrane-associated protein exhibits high affinity for aFGF and bFGF and contains no homology to other proteins including tyrosine or serine/threonine kinases. Sequencing of cDNA clones predicts a signal peptide, and a membrane-spanning region, suggesting the protein may function as an FGF receptor. Further details including the complete primary amino acid sequence and structural analysis will be described elsewhere. The significance of all three classes of FGF-binding proteins as well as their respective contributions to transducing FGF-mediated biological processes is not known. The recent identification of

TABLE 2. FGF-Binding Proteins

Binding Protein	Protein Structure	FGFs Bound	Affinity for FGF
cek 1 (flg)	Tyrosine kinase	bFGF, aFGF	100 pM[26]
cek 2	Tyrosine kinase	?	?
cek 3 (bek)	Tyrosine kinase	bFGF, k-FGF	100 pM[26]
cfr	Unknown membrane	bFGF, aFGF	≤ 1 nM (unpublished data)
Syndecan	Heparan/chondroitin-sulfate Proteoglycan	bFGF	≥ 10 nM (unpublished data)

these molecules should serve to greatly increase our understanding of their roles in FGF biology.

RESULTS

The presence of high levels of FGF and FGF-binding proteins in embryonic tissues strongly suggests that these factors play crucial roles in the developing organism. During chick embryogenesis, FGF binding can be detected as early as 2 days following fertilization (day 2) and remains relatively constant until organogenesis is completed by day 7 (FIGURE 1).[39] From day 7 to day 13 the FGF binding drops precipitously, and has disappeared from both heart and skeletal muscle by day 19, but is retained in brain, eye, and liver, although at lower levels than in early development.[39] Cross-linking of FGF to chick embryonic membrane FGF-binding proteins reveals at least two proteins that appear to decline in parallel with the loss of FGF binding.[39]

Because of the relatively high levels of FGF-binding proteins present during chick embryogenesis, we chose to use embryonic chick for the isolation of FGF-binding proteins. Two FGF-binding proteins have been purified from embryonic chick, a 120-kDa membrane-bound tyrosine kinase (bek or cek 1)[29] and a 150-kDa membrane-bound nontyrosine kinase.[33] Cross-linking of ^{125}I-aFGF or ^{125}I-bFGF to crude preparations of day 7 chick embryo membranes reveals a major product that migrates at 165 kDa and appears to bind both aFGF and bFGF (FIGURE 2).

To purify the 165-kDa product (150 kDa after subtraction of FGF mass) membranes were solubilized with 1% Triton X-100, applied to a wheat germ lectin affinity column, glycoproteins eluted with N-acetylglucosamine, and FGF-binding proteins purified over aFGF or bFGF affinity columns. Cross-linking of either ^{125}I-aFGF or ^{125}I-bFGF to the eluted fractions demonstrated purification of a major FGF-binding complex migrating at 165 kDa, as well as minor complexes migrating at 120 kDa and at the top of the separating gel.[33] Analysis of the protein content of FGF affinity purified fractions by coommassie/silver staining of sodium dodecyl sulfate-polyacrylamide gels revealed three major polypeptides migrating at 150, 70, and 45 kDa (cfr, p70, and p45). The 150-kDa protein contains N-linked carbohydrate, but did not contain detectable covalently linked heparan sulfate.[33] Neither p70 nor p45 contain detectable N-linked carbohydrate or heparan sulfate, suggesting they may be intracellular proteins that copurify with cef.[33] Since the FGF-binding protein is a component of a 165-kDa complex and is a membrane glycoprotein, the logical candidate protein for binding FGF was cfr (cysteine-rich FGF receptor).

Identification of FGF-binding proteins present in the purified pool was achieved via a ligand blotting technique. Fractions purified by aFGF-Sepharose affinity chromatography were subjected to sodium dodecyl sulfate-polyacrylamide gel electrophoresis (SDS-PAGE) at 4°C in the absence of a reducing agent. Fractions were then separated by SDS-PAGE, transferred to Immobilon (Millipore, Inc.) membranes, and incubated with ^{125}I-aFGF or ^{125}I-bFGF in the presence or absence of excess unlabeled aFGF, bFGF, and epidermal growth factor.[33] Two additional Immobilon strips were incubated with or without a monoclonal antibody to cfr and visualized using alkaline phosphatase (Lueddecke & Olwin, unpublished data). These results (FIGURE 3) unequivocally identify cfr as the FGF-binding protein present in the purified fractions. Additional experiments utilizing the monoclonal antibody to remove cfr from wheat germ lectin fractions suggest that cfr is the major FGF-binding protein present (Burrus and Olwin, unpublished data).

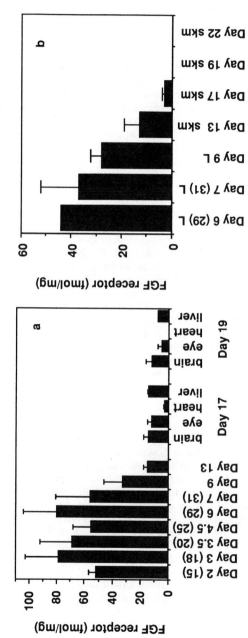

FIGURE 1. FGF receptor levels in embryonic chick. **a:** FGF receptor levels in whole chick embryo from day 2 through day 13 and in brain, eye, heart, and liver from day 17 and 19 embryonic membranes. **b:** FGF receptor levels in limbs (L) from day 6 to day 9 and in skeletal muscle (skm) from day 13 to day 22 (one day post-hatching). FGF receptor levels were determined by specific binding of ^{125}I-aFGF to staged embryos. Membrane assays contained 250 pM ^{125}I-aFGF (saturating) $\pm$ a 100-fold molar excess of unlabeled aFGF. Specific binding of ^{125}I-aFGF reported as fmol ^{125}I-aFGF bound per mg membrane protein is equivalent to FGF receptor per mg protein assuming stoichiometric interaction of ^{125}I-aFGF with FGF receptor. Numbers in parentheses refer to the developmental stage.[51] For each data point at least two independent extracts were analyzed in two separate assays, each performed in triplicate except for day 6 limb, where 1 extract was analyzed. For day 2 to day 7 embryos, 20 μg of protein was present per assay. For older embryos, assays were performed with 20 μg and when less than 10 fmol/mg bound ^{125}I-aFGF was detected, 50 μg of membrane protein was used. Data are plotted as the mean and standard deviation of these values. (Reprinted from Reference 39 with permission.)

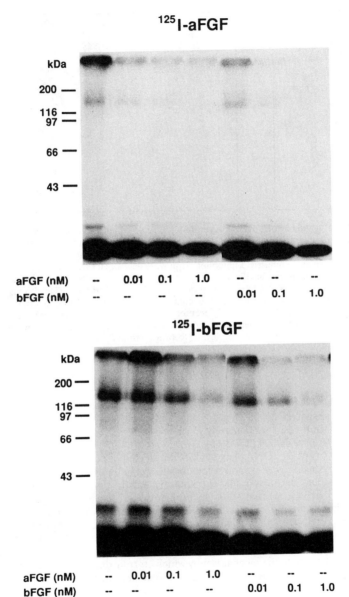

FIGURE 2. Cross-linking of ^{125}I-aFGF and ^{125}I-bFGF to chick membranes. Chick membranes (200 mg) were incubated with 100 pM ^{125}I-aFGF (top) or ^{125}I-bFGF (bottom) for 1 hr at 22°C and then covalently cross-linked with disuccinimidyl suberate as described.[33] Increasing concentrations of unlabeled aFGF or bFGF were used to specifically compete for binding of ^{125}I-FGF to FGF receptor. After completion of the cross-linking reaction, membranes were solubilized with 1% Triton X-100, separated by 7.5% SDS-PAGE, and subjected to autoradiography for 7 days. Molecular weight standards are indicated at the left. (Reprinted from Reference 33 with permission.)

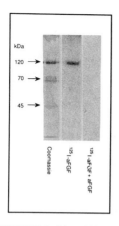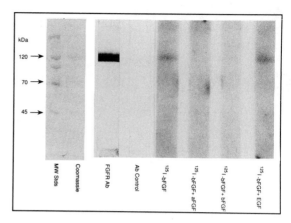

FIGURE 3. Ligand blotting of FGF receptor with ^{125}I-aFGF and ^{125}I-bFGF. Highly purified cfr fractions were separated by SDS-PAGE under nonreducing conditions and electroblotted onto Immobilon membranes at 4°C. Strips of Immobilon containing MW standards and cfr fractions were stained with Coommassie. Immobilon strips were developed as described previously.[33] The cfr polypeptide migrates at 120 kDa rather than 150 kDa when electrophoresed under nonreducing conditions. The cfr polypeptide binds ^{125}I-aFGF and ^{125}I-bFGF with high affinity. Unlabeled aFGF and bFGF, but not EGF, compete for binding of ^{125}I-FGF to cfr, demonstrating appropriate specificity for an FGF receptor. Two Immobilon strips used for the ligand blot were also incubated with and without a monoclonal antibody that immunoprecipitates noncovalently associated ^{125}I-FGF/cfr complexes. (Reprinted from Reference 33 with permission.)

DISCUSSION

We have isolated and characterized a 150-kDa polypeptide present in chick embryos that specifically binds aFGF and bFGF with high affinity. Starting with 40 dozen chick embryos, approximately 15 μg of purified cfr is recovered with an overall fold purification of 28,000 (TABLE 3).

Sequencing of cfr cDNAs has identified a signal peptide, N-linked carbohydrate attachment consensus sites, and a membrane spanning domain, but no tyrosine or serine/threonine conserved protein kinase catalytic domains. Thus, these data support our hypothesis that cfr is an FGF receptor, however, the mode of signal transduction from the extracellular to the intracellular milieu is unclear. The copurification of p70 and p45, which are most likely intracellular proteins, with cfr suggests these proteins form a complex that may be involved in FGF-mediated signal transduction. Purification of p70 and p45 and sequencing of their respective cDNAs may help to elucidate the mechanism by which cfr transduces FGF-mediated signals across the plasma membrane. Although we speculate that cfr function as an FGF

TABLE 3. Cfr Purification

Fraction	FGF receptor (μg)	FGFR (μg/mg protein)	Purification (fold)	Yield (%)
Membranes	38	0.015	1	100
Wheat germ lectin pool	42	5.6	470	110
aFGF-Sepharose pool	15	330	28,000	39

receptor, other plausible functions can be proposed for cfr. The protein could function as a "scavenger" or clearance receptor for extracellular FGF, assuming the heparan sulfate proteoglycans function solely as storage reservoir and are not responsible for clearance of FGF. Several groups have recently reported that bFGF and int-2 are targeted to the nucleus depending on the site of translational initiation. An additional report provides evidence that internalized FGF is delivered to the nucleus during the G1 phase of the cell cycle.[40] It is possible that cfr functions to internalize FGF for mediation of unknown FGF-dependent intracellular biological processes with cfr functioning in a manner analogous to steroid transport proteins. Finally, cfr may function as a specific transporter of FGF from the intracellular to the extracellular environment. This possibility seems least likely as cfr appears to undergo standard routing through the Golgi apparatus, evidenced by the presence of a defined signal peptide sequence and mature N-linked carbohydrate. It thus remains unclear how FGF, without a signal peptide, gains access to the Golgi lumen for binding and transport by cfr.

The relationship of cfr to other FGF-binding proteins is unknown. Two additional distinct classes of FGF-binding proteins include the transmembrane tyrosine kinases (three known members) and the heparan sulfate proteoglycans that reside both in the plasma membrane (syndecan) and in the extracellular matrix. Experiments are in progress to determine whether cfr associates with the tyrosine kinases or heparan sulfate proteoglycans. Obviously, the number of FGFs and FGF-binding proteins argues for a large number of complex interactions to generate the diversity of biological activities ascribed to the FGFs. Elucidation of the mechanisms involved will provide challenges for future research efforts.

REFERENCES

1. GREENHALGH, D. G., *et al.* 1990. PDGF and FGF stimulate wound healing in the genetically diabetic mouse. Am. J. Pathol. **136:** 1235–1246.
2. TSUBOI, R., Y. SATO & D. B. RIFKIN. 1990. Correlation of cell migration, cell invasion, receptor number, proteinase production, and basic fibroblast growth factor levels in endothelial cells. J. Cell. Biol. **110:** 511–517.
3. BURGESS, W. H. & T. MACIAG. 1989. The heparin-binding (fibroblast) growth factor family of proteins. Annu. Rev. Biochem. **58:** 575–606.
4. OLWIN, B. B. 1989. Heparin-binding growth factors and their receptors. Cytotechnology **2:** 351–365.
5. OLWIN, B. B. & S. D. HAUSCHKA. 1988. Cell surface fibroblast growth factor and epidermal growth factor receptors are permanently lost during skeletal muscle terminal differentiation in culture. J. Cell Biol. **107:** 761–769.
6. CLEGG, C. H., *et al.* 1987. Growth factor control of skeletal muscle differentiation: commitment to terminal differentiation occurs in G1 phase and is repressed by fibroblast growth factor. J. Cell Biol. **105:** 949–956.
7. LINKHART, T. A., C. H. CLEGG & S. D. HAUSCHKA. 1981. Myogenic differentiation in permanent clonal myoblast cell lines: regulation by macromolecular growth factors in the culture medium. Dev. Biol. **86:** 19–30.
8. KOHTZ, D. S., *et al.* 1989. Growth and partial differentiation of presumptive human cardiac myoblasts in culture. J. Cell Biol. **108:** 1067–1078.
9. NAVRE, M. & G. M. RINGOLD. 1989. Differential effects of fibroblast growth factor and tumor promoters on the initiation and maintenance of adipocyte differentiation. J. Cell Biol. **1:** 1857–1863.
10. KATO, Y. & M. IWAMOTO. 1990. Fibroblast growth factor is an inhibitor of chondrocyte terminal differentiation. J. Biol. Chem. **265:** 5903–5909.
11. GENSBURGER, C., G. LABOURDETTE & M. SENSENBRENNER. 1987. Brain basic fibroblast

growth factor stimulates the proliferation of rat neuronal precursor cells in vitro. FEBS Lett. **217:** 1–5.

12. UNSICKER, K., *et al.* 1987. Astroglial and fibroblast growth factors have neurotrophic functions for cultured peripheral and central nervous system neurons. Proc. Nat. Acad. Sci. USA **84:** 5459–5463.

13. RYDEL, R. E. & L. A. GREENE. 1987. Acidic and basic fibroblast growth factors promote stable neurite outgrowth and neuronal differentiation in cultures of PC12 cells. J. Neurosci. **7:** 3639–3653.

14. WALICKE, P., *et al.* 1986. Fibroblast growth factor promotes survival of dissociated hippocampal neurons and enhances neurite extension. Proc. Nat. Acad. Sci. USA **83:** 3012–3016.

15. MORRISON, R. S., *et al.* 1986. Basic fibroblast growth factor supports the survival of cerebral cortical neurons in primary culture. Proc. Nat. Acad. Sci. USA **83:** 7537–7541.

16. TOGARI, A., *et al.* 1985. The effect of fibroblast growth factor on PC12 cells. J. Neurosci. **5:** 307–316.

17. PATERNO, G. D., *et al.* 1989. Mesoderm-inducing properties of INT-2 and kFGF: two oncogene-encoded growth factors related to FGF. Development **106:** 79–83.

18. KIMELMAN, D. & M. KIRSCHNER. 1987. Synergistic induction of mesoderm by FGF and TGF-beta and the identification of an mRNA coding for FGF in the early *Xenopus* embryo. Cell **51:** 869–877.

19. SLACK, J. M., *et al.* 1987. Mesoderm induction in early *Xenopus* embryos by heparin-binding growth factors. Nature **326:** 197–200.

20. PRATS, H., *et al.* 1989. High molecular mass forms of basic fibroblast growth factor are initiated by alternative CUG codons. Proc. Nat. Acad. Sci. USA **86:** 1836–1840.

21. FLORKIEWICZ, R. Z. & A. SOMMER. 1989. Human basic fibroblast growth factor gene encodes four polypeptides: three initiate translation from non-AUG codons. Proc. Nat. Acad. Sci. USA **86:** 3978–3981.

22. BUGLER, B., F. AMALRIC & H. PRATS. 1991. Alternative initiation of translation determines cytoplasmic or nuclear localization of basic fibroblast growth factor. Mol. Cell. Biol. **11:** 573–577.

23. ACLAND, P., *et al.* 1990. Subcellular fate of the Int-2 oncoprotein is determined by choice of initiation codon. Nature **343:** 662–665.

24. MANSUKHANI, A., *et al.* 1990. A murine fibroblast growth factor (FGF) receptor expressed in CHO cells is activated by basic FGF and Kaposi FGF. Proc. Nat. Acad. Sci. USA **87:** 4378–4382.

25. PASQUALE, E. B. 1990. A distinctive family of embryonic protein–tyrosine kinase receptors. Proc. Nat. Acad. Sci. USA **87:** 5812–5816.

26. DIONNE, C. A., *et al.* 1990. Cloning and expression of two distinct high-affinity receptors cross-reacting with acidic and basic fibroblast growth factors. Embo J **9:** 2685–2692.

27. PASQUALE, E. B. & S. J. SINGER. 1989. Identification of a developmentally regulated protein–tyrosine kinase by using anti-phosphotyrosine antibodies to screen a cDNA expression library. Proc. Nat. Acad. Sci. USA **86:** 5449–5453.

28. RUTA, M., *et al.* 1989. Receptor for acidic fibroblast growth factor is related to the tyrosine kinase encoded by the fms-like gene (FLG). Proc. Nat. Acad. Sci. USA **86:** 8722–8726.

29. LEE, P. L., *et al.* 1989. Purification and complementary DNA cloning of a receptor for basic fibroblast growth factor. Science **245:** 57–60.

30. KIEFER, M. C., *et al.* 1990. Ligand-affinity cloning and structure of a cell surface heparan sulfate proteoglycan that binds basic fibroblast growth factor. Proc. Nat. Acad. Sci. USA **87:** 6985–6989.

31. SAUNDERS, S., *et al.* 1989. Molecular cloning of syndecan, an integral membrane proteoglycan. J. Cell Biol. **108:** 1547–1556.

32. RAPRAEGER, A., *et al.* 1985. The cell surface proteoglycan from mouse mammary epithelial cells bears chondroitin sulfate and heparan sulfate glycosaminoglycans. J. Biol. Chem. **260:** 11046–11052.

33. BURRUS, L. W. & B. B. OLWIN. 1989. Isolation of a receptor for acidic and basic fibroblast growth factor from embryonic chick. J. Biol. Chem. **264:** 18647–18653.

34. REID, H. H., A. F. WILKS & O. BERNARD. 1990. Two forms of the basic fibroblast growth

factor receptor–like mRNA are expressed in the developing mouse brain. Proc. Nat. Acad. Sci. USA **87:** 1596–1600.

35. SAKSELA, O. & D. B. RIFKIN. 1990. Release of basic fibroblast growth factor–heparan sulfate complexes from endothelial cells by plasminogen activator-mediated proteolytic activity. J. Cell Biol. **110:** 767–775.

36. RIFKIN, D. B. & D. MOSCATELLI. 1989. Recent developments in the cell biology of basic fibroblast growth factor. J. Cell Biol. **109:** 1–6.

37. MOSCATELLI, D. 1988. Metabolism of receptor-bound and matrix-bound basic fibroblast growth factor by bovine capillary endothelial cells. J. Cell Biol. **107:** 753–759.

38. SAKSELA, O., *et al.* 1988. Endothelial cell–derived heparan sulfate binds basic fibroblast growth factor and protects it from proteolytic degradation. J. Cell Biol. **107:** 743–751.

39. OLWIN, B. B. & S. D. HAUSCHKA. 1990. Fibroblast growth factor receptor levels decrease during chick embryogenesis. J. Cell Biol. **110:** 503–509.

40. BALDIN, V., *et al.* 1990. Translocation of bFGF to the nucleus is G1 phase cell cycle specific in bovine aortic endothelial cells. Embo J. **9:** 1511–1517.

41. LOBB, R. R. & J. W. FETT. 1984. Purification of two distinct growth factors from bovine neural tissue by heparin affinity chromatography. Biochemistry **23:** 6295–6299.

42. THOMAS, K. A., C. M. RIOS & S. FITZPATRICK. 1984. Purification and characterization of acidic fibroblast growth factor from bovine brain. Proc. Nat. Acad. Sci. USA **81:** 357–361.

43. BOHLEN, P., *et al.* 1984. Isolation and partial molecular characterization of pituitary fibroblast growth factor. Proc. Nat. Acad. Sci. USA **81:** 5364–5368.

44. DICKSON, C. & G. PETERS. 1987. Potential oncogene product related to growth factors. (Letter.) Nature **326.**

45. MOORE, R., *et al.* 1986. Sequence, topography and protein coding potential of mouse *int-2:* a putative oncogene activated by mouse mammary tumour virus. EMBO J. **5:** 919–924.

46. DELLI-BOVI, P., *et al.* 1988. Processing, secretion, and biological properties of a novel growth factor of the fibroblast growth factor family with oncogenic potential. Mol. Cell. Biol. **8:** 2933–2941.

47. ZHAN, X., *et al.* 1988. The human FGF-5 oncogene encodes a novel protein related to fibroblast growth factors. Mol. Cell. Biol. **8:** 3487–3495.

48. DE LAPEYRIERE, O., *et al.* 1990. Structure, chromosome mapping and expression of the murine Fgf-6 gene. Oncogene **5:** 823–832.

49. MARICS, I., *et al.* 1989. Characterization of the HST-related FGF.6 gene, a new member of the fibroblast growth factor gene family. Oncogene **4:** 335–340.

50. RUBIN, J. S., *et al.* 1989. Purification and characterization of a newly identified growth factor specific for epithelial cells. Proc. Nat. Acad. Sci. USA **86:** 802–806.

51. HAMBURGER, V. & H. L. HAMILTON. 1951. A series of normal stages in the development of the chick embryo. J. Morphol. **88:** 49–92.

52. RAPRAEYER, A. C., A. KRUFKA & B. B. OLWIN. 1991. Requirement of heparan sulfate for bFGF-mediated fibroblast growth and myoblast differentiation. Science **252:** 1705–1708.

New Observations on the Intracellular Localization and Release of bFGF[a]

DANIEL B. RIFKIN, NATALINA QUARTO, PAOLO
MIGNATTI, JOZEF BIZIK, AND DAVID MOSCATELLI

Department of Cell Biology
New York University Medical Center and
Raymond and Beverly Sackler Foundation Laboratory
550 First Avenue
New York, New York 10016

INTRODUCTION

Basic fibroblast growth factor (bFGF) is produced by many cells both *in vivo* and *in vitro* and induces a variety of responses in cells and tissues.[1,2] The growth factor has a number of properties that are similar to other competence factors such as binding to plasma membrane receptors that are tyrosine kinases, multipotent functions, and unique expression during development.[1,2] Basic FGF does have several unique properties that differentiate it from most other growth factors. One of these is the lack of a signal peptide normally required for vectorial translation into the endoplasmic recticulum and secretion. In fact, no evidence exists that bFGF is secreted. A second property of bFGF, shared with int-2, is the use of CUG codons for the initiation of translation of high molecular weight (HMW) forms of the molecule in addition to the AUG used to generate an 18-kDa form.[3–5] The subcellular distribution of these different forms of bFGF is not equivalent.[6]

These observations raise two important questions concerning the biology of FGF. They are, one, what is the mechanism directing the distribution of the different molecular weight forms of the protein to different subcellular compartments and, two, is bFGF released from living cells? We have approached these two questions in a series of experiments described below.

RESULTS AND DISCUSSION

The observation by Renko *et al.* that the HMW forms of bFGF are found predominantly in the nucleus,[6] while the lowest molecular weight form is found predominantly in the cytoplasm, suggested that the additional amino acid sequences found in HMW bFGF contain nuclear targeting sequences. To check this hypothesis, a chimeric cDNA was made encoding the amino-terminal extensions of the HMW bFGF fused to β-galactosidase, a bacterial protein.[7,8] When control or chimeric DNA was transfected into HeLa cells, the following observations were made. Transient expression of wild-type β-galactosidase in HeLa cells resulted in fluorescence only in the cytoplasm when the cells were observed by indirect immunofluorescence. In

[a]This work was supported by grants CA 34282 and CA 42229 from the National Institutes of Health, BC544C from the American Cancer Society, and the Italian Association for Cancer Research to P.M.

contrast, when the DNA encoding the NH_2-bFGF-β-galactosidase chimeric protein was transfected into HeLa cells, almost all of the fluorescence observed with anti-β-galactosidase antibodies was nuclear. When the distribution of β-galactosidase in immunoreactive cells was quantitated, greater than 95% of the positive cells transfected with the hybrid DNA construct showed nuclear staining, whereas 90% of the cells transfected with wild-type β-galactosidase DNA displayed cytoplasmic staining.

The distribution of β-galactosidase enzymatic activity in nuclear and cytoplasmic fractions in cells transfected with either wild-type β-galactosidase or the NH_2-bFGF-β-galactosidase constructs was also determined. Again, cells transfected with wild-type β-galactosidase DNA contained approximately 85% of the total enzymatic activity in the cytoplasm, while cells transfected with the chimeric DNA contained 80–85% of the enzymatic activity in the nuclear fractions. Finally, immunoblots of these fractions revealed β-galactosidase antigen in the nuclear fraction from cells transfected with the NH_2-bFGF-β-galactosidase DNA, while cells transfected with wild-type β-galactosidase DNA contained immunoreactive material in the cytoplasmic fractions. The immunoreactive material in the nuclear extracts of cells transfected with the chimeric DNA had a slightly higher molecular weight than wild-type β-galactosidase consistent with the presence of the additional bFGF sequences.

These experiments, plus others using additional DNA constructs of HMW bFGF sequences fused to pyruvate kinase, a eukaryotic cytoplasmic enzyme, all indicate that the amino-terminal amino acid sequences unique to HMW bFGF confer upon the protein its characteristic nuclear localization. It will be interesting to determine if nuclear bFGF performs unique functions. Consistent with this possibility are preliminary data indicating that cell lines expressing only HMW bFGF have a phenotype different from that observed with cells expressing only 18 kDa or all four forms of bFGF.

The second question, whether bFGF is released from living cells, has been difficult to answer using traditional approaches since it is essentially impossible to exclude the contribution of a small number of dead cells from any traditional culture system. A solution to this problem is to conduct experiments measuring the effects of altering bFGF levels on isolated, single living cells. Under these conditions all bFGF must come from the cell under observation, and the possible contribution from dying cells is eliminated. A positive result, the ability to modulate a cellular function by altering endogenous bFGF levels, would imply a true autocrine role of bFGF. If the modulation occurred outside the cell, a positive result would also imply that bFGF was released from viable cells.

We have attempted such an analysis.[9] We have used transfected NIH 3T3 cells which express high levels of bFGF.[10] The assay employed is the phagokinetic assay originally developed by Albrecht-Buchler.[11] In this assay cells are plated on coverslips coated with colloidal gold particles. As the cells move on the coverslips, they clear away the gold particles. After an appropriate period of time, the cells are fixed and the size of the tracks formed by the migrating cells can be quantitated with an image analyzer. The parental NIH 3T3 cells which contain almost no bFGF, a clone, B1, that expresses as much bFGF as SK-Hep 1 cells, or clone B3, which express at least 10-fold more bFGF than B1 cells, were plated at densities of 0.5 cells per 16-mm well insuring that the majority of wells with cells contained only one cell. The migration of the cells was found to be proportional to the amount of bFGF produced. NIH 3T3 cells moved the least, B1 cells moved more, and B3 cells moved the most. If exogenous bFGF was added to the cultures, the migration of the NIH 3T3 cells and the B1 cells was increased to the level observed with the B3 cells. No effect of bFGF addition was seen with the B3 cells. We conclude that the endogenous expression of bFGF is so high in these cells that their receptors are saturated. Previous data regarding these cells published by Moscatelli and Quarto[12] are consistent with this hypothesis. The size of the tracks was also found to be time dependent. Normally, the incubation time was limited to no more than 16 hours to insure that the cells did not divide. When anti-bFGF immunoglobulin G (IgG) was added to wells containing either NIH 3T3 or B1 cells, no effect on the tract size of the NIH 3T3 cells was observed, while the track size of the B1 cells was reduced to a size almost equivalent to that of NIH 3T3 cells. The addition of nonimmune IgG had no effect on cell movement. Pretreatment of the cells with suramin to remove any bound bFGF bound before the inhibition of the experiment had no effect on migration and demonstrated that the results were not due to bFGF carried over from prior treatment of the cells.

If we assume that the migratory tracks measured come from living cells and that the anti-bFGF IgG works from the outside of the cell, these results indicate that bFGF is released from living cells and can act as a true autocrine factor. The ability to measure quantitative responses to endogenous bFGF should allow the execution of additional experiments probing the actual mechanism by which bFGF is released. For example, the addition of drugs that block the action of the multidrug resistance protein can be used to test the hypothesis that bFGF released is via this transporter protein.

REFERENCES

1. RIFKIN, D. B. & D. MOSCATELLI. 1989. J. Cell Biol. **109:** 1–6.
2. BURGESS, W. H. & T. MACIAG. 1989. Annu. Rev. Biochem. **58:** 575–606.
3. PRATS, H., M. KAGHAD, A. C. PRATS, M. KLAGSBRUN, J. M. LELIAS, P. LIAU ZUN, P. CHALON, J. P. TAUBER, F. AMALRIC, J. A. SMITH & D. CAPUT. 1989. Proc. Nat. Acad. Sci. USA **86:** 1836–1840.
4. FLORKIEWICZ, R. Z. & A. SOMMER. 1989. Proc. Nat. Acad. Sci. USA **86:** 3978–3981.
5. ACLAND, P., M. DIXON, G. PETERS & C. DICKSON. 1990. Nature **343:** 662–665.
6. RENKO, M., N. QUARTO, T. MORIMOTO & D. B. RIFKIN. 1990. J. Cell. Physiol. **144:** 108–114.
7. KALDERON, D., B. ROBERTS, W. D. RICHARDSON & A. E. SMITH. 1984. Cell **89:** 499–509.
8. QUARTO, N., F. FINGER & D. B. RIFKIN. 1991. J. Cell Physiol. **147:** 311–318.
9. MIGNATTI, P., T. MORIMOTO & D. B. RIFKIN. Manuscript submitted.
10. QUARTO, N., D. TALARICO, A. SOMMER, R. FLORKIEWICZ, C. BASILICO & D. B. RIFKIN. 1989. Oncogene Res. **5:** 101–110.
11. ALBRECHT-BUEHLER, G. 1977. Cell **11:** 395–404.
12. MOSCATELLI, D. & N. QUARTO. 1989. J. Cell Biol. **109:** 2519–2527.

Sequestration and Release of Basic Fibroblast Growth Factor[a]

I. VLODAVSKY,[b,f] P. BASHKIN,[b] R. ISHAI-MICHAELI,[b]
T. CHAJEK-SHAUL,[c] R. BAR-SHAVIT,[b]
A. HAIMOVITZ-FRIEDMAN,[d] M. KLAGSBRUN,[e]
AND Z. FUKS[d]

[b]Department of Oncology
[c]Department of Medicine B
Hadassah University Hospital
Jerusalem 91120, Israel

[d]Department of Radiation Oncology
Memorial Sloan-Kettering Cancer Center
New York, New York 10021

[e]Department of Surgery
The Children's Hospital
Boston, Massachusetts 02115

INTRODUCTION

Fibroblast growth factors are a family of structurally related polypeptides characterized by high affinity to heparin.[1] They are highly mitogenic for vascular endothelial cells (EC) and are among the most potent inducers of neovascularization[2] and mesenchyme formation.[3] This gene family includes the prototypes aFGF (acidic fibroblast growth factor) and bFGF (basic FGF) which, unlike most other polypeptide growth factors, are primarily cell-associated proteins, consistent with the lack of a conventional signal sequence for secretion.[4] Biochemical and immunohistochemical studies performed *in vitro* and *in vivo* demonstrated that both factors are tightly adsorbed to the extracellular matrix (ECM), presumably by their avid affinity for heparinlike glycosaminoglycans (GAGs).[5-10] In fact, bFGF can be regarded as an ECM component required for supporting cell proliferation and differentiation.[11,12] The association between the FGF prototypes and heparan sulfate (HS) was found to protect these polypeptides from inactivation and proteolytic modification.[13] These results and the finding that anti-bFGF antibodies inhibit the proliferation of vascular EC in the absence of added bFGF[14] suggest that under certain conditions bFGF is secreted by a nontraditional pathway not involving a hydrophobic signal peptide. A possible, but yet unproved pathway for deposition of bFGF into ECM is by forming an intracellular complex with HS found in the cell cytoplasm and nucleus,[15] followed by insertion into the cell surface and/or liberation into the ECM. Despite the

[a]This work was supported by Public Health Service Grants CA-30289 (I.V.) and CA-52462 (Z.F.) awarded by the National Cancer Institute, Department of Health and Human Services; and by grants from the USA-Israel Binational Science Foundation and the German-Israel Foundation for Scientific Research and Development (I.V.).

[f]Address correspondence to Dr. Vlodavsky at the Department of Oncology, Hadassah Hospital, P.O.B. 12000, Jerusalem 91120, Israel.

ubiquity of FGF in normal tissues, EC proliferation in these tissues is usually very low, with turnover time measured in years.[16] This raises the question of how these EC growth factors are prevented from acting on the vascular endothelium continuously and in response to what signals they become available for stimulation of capillary EC proliferation. One possibility is that they are sequestered from their site of action by means of binding to HS in cell surfaces and ECM and saved for emergencies such as wound repair and neovascularization.[5]

In the present article we summarize results obtained by us and other investigators on: (1) ECM storage and distribution of bFGF in normal tissues; (2) interaction of bFGF with heparan sulfate in ECM and cell surfaces; (3) release of cell-surface- and ECM-bound bFGF by heparinlike molecules, enzymes, and intact cells; and (4) autocrine effect of bFGF in repair of radiation damage to endothelial cells.

BASIC FGF IS STORED WITHIN BASEMENT MEMBRANES
IN VITRO AND IN VIVO

Our studies on the control of cell proliferation and tumor progression by its local environment focused on the interaction of cells with the ECM produced by cultured corneal and vascular endothelial cells.[12,17,18] This ECM closely resembles the subendothelium in vivo in its morphological appearance and molecular composition. It contains collagens (mostly types III and IV, with smaller amounts of types I and V), proteoglycans (mostly heparan sulfate and dermatan sulfate proteoglycans, with smaller amounts of chondroitin sulfate proteoglycans), laminin, fibronectin, entactin and elastin.[12,17,18] We have demonstrated that EC and other cell types plated in contact with the subendothelial ECM no longer require the addition of soluble FGF in order to proliferate and express their differentiated functions.[17,18] They do, however, deposit bFGF into the subendothelial ECM.[6,19] In subsequent studies bFGF has been extracted from the subendothelial ECM produced in vitro[6] and from basement membranes of the cornea,[8] suggesting that ECM may serve as a reservoir for bFGF. Antisera directed against the internal and amino-terminal portions of bFGF were used to localize bFGF within frozen sections of whole bovine cornea. Basic FGF appeared to be concentrated in a fine line delineating the outer aspect of Bowman's membrane (FIGURE 1b). Intense bFGF staining was observed throughout the entire thickness of Descemet's membrane (FIGURE 1d). Stromal staining, when seen, was of low intensity and appeared mainly within the inner half of the corneal stroma.[8]

We characterized the distribution of bFGF in normal human tissues by immunohistochemical staining of unprocessed fresh frozen sections of various organs.[20] Expression of bFGF in normal human tissues was ubiquitously detected in the basement membranes of all size blood vessels, but was not found in epithelial basement membranes of all variety of tissues tested. Intensity and patterns of localization in blood vessels were consistent in various tissues, but varied among different regions of the vascular bed. Whereas homogeneous and intense immunoreactivity was observed in large and intermediate size blood vessels, heterogeneity of expression was found in capillaries, with the most intense immunoreactivity observed in the anastamosing sites of branching regions of capillary beds.[20] Strong staining for bFGF was also found in cardiac muscle fibers, smooth muscle cells of midsize blood vessels, the gut, and the myometrium. Basic FGF was also found in a subset of central nervous system neurons and cerebellar Purkinje cells, but not in glial cells.[20] The localization of bFGF in cardiac muscle and nerve cells suggests a role in stimulating myocardial collateral vascularization and neuronal development and regeneration.

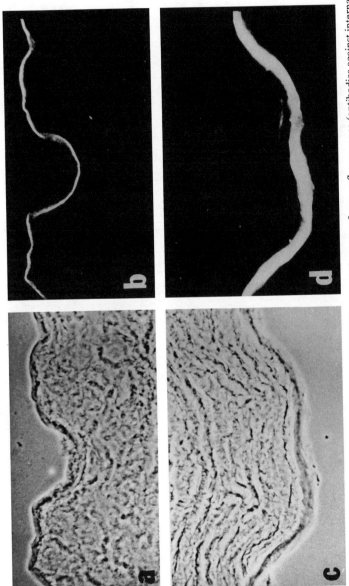

FIGURE 1. Storage in basement membranes of the cornea of bFGF-like growth factors. **a:** Immunofluorescence (antibodies against internal sequence of bFGF) (b, d) and corresponding phase contrast (a, c) micrographs of bFGF within frozen sections of normal bovine corneas. **b:** Outer aspect of Bowman's membrane. **d:** Entire thickness of Descemet's membrane.

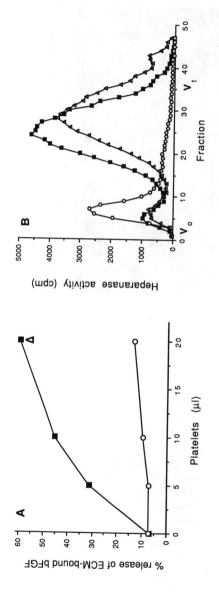

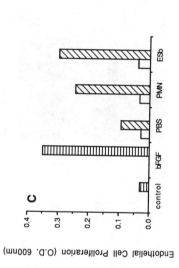

Immunohistochemical staining performed by other investigators revealed the localization of bFGF in basement membranes of diverse tissues of the rat fetus[9] and in nuclei, intercalated discs, and endomysium of muscle fibers of the bovine heart.[10] Substantial extracellular accumulation of bFGF around atrial, but not ventricular, myofibers was attributed to a more extensive endomysium in the atria.[10] Basic FGF was also identified in developing neovascular capillaries of the bovine retina[21] and in striated muscle cells of the chick embryo.[22]

EXTRACELLULAR MATRIX BINDING AND RELEASE OF bFGF

Scatchard analysis of ^{125}I-bFGF binding to ECM revealed that bFGF binds to ECM with an affinity (k_D = 610 nM) lower than that reported for binding of FGF to low-affinity, presumably heparinlike sites on cell surfaces.[23] It appears that bFGF binds specifically to HS in ECM and basement membranes, since up to 90% of the bound growth factor was displaced by heparin, HS, or HS-degrading enzymes (i.e. heparanase), but not by unrelated GAGs (i.e., chondroitin sulfate, keratan sulfate, hyaluronic acid) or enzymes (chondroitinases AC and ABC).[23,24] There was little or no binding of bFGF to ECM pretreated with heparanase, but there was no effect to pretreatment with chondroitinase ABC. Oligosaccharides derived from depolymerized heparin, and as small as the hexasaccharide, efficiently released the ECM-bound FGF. In contrast, there was little or no release of bFGF by totally desulfated heparin, N-desulfated heparin, or N-acetylated heparin,[23] indicating that N-sulfate groups of heparin are required for its high-affinity binding of bFGF. FGF released from ECM was biologically active, as indicated by its stimulation of EC proliferation. By using both in vitro and in vivo experimental systems, we have demonstrated that requirements for release of ECM-bound FGF were the same regardless of whether the FGF was exogenously added and bound to ECM, or was an endogenous constituent of intact basement membranes.[8,23,25]

FIGURE 2. Release of bFGF by platelets, neutrophils, and lymphoma cells. **A:** ECM-coated wells were incubated (3 hours, 24°C) with ^{125}I-bFGF (2.5×10^4 cpm/well). Unbound FGF was washed (×4) and the ECM incubated (2 hours, 37°C) with increasing amounts of platelets (1 µl = 1.5×10^6 platelets) in the absence (squares) or presence of either 10 µg/ml carrageenan lambda (circles), or a mixture of protease inhibitors (triangle). Released ^{125}I-bFGF was counted in a γ counter. Radioactivity released into the incubation medium is expressed as percent of total ECM-bound ^{125}I-FGF. **B:** Sulfate-labeled ECM was incubated (3 hours, 37°C, pH 6.8) with 2.5×10^8 platelets in the absence (squares) or presence of either 10 µg/ml carrageenan lambda (circles), or a mixture of protease inhibitors (triangles). Labeled degradation products released into the incubation medium were analyzed by gel filtration on Sepharose 6B, as described.[26] **C:** Neutrophils (PMN), ESb lymphoma cells, or phosphate-buffered saline (PBS) alone was incubated (3 hours, 37°C) on top of the inner side of bovine corneas (diagonal hatching), or in regular tissue culture plastic wells (white columns). Aliquots (20 µl) of the 0.5 ml incubation medium were added to sparsely seeded endothelial cells (EC) on days 2 and 4 after seeding and tested for stimulation of EC proliferation, as described.[25] EC proliferation was also measured in the absence (control) and presence (bFGF) of 5 ng/ml bFGF added on days 2 and 4 after seeding. Each point is the average of triplicate wells, and the variation in cell number (i.e., uptake of methylene blue) was less than 10%.

Heparanase Activity Expressed by Platelets, Neutrophils, and Metastatic
Tumor Cells Releases Active bFGF from ECM

An endoglycosidase (heparanase) that specifically degrades heparan sulfate was found to be a most efficient specific releaser of active bFGF from ECM.[8,23] Heparanase has been shown to participate in the extravasation of blood-borne tumor cells and activated cells of the immune system.[26–28] Our results suggest that this enzyme may also participate in tumor angiogenesis, through mobilization of ECM-bound EC growth factors. To investigate whether heparanase, expressed by various normal and malignant cells, is involved in release of bFGF from ECM, we first identified molecules (i.e., carrageenan lambda, N-acetylated heparin) that inhibit the enzyme but do not release the ECM-bound bFGF. Using these inhibitors, we demonstrated that heparanase activity expressed by platelets, neutrophils, and lymphoma cells is involved in release of active bFGF from ECM (FIGURE 2A) and Descemet's membrane of bovine corneas (FIGURE 2C).[25] Regardless of the source of heparanase and of whether release of bFGF was brought about by a pure enzyme, intact cells, or cell lysates, inhibition of FGF release correlated with inhibition of heparanase activity, measured by release from ECM of sulfate-labeled HS degradation products (FIGURE 2B).[25] Our results indicate that both endogenous and exogenously added bFGF are accessible to release by heparanase and that the released factor is active in promoting endothelial cell proliferation (FIGURE 2C). We suggest that heparanase activity expressed by tumor cells may not only function in cell migration and invasion,[26–28] but at the same time may also elicit an indirect neovascular response by means of releasing the ECM-resident FGF. Alterations in basement membrane structure and turnover that are associated with tumor progression may thus be responsible for the onset of angiogenic activity upon the transition of an *in situ* carcinoma from the prevascular to the vascularized state. Likewise, platelets, mast cells,[29] and activated cells of the immune system (i.e., macrophages, neutrophils, T lymphocytes) that are often attracted by tumor cells may indirectly stimulate tumor angiogenesis by means of their heparanase activity.[25] These cells may also elicit an angiogenic response in the process of inflammation and wound healing. Apart from HS degrading enzymes, ECM-bound bFGF is released by plasmin as a noncovalent complex with HS proteoglycan or GAG.[30] Basic FGF complexed to HS proteoglycan stimulates production of plasminogen activator (PA) by EC which may further facilitate release of bFGF from ECM.[30] PA may also stimulate release of ECM-bound FGF through stimulation of heparanase-mediated degradation of HS in ECM.[31]

Several studies and our own results indicate that μg quantities of heparin and HS inhibit the mitogenic activity of bFGF, but at the same time stabilize and protect the molecule from inactivation.[13] It is therefore conceivable that bFGF is stored in ECM in a highly stable but relatively inactive form, as also indicated by the highly stable ECM-resident growth-promoting activity, as compared to that of bFGF in a fluid phase. Release from ECM of bFGF as a complex with an HS fragment may yield a form of bFGF that is more stable than free bFGF and yet capable of binding the high-affinity plasma membrane receptors.[30] Moreover, bFGF complexed to HS fragments should diffuse through the stroma to the target cells more readily than free bFGF since bFGF-HS complexes do not bind to the ECM.[32] Recent studies indicate that heparin and HS are in fact required for binding of bFGF to high-affinity cell surface receptors. In these experiments bFGF did not bind to HS-deficient CHO mutant cells. Binding to high-affinity receptor sites was however fully restored by the addition of heparin and HS at concentrations as low as 20 ng/ml.[33] It is conceivable that binding of heparin or HS imposes on the bFGF molecule the conformation necessary for optimal interaction with its high-affinity cell surface receptor.

Since FGF is a pluripotent factor, restriction of its release at the vicinity of the target cell would ensure that its effect is not systemic.[5] Sequestration of FGF may also provide for a constant extracellular source of the angiogenic stimulus under conditions where the initial release of the growth factor may be as a bolus after cell death.[32] We propose that storage of EC growth factors in ECM prevents them from acting on the vascular endothelium, thus maintaining a very low rate of EC turnover and vessel growth. On the other hand, release from storage in ECM may elicit a localized EC proliferation and neovascularization in processes such as wound healing, inflammation, and tumor development.

bFGF is an ECM Component Required for Supporting Endothelial Cell Proliferation and Neuronal Differentiation

To investigate the involvement of the ECM-associated bFGF in its induction of cell proliferation and differentiation, we compared the ECM produced by PF-HR-9 mouse endodermal carcinoma cells, which do not synthesize bFGF, to ECM produced by PF-HR-9 cells transfected with the gene for bFGF.[11] PF-HR-9 cells secrete an underlying basement-membrane-like ECM which is not mitogenic for vascular EC. This ECM also failed to induce extension of neuronal processes in PC12 cells.[11] When lysates of HR9 cells were subjected to heparin-Sepharose chromatography, no growth factor activity for vascular EC was eluted with a salt gradient ranging from 0.2 to 3.0 M NaCl. In contrast, a clonal cell population (HR9/bFGF), derived by transfecting HR9 cells with bFGF expression vector, produced bFGF. ECM deposited by HR9 cells had little or no effect on both clonal growth of vascular EC and neurite outgrowth by PC12 cells. In contrast, ECM produced by PF-HR-9 cells that expressed bFGF induced both clonal growth of vascular EC and extensive neuronal differentiation of PC12 cells.[11] These effects were inhibited in the presence of anti-bFGF antibodies, indicating that bFGF is an ECM component required for supporting proliferation of EC and differentiation of PC12 cells.

The mechanism by which components of the ECM may modulate the activity and bioavailability of bFGF is the subject of current investigations. It was found that PK-A phosphorylates bFGF in the receptor binding domain and that the phosphorylated bFGF has a greater affinity for its high-affinity receptor.[34] When bFGF is associated with ECM proteins (i.e., fibronectin, laminin, collagen), the site of phosphorylation is masked and the mitogen is no longer a substrate for PK-A. In the presence of heparin, bFGF is phosphorylated by PK-A at a cryptic site that is not a PK-A consensus sequence.[34] The identification of a functional effect of heparin on the phosphorylation of bFGF, coupled with the demonstration of a novel interaction between bFGF, fibronectin, and laminin, emphasizes the potential role of components of the ECM in regulating the activity, stability, and storage of bFGF.[34] These results suggest that the bioavailability and function of bFGF are regulated by a complex array of biochemical interactions with the proteins, proteoglycans, and glycosaminoglycans present in the extracellular milieu and the cytoplasm.[34]

Release of Cell-Surface-Associated bFGF by Glycosyl-phosphatidylinositol-Specific Phospholipase C

Heparan sulfate proteoglycans (HSPGs) are ubiquitous constituents of mammalian cell surfaces.[35] They play a role in cellular interactions and in growth control during morphogenesis, cell differentiation, and proliferation.[35,36] A small proportion of the cell-surface-associated HSPG is anchored via a covalently linked glycosyl-phosphatidylinositol residue that has its fatty acyl chains buried in the lipid bilayer.[37] This HSPG can be released from the cell surface by treatment of cells with a glycosyl-phosphatidylinositol-specific phospholipase C (PI-PLC).[38] Exposure of cells (i.e., bovine aortic endothelial and smooth muscle cells) to PI-PLC resulted in release of growth-promoting activity which was neutralized by anti-bFGF antibodies. Under the same conditions there was no release of mitogenic activity from cells (i.e., 3T3 fibroblasts) that expressed little or no bFGF. Moreover, addition of PI-PLC to sparsely seeded vascular endothelial cells resulted in a marked stimulation of cell proliferation (FIGURE 3), but there was no mitogenic effect to PI-PLC on 3T3 fibroblasts. In other experiments (FIGURE 4), cells were first incubated with ^{125}I-bFGF, washed free of unbound bFGF, and exposed (1 hour, 37°C) to PI-PLC. About 6.5% of the cell-bound ^{125}I-bFGF was released by PI-PLC, as compared to 20% and 48% release induced by treatment with heparitinase and heparinase, respectively. In contrast, PI-PLC failed to release ^{125}I-bFGF from the subendothelial extracellular matrix, as compared to 35% and 60% release of the ECM-bound ^{125}I-bFGF by bacterial heparinase and heparitinase, respectively (FIGURE 4) (Bashkin *et al.*, submitted for publication). Cells were also metabolically labeled with $Na_2^{35}SO_4$, washed free of unincorporated radioactivity, and exposed to PI-PLC. Up to 75% of the sulfate-labeled material released in the presence of PI-PLC was precipitable by 0.05% cetylpiridinium chloride (CPC) in 0.6 M NaCl. Under these conditions CPC precipitates mostly HS glycosaminoglycans.[39] Exposure of cells to bacterial heparitinase released about twice as much sulfate-labeled material and mitogenic activity as

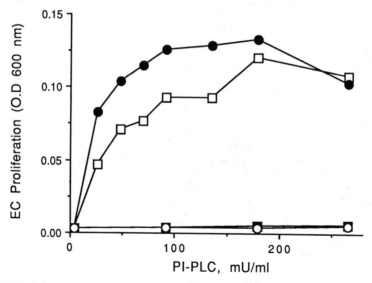

FIGURE 3. Mitogenic effect of PI-PLC on endothelial cells. Cells were seeded at a low density (200 cells/well of a 96-well plate) in 0.2 ml DMEM supplemented with 10% heat inactivated calf serum. Increasing amounts of PI-PLC (filled circles), recombinant PI-PLC (open squares), heparinase (filled squares), or chondroitinase ABC (open circles) were added on days 2 and 4. The number of cells was determined on day 8 by the methylene blue uptake assay.[25]

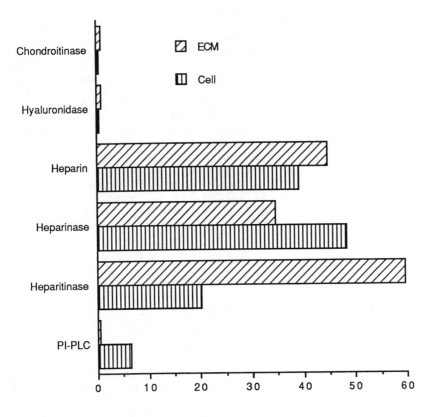

Release of bFGF (% of total bound)

FIGURE 4. PI-PLC-mediated release of cell-bound and ECM-bound ^{125}I-bFGF. ECM (diagonal hatching) and 3T3 fibroblasts (vertical hatching) were incubated with ^{125}I-FGF for 2 hours at 4°C. The ECM and cell monolayers were then washed with cold PBS to remove unbound ^{125}I-FGF and incubated (1 hour, 37°C) in the absence or presence of 1.4 U/ml PI-PLC, 0.2 U/ml heparitinase, 0.05 U/ml heparinase, 10 μg/ml low M_r heparin, 0.1 U/ml hyalurondiase, or 0.1 U/ml chondroitinase ABC. The supernatants were collected and counted for determination of released iodinated material.

compared to PI-PLC, indicating that a portion of the cell-surface-associated bFGF is bound to HSPG that are not anchored by the glycosyl-PI anchor. A transmembrane HSPG that binds bFGF has recently been cloned and characterized.[40] Our results indicate that a small proportion (3–8%) of the cellular content of bFGF is associated with glycosyl-PI-anchored HSPG. We suggest that bFGF associated with HS on the cell surface may exert both autocrine and paracrine effects, provided that the bFGF molecules are adequately presented to high-affinity cell surface receptor sites, following cleavage by enzymes such as PI-PLC and heparitinase (Bashkin *et al.*, submitted for publication). On the other hand, high-affinity receptors may not be accessible for FGF binding and signal transduction in the absence of low-affinity

HS-binding sites, as demonstrated in a recent study performed with HS-deficient CHO cells.[33] Since many cells that synthesize bFGF are capable of responding to bFGF through interaction with cell surface receptors,[1,2,32] it is conceivable that cellular bFGF may reach an extracytoplasmic compartment where it can bind to its high-affinity receptor and act as an autocrine growth and differentiation promoting factor. This possibility is supported by the findings that (1) 3T3 fibroblasts transfected with a bFGF cDNA fused to a signal sequence no longer bind exogenous bFGF[41] and (2) neutralizing anti-bFGF antibodies inhibit bFGF-dependent functions, such as cell migration, proliferation, and growth in soft agar in the absence of exogenously added bFGF.[14] However, no defined mechanism for release of bFGF has been described. HSPG released by PI-PLC is internalized via a cell surface receptor that recognizes myo-inositol phosphate. The internalized HSPG is processed by a nonlysosomal pathway, and a portion of the HS side chains may then be transported into the nucleus.[38] PI-PLC may thus also function in translocation of bFGF from the cell surface into the nucleus, where it may exert direct effects on genomic DNA. In fact, preferential localization of bFGF in the cell nucleus has been demonstrated in several cell systems.[42]

Sequestration of bFGF by Heparinlike Molecules in the Vessel Wall

Our studies on the fate of bFGF administered into the blood of rats revealed that bFGF is rapidly sequestered by heparinlike molecules in the luminal surface of the vascular endothelium. FIGURE 5 demonstrates that as early as one minute after a bolus administration of ^{125}I-bFGF, only 13–15% of the initial plasma concentration of bFGF remained in the circulation, as compared to about 50% remaining when heparin was administrated 2–5 min prior to the ^{125}I-bFGF. Similar results were obtained by Whalen et al.[43] Administration of heparin 5 minutes after the ^{125}I-bFGF resulted in rapid restoration of the plasma concentration of bFGF to the level measured in animals that received heparin prior to bFGF. At 10 minutes, the plasma concentration of bFGF was four- to fivefold higher in the presence than in the absence of heparin, regardless of whether the heparin was administered prior to or after the ^{125}I-bFGF. These results suggest that a large proportion of the bFGF in the bloodstream is rapidly sequestered by heparinlike molecules in the luminal surface of the vessel wall and can be released by heparin, in a manner similar to that observed with ^{125}I-bFGF bound to cultured vascular endothelial cells or to their underlying ECM. Analysis of the distribution of ^{125}I-bFGF in various organs revealed that bFGF was predominantly cleared by the liver and to a lesser extent by the kidneys (50% and 15% of the amount administered, respectively). Immunoprecipitation of tissue extracts followed by sodium dodecyl sulfate-polyacrylamide gel electrophoresis (SDS-PAGE) and autoradiography revealed the presence of intact ^{125}I-bFGF in both organs.

AUTOCRINE EFFECT OF bFGF IN REPAIR OF RADIATION DAMAGE IN ENDOTHELIAL CELLS

Radiation injury to slowly proliferating normal tissues is characterized in early phases by microvascular damage, while atrophy and fibrosis are prominent features of late radiation damage.[44] Ultrastructural studies suggested that capillary endothelial cells are the most sensitive targets for irradiation in the vessel wall.[44] Postradiation proliferation of endothelial and smooth muscle cells was observed in small-caliber arteriols, followed by general thickening of the vessel wall.[44] We have demonstrated that irradiation (20–60 cGy) of EC from several origins resulted in a dose- and time-dependent release of mitogenic factors into the culture medium.[45] Although the cultures exhibited some cell loss due to cell death within the first 48–72 hours after irradiation, our data indicated that the released mitogenic activity was mainly due to an enhanced *de novo* synthesis of growth-promoting factors.[45] Mitogenic assays performed in the presence of neutralizing anti-bFGF and anti-PDGF antibodies revealed that approximately 20% of the mitogenic activity was due to PDGF and 70–80% to bFGF.[45] Northern blot analysis revealed a 5.6- and 4.7-fold

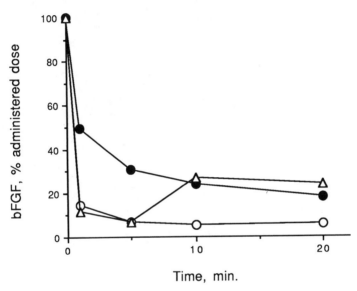

FIGURE 5. Effect of heparin on the rate of bFGF disappearance from the bloodstream. ^{125}I-bFGF was administered as a bolus into the bloodstream of rats (*n* = 6) with (filled circles) or without (open circles) heparin (50 U/rat). Samples of blood (0.3 ml) were drawn from the arterial line at the indicated time intervals and counted in a gamma counter. Some rats received heparin (50 U, intravenously) 5 minutes after the administration of ^{125}I-bFGF (triangles). Data are presented as percentage of the expected blood concentration of ^{125}I-bFGF at zero time.

increase of the 3.7 and 7.0 kilobase (kb) species of the bFGF-specific mRNA, respectively, within 6 hours after delivery of a single dose of 400 cGy.[46] While release of bFGF from irradiated EC may occur in response to perturbation of the cell membrane, the actual mechanism of its secretion remains unknown. The released FGF may participate in the abnormal proliferation of EC reported to obliterate the lumen of small-caliber arteriols in various organs. It may also initiate the postradiation fibrosis observed in chronic radiation damage in normal tissues.[44] In addition, the secreted FGF may have a critical role in the repair of radiation-induced damage in the irradiated endothelial cells themselves. We have demonstrated that soluble and ECM-bound bFGF induce radiation damage repair in EC. In fact, radiation induced in bovine aortic EC a complete cycle of an autoregulated damage-repair pathway, initiated by damage to cellular DNA and followed by an increase in bFGF-specific mRNA, stimulation of bFGF synthesis, and its secretion into the medium. The newly synthesized bFGF stimulates repair of potentially lethal damage (PLDR) acting via an extracellular autocrine loop (inhibitable by anti-bFGF antibodies) and leading to recovery of EC and possibly smooth muscle cells from radiation lesions resulting in restoration of their proliferative capacity.[46]

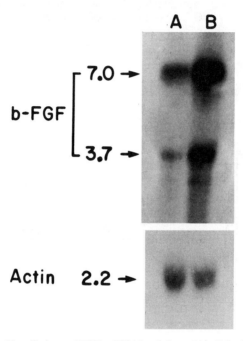

FIGURE 6. Effect of irradiation on bFGF mRNA levels in endothelial cells. Confluent bovine aortic endothelial cells were irradiated with a single dose of 400 cGy. Six hours later the cells were subjected to Northern blot analysis using the pJJ11-1 bFGF cDNA probe. The blots were also probed with a human γ-actin probe to normalize for the amounts of mRNA loaded in each lane.[46] **Lane A:** Control unirradiated cells. **Lane B:** EC irradiated with a single dose of 400 cGy.

CONCLUSIONS

1. Basic FGF is stored in the subendothelial ECM produced by cultured endothelial cells as well as in basement membranes of the cornea and blood vessels *in vivo*.
2. Basic FGF is an ECM component required for supporting endothelial cell proliferation and neuronal differentiation.
3. Basic FGF binds to heparan sulfate and heparinlike molecules in ECM and cell surfaces. It can be released both *in vitro* and *in vivo* by various heparin species and by heparin- and heparan sulfate–degrading enzymes.
4. Platelets, neutrophils and tumor cells release active bFGF from ECM and

basement membranes by means of their heparanase activity. Heparanase may thus function in both tumor cell invasion and neovascularization.

5. Some of the cell-surface- but not ECM-bound bFGF is released in an active form by phosphatidylinositol-specific phospholipase C (PI-PLC), suggesting association with glycosyl-PI-anchored heparan sulfate.

6. Exposure of sparsely seeded vascular EC to PI-PLC results in a marked stimulation of cell proliferation, suggesting that bFGF associated with HS on the cell surface may exert an autocrine effect provided that the bFGF molecules are released and properly presented to high-affinity cell surface receptor sites.

7. Radiation induces in endothelial cells an increased synthesis of the bFGF mRNA and protein, followed by secretion of bFGF into the medium. The secreted bFGF led to recovery of EC from potentially lethal radiation damage as revealed by restoration of their clonogenic capacity.

Our results indicate that bFGF is sequestered from its site of action by binding to heparan sulfate and heparinlike molecules in the ECM and cell surface. It is thus protected and reserved for induction of localized neovascularization upon its release by various enzymes (i.e., heparanase, heparinase, PI-PLC, plasmin) or heparinlike molecules. Release of intracellular bFGF may occur under pathological conditions (i.e., radiation) and the released factor may stimulate repair of a potentially lethal damage.

REFERENCES

1. BURGESS, W. H. & T. MACIAG. 1989. Annu. Rev. Biochem. **58:** 575–606.
2. FOLKMAN, J. & M. KLAGSBRUN. 1987. Science **235:** 442–447.
3. KIMELMAN, D. & M. KIRSCHNER. 1987. Cell **51:** 869–877.
4. ABRAHAM, J. A., A. MERGIA, J. L. WHANG, A. TUMOLO, J. FRIEDMAN, A. HJERRILD, D. GOSPODAROWICZ & C. FIDDES. 1986. Science **233:** 545–548.
5. BAIRD, A. & P. A. WALICKE. 1989. Brit. Med. Bull. **45:** 438–452.
6. VLODAVSKY, I., J. FOLKMAN, R. SULLIVAN, R. FRIDMAN, R. ISHAI-MICHALI, J. SASSE & M. KLAGSBRUN. 1987. Proc. Nat. Acad. Sci. USA **84:** 2292–2296.
7. WEINER, H. L. & J. L. SWAIN. 1989. Proc. Nat. Acad. Sci. USA **86:** 2683–2687.
8. FOLKMAN, J., M. KLAGSBRUN, J. SASSE, M. WADZINSKI, D. INGBER & I. VLODAVSKY. 1988. Am. J. Pathol. **130:** 393–400.
9. GONZALEZ, A-M., M. BUSCAGLIA, M. ONG & A. BAIRD. 1990. J. Cell Biol. **110:** 753–765.
10. KARDAMI, E. & R. R. FANDRICH. 1989. J. Cell Biol. **109:** 1865–1875.
11. ROGELJ, S., M. KLAGSBRUN, R. ATZMON, M. KUROKAWA, A. HAIMOVITZ, Z. FUKS & I. VLODAVSKY. 1989. J. Cell Biol. **109:** 824–831.
12. VLODAVSKY, I., G. KORNER, R. ISHAI-MICHAELI, P. BASHKIN, R. BAR-SHAVIT & Z. FUKS. 1990. Cancer Metastasis Rev. **9:** 203–226.
13. SAKSELA, O., D. MOSCATELLI, A. SOMMER & D. B. RIFKIN. 1988. J. Cell Biol. **107:** 743–751.
14. SATO, Y. & D. B. RIFKIN. 1988. J. Cell Biol. **107:** 1199–1205.
15. ISHIHARA, M., N. S. FEDARKO & H. E. CONARD. 1987. J. Biol. Chem. **262:** 4708–4716.
16. DENEKAMP, J. 1984. Prog. Appl. Microcirc. **4:** 28–38.
17. GOSPODAROWICZ, D., D. DELGADO & I. VLODAVSKY. 1980. Proc. Nat. Acad. Sci. USA **77:** 4094–4098.
18. VLODAVSKY, I., G. M. LUI & D. GOSPODAROWICZ. 1980. Cell **19:** 607–616.
19. VLODAVSKY, I., R. FRIDMAN, R. SULLIVAN, J. SASSE & M. KLAGSBRUN. 1987. J. Cell Physiol. **131:** 402–408.
20. CARDON-CARDO, C., I. VLODAVSKY, A. HAIMOVITZ-FRIEDMAN, D. HICKLIN & Z. FUKS. 1990. Lab. Invest. **63:** 832–840.

21. HANNEKEN, A., G. A. LUTTY, D. S. MCLEAD, F. ROBEY, A. K. HARVEY & L. M. HJELMELAND. 1989. J. Cell. Physiol. **138:** 115–120.
22. SILVERSTEIN, J., S. A. CONSIGLI, K. M. LYSER & C. VER PAULT. 1989. J. Cell Biol. **108:** 2459–2466.
23. BASHKIN, P., M. KLAGSBRUN, S. DOCTROW, C-M. SVAHN, J. FOLKMAN & I. VLODAVSKY. 1989. Biochemistry **28:** 1737–1743.
24. JEANNY, J. C., N. FAYEIN, M. MOENNER, B. CHEVALLIER, D. BARRITAULT & Y. COURTOIS. 1987. Exp. Cell Res. **117:** 63–72.
25. ISHAI-MICHAELI, R., A. ELDOR & I. VLODAVSKY. 1990. Cell Regul. **1:** 833–842.
26. VLODAVSKY, I., Z. FUKS, M. BAR-NER, Y. ARIAV & V. SCHIRRMACHER. 1983. Cancer Res. **43:** 2704–2711.
27. NAKAJIMA, M., T. IRIMURA & G. L. NICOLSON. 1988. J. Cell. Biochem. **36:** 157–167.
28. NAPARSTEK, Y., I. R. COHEN, Z. FUKS & I. VLODAVSKY. 1984. Nature **310:** 241–243.
29. BASHKIN, P., E. RAZIN, A. ELDOR & I. VLODAVSKY. 1990. Blood **75:** 2204–2212.
30. SAKSELA, O. & D. B. RIFKIN. 1990. J. Cell Biol. **110:** 767–775.
31. BAR-NER, M., M. MAYER, V. SCHIRRMACHER & I. VLODAVSKY. 1986. J. Cell. Physiol. **128:** 299–307.
32. RIFKIN, D. B. & D. MOSCATELLI. 1989. J. Cell Biol. **109:** 1–6.
33. YAYON, A., M. KLAGSBRUN, D. E. ESKO, P. LEDER & D. M. ORNITZ. Cell **64:** 841–848.
34. FEIGE, J-J., J. D. BRADEY, K. FRYBURG, J. FARRI, L. C. COUSENS, P. J. BARR & A. BAIRD. 1989. J. Cell Biol. **109:** 3105–3114.
35. HOOK, M., L. KJELLEN, S. JOHANSSON & J. ROBINSON. 1984. Annu. Rev. Biochem. **53:** 847–869.
36. SAUNDERS, S., M. JALKANEN, S. O'FARRELL & M. BERNFIELD. 1989. J. Cell Biol. **108:** 1547–1556.
37. LOW, G. L. & A. R. SALTIEL. 1988. Science **239:** 268–275.
38. ISHIHARA, M., N. S. FEDARKO & H. E. CONRAD. 1987. J. Biol. Chem. **262:** 4708–4716.
39. RODEN, L., J. R. BAKER, J. A. CITONELLI & M. B. MATHEWS. 1972. Methods Enzymol. **28:** 73–140.
40. KIEFER, M. C., J. C. STEPHANS, K. CRAWFORD, K. OKINO & P. J. BARR. 1990. Proc. Nat. Acad. Sci. USA **87:** 6985–6989.
41. YAYON, A. & M. KLAGSBRUN. 1990. Proc. Nat. Acad. Sci. USA **87:** 5346–5350.
42. BALDIN, V., A. M. ROMAN, I. BOSC-BIERNE, F. AMALRIC & G. BOUCHE. 1990. EMBO J. **9:** 1511–1517.
43. WHALEN, G. F., Y. SHING & J. FOLKMAN. 1989. Growth Factors **1:** 157–164.
44. FAJARDO, L. F. & M. BERTHRONG. 1988. Pathol. Annu. **23:** 297–330.
45. WITTE, L., Z. FUKS, A. HAIMOVITZ-FRIEDMAN, I. VLODAVSKY, D. S. GOODMAN & A. ELDOR. 1989. Cancer Res. **49:** 5066–5072.
46. HAIMOVITZ-FRIEDMAN, A., I. VLODAVSKY, A. CHAUDHURI, L. WITTE & Z. FUKS. 1991. Cancer Res. **51:** 2552–2558.

Basic Fibroblast Growth Factor in the Mature Brain and Its Possible Role in Alzheimer's Disease

CARL W. COTMAN AND FERNANDO GÓMEZ-PINILLA

Department of Psychobiology
University of California Irvine
Irvine, California 92717

INTRODUCTION

A fine balance seems to exist in certain brain systems between the adaptive functions, broadly termed plasticity, and the initiation of brain pathology leading to neuronal degeneration. Each process consists of a variety of different molecular cascades that share certain events, several of which are regulated at least in part by basic fibroblast growth factor (bFGF). In this chapter, we will describe relevant data on the anatomical organization of bFGF in the rodent and human brain, the role of bFGF in normal brain function and synaptic plasticity, and its possible role in an age-related neurodegenerative disease, Alzheimer's disease (AD). In this disease it appears that molecular plasticity processes that are initially adaptive may ironically become part of pathological processes.

One of the most important adaptive or plastic functions of the nervous system is that of learning and memory. A particular type of memory called data memory (the learning of lists and facts) involves a specific set of brain circuitries.[1] The definition of these circuitries has evolved from studies in several areas, including lesion studies in rodents, studies on brain imaging in man, and behavioral studies on man with specific types of brain damage. These highly defined circuitries involve the limbic system and an integrative type of association cortex, the entorhinal cortex. It is also known that a cholinergic input from the basal forebrain which is responsive to bFGF and nerve growth factor (NGF) amplifies and modulates the workings of this circuit. Significantly, current studies show that areas involved in data memory are particularly vulnerable to injury and insult, including but not limited to stroke, epilepsy, anoxia, and AD. Our recent research has focused on the possible role of bFGF in data memory and in the degeneration of select neurons in this system due to insult and injury.

The first evidence that bFGF was a neurotrophic factor in brain was reported simultaneously by Morrison et al.[2] and Walicke et al.[3] These groups showed that bFGF added in small quantities to cultured hippocampal cells increased neuronal survival and the number of process-bearing neurons. The effect of bFGF lasted for several days, and various types of brain neurons tested were supported by bFGF.[4] Further work demonstrated that glycosaminoglycans (i.e., heparin and heparan sulfate) interact synergistically with bFGF to increase process outgrowth.[5]

Following studies on cells in culture, various investigators examined the role of bFGF on neuronal survival and growth in vivo. For example, Anderson et al.[6] studied bFGF in a defined neural system, the septal hippocampal system, using two-month-old and two-year-old Sprague-Dawley rats. Septal neurons project to the hippocampus and disruption of this projection causes loss of target-derived trophic support

221

and neuronal degeneration. Septal hippocampal neurons were transected and bFGF was infused into the ventricles. bFGF enhanced neuronal survival of cholinergic neurons in young and old rats. These and similar findings *in vitro* and *in vivo* have paved the way for future work on bFGF in brain.

CELLULAR DISTRIBUTION OF bFGF IN THE NORMAL RAT BRAIN

Recently we used a monoclonal antibody that specifically recognizes the conformation of bFGF associated with its biological activity[7] to study the organization and cellular locus of bFGF in brain. The main locus of bFGF immunoreactivity appeared to be in small cells which resembled astrocytes. Double-staining immunohistochemistry with antibodies to glial fibrillary acidic protein (GFAP), a specific marker for astrocytes, clearly identified and confirmed the nature of these cells. A few bFGF immunostained cells that did not co-stain with anti-GFAP antibody also showed positive immunoreactivity for anti–leucocyte common antigen (LCA) antibody, a marker for macrophage line cells, which include microglia.[8] Other cells such as periventricular cells and perivascular endothelial cells also showed bFGF immunoreactivity, in general agreement with the recognized role of bFGF as a potent angiogenic factor.[9,10]

Only select neuronal elements within particular brain areas showed strong bFGF immunoreactivity. Neurons of CA2 subfield of the hippocampal formation stained throughout the rostrocaudal extent of the hippocampus (FIGURE 1). Neurons in CA2 showed strong bFGF immunoreactivity around the perinuclear area and weak staining in the cytoplasm and processes proximal to the soma. The presence of bFGF in CA2 neurons is quite interesting because this neuronal group is one of the most resistant to aging and epilepsy, and suggests that bFGF may contribute to this property.

Basic FGF-positive neurons were also shown in the septohippocampal nucleus, mainly visualized in horizontal brain sections. Cingulate cortex neurons stained strongly throughout most of their rostrocaudal extents. In the cerebellum, somas of Purkinje cells and neurons within the granular layer showed bFGF immunoreactivity. Deep nuclei of the cerebellum also showed bFGF immunoreactivity. In the brain stem there was strong bFGF immunoreactivity within the facial nerve nucleus and the motor and spinal components of the trigeminal nucleus.

Our results on the locus of bFGF in the brain are partially in disagreement with previous studies by Pettmann *et al.*[11] According to these authors, bFGF is localized exclusively in neurons in the normal brain. In our study, however, only a restricted number of neuronal populations showed bFGF immunoreactivity in the adult rat brain. Possible sources of disagreement between our studies and their studies may be due to differences in the antibodies used and/or differences in the age of the animals since these investigators used immature rats whereas we used adult rats.

REGULATION OF bFGF FOLLOWING LESIONS

Trophic factors may not only participate in the maintenance of the normal cytoarchitecture of the brain but may also play a role on its response to injury. When damaged, neurons appear to react in a manner dependent on the type of injury. For example, neurons that are axotomized generally dedifferentiate and degenerate while the remaining healthy neurons may undergo a sprouting reaction. Neuro-

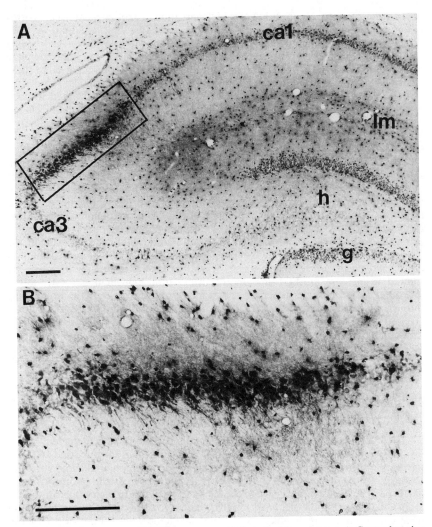

FIGURE 1. Typical pattern of bFGF immunoreactivity in the hippocampus. **A:** Coronal section showing bFGF-positive astrocytes widely distributed in the hippocampus while neuronal bFGF is restricted to the CA2 region (box). The inset (B) shows higher magnification of the CA2 region from A. bFGF was concentrated in the nucleus and within proximal dendritic processes of pyramidal neurons. CA1 and CA3 regions of the hippocampus are labeled as a reference; lm, lacunosum moleculare; h, hilus of the DG; G, granular layer of the DG. Calibration bars (A and B): 100 mm. (From Gómez-Pinilla et al.[16] with permission.)

trophic factors may participate in this response to injury by triggering growth and preventing degeneration. Indeed, previous work has suggested that the sprouting reaction may involve the induction of select neurotrophic factors which increase after injury.[12] bFGF is a good candidate since it acts on a large variety of neurons and, as previous work suggests, bFGF immunoreactivity increases near the site of injury.[13]

In order to study responses in areas of denervation or axotomy, entorhinal cortex lesion studies were performed which cause loss of synapses in the dentate gyrus of the hippocampal formation. After a period beginning within 5–7 days and ending within two or three months, axons sprout, forming a new, virtually complete complement of synapses. In aged animals this process begins after a period of slight delay, but appears to eventually be as effective as it is in younger animals. This overall process and the individual inputs that contribute have been established through quantitative electron microscopy, various light microscopic methods, and biochemical markers. For example, after unilateral lesioning there is a sprouting of cholinergic input from FGF/NGF-sensitive septal cells (for review, see Cotman and Anderson).[14]

After a unilateral entorhinal cortex lesion, the hippocampus ipsilateral to a lesion showed an enhancement of bFGF immunoreactivity in the dentate gyrus (DG) outer molecular layer where the entorhinal fibers terminate. The lesion effects on bFGF immunoreactivity were expressed as a relative increase in the number of bFGF astrocytes, an increase in relative intensity of the bFGF immunoreactivity within astrocytes and a relative increase of the bFGF immunoreactivity in the surrounding extracellular matrix, as compared to the contralateral side. The increase in the number of astrocytes was already evident at postlesion day 2, reached a maximum by day 7, and decreased to about normal levels by day 14. Computer densitometry showed an increase in bFGF immunoreactivity within individual astrocytes in the DG outer molecular layer reaching significance by postlesion day 7 and remaining through postlesion day 14, relative to the contralateral side and untreated controls (FIGURE 2). The extracellular matrix surrounding bFGF-positive astrocytes also showed an increase in bFGF immunoreactivity with a similar time course. Parallel serial sections stained for AChE histochemistry, a marker of axon sprouting by septal cells in the hippocampus,[15] also showed an enhancement of the staining in the DG outer molecular layer. This intensification had a similar time course to that observed with anti-bFGF antibody.[16] Astrocytes surrounding lesion areas showed increased bFGF immunoreactivity in their processes, and their cell bodies appeared to be enlarged and stained darker as compared to bFGF immunoreactive astrocytes of normal rats.

The fact that an increase in bFGF immunoreactivity was detected in the extracellular matrix of the dentate gyrus of the lesion side suggests that bFGF is released from astrocytes and becomes available to other neural elements. Our observations that normal astrocytes show bFGF in the nucleus and perinuclear area and hyperreactive astrocytes around the lesion show bFGF in the cytoplasm suggest a possible translocation of bFGF towards secretion pathways. The mechanism of release of bFGF is largely unknown but probably undergoes a series of posttranslational processing and comparmentalizations in the cytoplasm.[17] In addition, evidence indicates that bFGF reaches the cell surface in a dynamic way, not exclusively as the result of cell death.[18]

POSSIBLE ROLE OF bFGF IN ALZHEIMER'S DISEASE

AD specifically causes neuronal degeneration in select brain areas including the entorhinal cortex, the main input into the hippocampus. Neuronal degeneration in these areas has been described in detail by Hyman and colleagues.[19] The course of degeneration is such that the neurons of the entorhinal cortex (layer II and III) projecting into the hippocampus are among the first affected. As described above, studies suggest that axon sprouting might occur in AD. But while degeneration is a prominent feature of AD, reactive growth in both neurons and glial reactions is exhibited in the disease as well.[20]

Light microscopy can be used for the study of the terminal fields of the cholinergic input in postmortem human normal and AD brain. In the dentate gyrus of the normal brain there is a light pattern of AChE staining from cholinergic input; whereas in AD brain the pattern is intensified in the denervated zone, as predicted from animal models. We suggest that during the early course of the disease, when cells are lost, the remaining cells sprout connections in a compensatory fashion in order for the circuits to be maintained in some state of completion.[14]

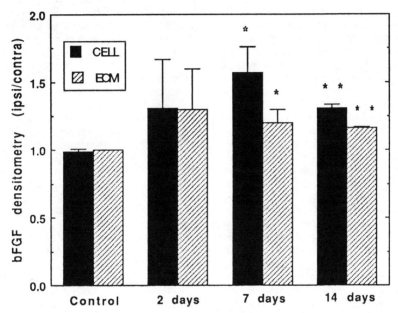

FIGURE 2. Graph illustrating the effects of entorhinal cortex lesions on the relative intensity of bFGF immunoreactivity within the DG outer molecule layer. Computer densitometry readings were taken from individual astrocytes and the surrounding extracellular matrix (ECM) from the immunohistochemical material. The graph shows the ratio of optical density on the ipsilateral (lesioned) side versus the contralateral (control) side. There is a relative increase in bFGF immunoreactivity as compared to intact controls in both astrocytes and extracellular matrix in the dentate gyrus ipsilateral to the lesion, reaching significance by postlesion day 7 ($p < 0.05$) and remaining through day 14 ($p < 0.01$). Error bars are standard error of the mean (*$p < 0.05$; **$p < 0.01$, student t-test). (From Gómez-Pinilla et al.[16] with permission.)

POSSIBLE NEURITIC SPROUTING IN PLAQUE FORMATION

It appears that at least some sprouting occurs in AD and it may be associated with an induction of bFGF. One of the neuropathological hallmarks of AD is the presence of senile plaques in selected brain areas. Plaques are extracellular deposits that consist primarily of a polypeptide product called amyloid. These amyloid deposits appear to consist of an insoluble matrix of protein derived from a larger precursor protein, the amyloid precursor protein. Abnormal proteolysis of this larger protein appears to produce a small amino acid fragment of approximately forty-two amino acids as a byproduct that becomes insoluble and forms a beta-pleated sheet structure, also known as β-amyloid. Plaques also have certain neuritic and cellular involvements, such as astrocytes and microglia. In the observation of our AD material we noticed the presence of small round cholinesterase-positive structures surrounded by astrocytes that were not predicted from animal models. As revealed by thioflavin staining, a specific marker for plaques, these cholinesterase-rich deposits corresponded to plaques. They appeared to form along the areas of interface between degeneration and neuronal sprouting. We suggested that there might be an abortive turning of the sprouting reactions into plaque formation.[20] This hypothesis did not agree with the prevailing theory at the time that plaque formation was due to a collection of degeneration products that accumulated into a massive deposition of insoluble debris. We were further convinced that such a hypothesis might explain the data, however, when we were able to show that certain antibodies specific to the cytoskeletal neurons such as phosphorylated neurofilament antibody stained processes that seemed to be entering and aborizing in the plaques.

In the earlier literature, some reports had suggested that sprouting occurred around plaques. In fact, Ramon y Cajal in 1928 had suggested that sprouts were attracted to plaques by some neurotrophic factor,[21] which was a remarkable insight considering that at the time there was little knowledge regarding the role of neurotrophic factors in brain. Having determined, then, that growth factors might play some role in sprouting in plaques, we proceeded to attempt to identify specific molecular mechanisms that could contribute to such aberrant growth. It is known from cell cultures studies that neuritic growth requires both neurotrophic factor(s) and a suitable substrate(s). We began studying amyloid for possible activity and found that β-amyloid stimulated neural survival and neuritic growth in cultured hippocampal neurons. In order to achieve that effect, however, the concentrations were too high to consider β-amyloid a suitable long-term growth factor in itself, although it might be a participant in the process. Further, β-amyloid did not maintain neurons in culture for more than a few days.[22,23]

In order to explore a possible mechanism that increases bFGF availability to plaques, we performed unilateral entorhinal cortex ablation in rats and tested both hippocampi for bFGF immunoreactivity. These experiments could determine whether denervation induces an increase in astrocytic or neuronal bFGF. The dentate gyrus receives prominent afferents from the entorhinal cortex and the septum. Destruction of entorhinal input in rats[14,15] or neuronal cell death in AD[24] evokes a powerful sprouting of cholinergic septal afferents in the dentate gyrus molecular layer. Results showed an increase in bFGF immunoreactivity in the dentate gyrus outer molecular layer and hilus, ipsilateral to the lesion, in a pattern similar to the sprouting septal fibers found after such lesions (FIGURE 3A and B). Comparison with the AD material showed a similar pattern of bFGF immunoreactivity in the outer molecular layer of hippocampus (FIGURE 3C and D), suggesting that loss of entorhinal input may contribute to the stimulus leading to a local increase of bFGF in the dentate gyrus.

POSSIBLE ROLE OF bFGF IN PLAQUE BIOGENESIS

In our search for known trophic factors that could induce sprouting of neurites into plaques, we turned our attention to bFGF. Accordingly, we have shown that plaques in the dentate gyrus along the sprouting zone were bFGF immunoreactive, as was the area surrounding them[25] (FIGURE 4). At higher magnification, we observed what appeared to be cells around the plaques that were particularly immunoreactive. Double-label immunocytochemistry showed that bFGF immuno-positive cells around bFGF immunostained plaques were actually astrocytes (FIG-

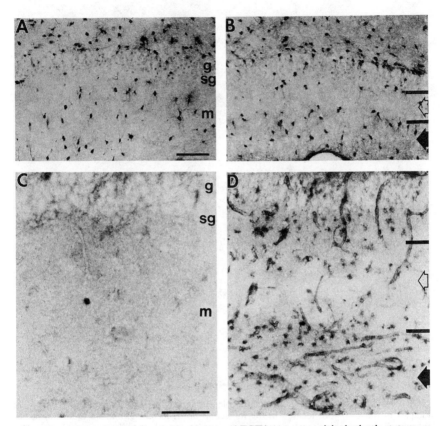

FIGURE 3. Effect of a loss of entorhinal input on bFGF immunoreactivity in the dentate gyrus in rats or in AD. Following entorhinal cortex ablation in rats, there was an increase of bFGF immunoreactivity in the hippocampus ipsilateral to the lesion (B) as compared to the normal hippocampus (A). There was an increase in bFGF immunoreactivity in the outer molecular layer (dark arrow) and a clearing zone between inner and outer molecular layers (hollow arrow). A similar pattern of bFGF immunoreactivity was observed in the dentate gyrus molecular layer of AD hippocampus (D) as compared to normal control hippocampus (C). This local increase in bFGF immunoreactivity in the dentate gyrus after loss of entorhinal input may represent a mechanism responsible for the availability of bFGF in AD plaques. g, granule cell layer; sg, supragranule layer; m, molecular layer. Calibration bars (A and B): 100 μm; (C and D): 100 μm. (From Gómez-Pinilla *et al.*[24] with permission.)

URE 4A). Next we asked, what might be triggering the appearance of bFGF in and around plaques? Given data described above that bFGF immunoreactivity increased within astrocytes after entorhinal lesions in rodents, it follows that denervation may serve to trigger a molecular cascade that causes astrocytes to synthesize more bFGF.

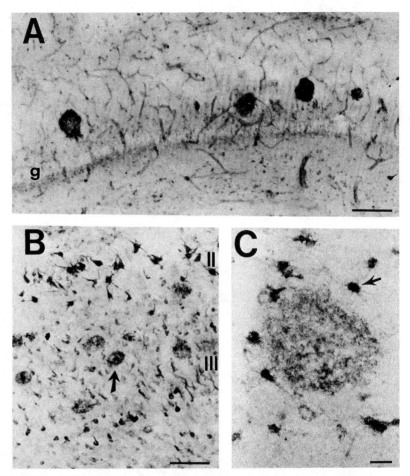

FIGURE 4. bFGF immunoreactivity in AD plaques in the DG and the entorhinal cortex. **A:** Plaques in the DG were located in the molecular layer, mostly within the middle zone. **B:** Plaques in the entorhinal cortex were widely distributed and stained more weakly than DG plaques in the same section. Some plaques contained a denser core of bFGF immunoreactivity (arrow). **C:** High magnification of an entorhinal cortex plaque. Astrocytelike bFGF-positive cells (arrow) were distributed around plaques. g, granule cell layer; II, and III, cell layers in entorhinal cortex. Calibration bars (A and B): 100 mm; (C): 10 mm. (From Gómez-Pinilla *et al.*[24] with permission.)

Thus, it appears that within this microenvironment glial cells are stimulated to produce increased levels of bFGF that may, in effect, "trick" fibers into growing into plaques.

But if this is the case, there should be some molecule involved that both attracts bFGF and also perhaps provides a proper substrate to stimulate fiber growth. It is well known that heparan sulfate can bind to bFGF. Further, as mentioned above, Walicke had shown that heparan sulfate can act synergistically with bFGF to stimulate neuritic growth in culture.[5] In agreement with these studies we also found that culture dishes coated with heparan sulfate were able to greatly stimulate process growth and increase neuronal survival. Then, using an antibody to heparan sulfate we showed that many plaques in AD do in fact stain positive for heparan sulfate, in agreement with previous suggestions. Thus, heparan sulfate within plaques may accumulate bFGF produced by local cellular reactions. Whether all plaques show a colocalization of heparan sulfate and bFGF is unknown; but, based on preliminary findings, it appears that the plaque population is heterogeneous. Thus, at least in some but perhaps not all plaques the potential exists for misdirected neuronal growth, since there exist both neurotrophic factor(s) and suitable substrate(s).

CONCLUSION

Accumulating evidence suggests that bFGF in concert with other growth factors regulates neuronal differentiation during development and also participates in synaptic regeneration in the mature and aged brain. It is also significant that partial denervation causes some reactive neurons to reexpress developmentally regulated genes, e.g., α-tubulin, as part of the sequel of events in the sprouting-regeneration process.[26] Such events may also be regulated by FGFs.

In the brain of patients who have died of AD there is also evidence of reactive neuronal sprouting probably triggered as a result of entorhinal cell loss which occurs early in the disease. Such sprouting may help maintain input to offset cell loss. However, in the AD brain morphological data suggest some sprouting appears to be misdirected into plaques. In order to understand this process, we attempted to identify the presence of growth factor(s) and suitable substrate(s). We found that plaques contain bFGF and heparan sulfate–like molecules. It appears as if bFGF, probably produced by astrocytes as part of an injury-induced growth factor cascade, accumulates in plaques. This accumulation itself may arise as a result of the presence of heparan sulfate or heparan sulfate-amyloid complexes in the plaque environment. Thus, the presence of a neurotrophic factor and a suitable substrate may cause altered growth.

It is ironic that the initial and/or subsequent mechanisms that promote growth and slow degeneration somehow contribute to the disorganization of the environment. bFGF in plaques may activate a number of other processes, some of which may further add to the growth of plaques. For example, it appears that bFGF can enhance the production of the amyloid precursor protein in C6 glioma cells[27] and in primary cultures of astrocytes. Further, it appears that β-amyloid can increase the production of bFGF by astrocytes.[28] Such feedback cycles may produce molecular cascades that enhance the progression of pathology. As such, this is but one other example of the role bFGF may play in the fine balance between maintaining healthy brain function and potentially contributing to maladaptive function and pathology.

REFERENCES

1. COTMAN, C. W. & G. LYNCH. 1989. The neurobiology of learning and memory. Cognition **33:** 201–241.
2. MORRISON, R. S., A. SHARMA, J. DE VELLIS & R. A. BRADSHAW. 1986. Basic fibroblast growth factor supports the survival of cerebral cortical neurons in primary culture. Proc. Nat. Acad. Sci. USA **83**(19): 7537–7541.
3. WALICKE, P. A., W. M. COWAN, N. UENO, A. BAIRD & R. GUILLEMIN. 1986. Fibroblast growth factor promotes survival of dissociated hippocampal neurons and enhances neurite extension. Proc. Nat. Acad. Sci. USA **83**(9): 3012–3016.

4. WALICKE, P. A. 1988. Basic and acidic fibroblast growth factors have trophic effects on neurons from multiple CNS regions. J. Neurosci. **8**(7): 2618–2627.

5. WALICKE, P. A. 1988. Interactions between basic fibroblast growth factor (FGF) and glycosoaminoglycans in promoting neurite outgrowth. Exp. Neurol. **102**: 144–148.

6. ANDERSON, K. J., D. DAM, & C. W. COTMAN. 1988. Basic fibroblast growth factor prevents death of cholinergic neurons in vivo. Nature **332**: 360–361.

7. MATSUZAKI, K., Y. YOSHITAKE, Y. MATUO, H. SASAKI & K. NISHIKAWA. 1989. Monoclonal antibodies against heparin-binding growth factor II/basic fibroblast growth factor that block its biological activity: invalidity of the antibodies for tumor angiogenesis. Proc. Nat. Acad. Sci. USA **86**: 9911–9915.

8. MCGEER, P. L., H. AKIYAMA, S. ITAGAKI & E. G. MCGEER. 1989. Immune system response in Alzheimer's disease. Can. J. Neurol. Sci. **16**: 516–527.

9. FOLKMAN, J. & M. KLAGSBRUN. 1987. Angiogenic factors. Science. **235**: 442–446.

10. SCHWEIGERER, L., G. NEUFIELD, J. FRIEDMAN, J. A. ABRAHAM, J. C. FIDDES & D. GOSPODAROWICZ. 1987. Capillary endothelial cells express basic fibroblast growth factor, a mitogen that promotes their own growth. Nature **325**: 257–259.

11. PETTMANN, B., G. LABOURDETTE, M. WEIBEL & M. SENSENBRENNER. 1986. The brain fibroblast growth factor (FGF) is localized in neurons. Neurosci. Lett. **68**: 175–180.

12. NIETO-SAMPEDRO, M., E. R. LEWIS, C. W. COTMAN, M. MANTHORPE, S. D. SKAPER, G. BARBIN, F. M. LONGO & S. VARON. 1982. Brain injury causes a time-dependent increase in neurotrophic activity at the lesion site. Science **221**: 860–861.

13. FINKLESTEIN, S. P., P. J. APOSTOLIDES, C. G. CADAY, J. PROSSER, M. F. PHILIPS & M. KLAGSBRUN. 1988. Increased basic fibroblast growth factor (bFGF) immunoreactivity at the site of focal brain wounds. Brain Res. **460**: 253–259.

14. COTMAN, C. W. & K. J. ANDERSON. 1986. Synaptic plasticity and functional stabilization in the hippocampal formation: Possible role of Alzheimer's disease. *In* Physiologic Basis for Functional Recovery in Neurological Disease. S. Waxman, Ed.: 313–336. Raven Press. New York, N.Y.

15. COTMAN, C. W., D. A. MATHEWS, D. TAYLOR & G. LYNCH. 1973. Synaptic rearrangement in the dentate gyrus: histochemical evidence of adjustments after lesions in immature and adult rats. Proc. Nat. Acad. Sci. USA **70**: 3473–3477.

16. GÓMEZ-PINILLA, F., J. K.-W. LEE & C. W. COTMAN. Basic FGF in adult rat brain: cellular distribution and response to entorhinal lesion and fimbria-fornix transection. J. Neurosci. (In press.)

17. AMALRIC, F., V. BALDIN, I. BOSC-BIERNE, B. BUGLER, B. COUDERC, N. GUYADER, B. PATRY, A. ROMAN, & G. BOUCHE. Nuclear translocation of basic fibroblast growth factor. Ann. N.Y. Acad. Sci. (This volume.)

18. FLORKIEWICZ, R. S., A. BAIRD & M. GONZALEZ. Multiple forms of bFGF: differential nuclear and cell surface localization. Growth Factors. (In press.)

19. HYMAN, B. T., G. W. VAN HOESEN, A. R. DAMASIO & C. L. BARNES. 1984. Alzheimer's disease: cell-specific pathology isolates the hippocampal formation. Science **225**: 1168–1170.

20. GEDDES, J. W., K. J. ANDERSON & C. W. COTMAN. 1986. Senile plaques as aberrant sprout stimulating structures. Exp. Neurol. **94**: 767–776.

21. CAJAL, R. S. 1928. Degeneration and Regeneration of the Nervous System. Oxford University Press. Oxford, England. (Translated by R. M. May.)

22. WHITSON, J. S., D. J. SELKOE & C. W. COTMAN. 1989. Amyloid β-protein enhances the survival of hippocampal neurons in vitro. Science **243**: 1488–1490.

23. WHITSON, J. S., C. G. GLABE, E. SHINTANI, A. ABCAR & C. W. COTMAN. 1990. β-amyloid protein promotes neuritic branching in hippocampal cultures. Neurosci. Lett. **110**: 319–324.

24. GEDDES, J. W., D. T. MONAGHAN, C. W. COTMAN, I. T. LOTT, R. C. KIM & H. C. CHUI. 1985. Plasticity of hippocampal circuitry in Alzheimer's disease. Science **230**: 1179–1181.

25. GÓMEZ-PINILLA, F., B. J. CUMMINGS & C. W. COTMAN. 1990. Induction of basic fibroblast growth factor in Alzheimer's disease pathology. NeuroReport **1**: 211–214.

26. GEDDES, J. W., J. WONG, B. H. CHOI, R. C. KIM, C. W. COTMAN & F. D. MILLER. 1990. Increased expression of an embryonic growth-associated mRNA in Alzheimer's disease. Neurosci. Lett. **109:** 54–61.
27. QUON, D., R. CATALANO & B. CORDELL. 1990. Fibroblast growth factor induces beta-amyloid precursor mRNA in glial but not neuronal cultured cells. Biochem. Biophys. Res. Commun. **167**(1): 96–102.
28. ARAUJO, D. & C. W. COTMAN. β-amyloid stimulates glial cells in vitro to produce growth factors that accumulate in senile plaques in Alzheimer's disease. Brain Res. (Submitted.)

Fibroblast Growth Factors in Normal and Malignant Melanocytes[a]

RUTH HALABAN,[b] YOKO FUNASAKA,[b] PAULINE LEE,[c]
JEFFREY RUBIN,[d] DINA RON,[e]
AND DANIEL BIRNBAUM[f]

MITOGENIC FACTORS FOR MELANOCYTES

Studies of normal pigment cell growth were severely hampered until 1982 when Eisinger and Marko showed that these cells proliferate in culture in the presence of the phorbol ester TPA (12-O-tetradecanoyl phorbol-13-acetate) plus the cyclic adenosine monophosphate (cAMP) stimulator cholera toxin.[1] The search then began for natural mitogens, and the combined experience showed that normal human melanocytes in culture did not respond to a vast number of growth factors, including melanotropin (MSH, melanocyte stimulating factor), nerve growth factor, epidermal growth factor (EGF), or platelet-derived growth factor (PDGF).[2-4] A breakthrough in the search was the identification of basic fibroblast growth factor (bFGF)[5-7] as a natural mitogen for melanocytes.[8] Recently we have shown that other peptides from the fibroblast growth factor family, acidic FGF (aFGF),[9] K-FGF/*hst*,[10-12] FGF-5,[13] and FGF-6[14] are likewise melanocyte mitogens.[3,4,15] The two unrelated ligands hepatocyte growth factor (HGF)[16-18] and mast cell growth factor (MGF), known also as the *kit*-ligand and by other designations,[19-26] have recently been added to the list of melanocyte mitogens (Reference 18, and Funasaka *et al.,* manuscript submitted). Like TPA and bFGF in chemically defined medium, each of these growth factors when added alone is an effective melanocyte mitogen only in the presence of substances that increase intracellular levels of cAMP. However, this indispensable synergism can be achieved also by the simultaneous addition of two factors such as bFGF plus HGF, but not MGF (our unpublished results). Two recently discovered ligands, keratinocyte growth factor (KGF)[27] and vascular endothelial growth factor (VEGF)[28] are not mitogenic to melanocytes (our unpublished observations). While the role of bFGF in melanocyte proliferation and malignant transformation has been documented extensively, the studies on the other FGFs, MGF, and HGF are in their initial stages, because these peptides have been purified and cloned only recently.

[a]This work was supported by Public Health Service grants 5 R29 CA44542, 1 RO1 CA 04679, and 5 PO1 AR25252.

[b]Department of Dermatology, Yale University School of Medicine, 500 LCI, P.O. Box 333, New Haven, Connecticut 06510.

[c]Scripps Clinic and Research Institute, Department of Molecular and Experimental Medicine, BCR7, Room SR 305B, 10666 Torrey Pines Road, La Jolla, California 92037.

[d]Laboratory of Cellular and Molecular Biology, Building 37, Room IE24, National Cancer Institute, Bethesda, Maryland 20892.

[e]Department of Biology, Technion, Technion City, Haifa 32000, Israel.

[f]Institut Paoli-Calmettes, Unité 119 de l'INSERM, 27, Boulevard Leï Roure, 13009 Marseille, France.

BASIC FGF

bFGF is mitogenic to melanocytes in culture and, *in situ,* appears to be provided by basal keratinocytes.[29,30] Normal melanocytes, unlike other cell types that respond to bFGF, do not produce bFGF on their own.[15] Basic FGF is expressed neither at the mRNA nor at the protein level.[4,15] In fact, melanocytes do not produce any mitogen that stimulates their own proliferation, the reason they fail to proliferate in regular serum-supplemented or chemically defined media suitable for other cell types, including metastatic melanoma cells. In contrast, other skin cells, such as proliferating fibroblasts and keratinocytes, produce bFGF and, in coculture with melanocytes, maintain melanocyte viability.[31]

Basic FGF is the main factor that stimulates growth of melanocytes in melanocytic lesions. Messenger RNA for bFGF can be detected in dysplastic and dermal nevi, in primary and metastatic melanomas by *in situ* hybridization with a bFGF riboprobe.[30] The presence of an active mitogen was verified in the tissue of freshly excised dermal nevi (our unpublished results) and in advanced primary and metastatic melanoma cells in culture.[4,8,15,32-35] Melanoma cells depend on this intrinsic mitogen because internalized neutralizing anti-bFGF antibodies[15] and antisense, but not sense, oligodeoxynucleotides targeted against human bFGF mRNA[36] inhibit growth. Antisense oligodeoxynucleotides to nerve growth factor (β-NGF) and insulinlike growth factor I (IGF-I) mRNA had no effect,[36] suggesting that growth factors other than bFGF that are also produced by melanoma cells do not give autocrine growth advantage to melanomas.

Indirect evidence for the importance of bFGF in melanoma growth is the inhibition of proliferation of a human melanoma cell line by a monoclonal antibody against the active site of urokinase-type plasminogen activator (u-PA).[37] Because one of the functions of bFGF is to increase the synthesis of plasminogen activator,[38] inhibition of melanoma growth by anti-u-PA antibody suggests that bFGF in melanomas acts in part via induction of this serine protease.

INTERNAL VERSUS EXTERNAL AUTOCRINE LOOP

The autostimulation of melanoma cells with bFGF cannot be explained by a classical autocrine loop in which a factor is secreted and in turn binds to cell surface receptors as an extracellular growth factor would. Significant levels of bFGF were detected in the conditioned media of only 3 out of 10 melanomas tested[4]; in the other cases the bFGF was cell associated.[4,8,15,32-35] In contrast to classical autocrine loops (see Reference 39 for review), the autostimulation of melanoma cells can be inhibited by bFGF-neutralizing antibodies only if they are injected into the cells but not by the simple addition of antibodies to the culture medium.[15]

Other effects of bFGF suggest that in melanomas this growth factor acts through an intracellular loop of autostimulation. Immortalized murine melanocytes expressing bFGF constitutively via infection with a recombinant retroviral vector, became, as expected, independent of exogenous bFGF.[40] Not expected was the associated loss of all differentiated functions, such as tyrosinase (the key enzyme in melanin synthesis), melanosomes (the subcellular organelles in which melanin is normally deposited), pigmentation, and dendrite formation. The transformed melanocytes appeared in culture like fibroblasts or endothelial cells,[40] and the bFGF accumulated in the cytoplasm (Neufeld G., unpublished results of immunostaining). Loss of differentiated functions in culture is also common in cells from advanced primary and metastatic melanoma lesions that express intrinsic bFGF (Reference 41 and our unpublished results).

Differentiation of melanocytes is dependent on continuous cell surface stimulation, suggesting that internal bFGF activates a pathway that effectively maintains proliferation but not differentiation. Exogenous bFGF does not suppress differentiation. In contrast, it induces the melanocytic phenotype as shown with neural-crest-derived cells from embryonic quail dorsal root ganglia.[42] bFGF was fully effective only in the presence of TPA, suggesting that a second factor was needed to fully potentiate the bFGF effect on differentiation. Continuous stimulation of normal human melanocytes with external bFGF plus dbcAMP maintains their differentiated functions. In fact, any other substance that stimulates the proliferation of normal and malignant melanocytes on addition to the cultures also stimulates the production of melanocyte-specific mRNAs, tyrosinase, and melanin.[41,43,44]

FGFs AND MALIGNANT TRANSFORMATION

In vivo, the growth of bFGF-expressing melanocytes is restrained by environmental factors with concomitant induction of differentiated functions. Expression of intrinsic bFGF alone is insufficient to confer the malignant phenotype, because bFGF-transformed melanocytes do not grow as tumors in syngeneic or nude mice.[40] Instead, they regain their differentiated functions, including the ability to produce melanin. They appear in all respects like normal melanocytes. Most of the melanocytes in dysplastic nevi that expressed bFGF[30] were clearly not malignant. In patients with the dysplastic nevus syndrome who may have as many as 300 nevi, a single cell in only one dysplastic nevus suffices to give rise to a melanoma.[45]

Spontaneous expression of bFGF in melanomas is probably due to gene activation and not amplification or rearrangement (Reference 46 and unpublished results). Basic FGF has been mapped to human chromosome 4q26-q27,[47,48] which is not among the chromosomes known to be affected in melanomas.[49,50] In normal melanocytes, the expression of bFGF is suppressed, a process that might be important for maintaining the differentiated state.

Southern blot analysis of other genes from the FGF family has shown that 2 out of 15 melanomas contained amplification of the linked FGF-related genes *int-2* and K-FGF/*hst* (Reference 46 and FIGURE 1). However, in one such melanoma, YU SIT1, with 12-fold amplification of *int-2* and K-FGF genes, both known to encode secreted factors, there was no mitogenic activity in the medium (Reference 4 and TABLE 1) and Northern blot analysis showed that the *int-2* and K-FGF/*hst* genes were not expressed. All the cell-associated mitogenic activity was neutralized by anti-bFGF antibodies specific for this peptide (Reference 4 and TABLE 2). These results suggest that the amplification of the *int-2* and K-FGF/*hst* genes was of no consequence to the development of the melanomas.

FGF RECEPTORS

Basic and acidic FGF bind to proteins of 110–125 and 145 kDa and induce tyrosine phosphorylation of proteins of the same sizes.[51-55] In addition they phosphorylate proteins of 145, 90, 42, and 46 kDa.[54,55] The 145-kDa protein was identified as phospholipase Cγ.[56,57] A tyrosine-kinase domain on the receptor was recently confirmed by the cloning and DNA sequencing of four receptor genes from the FGF-R

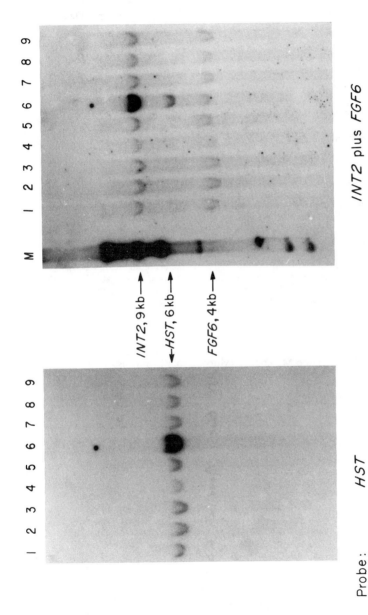

Probe: *HST*

INT2 plus *FGF6*

FIGURE 1. Southern blot hybridization of DNA from human metastatic melanomas. The following cell lines were used: lane 1, YU BEM11; lane 2, YU MAC12; lane 3, YU SAC2; lane 4, SK MEL 23; lane 5, SK MEL 37; lane 6, YU SIT1; lane 7, YU WIL4; lane 8, WM 9; lane 9, YU ZAZ6. DNA was digested with Eco RI, and the Southern blot was hybridized first with an *hst* probe (left panel). The same filter was then used to hybridize with *int-2* and *FGF-6* DNA probes (right panel) as described in Reference 46. Notice amplification of the linked *int-2* and K-FGF/*hst* genes in YU SIT1 (lane 6), but normal copy number of FGF-6. The YU series of melanomas was established in culture at Yale University by RH.

TABLE 1. Melanocyte Mitogen of YU SIT1 Is Not Secreted and the One Secreted by Two Other Melanomas Is bFGF[a]

Additions	³H-Thymidine Incorporation (cpm/well per 3 hours)
hrbFGF	19,600
dbcAMP	1,450
hrbFGF + dbcAMP	38,000
Conditioned media	
YU S1T1 + dbcAMP	1,000
YU DAN3	1,200
YU DAN3 + dbcAMP	13,000
YU DAN3 + dbcAMP + anti-bFGF-(1-24)-ab	1,600
SW1614	2,340
SW1614 + dbcAMP	23,000
SW1614 + anti-bFGF-(1-24)-ab	2,400

[a]YU SIT1, YU DAN3 and SW1614 are human metastatic melanoma cell lines. Mitogenicity of conditioned media was tested on human melanocytes derived from newborn foreskins seeded in chemically defined medium (PC-1, Ventrex Laboratories, Inc. Portland, Me.) in 24-well cluster plates (approximately 30,000 cells/well). Additions were 1 ng/ml hrbFGF (human recombinant basic fibroblast growth factor, Chiron), 1.0 mM dbcAMP (dibutyryl cyclic adenosine monophosphate); PC-1 conditioned medium from melanomas, 3 days, diluted 1:1 with fresh PC-1 medium. Neutralizing antibodies to bFGF [anti-bFGF-(1-24)-ab], 1:100 dilution, were added directly to experimental media. ³H-thymidine was incorporated during the last 3 hours of the 24-hour incubation period. Data are means of duplicate wells. Notice that like bFGF, the melanocyte mitogen in melanoma-conditioned media acts in synergy with dbcAMP and is completely neutralized by anti-bFGF-antibodies. Antibodies to hepatocyte growth factor were without effect (data not shown).

TABLE 2. The Cell-Associated Melanocyte Mitogen in YU SIT1 is bFGF[a]

Additions	³H-Thymidine Incorporation (cpm/well per 3 hours)
None	230
hrbFGF	1,290
dbcAMP	440
hrbFGF + dbcAMP	20,200
Cell extracts plus dbcAMP	
YU SIT1	41,100
YU SIT1 + anti-bFGF-(1-24)-ab	1,100
NIH 3T3	9,620
NIH 3T3 + anti-bFGF-(1-24)-ab	4,500
NIH 3T3-FGF-5	24,700
NIH 3T3-FGF-5 + anti-bFGF-(1-24)-ab	17,300
NIH 3T3-FGF-6	50,700
NIH-3T3-FGF-6 + anti-bFGF-(1-24)-ab	47,000

[a]Melanoma cells as in legend to TABLE 1. NIH 3T3-FGF-5 and FGF-6 are fibroblast lines transformed with the respective cDNA clones. They were obtained from Drs. M. Goldfarb and D. Birnbaum, respectively. Additions were 100 μg protein/ml cell extract; 1:100 dilution of anti-bFGF-(1-24)-ab. Other details as in legend to TABLE 1. Data are means of duplicate wells. Notice that anti-bFGF antibodies neutralized 97% of the melanocyte mitogen associated with the melanoma cells. NIH 3T3-FGF-5 and -FGF-6 lost only the mitogenic activity inherent in the parental cell line which expresses low levels of bFGF. The other mitogen in NIH 3T3 is HGF.[18]

family.[58-71] These four belong to the immunoglobulin (Ig) superfamily of proteins. Receptor isoforms with two or three immunoglobulinlike extracellular domains were identified.[61,64-66,70,71] Two of the human bFGF-R (*flg* isoform) bind basic and acidic FGF with similar affinity,[61] and possibly by K-FGF/*hst*,[67] with concomitant activation of the signal transduction pathway. The murine *bek* homologue with two IgG-like domain binds KGF and acidic FGF with high affinity and bFGF with approximately $20\times$ lower affinity.[70] In addition, an mRNA species encoding only the extracellular domain of the FGF receptor has been identified,[61,70] suggesting the existence of a secreted FGF-binding protein.

Basic FGF receptor kinase activity in melanomas is critical for proliferation because synthetic bFGF peptides that span the receptor binding domain and act as bFGF antagonists inhibit melanoma cell growth, whereas peptides from nonrelevant domains do not have that effect.[15] In addition, internalized antiphosphotyrosine antibodies also inhibit melanoma cell growth,[15] suggesting inhibition of bFGF receptor activity. In agreement with this conclusion is our recent finding that the bFGF receptor is constitutively phosphorylated in a metastatic human melanoma cell line (FIGURE 2). In this case the bFGF receptor appears to be a 125-kDa protein, compatible with the two-loop form of receptor. To identify FGF-R transcripts, we amplified mRNA from normal melanocytes and several melanomas by the polymerase chain reaction (PCR) using a set of primers highly specific for the tyrosine-kinase domain of *flg* and then a second set from sequences flanking the deletion in the two-loop form of *flg*.[70] Ethidium bromide staining of the amplified products showed that the two-loop form of *flg* was predominantly expressed in both cell types. However, the expression of other *flg*-related genes cannot yet be ruled out. The FGF receptor genes *bek* and *flg* are not amplified or rearranged in the series of melanomas presented in FIGURE 1, and the FGF receptor protein is not expressed at abnormally high levels, suggesting that the uncontrolled growth is not associated with overexpression of this receptor kinase.

OTHER RECEPTORS

Because continuous activation of FGF receptors by internal bFGF is not sufficientto confer tumorigenicity, it is likely that other receptor kinases collaborate with the FGF receptor kinase to transform melanocytes to melanomas. The two unrelated mitogens for melanocytes mentioned at the beginning, MGF and HGF, stimulate receptors with tyrosine-kinase that are known as c-*kit* and *met,* respectively.[19-26,72,73] It was of particular interest to find out the role these two receptors play, if any, in melanomas because *v-kit,* the oncogenic form of c-*kit,* induces sarcomas in cats,[74] and a chimeric molecule composed of the extracellular portion of the EGF receptor and the transmembrane and cytoplasmic domain of c-*kit* confers an EGF-dependent transformed phenotype on NIH-3T3 cells.[75] *Met,* on the other hand, was first identified as an oncogene in a carcinogen-treated human osteosarcoma cell line, by DNA transfection of NIH-3T3 cells.[76]

We found that MGF-activated c-*kit* kinase is one of the most prominent receptor kinases in normal human melanocytes, phosphorylating several substrate proteins. However, there was no stimulation of DNA synthesis in response to MGF in one primary and 7 metastatic melanoma cell lines tested, and tyrosine phosphorylation of p145[c-*kit*] could not be detected or was present at extremely low levels in only two cell lines (Funasaka *et al.,* manuscript in preparation). Further analysis of c-*kit* receptor kinase in melanomas should reveal whether or not this abnormality contributes to the malignant state. In contrast, *met* was expressed in all the melanoma cell lines

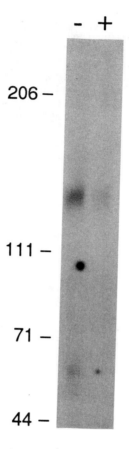

ip: α bFGF-R
ib: α p-Tyr

FIGURE 2. Basic-FGF receptor in metastatic melanoma YU PAC7. Deprived culture (−) or cultures stimulated with bFGF for 10 minutes (+) were harvested, and cell extracts were subjected to immunoprecipitation with anti-FGF-receptor antibodies prepared against the synthetic peptide spanning the intracellular juxtamembrane domain deduced from the nucleotide sequence of *flg* (See references 58 and 61). Eluted proteins were subjected to polyacrylamide gel electrophoresis, transferred to nitrocellulose membrane, and immunoblotted with antiphosphotyrosine antibodies (Upstate Biotechnology, Inc. Lake Placid, N.Y.). Antibody-antigen complexes were visualized by radioautography of bound ^{125}I-protein A. Notice that the level of phosphotyrosine in the receptor was similar whether or not the cells were treated with bFGF.

tested and was stimulated by HGF in a fashion similar to that of normal melanocytes. Because *met* is not constitutively activated in melanomas, it is possible that external HGF promotes melanoma metastasis in synergy with internal FGF. Preferential metastasis to the liver was shown to be the result of an unidentified growth factor produced by hepatocytes.[77] Because this growth factor stimulated only the cell line that was cloned through several passages from liver metastasis, but not the one cloned from lung metastasis, it is possible that the selection resulted in enrichment of *met*-expressing melanoma cells and that the unidentified growth factor was HGF.

ACKNOWLEDGMENTS

We thank Dr. Gisela Moellmann for critically reading the manuscript and for helpful suggestions, and Jack Schreiber for technical help.

REFERENCES

1. EISINGER, M. & O. MARKO. 1981. Selective proliferation of normal human melanocytes *in vitro* in the presence of phorbol ester and cholera toxin. Proc. Nat. Acad. Sci. USA **79:** 2018–2022.
2. HALABAN, R. 1988. Responses of cultured melanocytes to defined growth factors. Pigment Cell Res. **1:** 18–26.
3. HALABAN, R. 1990. Growth regulation in normal and malignant melanocytes. *In* Human Melanoma. S. Ferrone, Ed.: 3–14. Springer-Verlag. Berlin, Heidelberg & New York.
4. HALABAN, R. 1991. The regulatory role of fibroblast growth factor in melanomas. *In* Melanoma 5 (Cancer Treatment and Research Series). L. Nathanson, Ed. Kluwer Academic Publishers. Norwell, Mass.
5. ABRAHAM, J. A., A. MERGIA, J. L. WHANG, A. TUMOLO, J. FRIEDMAN, K. A. HJERRILD, D. GOSPODAROWICZ & J. C. FIDDES. 1986. Nucleotide sequence of a bovine clone encoding the angiogenic protein, basic fibroblast growth factor. Science **233:** 545–548.
6. BURGESS, W. H. & T. MACIAG. 1989. The heparin-binding (fibroblast) growth factor family of proteins. Annu. Rev. Biochem. **58:** 575–606.
7. RIFKIN, D. & D. MOSCATELLI. 1989. Recent developments in the cell biology of basic fibroblast growth factor. J. Cell Biol. **109:** 1–6.
8. HALABAN, R., S. GHOSH & A. BAIRD. 1987. bFGF is the putative natural growth factor for human melanocytes. In Vitro Cell. Dev. Biol. **23:** 47–52.
9. JAYE, M., J. T. HOWK, W. BURGESS, G. A. RICCA, I.-M. CHIU, M. W. RAVERA, S. J. O'BRIEN, W. S. MODI, T. MACIAG & W. N. DROHAN. 1986. Human endothelial cell growth factor: cloning nucleotide sequence, and chromosome localization. Science **233:** 541–545.
10. DELLI BOVI, P. & C. BASILICO. 1987. Isolation of a rearranged human transforming gene following transfection of Kaposi sarcoma DNA. Proc. Nat. Acad. Sci. USA **84:** 5660–5664.
11. DELLI BOVI, P., A. M. CURATOLA, F. G. KERN, A. CRECO, M. ITTMANN & C. BASILICO. 1987. An oncogene isolated by transfection of Kaposi's sarcoma DNA encodes a growth factor that is a member of the FGF family. Cell **50:** 729–737.
12. TAIRA, M., T. YOSHIDA, K. MIYAGAWA, H. SAKAMOTO, M. TERADA & T. SUGIMURA. 1987. cDNA sequence of human transforming gene *hst* and identification of the coding sequence required for transforming activity. Proc. Nat. Acad. Sci. USA **84:** 2980–2984.
13. ZHAN, X., B. BATES, X. HU & M. GOLDFARB. 1988. The human FGF-5 oncogene encodes a novel protein related to fibroblast growth factors. Mol. Cell. Biol. **8:** 3487–3495.
14. MARICS, I., J. ADELAIDE, F. RAYBAUD, M. G. MATTEI, F. COULIER, J. PLANCHE., O. DE LAPEYRIERE & D. BIRNBAUM. 1989. Characterization of the *HST*-related *FGF*.6 gene, a new member of the fibroblast growth factor family. Oncogene **4:** 335–340.
15. HALABAN, R., B. S. KWON, S. GHOSH, P. DELLI BOVI & A. BAIRD. 1988. bFGF as an autocrine growth factor for human melanomas. Oncogene Res. **3:** 177–186.
16. NAKAMURA, T., T. NISHIZAWA, M. HAGIYA, T. SEKI, M. SHIMONISHI, A. SUGIMURA, K. TASHIRO & S. SHIZIMU. 1989. Molecular cloning and expression of human hepatocyte growth factor. Nature **342:** 440–443.

17. TASHIRO, K., M. HAGIYA, T. NISHIZAWA, T. SEKI, M. SHIMONISHI, S. SHIMIZU & T. NAKAMURA. 1990. Deduced primary structure of rat hepatocyte growth factor and expression of the mRNA in rat tissues. Proc. Nat. Acad. Sci. USA **87**: 3200–3204.
18. RUBIN, J. S., A. M.-L. CHAN, D. P. BOTTARO, W. H. BURGESS, W. G. TAYLOR, A. C. CECH, D. W. HIRSCHFELD, J. WONG, T. MIKI, P. W. FINCH & S. A. AARONSON. 1991. A broad spectrum human lung fibroblast-derived mitogen is a variant of hepatocyte growth factor. Proc. Nat. Acad. Sci. USA **88**: 415–419.
19. WILLIAMS, D. E., J. EISENMAN, A. BAIRD, C. RAUCH, K. VAN NESS, C. J. MARCH, L. S. PARK, U. MARTIN, D. Y. MOCHIZUKI, H. S. BOSWELL, G. M. BURGESS, D. COSMAN & S. D. LYMAN. 1990. Identification of a ligand for the c-*kit* proto-oncogene. Cell **63**: 167–174.
20. COPELAND, N., D. J. GILBERT, B. C. CHO, P. J. DONOVAN, N. A. JENKINS, D. COSMAN, D. ANDERSON, S. D. LYMAN & D. E. WILLIAMS. 1990. Mast cell growth factor maps near the steel locus on mouse chromosome 10 and is deleted in a number of steel alleles. Cell **63**: 175–183.
21. FLANAGAN, J. G. & P. LEDER. 1990. The *kit* ligand: a cell surface molecule altered in steel mutant fibroblasts. Cell **63**: 185–194.
22. ZSEBO, K. M., J. WYPYCH, I. K. MCNIECE, H. S. LU, K. A. SMITH, S. B. KARKARE, R. K. SADCHEV, V. N. YUSCHENKOFF, N. C. BIRKETT, L. R. WILLIAMS, V. N. SATYAGAL, W. TUNG, R. A. BOSSELMAN, E. A. MENDIAZ & K. E. LANGLEY. 1990. Identification, purification, and biological characterization of hematopoietic stem cell factor from buffalo rat liver–conditioned medium. Cell **63**: 195–201.
23. MARTIN, F. H., S. V. SUGS, K. E. LANGLEY, H. S. LU, J. TING, K. H. OKINO, C. F. MORRIS, I. K. MCNIECE, F. W. JACOBSEN, E. A. MENDIAZ, N. C. BIRKETT, K. A. SMITH, M. J. JOHNSON, V. P. PARKER, J. C. FLORES, A. C. PATEL, E. FISHER, H. O. ERJAVEC, C. H. HERRERA, J. WYPYCH, R. K. SACHDEV, J. A. POPE, I. LESLIE, D. WEN, C.-H. LIN, R. L. CUPPLES & K. ZSEBO. 1990. Primary structure and functional expression of rat and human stem cell factor DNAs. Cell **63**: 203–211.
24. ZSEBO, K., D. A. WILLIAMS, E. N. GEISSLER, V. C. BROUDY, F. H. MARTIN, H. L. ATKINS, R.-Y. HSU, N. C. BIRKETT, K. H. OKINO, D. C. MURDOCK, F. W. JACOBSEN, K. E. LANGLEY, K. A. SMITH, T. TAKEISHI, B. M. CATTANACH, S. J. GALLI & S. SUGGS. 1990. Stem cell factor is encoded at the *Sl* locus of the mouse and is the ligand for the c-*kit* tyrosine kinase receptor. Cell **63**: 213–224.
25. HUANG, E., K. NOCKA, D. R. BEIER, T.-Y. CHU, J. BUCK, H.-W. LAHM, D. WELLNER, P. LEDER & P. BESMER. 1990. The hematopoietic growth factor KL is encoded by the *Sl* locus and is the ligand of the c-*kit* receptor, the gene product of the *W* locus. Cell **63**: 225–233.
26. ANDERSON, D. M., S. D. LYMAN, A. BAIRD, J. M. WIGNALL, J. EISENMAN, C. RAUCH, C. J. MARCH, H. S. BOSWELL, S. D. GIMPEL, D. COSMAN & D. E. WILLIAMS. 1990. Molecular cloning of mast cell growth factor, a hematoprotein that is active in both membrane-bound and soluble forms. Cell **63**: 235–243.
27. FINCH, P. W., J. S. RUBIN, T. MIKI, D. RON & S. A. AARONSON. 1989. Human KGF is FGF-related with properties of a paracrine effector of epithelial cell growth. Science **245**: 752–755.
28. CONN, G., M. L. BAYNE, D. D. SODERMAN, P. W. KWOK, K. A. SULLIVAN, T. M. PALISI, D. A. HOPE & K. A. THOMAS. 1990. Amino acid and cDNA sequences of a vascular endothelial cell mitogen that is homologous to platelet-derived growth factor. Proc. Nat. Acad. Sci. USA **87**: 2628–2632.
29. SCHULZE-OSTHOFF, K. W. RISAU, E. VOLLMER & S. CLEMENS. 1990. *In situ* detection of basic fibroblast growth factor by highly specific antibodies. Am. J. Pathol. **173**: 85–92.
30. SCOTT, G., M. STOLER, S. SARKAR & R. HALABAN. 1991. Localization of basic fibroblast growth factor mRNA in melanocytic lesions by *in situ* hybridization. J. Invest. Dermatol. **96**: 318–322.
31. HALABAN, R., R. LANGDON, N. BIRCHALL, C. CUONO, A. BAIRD, G. SCOTT, G. MOELL-MANN & J. McGUIRE. 1988. Basic fibroblast growth factor of keratinocytes is a natural mitogen for normal human melanocytes. J. Cell Biol. **107**: 1611–1619.
32. EISINGER, M., O. MARKO, S.-L. OGATA & L. J. OLD. 1985. Growth regulation of human

melanocytes: mitogenic factors in extracts of melanoma, astrocytoma, and fibroblast cell lines. Science **229:** 984–986.

33. LOBB, R., J. SASSE, R. SULLIVAN, Y. SHING, P. D'AMORE, J. JACOBS & M. KLAGSBRUN. 1986. Purification and characterization of heparin-binding endothelial cell growth factors. J. Biol. Chem. **261:** 1924–1926.

34. MOSCATELLI, D., M. PRESTA, J. JOSEPH-SILVERSTEIN & D. B. RIFKIN. 1986. Both normal and tumor cells produce basic fibroblast growth factor. J. Cell. Physiol. **129:** 273–276.

35. OGATA, S., Y. FURUHASHI & M. EISINGER. 1987. Growth stimulation of human melanocytes: identification and characterization of melanoma-derived growth factor (M-McGF). Biochem. Biophys. Res. Commun. **146:** 1204–1211.

36. BECKER, D., C. B. MEIER & M. HERLYN. 1989. Proliferation of human malignant melanomas is inhibited by antisense oligodeoxynucleotides targeted against basic fibroblast growth factor. EMBO J. **8:** 3685–3691.

37. KIRCHHEIMER, J. C., J. WOJTA, G. CHRIST & B. R. BINDER. 1989. Functional inhibition of endogenously produced urokinase decreases cell proliferation in a human melanoma cell line. Proc. Nat. Acad. Sci. USA **86:** 5424–5428.

38. SATO, Y. & D. B. RIFKIN. 1988. Autocrine activities of basic fibroblast growth factor: regulation of endothelial cell movement, plasminogen activator synthesis and DNA synthesis. J. Cell Biol. **107:** 1199–1205.

39. BROWDER, T. M., C. E. DUNBAR & A. W. NIENHUIS. 1989. Private and public autocrine loops in neoplastic cells. Cancer Cells **1:** 9–17.

40. DOTTO, G., S. GHOSH, G. MOELLMANN, A. B. LERNER & R. HALABAN. 1989. Transformation of melanocytes with basic fibroblast growth factor cDNA and oncogenes and selective suppression of the transformed phenotype in a reconstituted cutaneous environment. J. Cell Biol. **109:** 3115–3128.

41. HALABAN, R., S. H. POMERANTZ, S. MARSHALL, D. T. LAMBERT & A. B. LERNER. 1983. Regulation of tyrosinase in human melanocytes grown in culture. J. Cell Biol. **97:** 480–488.

42. STOCKER, K. M., L. SHERMAN, S. REES & G. CIMENT. 1991. Basic FGF and TFG-β1 influence commitment to melanogenesis in neural crest–derived cells of avian embryos. 1991. Development **111:** 635–645.

43. KWON, B. S., R. HALABAN, G. S. KIM, L. USACK, S. POMERANTZ & A. K. HAQ. 1987. A melanocyte-specific cDNA clone whose expression is inducible by melanotropic and isobutylmethyl xanthine. Mol. Biol. Med. **4:** 339–355.

44. KWON, B. S., M. WAKULCHIK, A. K. HAQ, R. HALABAN & D. KESTLER. 1988. Sequence analysis of murine tyrosinase cDNA and the effect of melanotropin on its gene expression. Biochem. Biophys. Res. Commun. **153:** 1301–1309.

45. GREENE, M. H., W. H. CLARK, JR., M. A. TUCKER, K. H. KRAEMER, D. E. ELDER & M. C. FRASER. 1985. High risk of malignant melanoma in melanoma-prone families with dysplastic nevi. Ann. Intern. Med. **102:** 458–465.

46. ADELAIDE, J., M.-G. MATTEI, I. MARICS, F. RAYBAUD, J. PLANCHE, O. DE LAPEYRIERE & D. BIRNBAUM. 1988. Chromosomal localization of the *hst* oncogene and its co-amplification with the *int*.2 oncogene in human melanoma. Oncogene **2:** 413–416.

47. MERGIA, A., R. EDDY, J. A. ABRAHAM, J. C. FIDDES & T. B. SHOWS. 1986. The genes for basic and acidic fibroblast growth factors are on different human chromosomes. Biochem. Biophys. Res. Commun. **138:** 644–651.

48. LAFAGE-POCHITALOFF, M., F. GALLAND, J. SIMONETTI, H. PRATS, M.-G. MATTEI & D. BIRNBAUM. 1990. The human basic fibroblast growth factor gene is located on the long arm of chromosome 4 bands q26-q27. Oncogene Res. **5:** 241–244.

49. BALABAN, G., M. HERLYN, D. GUERRY IV, R. BARTOLO, H. KOPROWSKI, W. H. CLARK & P. C. NOWELL. 1984. Cytogenetics of human malignant melanoma and premalignant lesions. Cancer Genet. Cytogenet. **11:** 429–439.

50. COWAN, J. M., R. HALABAN & U. FRANKCE. 1988. Cytogenetic analysis of melanocytes from premalignant nevi and melanomas. J. Nat. Cancer Inst. **80:** 1159–1164.

51. NEUFELD, G. & D. GOSPODAROWICZ. 1985. The identification and partial characterization of the fibroblast growth factor receptor of baby hamster kidney cells. J. Biol. Chem. **260:** 13860–13868.

52. HUANG, S. S. & J. S. HUANG. 1986. Association of bovine brain–derived growth factor receptor with protein tyrosine kinase activity. J. Biol. Chem. **261:** 9568–9571.

53. IMAMURA, T., Y. TOKITA & Y. MITSUI. 1988. Purification of basic FGF receptors from rat brain. Biochem. Biophys. Res. Comm. **155:** 583–590.

54. COUGHLIN, S. R., P. J. BARR, L. S. COUSENS, L. J. FRETTO & L. T. WILLIAMS. 1988. Acidic and basic fibroblast growth factors stimulate tyrosine kinase activity *in vivo*. J. Biol. Chem. **263:** 988–993.

55. FRIESEL, R., W. H. BURGESS & T. MACIAG. 1989. Heparin-binding growth factor 1 stimulates tyrosine phosphorylation in NIH 3T3 cells. Mol. Cell. Biol. **9:** 1857–1865.

56. BURGESS, W. H., C. A. DIONNE, J. KAPLOW, R. MUDD, R. FRIESEL, A. ZILBERSTEIN, J. SCHLESSINGER & M. JAYE. 1990. Characterization and cDNA cloning of phospholipase C-γ, a major substrate for heparin-binding growth factor 1 (acidic fibroblast growth factor)–activated tyrosine kinase. Mol. Cell. Biol. **10:** 4770–4777.

57. CUADRADO, A. & C. J. MOLLOY. 1990. Overexpression of phospholipase C-γ in NIH 3T3 fibroblasts results in increased phosphatidylinositol hydrolysis in response to platelet-derived growth factor and basic fibroblast growth factor. Mol. Cell. Biol. **10:** 6069–6072.

58. LEE, P. L., D. E. JOHNSON, L. S. COUSENS, V. A. FRIED & L. T. WILLIAMS. 1989. Purification and complementary DNA cloning of a receptor for basic fibroblast growth factor. Science **245:** 57–60.

59. RUTA, M., W. BURGESS, D. GIVOL, J. EPSTEIN, N. NEIGER, J. KAPLOW, G. CRUMLEY, C. DIONNE, M. JAYE & J. SCHLESSINGER. 1989. Receptor for acidic fibroblast growth factor is related to the tyrosine kinase encoded by the *fms*-like gene (FLG). Proc. Nat. Acad. Sci. USA **86:** 8722–8726.

60. PASQUALE, E. B. & S. J. SINGER. 1989. Identification of a developmentally regulated protein-tyrosine kinase by using anti-phosphotyrosine antibodies to screen a cDNA expression library. Proc. Nat. Acad. Sci. USA **86:** 5449–5453.

61. JOHNSON, D. E., P. L. LEE, J. LU & L. T. WILLIAMS. 1990. Diverse forms of a receptor for acidic and basic fibroblast growth factors. Mol. Cell. Biol. **10:** 4728–4736.

62. BOTTARO, D. P., J. S. RUBIN, D. RON, P. W. FINCH, C. FLORIO & S. A. AARONSON. 1990. Characterization of the receptor for keratinocyte growth factor. J. Biol. Chem. **265:** 12767–12770.

63. SAFRAN, A., A. AVIVI, A. ORR-URTEREGER, G. NEUFELD, P. LONAI, D. GIVOL & Y. YARDEN. 1990. The murine *flg* gene encodes a receptor for fibroblast growth factor. Oncogene **5:** 635–643.

64. HATTORI, Y., O. ODAGIRI, H. NAKATANI, K. MIYAGAWA, K. NAITO, H. SAKAMOTO, O. KATOH, T. YOSHIDA, T. SUGIMURA & M. TERADA. 1990. K-*sam*, an amplified gene in stomach cancer, is a member of the heparin-binding growth factor receptor genes. Proc. Nat. Acad. Sci. USA **87:** 5983–5987.

65. DIONNE, C. A., G. CRUMLEY, B. FRAÇOISE, J. M. KAPLOW, G. SEARFOSS, M. RUTA, W. H. BURGESS, M. JAYE & J. SCHLESSINGER. 1990. Cloning and expression of two distinct high-affinity receptors cross-reacting with acidic and basic fibroblast growth factors. EMBO J. **9:** 2685–2692.

66. REID, H. H., A. F. WILKS & O. BERNARD. 1990. Two forms of the basic fibroblast growth factor receptor-like mRNA are expressed in the developing mouse brain. Proc. Nat. Acad. Sci. USA **87:** 1596–1600.

67. MANSUKHANI, A., D. MOSCATELLI, D. TALARICO, V. LEVYTSKA & C. BASILICO. 1990. A murine fibroblast growth factor (FGF) receptor expressed in CHO cells is activated by basic FGF and Kaposi FGF. Proc. Nat. Acad. Sci. USA **87:** 4378–4382.

68. PARTANEN, J., T. P. MAKELA, R. ALITALO, H. LEHVASLAIHO & K. ALITALO. 1990. Putative tyrosinase kinases expressed in K-562 human leukemia cells. Proc. Nat. Acad. Sci. USA **87:** 8913–8917.

69. MIKI, T., T. P. FLEMING, D. P. BOTTARO, J. S. RUBIN, D. RON & S. A. AARONSON. 1991. Expression cDNA cloning of the KGF receptor by creation of a transforming autocrine loop. Science **251:** 72–75.

70. EISEMAN, A., J. A. AHN, G. GRACIANI, S. R. TRONICK & D. RON. 1991. Alternative splicing generates at least five different isoforms of the human basic-FGF receptor. Oncogene **6:** 1195–1202.

71. KEEGAN, K., D. E. JOHNSON, L. T. WILLIAMS & M. J. HAYMAN. 1991. Isolation of an additional member of the fibroblast growth factor receptor family, FGFR-3. Proc. Nat. Acad. Sci. USA **88:** 1095–1099.

72. YARDEN, Y., W.-J. KUANG, T. YANG-FENG, L. COUSSENS, S. MUNEMITSU, T. J. DULL, E. CHEN, J. SCHLESSINGER, U. FRANCKE & A. ULLRICH. 1987. Human proto-oncogene c-*kit:* a new cell surface receptor tyrosine kinase for an unidentified ligand. EMBO J. **6:** 3341–3351.

73. BOTTARO, D. P., J. S. RUBIN, D. L. FALETTO, A. M.-L. CHAN, T. E. KMIECIK, G. P. VAN DE WOUDE & S. A. AARONSON. 1991. Identification of the HGF receptor as the c-*met* protooncogene product. Science **251:** 802–804.

74. QIU, F., P. RAY, K. BROWN, P. E. BARKER, S. JHANWAR, F. H. RUDDLE & P. BESMER. 1988. Primary structure of c-*kit:* relationship with the CSF-1/PDGF receptor kinase family-oncogenic activation of v-*kit* involves deletion of extracellular domain and c-terminus. EMBO J. **7:** 1003–1011.

75. LEV, S., Y. YARDEN & D. GIVOL. 1990. Receptor functions and ligand-dependent transforming potential of a chimeric *kit* proto-oncogene. Mol. Cell. Biol. **10:** 6064–6068.

76. COOPER, C. S., M. PARK, D. G. BLAIR, M. A. TAINSKY, K. HUEBNER, C. M. CROCE & G. F. VAN DE WOUDE. 1984. Molecular cloning of a new transforming gene from a chemically transformed human cell line. Nature **311:** 29–33.

77. SARGENT, N. S., M. OESTREICHER, H. HAIDVOGL, H. M. MADNICK & M. M. BURGER. 1988. Growth regulation of cancer metastases by their host organ. Proc. Nat. Acad. Sci. USA **85:** 7251–7255.

Basic Fibroblast Growth Factor in Cultured Cardiac Myocytes[a]

ELISSAVET KARDAMI, LEI LIU, AND
BRADLEY W. DOBLE

St. Boniface General Hospital Research Center and
Department of Physiology
University of Manitoba
351 Tache Avenue
Winnipeg, Manitoba, Canada R2H 2A6

INTRODUCTION

Basic fibroblast growth factor (bFGF) is a multifunctional polypeptide which has been detected in all tissues examined so far.[1,2] Low (16–19 kDa) and high (over 20 kDa) molecular weight forms of bFGF have been characterized.[3–7] Originally considered primarily as a mitogen, bFGF is now known to have several other properties such as the ability to stimulate or inhibit cellular differentiation,[1,2] to promote cell migration,[8] to affect cell attachment,[9] and to induce mesoderm formation in embryonic development.[10–12] *In vivo* bFGF promotes tissue repair[13,14] and angiogenesis.[1,2]

Basic FGF belongs to the larger FGF family of related heparin-binding proteins which share 30–55% amino acid sequence homology.[1,2] The response of cells to FGFs is mediated by binding to and activation of specific cell surface receptors.[15,16] Basic FGF can be found in extracellular matrices[17] as well as intracellularly.[18] However the mode of secretion of this factor to the environment is not known since there is no signal peptide in the sequence of bFGF.[19] Gentle mechanical injury of cells has been reported to induce release of bFGF.[20]

We have shown that the vertebrate heart is rich in bFGF, especially in the atria,[21,22] and that this factor is mitogenic for young cardiomyocytes and nonmuscle cells.[25] Basic FGF therefore may play a role in cardiac repair processes. Here we report the release of bFGF by cultured cardiomyocytes after gentle disruption and on the localization of this factor in cultured immature cardiomyocytes. Our data suggest a continuous participation of bFGF in myocyte function and the cell cycle as well as involvement in myocyte response to injury.

MATERIALS AND METHODS

Antibodies

Antisera S1 and S2 were raised in rabbits against a synthetic peptide containing residues 1–24 of the truncated, 146 amino acid, bovine brain bFGF[19] conjugated to keyhole limpet hemocyanin, as described.[22,23] Characterization of both antisera has

[a]This work was supported by a grant from the Heart and Stroke Foundation of Canada. E. K. has a Scholarship Award from the Medical Research Council of Canada and L. L. a Studentship from the Manitoba Health Research Council.

been reported.[23,24] Briefly, both S1 and S2 are highly potent and specific for native or denatured bFGF, in *in vitro* assays or *in situ*. A monoclonal antibody specific for striated muscle myosin (no. 52) is a gift from Dr. R. Zak (University of Chicago, Ill.).

Construction of immunoaffinity columns with the immunoglobulin G (IgG) from either antiserum S1 or S2 and use of these columns for isolation of bFGFs from tissue extracts are as previously described.[24]

Cell Cultures

Cardiac myocytes were obtained from the ventricles of 7-day-old chick embryos or 1-day-old rat pups (Sprague Dawley) as described previously.[25] Cells were plated on collagen-coated coverslips or chamber slides and grown in 2% fetal calf serum (heat-inactivated) in a 1:1 mixture of Ham's F-12 medium and Eagle's minimal essential medium (MEM).

Cell Extracts and "Conditioned" Media

Confluent cardiomyocyte cultures were rinsed three times with MEM and then scraped gently in phosphate-buffered saline (PBS; 2 ml per 100-mm plate), one week after plating. Disrupted cells were left to "condition" the PBS for 15 minutes at room temperature. Subsequently cells were pelleted and lysed by freezing and thawing (three cycles) in column buffer (0.6 M NaCl, 10 mM TrisHCl pH 7.0; 1 ml per 100-mm plate). PBS and column buffer contained protease inhibitors (1 mM phenylmethylsulphonyl fluoride, 5 μg/ml leupeptin, 5 μg/ml pepstatin). Insoluble residue was removed by centrifugation (100,000 × g, 30 minutes).

Heparin-Sepharose Fractionation

Soluble cell lysates and "conditioned" PBS were separately mixed with heparin-sepharose beads (40 μl settled beads per 100-mm plate) for 30 minutes at room temperature. Unbound material was removed by aspiration, and the beads washed twice with column buffer and twice with 1.1 M NaCl in 10 mM TrisHCl, pH 7.0. Beads were then washed with 0.1 M NaCl and bound peptides eluted by boiling in two volumes of double-strength gel sample buffer.[6]

Gel Electrophoresis

Sodium dodecyl sulfate-polyacrylamide gel electrophoresis (SDS-PAGE) was performed according to Laemmli[26] in 12.5% polyacrylamide gels. Recombinant human bFGF (Upstate Biotechnology, Inc., Lake Placid, N.Y.) was included as positive control, at 5 ng/lane. *Immunoblotting* was as previously described.[22-24] Proteins were transferred electrophoretically from SDS-PAGE gels onto Immobilon-P membranes (Millipore, Ontario, Canada). After blocking of the nonspecific protein binding sites, blots were incubated with serum S2 (1:10,000) overnight. Antigen-antibody complexes were visualized by incubation with ^{125}I-protein A (Amersham Corp.) and autoradiography as described.[22,23]

Immunofluorescence

Before processing for indirect immunofluorescence labeling, cultured cells were rinsed with PBS and then fixed by method A (1% paraformaldehyde in PBS, 10 minutes, 4°C), or method B (3–4% paraformaldehyde in PBS, 30 minutes, 4°C), or method C (1% paraformaldehyde in PBS, 10 minutes; 50% methanol in PBS, 10 minutes; 50% methanol, 50% acetone, 10 minutes; 4°C). Fixed cells were incubated overnight with antisera S1 or S2 (1:2000); methods and reagents used are as described in detail in previous reports.[22–24]

RESULTS AND DISCUSSION

Antisera S1 and S2 were both raised against the N-terminal region of the truncated (146 amino acids) bovine brain bFGF. Both antisera as well as their respective affinity purified IgG were shown to be highly specific for SDS-denatured or native purified bFGF.[22–24] To examine the ability of S1 and S2 to react with bFGF in tissues, an extract from bovine pituitaries was absorbed either through immunoaffinity columns constructed with purified IgG from sera S1 and S2 or through a heparin-sepharose column. An equivalent amount of extract was also absorbed through an affinity column constructed with IgG from a nonimmune rabbit serum. Specifically retained peptides from these columns were analyzed by Western blotting for bFGF: heparin-sepharose-bound pituitary bFGF is shown in FIGURE 1, lane 1. Similar amounts of native bFGF were retained by both immunoaffinity columns (FIGURE 1, lanes 2 and 3). Both amino-extended (over 20 kDa) and truncated (18 kDa) forms of pituitary bFGF were retained equally well by either immunoaffinity column [FIGURE 1: compare lane 2 (S1) with lane 3 (S2)] as by heparin-sepharose (FIGURE 1, lane 1). Basic FGF was not retained by the control IgG column (FIGURE 1, lane 4).

The immunofluorescence staining pattern elicited by both types of antisera and IgG preparations diminishes dramatically after absorption with immobilized recombinant bFGF[22–24] (see also FIGURE 4d), confirming that these antibodies are specific for epitopes present in the intact bFGF molecule or closely related peptides.

Serum S1 has been shown to detect bFGF-like peptides predominantly in association with the cellular periphery, including the intercalated disc region of adult cardiomyocytes in tissue sections.[23] This staining pattern is also shown here (FIGURE 2a) for comparative purposes. Pericellular staining reflects most likely presence of bFGF in the basement membrane, consistent with findings from other investigators.[27,28] In the cytoplasm, S1 detects bFGF mainly along the Z discs, which are dense structures separating one sarcomere from the next (FIGURE 2a, arrows). Muscle intermediate filaments show a similar cytoplasmic pattern of localization.[29] Serum S1 has also been shown to variably stain muscle nuclei, and different affinity purified IgG preparations derived from S1 can detect bFGF either intranuclearly or in the nuclear membrane.[23] Presence of bFGF in the nucleus has now been confirmed for several other cell types.[30–32]

Antiserum S1 was used to localize bFGF in neonatal cardiac myocytes in culture. To ensure preservation and/or access to extracellular as well as intracellular sites on the myocytes several cell-fixation protocols were employed. Fixation according to method A preserved predominantly the extracellular surfaces, which, in the case of myocytes, stained strongly for bFGF (FIGURE 2b, large arrowhead). Myocytes were identified by phase contrast optics (presence of cross-striations) or double immunolabeling for myosin (not shown here). FIGURE 2b shows a brightly labeled myocyte forming contacts with another cell which, according to the criteria mentioned above, is presumed to be of nonmuscle origin. Fixation with a hyperosmotic fixative (method B) induced rupture of the myocyte membrane and thus access to the myofibrillar cytoplasm: FIGURE 2c and d show three such cardiomyocytes forming contacts with each other. Antimyosin labeling clearly highlights the sarcomeres and verifies the identity of these cells (FIGURE 2c). The outer surface of a fraction of the cell membrane has been preserved, staining intensely for bFGF (FIGURE 2d, thick arrows), but not (as expected) for myosin (FIGURE 2c, thick arrows). The myocyte cytoplasm was also labeled (FIGURE 2d), although not as intensely as the external surfaces. As observed for myocytes *in situ* (FIGURE 2a), bFGF localizes along what appears to be the Z discs of the myofibrils (compare sarcomeric staining with antimyosin with the bFGF localization).

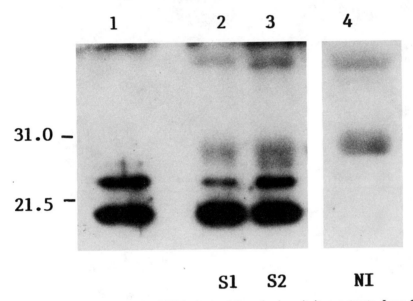

FIGURE 1. Western blotting for bFGF obtained from bovine pituitary extracts. **Lane 1:** Heparin-bound peptides. **Lane 2:** Peptides bound to (S1-IgG)-affinity column. **Lane 3:** Peptides bound to (S2-IgG)-affinity column. **Lane 4:** Peptides bound to nonimmune serum–(NI-IgG)-affinity column.

Antiserum S2 was also used to localize bFGF in cardiac cells *in vivo*. By immunofluorescence of tissue sections, S2 is shown to elicit an intense punctate staining of the intercalated disc area (FIGURE 3a, thick arrow). By immunoelectron microscopy, S2 was actually found to detect bFGF in cardiac gap junctions,[24] which form part of the intercalated disc. In addition this antiserum labels myofibrillar Z lines (FIGURE 3a, small arrows). Immunolabeling of myocytes with S2 is not as extensive as that produced by S1, suggesting that S2 recognizes only a subset of the bFGF epitopes recognized by S1 [compare FIGURE 3a (S2) with FIGURE 2a (S1)]. Serum S2 does not detect bFGF in the basement membrane (FIGURE 3a). Differences in immunostaining patterns with antibodies directed at different domains of the N terminal of bFGF have been previously noted.[23]

As shown for myocytes in tissue sections, antiserum S2 does not stain the external surface of these cells in culture but detects bFGF in regions of intercellular contact

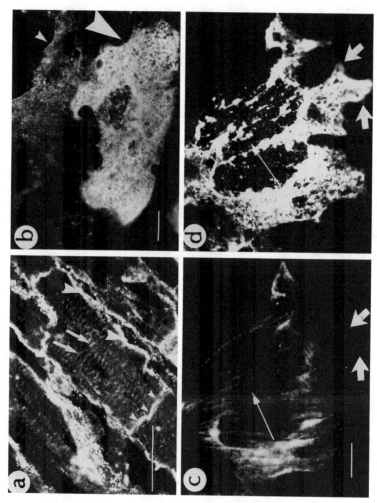

FIGURE 2. Indirect immunofluorescence for bFGF, using antiserum S1. **a:** Longitudinal section from adult rat left ventricle. **b:** Neonatal cardiac cells in culture, fixed by method A. Large and small arrowheads point to a myocyte and a nonmuscle cell, respectively. **c** and **d:** Double immunofluorescence for myosin and bFGF, respectively, in neonatal cardiomyocytes, fixed by method B. Short arrows indicate the external surface of the myocytes, while long arrows denote the myofibrillar cytoplasm. Scale bar represents 20 μm.

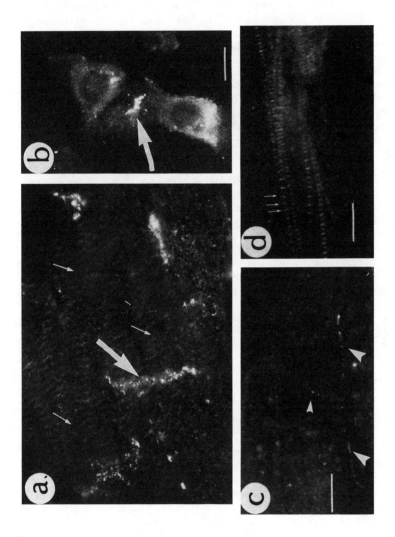

FIGURE 3. Indirect immunofluorescence for bFGF, using antiserum S2. **a:** Longitudinal section from adult rat ventricle. Arrows and arrowheads indicate intercalated discs and nuclei, respectively. **b:** Fetal cardiomyocytes, 4 hours after plating, fixed by method A. Arrow indicates intercellular junctions. **c** and **d:** Fetal myocytes, 1 week after plating, fixed by method C. Arrowheads or arrows indicate gap-junction-like or Z-line-like straining patterns, respectively.

(presumably gap junctions; FIGURE 3b and c). To ensure preservation of myofibrillar structure as well as membrane permeabilization, myocytes were fixed following method C. FIGURE 3d shows that serum S2 clearly detects bFGF in association with the myofibril Z discs and to some extent with the middle of the sarcomere, the M line.

Antiserum S2 was used to localize bFGF in near-confluent, proliferative cardiac myocyte cultures (FIGURE 4b). These cultures were also counterstained with a fluorescent nuclear stain (FIGURE 4a). Myocytes in these cultures were clearly identified not only by the anti-bFGF staining of the myofibril Z discs but also by the smaller size and round shape of their nuclei, compared to the larger cigar-shaped nuclei of fibroblastic cells (FIGURE 4a). S2 produced strong punctate nuclear staining in approximately 10% of the nuclei of interphase myocytes (FIGURE 4c, small arrows). The majority of nuclei did not stain significantly for bFGF (FIGURE 4c, arrowheads). Total loss of nuclear or myofibrillar labeling is observed when antiserum S2 is absorbed with immobilized recombinant bFGF (FIGURE 4d). To explain the nuclear anti-bFGF staining of a fraction of nuclei we suggest that S2 detects increases in nuclear bFGF in myocytes that are actively synthesizing DNA and are about to undergo mitotic division.

We compared the pattern of nuclear bFGF distribution in interphase cells with that in myocytes at the early stages of mitosis, after the dissolution of the nuclear envelope (FIGURE 5). In all myocytes in prophase a strong punctate anti-bFGF staining is discerned, similar in intensity and overall shape to the pattern of interphase nuclei [FIGURE 5, compare d (prophase) to b (interphase)]. In addition anti-bFGF labeling appears to occur only in partial association with the chromosomes in all prophase myocytes examined (compare the anti-bFGF labeling in FIGURE 5d with the chromosomal DNA-staining in FIGURE 5c). The strong anti-bFGF staining of the nuclear region in prophase cells offers support to our suggestion that the strongly staining interphase nuclei shown in FIGURE 4c belong to cells traversing a stage of the cell cycle that immediately precedes prophase, namely, the S or G2 phase. The close association of anti-bFGF and chromosomal staining persists in metaphase (FIGURE 4c, central arrow) but disappears completely in anaphase or telophase; a detailed description of bFGF localization during mitosis will be presented elsewhere (Liu and Kardami, in preparation).

Confluent cultured cardiomyocytes were scraped gently in PBS which was thus "conditioned" by these cells during a brief incubation at ambient temperature. This approach provides an *in vitro* model of cellular damage induced by mild mechanical disruption of the cell membrane and results in the release of bFGF-like factors into the medium.[20] We therefore analyzed bFGF present in cell lysates as well as bFGF released into the medium after this treatment. FIGURE 6 shows that both basal medium (lanes 2 and 3) and cell lysates (lanes 4 and 5) contained bFGF. Basic FGF released by mechanically disrupted (scraped) cells migrated with an apparent M_r of 18.5 KDa (indistinguishable from human recombinant bFGF used as positive control; FIGURE 6, lanes 2 and 3), while bFGF from cell lysates (cell-associated) was somewhat smaller (17 kDa; FIGURE 6, lanes 4 and 5) and presumably represents a truncated version of the larger form. These results could be explained by the preferential release or "secretion" of the larger bFGF form. Alternatively, the shorter, cell-associated bFGF may be the result of limited proteolysis (despite the protease inhibitors used) by enzymes released during cell lysis. In any case, these results demonstrate the presence of bFGF in cultured cardiac myocytes and offer additional support to the immunocytochemical detection of this growth factor. Furthermore, release of bFGF by mechanically disrupted myocytes *in vitro* suggests that damaged adult cardiac muscle (as would occur for instance after an ischemic

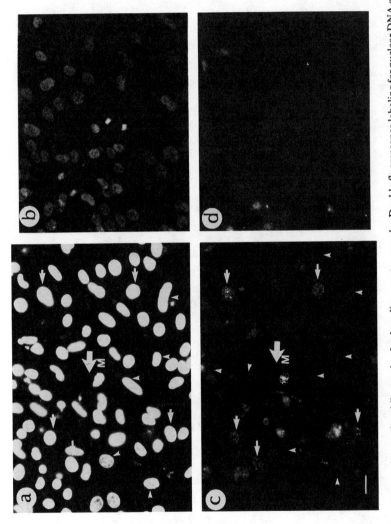

FIGURE 4. Near-confluent cultures of rapidly growing fetal cardiomyocytes. **a** and **c:** Double fluorescence labeling for nuclear DNA and for bFGF (S2), respectively. Short arrows indicate interphase nuclei that stain intensely for bFGF. Arrowheads indicate some of the nuclei that do not stain for bFGF. Thick central arrow points to a mitotic myocyte in metaphase (M). **b** and **d:** Double fluorescence labeling for nuclear DNA and for non-bFGF reactivities ("absorbed" S2), respectively. Scale bars represent 20 μm.

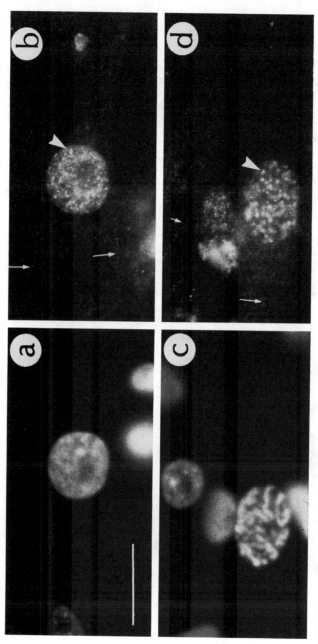

FIGURE 5. Near-confluent cultures of proliferating fetal cardiomyocytes. **a** and **b:** Double fluorescence labeling of interphase cells for nuclear DNA and bFGF (S2), respectively. **c** and **d:** Double fluorescence labeling of prophase cells for nuclear DNA and bFGF (S2), respectively. Arrows indicate myofibrillar structures in the cytoplasm. Arrowheads point to the punctate nuclear staining.

episode or extreme mechanical exertion) would similarly release bFGF to the environment and thus promote tissue repair or fibrosis.

Basic FGF localization in fetal cardiac myocytes in culture was qualitatively similar to that observed for adult myocytes in tissue sections and indicated a continuous involvement with myocardial function or functions. Cellular growth and contractile function are interlinked in cardiac myocytes. Presence of a growth-promoting factor in association with myofibril Z discs suggests a potential involvement of bFGF with the transformation of mechanical stresses to a chemical trigger for growth. Presence of bFGF in specialized intercellular junctions suggests participation in intercellular communication, which is in turn implicated not only in growth control but also in synchronous contractility.[33] Involvement of bFGF in intercellular

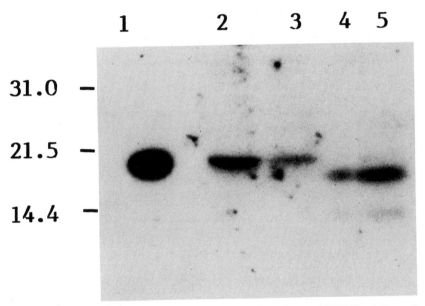

FIGURE 6. Western blotting for bFGF. **Lane 1:** Recombinant human bFGF (5 ng). **Lanes 3** and **4:** Heparin-binding peptides from myocyte "conditioned" medium. **Lanes 4** and **5:** Heparin-binding peptides from myocyte lysates.

signaling has already been inferred from its morphogenetic effects in early embryonic development.[10–12] Finally, the pattern of nuclear localization of bFGF suggested a role in regulating the cell cycle and possibly an involvement in the mechanics of mitotic division.

SUMMARY

Distribution of basic-fibroblast-growth-factor-like peptides in immature cultured cardiac myocytes was investigated using specific antisera and immunolocalization. Basic FGF was detected in association with the external surface of the cell membrane, with specialized intercellular junctions and with the myofibril Z lines in the cytoplasm. Intense, punctate nuclear anti-bFGF labeling was observed in a fraction of interphase myocytes of near-confluent, proliferating cultures. This staining pattern persisted even after the dissolution of the nuclear envelope in prophase myocytes. The pattern of cellular localization of bFGF indicates a continuous participation of this factor in myocyte physiology as well as a role in the cell cycle. Furthermore, the identification of bFGF not only in cell lysates but also in culture media after gentle mechanical disruption suggests that cardiac myocytes may release bFGF *in vivo* following tissue damage.

ACKNOWLEDGMENTS

We thank Dr. R. Zak (University of Chicago, Ill.) for the antibodies to striated myosin.

REFERENCES

1. KLAGSBRUN, M. 1989. The fibroblast growth factor family: structural and biological properties. Prog. Growth Factor Res. **1:** 207–236.
2. RIFKIN, D. B. & D. MOSCATELLI. 1989. Recent developments in the cell biology of basic fibroblast growth factor. J. Cell Biol. **109:** 1–6.
3. KLAGSBRUN, M., S. SMITH, R. SULLIVAN, Y. SHING, S. DAVIDSON, J. A. SMITH & J. SASSE. 1987. Multiple forms of basic fibroblast growth factor: amino terminal cleavages by tumor cell- and brain cell–derived proteinases. Proc. Nat. Acad. Sci. USA **84:** 1839–1843.
4. BRIGSTOCK, D. R., M. KLAGSBRUN, J. SASSE, P. A. FARBER & N. IBERG. 1990. Species-specific high molecular weight forms of basic fibroblast growth factor. Growth Factors **4:** 45–52.
5. MOSCATELLI, D., J. JOSEPH-SILVERSTEIN, R. MANEJIAS & D. B. RIFKIN. 1987. M_r 25 000 heparin-binding protein from guinea pig brain is a high molecular weight form of basic fibroblast growth factor. Proc. Nat. Acad. Sci. USA **84:** 5778–5782.
6. DOBLE, B. W., R. R. FANDRICH, L. LIU, R. R. PADUA & E. KARDAMI. 1990. Calcium protects pituitary basic FGFs from limited proteolysis by co-purifying proteases. Biochem. Biophys. Res. Comm. **173:** 1116–1122.
7. FLORKIEWICZ, R. & A. SOMMER. 1989. Human basic fibroblast growth factor encodes four polypeptides: three initiate translation from non-AUG codons. Proc. Nat. Acad. Sci. USA **86:** 3978–3981.
8. PRESTA, M., D. MOSCATELLI, J. JOSEPH-SILVERSTEIN & D. B. RIFKIN. 1986. Purification from a human hepatoma cell line of a basic fibroblast growth factor–like molecule that stimulates capillary endothelial cell plasminigen activator production, DNA synthesis, and migration. Mol. Cell. Biol. **6:** 4060–4066.
9. SCHUBERT, D., N. LING & A. BAIRD. 1987. Multiple influences of a heparin-binding growth factor in neuronal development. J. Cell Biol. **104:** 635–643.
10. SLACK, J. M. W., B. J. DARLINGTON, J. K. HEATH & S. F. GODSAVE. 1987. Mesoderm induction in early *Xenopus* embryos by heparin-binding growth factors. Nature London **326:** 197–200.
11. KIMMELMAN, D. & M. KIRSHNER. 1987. Synergistic induction of mesoderm by FGF and TGF-β and the identification of an mRNA coding for FGF in the early *Xenopus* embryo. Cell **51:** 869–877.
12. PATERNO, G. D., L. L. GILESPIE, M. S. DIXON, J. W. M. SLACK & J. K. HEATH. 1989. Mesoderm-inducing properties of INT-2 and kFGF: two oncogene-encoded growth factors related to FGF. Development **106:** 79–83.
13. CUEVAS, P., J. BURGOS & A. BAIRD. 1988. Basic fibroblast growth factor promotes cartilage repair in vivo. Biochem. Biophys. Res. Commun. **156:** 611–618.

14. TSUBOI, R. & D. B. RIFKIN. 1990. Recombinant basic fibroblast growth factor stimulates wound healing-impaired db/db mice. J. Exp. Med. **172:** 245–251.
15. NEUFELD, G. & D. GOSPODAROWICZ. 1985. The identification and partial characterization of the fibroblast growth factor receptor by baby hamster kidney cells. J. Biol. Chem. **260:** 13860–13868.
16. DIONNE, C. A., G. CRUMLEY, F. BELLOT, J. M. KAPLOW, G. SEARFOSS, M. RUTA, W. H. BURGESS, M. JAYE & J. SCHLESSINGER. 1990. Cloning and expression of two distinct high-affinity receptors cross-reacting with acidic and basic fibroblast growth factors. EMBO J. **9:** 2685–2682.
17. VLODASKY, I., J. FOLKMAN, R. SULLIVAN, R. FRIDMAN, R. ISHAI-MICHAELI, J. SASSE & M. KLAGSBRUN. 1987. Endothelial cell–derived basic fibroblast growth factor: synthesis and deposition into subendothelial extracellular matrix. Proc. Nat. Acad. Sci. USA **84:** 2292–2296.
18. JOSEPH-SILVERSTEIN J., S. A. CONSIGLI, K. M. LYSER & C. VER PAULT. 1989. Basic fibroblast growth factor in the chick embryo: immunolocalization to striated muscle cells and their precursors. J. Cell Biol. **108:** 2459–2466.
19. ESCH, F., A. BAIRD, N. LING, N. UENO, F. HILL, L. DENOROY, R. KLEPPER, D. GOSPO-DAROWICZ, P. BOHLEN & R. GUILLEMIN. 1985. Primary structure of bovine pituitary basic fibroblast growth factor (FGF) and comparison with the amino-terminal sequence of bovine brain acidic FGF. Proc. Nat. Acad. Sci. USA **82:** 6507–6511.
20. MCNEIL, P. L., L. MUTHUKRISHNAN, E. WARDER & P. D'AMORE. 1989. Growth factors are released by mechanically wounded endothelial cells. J. Cell Biol. **109:** 811–822.
21. KARDAMI, E. & R. R. FANDRICH. 1989. Heparin-binding mitogens in the heart: in search of origin and function. *In* Molecular Biology of Muscle Development. UCLA Symposia New Series: 315–325. Alan R. Liss, Inc. New York, N.Y.
22. KARDAMI, E. & R. R. FANDRICH. 1989. Basic fibroblast growth factor in atria and ventricles of the vertebrate heart. J. Cell Biol. **109:** 1865–1875.
23. KARDAMI, E., L. J. MURPHY, L. LIU, R. R. PADUA & R. R. FANDRICH. 1990. Growth Factors **4:** 69–80.
24. KARDAMI, E., R. STOSKI, B. W. DOBLE, T. YAMAMOTO, E. L. HERTZBERG & J. I. NAGY. Evidence for the association of basic fibroblast growth factor with cardiac gap junctions. (Submitted for publication.)
25. KARDAMI, E. 1990. Stimulation and inhibition of cardiac myocyte proliferation in vitro. J. Mol. Cell. Biochem. **92:** 124–129.
26. LAEMMLI, U. K. 1970. Cleavage of structural proteins during the assembly of the head of bacteriophage T4. Nature (London) **227:** 680–685.
27. GONZALEZ, A-M., M. BUSCAGLIA, M. ONG & A. BAIRD. 1990. Distribution of basic fibroblast growth factor in the 18-day rat fetus: localization in the basement membranes of diverse tissues. J. Cell Biol. **110:** 753–765.
28. CASSCELLS, W., E. SPEIR, J. SASSE, M. KLAGSBRUN, P. ALLEN, M. LEE, B. CALVO, M. CHIBA, L. HAGGROTH, J. FOLKMAN & S. E. EPSTEIN. 1990. Isolation, characterization and localization of heparin-binding growth factors in the heart. J. Clin. Invest. **85:** 433–441.
29. GRANGER, B. L. & E. LAZARIDES. 1979. Desmin and vimentin coexist at the periphery of the myofibril Z-disc. Cell **18:** 1053–1063.
30. RENKO, M., N. QUARTO, T. MORIMOTO & D. B. RIFKIN. 1990. Nuclear and cytoplasmic localization of different basic fibroblast growth factor species. J. Cell. Physiol. **144:** 108–114.
31. TESSLER, S. & G. NEUFELD. 1990. Basic fibroblast growth factor accumulates in the nuclei of various bFGF-producing cell types. J. Cell. Physiol. **145:** 310–317.
32. BALDIN, V., A-M. ROMAN, I. BOSC-BIERNE, F. AMALRIC & G. BOUCHE. 1990. Transloca-tion of bFGF to the nucleus is G_1 phase cell cycle specific in bovine aortic endothelial cells. EMBO J. **9:** 1511–1517.
33. FORBES, M. S. 1989. Gap junctions in vertebrate myocardium. *In* Cell Interactions and Gap Junctions. N. Sperelakis & W. C. Cole, Eds. **2:** 3–17. CRC Press, Inc. Boca Raton, Fla.

The Role of Fibroblast Growth Factor in Eye Lens Development[a]

J. W. McAVOY, C. G. CHAMBERLAIN, R. U. DE IONGH,
N. A. RICHARDSON, AND F. J. LOVICU

Department of Histology and Embryology
University of Sydney
Sydney, New South Wales, Australia 2006

The way in which cells and tissues differentiate and become integrated temporally and spatially into organs and organ systems is a fundamental question in developmental biology. It appears that these events depend on inductive cell and tissue interactions; consequently, considerable attention is now being focused on the molecules mediating such interactions. Members of the fibroblast growth factor (FGF) family of molecules are at the forefront of interest in this area since they have been shown to have potent inducing effects on a wide range of processes in a variety of species.[1-4] In our laboratory we have been studying mechanisms controlling eye lens development, and here we review evidence that FGF plays a central role in inducing events in lens morphogenesis and growth.

THE OPTIC CUP AND LENS FIBER INDUCTION

The lens arises from ectoderm that becomes associated with an outpocketing of neuroectoderm, the optic vesicle; close contact between these tissues coincides with the initiation of lens morphogenesis from ectoderm (FIGURE 1A and B). The period of intimate physical association between these two tissues (FIGURE 1A–D) appears to be important in lens morphogenesis.[5] At later stages (FIGURE 1E and F), two forms of lens cells arise; fiber cells eventually make up the bulk of the lens, while a monolayer of epithelial cells covers the anterior surface of the fibers (FIGURE 1F). This gives the lens a distinctive polarity. When morphogenesis is complete, the lens continues to grow by proliferation of epithelial cells mainly in the germinative zone just above the lens equator. Progeny of these divisions just below the lens equator in the transitional zone elongate and differentiate into fibers (FIGURE 2).

There is convincing evidence that lens polarity is influenced by the optic cup and the retina, which differentiates from it. In their classic experiment using chick embryos, Coulombre and Coulombre turned the lens through 180° so that the epithelial cells that normally faced the cornea then faced the retina.[6] In this new environment the epithelial cells elongated and gave rise to fiber cells. Comparable experiments using mouse eyes gave similar results and also showed that growth of the lens depended on the presence of the retina.[7]

To study the molecular basis of the inductive interaction between retina and lens, we developed a lens epithelial explant culture system using lenses from neonatal rats.

[a]The work was supported by grants from NEI (R0I 03177) Department of Health Education and Welfare USA, National Health and Medical Research Council of Australia, the Australian Research Council, the Australian Research Grants Scheme, and the Medical Foundation, University of Sydney.

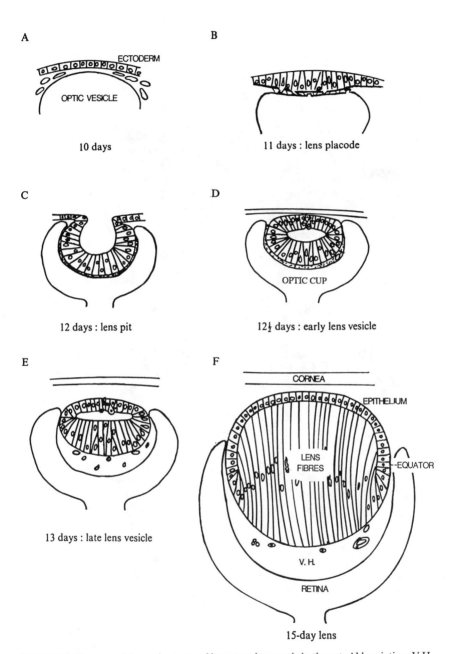

FIGURE 1. Diagram of the main stages of lens morphogenesis in the rat. Abbreviation: V.H., vitreous humor.

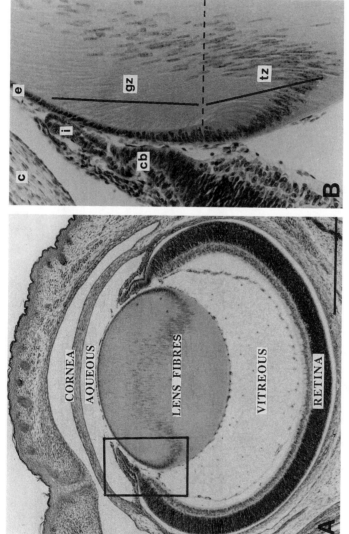

FIGURE 2. Structure of the eye in the 20-day rat fetus. **A:** The lens in relationship to other ocular tissues. The lens is composed largely of fiber cells aligned in anteroposterior direction, with a monolayer of epithelial cells on their anterior surface (cf. FIGURE 1F). **B:** Higher magnification of the equatorial region of the lens indicated by the square in A. This region of the lens can be divided into two broadly defined zones; the germinative zone (gz) of the epithelium (e) anterior to the lens equator (dashed line), where most cell proliferation occurs, and the transitional zone (tz) posterior to the equator, where epithelial cells elongate and differentiate into fiber cells. Functionally, there is some overlap between the two zones. The optic cup in this region has differentiated into the two epithelial layers of both the ciliary body (cb) and iris (i). Stained with hematoxylin-phloxine. The bar indicates 500 μm (A), 170 μm (B).

In the rat, β- and γ-crystallins serve as markers for lens fiber cell differentiation, because they are not found in the epithelial cells *in vivo* but appear sequentially as the cells elongate and differentiate into fiber cells[8,9] (FIGURE 3). In contrast, α-crystallin is present in all lens cells. We therefore used the specific crystallin content of the explants as a measure of fiber induction. We showed that the retina induces the onset of fiber-specific β- and γ-crystallin accumulation in lens epithelial explants.[10] Subsequent experiments with retina-conditioned medium showed clearly that a soluble factor was responsible for controlling these events[11] and that the events were characteristic of fiber differentiation at both the molecular[12] and morphological[13] levels.

Therefore there is strong evidence that the retina contains a factor or factors that induce fiber differentiation. We hypothesize that, during morphogenesis, only posterior cells of the lens vesicle form primary fibers since they are the only lens cells exposed to the inducing factor(s) in the optic cup (see FIGURE 1D–F). Subsequently, progeny of epithelial cell divisions that migrate or are displaced below the equator (see McAvoy)[14] elongate and differentiate into fiber cells (FIGURE 2). We suggest that an inducing signal from the optic cup is also involved in this process.

ROLE OF FGF IN LENS FIBER INDUCTION

To identify the inducing factor(s) in the retina, we carried out conventional chromatography of retina-conditioned medium and achieved partial purification of a soluble fiber differentiation factor.[15] However, a breakthrough in our efforts to identify and purify this factor came with the application of cation exchange and heparin affinity chromatography to extracts of retina.[16] The latter technique had been found by others to be very useful in the purification of growth factors such as FGF (reviewed by Baird *et al.*[17]; Lobb *et al.*[18]). We used standard procedures for isolating acidic FGF (aFGF) and basic FGF (bFGF) from retina and brain and showed that, in our lens epithelial explant culture system, these preparations promoted β- and γ-crystallin accumulation and morphological changes consistent with fiber differentiation[16,19] (FIGURE 4). We also showed that lens epithelial explants responded in a dose-dependent manner and that under our assay conditions bFGF was approximately five times more potent than aFGF (FIGURE 5).

FGF INDUCES PROLIFERATION, MIGRATION, AND FIBER DIFFERENTIATION

In order to analyze in detail the responses of individual lens epithelial cells to FGF, we developed a method for labeling cells.[14,20] The central region of explants was divided into nine squares, and single cells in each square were injected with fluorescein isothiocyanate (FITC) dextran. Cells were visualized by fluorescence microscopy and were monitored for up to 5 days after exposure to bFGF (FIGURE 6). This allowed analysis of the interrelation between various responses to bFGF in individual cells (FIGURE 7). Some cells (about 25%) divided in response to FGF, mainly within the first 3 days. However, whether or not they divided, all labeled cells responded to FGF by migrating and elongating. Maximal migration occurred during the first day of culture, and maximal elongation was achieved by day 4. This study showed that FGF induced three responses in lens epithelial cells: proliferation, migration, and fiber elongation.[20]

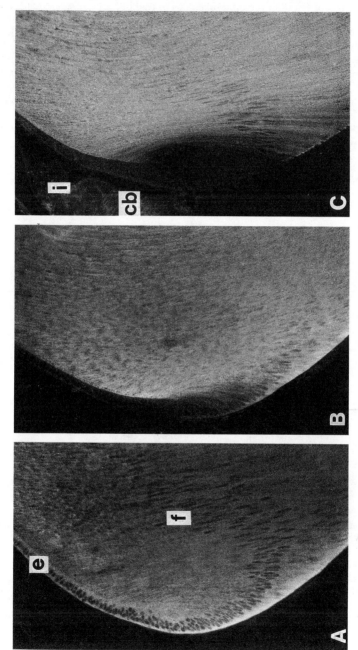

FIGURE 3. Typical distribution of crystallins in sections of the neonatal lens. **A:** α-Crystallin is detected in all lens epithelial (e) and fiber (f) cells. **B:** β-crystallin is first detected in elongating fiber cells just below the lens equator (see FIGURE 2B) and is not detected in the lens epithelium. **C:** γ-Crystallin is first detected in young fiber cells in the lens cortex and is not detected in the lens epithelium. Crystallins are not detected outside the lens, e.g., in iris (i) or ciliary body (cb). Magnification ×170.

To compare the dose dependence of the three different responses of lens cells in explants, experiments were carried out using a range of doses of bFGF. Proliferation was assessed by measuring [³H]thymidine incorporation, cell migration was measured by monitoring the movement of FITC-dextran-labeled cells by fluorescence microscopy (as described above), and fiber differentiation was assessed on the basis of β-crystallin accumulation. This study showed that as the concentration of FGF was increased, the first response of the cells was proliferation. A higher concentration of FGF was required to stimulate cell migration, and an even higher concentration to induce accumulation of β-crystallin. The half-maximal concentrations for these three responses were 150 pg/ml, 3 ng/ml, and 40 ng/ml, respectively (FIGURE

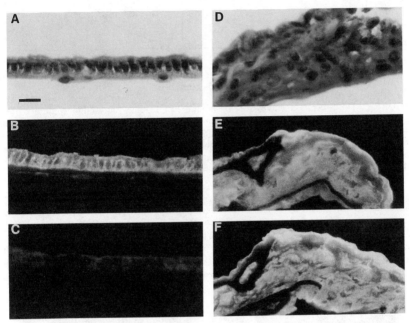

FIGURE 4. Immunofluorescent localization of α- and β-crystallin in lens epithelial explants cultured for 5 days with or without FGF. Control medium (A–C) or medium containing a saturating dose (550 ng/ml) of aFGF (D–F) was used. Sections from the same explant were stained with hematoxylin-phloxine (A,D) or used to localize α-crystallin (B,E) or β-crystallin (C,F). The bar indicates 20 μm.

8). The three responses were not mutually exclusive, although each appeared to be at least partially inhibited at concentrations of FGF greater than the optimal for that response.[20]

To confirm that bFGF was the molecule in our preparation responsible for stimulating proliferation, migration, and differentiation, immunoneutralization studies were carried out using a bFGF-specific antibody. This antibody inhibited all three responses by 90–93% (TABLE 1), thus providing good evidence that they were all due to bFGF and not due to contamination of our bFGF preparations by other factors.[19,20]

We also examined in detail the structural changes that are induced in explants by FGF, using transmission and scanning electron microscopy.[21] We showed that cells in

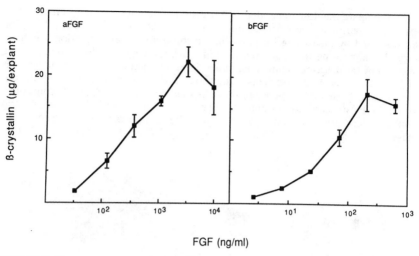

FGF (ng/ml)

FIGURE 5. Dose-response curves for aFGF and bFGF. The β-crystallin content of lens epithelial explants was determined after culturing for 5 days in the presence of FGF at the concentrations indicated. Each point represents the mean ± standard error of the mean (SEM) of data derived from three culture dishes. Explants cultured for 5 days in medium without FGF contained only 0.04–0.15 μg β-crystallin/explant. (From Chamberlain and McAvoy.[19] Reproduced by permission of Harwood Academic Publishers GmbH.)

explants were induced by either aFGF or bFGF to undergo the structural changes characteristic of fiber differentiation in the intact lens. The most obvious morphological change induced by FGF was an increase in explant thickness (FIGURE 9). Shortly after its addition to the cultures, FGF induced cell migration. This resulted in cells extending and migrating over other cells that still retained normal epithelial attachment to the capsule. The cells then underwent the following fiber-specific structural

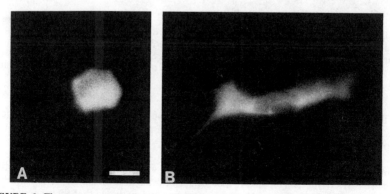

FIGURE 6. Fluorescence micrographs of FITC-dextran-labeled cells cultured for 24 hours with or without FGF. Control medium (A) or medium containing bFGF, 125 ng/ml (B), was used. (From McAvoy and Chamberlain.[20] Reproduced by permission of Company of Biologists Ltd.) The bar indicates 10 μm.

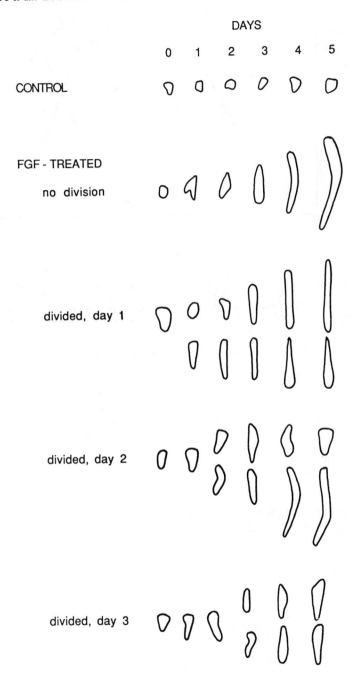

FIGURE 7. Diagram representing typical changes in the morphology of cells in explants cultured for 5 days in control medium or in medium supplemented with bFGF, 125 ng/ml. (From McAvoy and Chamberlain.[20] Reproduced by permission of Company of Biologists Ltd.)

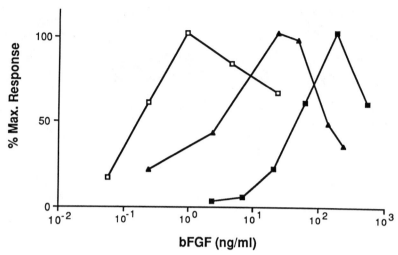

bFGF (ng/ml)

FIGURE 8. The dose dependence of the different responses of lens epithelial explants to bFGF; open squares, proliferation assessed by measuring the incorporation of [³H]thymidine over a 6-hour period, commencing 18 hours after the addition of FGF; triangles, migration assessed by injecting individual cells with FITC-dextran and recording their position just before the addition of FGF and again 24 hours later; filled squares, fiber differentiation assessed by measuring β-crystallin accumulation 5 days after addition of FGF. In each case, FGF at the concentrations indicated was added on day 0 and explants were cultured without replacing the medium. Derived from data presented by McAvoy and Chamberlain.[20]

changes (FIGURES 10 and 11): (1) cell elongation, (2) a reduction in cytoplasmic organelles, (3) the formation of specialized cell-cell junctions, including fingerlike processes and fingerprints, ball and socket junctions, tonguelike flaps and imprints, and gap junctions, and (4) denucleation. All these changes are associated with fiber differentiation *in vivo*.

In summary, these studies show that FGF stimulates proliferation, migration, and fiber differentiation at both the morphological and molecular levels (FIGURE 9)

TABLE 1. Inhibition of the Action of bFGF by an Antibody Specific for bFGF[a]

Control[b]	FGF[b]	FGF + Antibody[c]	Inhibition (%)
[³H]Thymidine incorporation (d.p.m. $\times 10^{-3}$/explant):			
0.45 ± 0.05	1.94 ± 0.24	0.60 ± 0.18[d]	90
Migration (μm/24 hours):			
25.6 ± 4.4	69.5 ± 7.3	28.5 ± 7.2[d]	93
β-Crystallin (μg/explant):			
0.06 ± 0.04	16.7 ± 1.1	1.2 ± 0.4[d]	93

[a]Summary of data derived from Chamberlain and McAvoy[19] and McAvoy and Chamberlain.[20] The concentration of bFGF in the assay was 1 ng/ml for thymidine incorporation, measured over a 21–24 hour period; 40 ng/ml for migration, measured over a 0–24 hour period; and 125 ng/ml for β-crystallin, measured after 5 days of culture.
[b]The medium contained nonimmune immunoglobulin G (IgG).
[c]The medium contained anti-FGF IgG.
[d]This value is significantly lower than the corresponding value for FGF alone, $p < 0.001$.

and that these three responses occur sequentially as the concentration of FGF is increased (FIGURE 8). Thus, the concentration of FGF in the culture medium influences not only the magnitude but also the nature of the response of lens epithelial cells. We believe that this information is highly significant for understanding the behavior of lens cells in the eye. It has led us to hypothesize that different levels of stimulation by FGF in different regions of the lens play an important role in determining the characteristic polarity and growth patterns of the lens. For example, a low level of FGF stimulation above the lens equator may only be sufficient to stimulate epithelial cell proliferation, whereas a high level of FGF in vitreous below the equator may be enough to stimulate fiber differentiation.

Control   FGF-treated

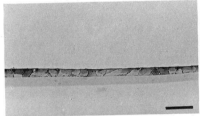

*proliferation

*migration

*fibre   differentiation
-elongation
-structural   specialisations
-increased   α-crystallin
-appearance of β- and γ-crystallins

FIGURE 9. A summary of the responses of lens epithelial explants to FGF, including micrographs of sections of lens epithelial explants to illustrate morphological changes. Stained with azurII-methylene blue. Scale bar indicates 20 μm.

LOCALIZATION OF FGF IN THE EYE

To test the hypothesis that FGF plays an important role in lens development, we required detailed information about the distribution of FGF in the eye, particularly in relation to the lens. There are now a number of reports that the ocular media that bathe the lens contain FGF. Interestingly, Caruelle *et al.* reported that bovine vitreous contains a substantially higher concentration of aFGF (295 ng/g) than aqueous (35 ng/g).[22] bFGF has also been detected in vitreous[17] and aqueous[23] of humans and in vitreous of chicks;[24] however, it appears that, in chick vitreous at least, most of the FGF-like mitogenic activity is due to aFGF.[24] In a recent immunohistochemical analysis using an antibody against a synthetic peptide corresponding to the first 24 amino acids of bovine bFGF (as used by Gonzales *et al.*),[25] we localized

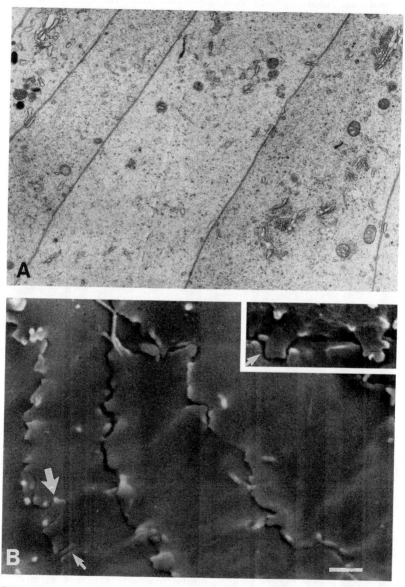

FIGURE 10. A: Transmission electron micrograph. Explant cultured with bFGF for 4 days. Fibers are arranged in parallel. Longitudinal sections through fibers demonstrate sparse organelles. **B:** Scanning electron micrograph (SEM). Fiber cells are packed in parallel on the explant surface. The cells have a smooth broad surface, and at their lateral margins fingerlike processes (small arrow) and tonguelike flaps (large arrow) overlap neighboring cells. The inset shows tonguelike flaps, one fits into a complementary imprint (arrow) on an adjacent cell. The bar indicates 1.0 μm (A), 0.5 μm (B), 0.4 μm (inset).

strong bFGF-like reactivity in the lens capsule of fresh-frozen sections of the 20-day embryonic rat eye (FIGURE 12B; de Iongh and McAvoy).[26] This localization, which required pretreatment of sections with hyaluronidase, is consistent with the findings of others who showed that, using sections pretreated with hyaluronidase[25] or protease,[27] bFGF was localized to various extracellular matrices of the body. Interestingly, different localization was obtained for aFGF. Using two monoclonal antibodies specific for aFGF and fresh-frozen sections of the 20-day embryonic rat eye, we localized aFGF-like reactivity in a band of cells around the lens equator (FIGURE 12C and D; de Iongh and McAvoy).[26] This band of cells includes the germinative zone where most of the cell proliferation occurs and the transitional zone where epithelial cells elongate and differentiate into fibers (see McAvoy[8,28] for details of spatial patterns of proliferation and fiber elongation).

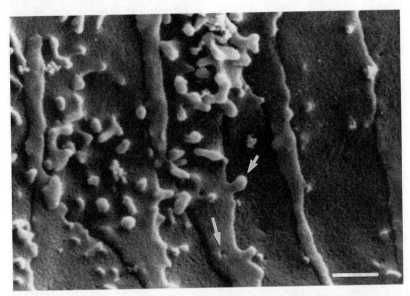

FIGURE 11. SEM, of explant cultivated with aFGF for 6 days. Surface processes from the fibers are numerous in some regions and are generally fingerlike or have typical ball and short stalk shape (short arrow). Fingerprints or sockets are also prominent on the fiber surface (long arrow). The bar indicates 0.5 μm.

Since our *in vitro* studies have shown that lens cells are induced to proliferate, migrate, and differentiate into fibers at low, medium, and high FGF concentrations respectively,[20] it is possible that aFGF associated with the lens cells influences these events *in vivo* by an autocrine mechanism. A paracrine component is probably also involved since the inner epithelium of the ciliary and iridial regions of the retina, which directly apposes the equatorial region of the lens, also fluoresces strongly for aFGF (FIGURE 12C and D). Moreover, the relatively high concentration of aFGF known to be present in vitreous[22] may be responsible for inducing lens cells below the equator (in the transitional zone) to differentiate into fibers. Therefore it seems likely that the action of FGF in determining spatial patterns of proliferation,

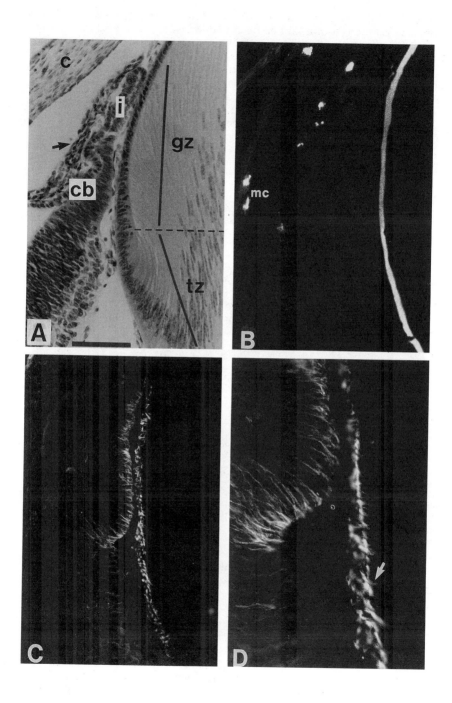

migration, and fiber differentiation in the lens *in vivo* may involve both autocrine and paracrine mechanisms.

DOES THE RESPONSE OF LENS EPITHELIAL CELLS TO FGF CHANGE WITH AGE?

Once the lens has formed, new fiber cells are laid down at the lens equator continuously throughout life. To investigate whether FGF plays a role in inducing lens fiber differentiation in later life, lens epithelial explants from 3-, 10-, 21-, 100-, and 175-day-old rats were cultured in the presence of bFGF. (Previously we had used only 3-day-old rats.) To ensure consistency between preparations, explants were prepared from comparable regions of the central lens epithelium at all ages. The crystallin composition of explants was analyzed by immunofluorescence and ELISA methods. This study revealed an age-related decline in responsiveness to FGF. Fiber-specific γ-crystallin was only detected in explants from neonates, and fiber-specific β-crystallin accumulation progressively diminished with increasing age. In fact, in explants from adult rats, central epithelial cells did not accumulate any β-crystallin in response to FGF at concentrations known to have significant effects on explants from neonates.[29]

Since we hypothesize that FGF plays a central role in fiber induction in rats (and probably in other mammals), this study[29] left us with the question: how do new secondary fibers arise in rats over 14 weeks of age? It occurred to us that cells in different regions of the lens epithelium may vary in their responsiveness to FGF. The lens epithelium can be subdivided into two main regions, a central region and a peripheral region nearer the lens equator. In the study of age-related changes described above,[29] explants were taken from the central region of the epithelium. However, the peripheral region includes the germinative zone and autoradiography has shown that it is cells in this region near the equator that elongate into secondary fibers *in situ* (see McAvoy).[14] It seemed likely that cells in this region could be more responsive to FGF than cells in the central region. Therefore we extended our studies to include the investigation of cells in the peripheral region of lens epithelial explants. Firstly, we cultured peripheral or central regions of explants from 3-day-old rats with a range of concentrations of bFGF for 5 days and measured their β-crystallin content. Both central and peripheral explants responded to FGF in a dose-dependent manner, but the concentration required for half-maximal stimula-

FIGURE 12. Immunolocalization of FGF in lens, ciliary body, and iris of the 20-day rat fetus. **A:** Hematoxylin-stained section showing the same equatorial region of the lens as that shown in B and C. The germinative zone (gz), transitional zone (tz), and equator (dotted line) of the lens are indicated. Both ciliary body (cb) and iris (i) are composed of a double layer of epithelial cells (the ciliary and iridial regions of the retina) and a vascular stroma (arrow). The cornea (c) is also indicated. **B:** bFGF. Hyaluronidase-pretreated section showing bFGF-like immunoreactivity in the lens capsule and mast cells (mc), in the cornea, and along the lateral surface of the iris. **C:** aFGF. In the lens, strong aFGF-like reactivity is present in equatorial epithelial cells in the germinative zone and in incipient fiber cells in the transitional zone. Only the inner epithelial layer of the ciliary and iridial retina is immunoreactive. **D:** aFGF. The same section as in C at higher magnification. The immunofluorescent label is clearly localized along the lateral borders of the inner epithelial cells of the ciliary retina, and in the lens the label is localized mainly to the basal parts of cells with occasional labeling along their lateral surfaces (arrow). The bar indicates 100 μm (A, B, C) and 40 μm (D).

tion was lower for peripheral cells than for central cells (7 ng/ml compared with 38 ng/ml). We then compared the ability of peripheral explants from 3-, 21-, 100-, and 175-day-old rats to undergo fiber differentiation in response to FGF.[30] The crystallin composition of explants was analyzed by immunofluorescence and quantified by ELISAs. There was an age-related decrease in responsiveness to FGF, as already observed for central explants; however, unlike central explants, peripheral explants of all ages studied still retained the ability to respond to FGF by accumulating fiber-specific β-crystallin to some extent. Therefore the distribution of FGF in the eye may be important for maintaining lens polarity and growth patterns throughout life as well as for establishing them during morphogenesis.

In addition, these age-related changes in responsiveness of lens cells to FGF may be important for regulation of lens growth. The lens continues to grow throughout life. In rats, the fastest growth rate is around birth[31] and the growth rate decreases with age. Based on the results presented here we suggest that this may be attributed, at least in part, to a progressive slowing down in responsiveness to FGF with increasing age.

INSULIN/IGF POTENTIATES FGF-INDUCED FIBER DIFFERENTIATION

As our knowledge of growth factors expands, it is becoming increasingly clear that the response induced by a growth factor can be significantly influenced by the presence of other growth factors. Besides FGF, a number of other growth factors have been detected in the eye and shown to influence lens cells (reviewed by McAvoy and Chamberlain).[4] The insulin/insulinlike growth factor (IGF) family is of particular interest. Experiments with epithelial explants from chick lenses have shown that events in fiber differentiation are induced by insulin[32] and chicken vitreous humor.[33] Lentropin, the protein in vitreous humor that stimulates fiber differentiation events, has been characterized and shown to be related to IGF-1.[33] Both insulin and IGF receptors are present on chick lens epithelial cells.[34,35]

Recently we began to investigate the effects of insulin/IGF on rat lens explants. So far we have shown that insulin or IGF-1, in the absence of FGF, can induce traces of β-crystallin accumulation in explants, almost exclusively in cells in the peripheral region.[36] They also stimulate [³H]thymidine incorporation (Williams and McAvoy, unpublished), as previously reported by others who used whole lenses in culture.[37] However when FGF is combined with insulin (or IGF-1), the typical FGF-induced responses are substantially enhanced.[36] Both β- and γ-crystallin accumulation is affected synergistically, the effect being greater for γ- than for β-crystallin, and epithelial cells in both the central and peripheral regions of the explant participate in the synergistic response (FIGURE 13). We have also shown that combining FGF and insulin produces a synergistic effect on [³H]thymidine incorporation (Williams and McAvoy, unpublished).

Since IGF is present in ocular media (see McAvoy and Chamberlain),[4] these findings have important implications. For example, in terms of the model for FGF-regulated lens differentiation described earlier, much lower levels of FGF in vitreous would be capable of inducing optimal fiber differentiation if a member(s) of the insulin/IGF growth factor family was also present. It is also interesting that cells in peripheral, but not central, regions of explants can accumulate some fiber-specific β-crystallin in response to insulin or IGF-1 alone. The greater responsiveness of cells in the periphery of epithelial explants may reflect the fact that they contain some aFGF; immunolocalization shows aFGF-like reactivity in the germinative zone of the lens epithelium (FIGURE 12C) which would be included in peripheral explants. Thus it is possible that the addition of insulin/IGF may in some way potentiate the action of FGF in this region leading to the accumulation of some β-crystallin.

SUMMARY

In this review we have presented evidence that FGF plays an important role in inducing events in lens morphogenesis and growth. Our studies show that FGF

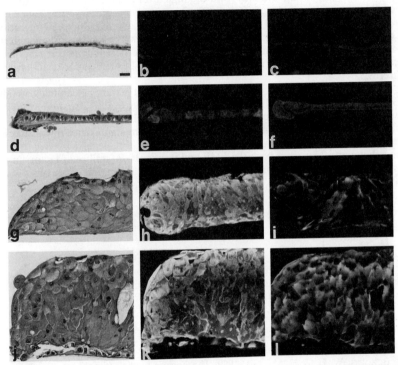

FIGURE 13. Immunofluorescent localization of β- and γ-crystallins in central epithelial explants cultured with insulin, FGF, or FGF plus insulin. Explants were trimmed to remove the peripheral region and cultured for 10 days with control medium (a, b, c) or with medium containing 2 μg/ml insulin (d, e, f) or 20 ng/ml bFGF (g, h, i) or a combination of these two factors (j, k, l). Serial sections were stained with hematoxylin-phloxine (a, d, g, j) or used to localize β-crystallin (b, e, h, k) or γ-crystallin (c, f, i, l). α-Crystallin was present in all cells in both control and growth-factor-treated explants. The bar indicates 20 μm.

stimulates lens epithelial cells in explants to proliferate, migrate, and differentiate into fibers at low, medium, and high concentrations, respectively. This has some important implications for understanding the behavior of lens cells in the eye.

The fact that aFGF is detected in the equatorial region of the lens where cells are actively proliferating, possibly migrating, and differentiating into fibers suggests that these events may be under autocrine control *in vivo,* at least to some extent. Because

FGF is also present in the ciliary and iridial region of retina and in the vitreous, paracrine control may also be involved.

Cell proliferation, fiber differentiation, and (possibly) cell migration occur in characteristic spatial patterns that are related to distinct compartments of the lens. We suggest that cells in the germinative zone receive only a low level of FGF stimulation arising from the cells themselves and possibly also from the ciliary and iridial regions of the retina but, whatever the source, this is only sufficient to stimulate proliferation. Lens epithelial cells that migrate or are displaced into the transitional zone below the lens equator receive some FGF from these sources but in addition receive a strong stimulus from the high level of FGF in the vitreous; thus, fiber differentiation is induced. Cells at the junction between these two zones may receive an intermediate level of FGF stimulation, sufficient to induce cell migration. In essence, we are proposing that, in the eye, FGF acts as a lens morphogen in the sense that different levels of FGF stimulation elicit different lens cell responses. Hence its characteristic distribution in the eye establishes lens polarity and growth patterns.

Since FGF has an inductive effect on lens cells from mature age animals, we also propose that this specific distribution of FGF in the eye is also important for maintenance of a normal lens throughout life. Finally the synergistic effects of insulin/IGF on the FGF-induced responses highlight the importance of considering the distribution of members of the insulin/IGF family of molecules *in vivo*. Mechanisms that control levels of both the FGF and insulin/IGF families of factors in the eye are probably of crucial importance in the formation and maintenance of a normal lens.

ACKNOWLEDGMENTS

The authors thank Dr. A. Baird for supplying the bFGF antiserum, Dr. I. Hendry for providing monoclonal antibodies to aFGF, and Homebush State Abattoirs and Riverstone Meat Co. Pty. Ltd. for providing bovine brains. Roland Smith and Fay Lau helped with the preparation of the manuscript.

REFERENCES

1. BURGESS, W. H. & T. MACIAG. 1989. The heparin-binding (fibroblast) growth factor family of proteins. Annu. Rev. Biochem. **58:** 575–606.
2. KLAGSBRUN, M. 1989. The fibroblast growth factor family: structural and biological properties. Prog. Growth Factor Res. **1:** 207–235.
3. RIFKIN, D. B. & D. MOSCATELLI. 1989. Recent developments in the cell biology of basic fibroblast growth factor. J. Cell Biol. **109:** 1–6.
4. MCAVOY, J. W. & C. G. CHAMBERLAIN. 1990. Growth factors in the eye. Prog. Growth Factor Res. **2:** 29–43.
5. PARMIGIANI, C. & J. MCAVOY. 1984. Localisation of laminin and fibronectin during rat lens morphogenesis. Differentiation **28:** 53–61.
6. COULOMBRE, J. L. & A. J. COULOMBRE. 1963. Lens development: fibre elongation and lens orientation. Science **142:** 1489–1490.
7. YAMAMOTO, Y. 1976. Growth of lens and ocular environment: role of neural retina in the growth of mouse lens as revealed by an implantation experiment. Dev. Growth Differ. **18:** 273–278.
8. MCAVOY, J. W. 1978. Cell division, cell elongation and distribution of α-, β-, and γ-crystallins in the rat lens. J. Embryol. Exp. Morphol. **44:** 149–165.

9. YANCEY, S. B., K. KOH, J. CHUNG & J.-P. REVEL. 1988. Expression of the gene for main intrinsic polypeptide (MIP): separate spatial distributions of MIP and β-crystallin gene transcripts in rat lens development. J. Cell Biol. **106**: 705–714.
10. McAVOY, J. W. 1980. β- and γ-crystallin synthesis in rat lens epithelium explanted with neural retina. Differentiation **17**: 85–91.
11. McAVOY, J. W. & V. T. P. FERNON. 1984. Neural retinas promote cell division and fibre differentiation in lens epithelial explants. Curr. Eye Res. **3**: 827–834.
12. CAMPBELL, M. T. & J. W. McAVOY. 1984. Onset of fibre differentiation in cultured rat lens epithelium under the influence of neural retina–conditioned medium. Exp. Eye Res. **39**: 83–94.
13. WALTON, J. & J. McAVOY. 1984. Sequential structural response of lens epithelium to retina-conditioned medium. Exp. Eye Res. **39**: 217–229.
14. McAVOY, J. W. 1988. Cell lineage analysis of lens epithelial cells induced to differentiate into fibres. Exp. Eye Res. **47**: 869–883.
15. CAMPBELL, M. T. & J. W. McAVOY. 1986. A lens fibre differentiation factor from calf neural retina. Exp. Cell Res. **163**: 453–466.
16. CHAMBERLAIN, C. G. & J. W. McAVOY. 1987. Evidence that fibroblast growth factor promotes lens fibre differentiation. Curr. Eye Res. **6**: 1165–1168.
17. BAIRD, A., F. ESCH, P. MORMEDE, N. UENO, N. LING, P. BOHLEN, S.-Y. LING, W. B. WEHRENBERG & R. GUILLEMIN. 1986. Molecular characterization of fibroblast growth factor: distribution and biological activities in various tissues. Recent Prog. Horm. Res. **42**: 143–205.
18. LOBB, R. R., J. W. HARPER & J. W. FETT. 1986. Purification of heparin-binding growth factors. Anal. Biochem. **154**: 1–14.
19. CHAMBERLAIN, C. G. & J. W. McAVOY. 1989. Induction of lens fibre differentiation by acidic and basic fibroblast growth factor (FGF). Growth Factors **1**: 125–134.
20. McAVOY, J. W. & C. G. CHAMBERLAIN. 1989. Fibroblast growth factor (FGF) induces different responses in lens epithelial cells depending on its concentration. Development **107**: 221–228.
21. LOVICU, F. J. & J. W. McAVOY. 1989. Structural analysis of lens epithelial explants induced to differentiate into fibres by fibroblast growth factor (FGF). Exp. Eye Res. **49**: 479–494.
22. CARUELLE, D., B. GROUX-MUSCATELLI, A. GAUDRIC, C. SESTIER, G. COSCAS, J. P. CARUELLE & D. BARRITAULT. 1989. Immunological study of acidic fibroblast growth factor (aFGF) in the eye. J. Cell. Biochem. **39**: 117–128.
23. TRIPATHI, R. C., C. B. MILLARD, B. J. TRIPATHI & V. REDDY. 1988. A molecule resembling fibroblast growth factor in aqueous humor. Am. J. Ophthalmol. **106**: 230–231.
24. MASCARELLI, F., D. RAULAIS, M. F. COUNIS & Y. COURTOIS. 1987. Characterization of acidic and basic fibroblast growth factors in brain, retina, and vitreous of the chick embryo. Biochem. Biophys. Res. Commun. **146**: 478–486.
25. GONZALEZ, A-M., M. BUSCAGLIA, M. ONG & A. BAIRD. 1990. Distribution of basic fibroblast growth factor in the 18-day rat fetus: localization in the basement membranes of diverse tissues. J. Cell Biol. **110**: 753–765.
26. DE IONGH, R. & J. W. McAVOY. Distribution of acidic and basic fibroblast growth factors (FGF) in the foetal rat eye: implications for lens development. (Submitted to Growth Factors.)
27. FOLKMAN, J., M. KLAGSBRUN, J. SASSE, M. WADZINSKI, D. INGBER & I. VLODAVSKY. 1988. A heparin-binding angiogenic protein—basic fibroblast growth factor—is stored within basement membrane. Am. J. Pathol. **130**: 393–400.
28. McAVOY, J. W. 1978. Cell division, cell elongation and the coordination of crystallin gene expression during lens morphogenesis in the rat. J. Embryol. Exp. Morphol. **45**: 271–281.
29. RICHARDSON, N. A. & J. W. McAVOY. 1990. Age-related changes in fibre differentiation of rat lens epithelial explants exposed to fibroblast growth factor. Exp. Eye Res. **50**: 203–211.
30. RICHARDSON, N. A., J. W. McAVOY & C. G. CHAMBERLAIN. Age of rats affects response of lens epithelial explants to fibroblast growth factor. (Submitted to Exp. Eye Res.)

31. PARMIGIANI, C. M. & J. W. McAVOY. 1989. A morphometric analysis of the development of the rat lens capsule. Curr. Eye Res. **8:** 1271–1277.

32. PIATIGORSKY, J. 1973. Insulin initiation of lens fiber differentiation in culture: elongation of embryonic lens epithelial cells. Dev. Biol. **30:** 214–216.

33. BEEBE, D. C., M. H. SILVER, K. S. BELCHER, J. J. VAN WYK, M. E. SVOBODA & P. S. ZELENKA. 1987. Lentropin, a protein that controls lens fiber formation, is related functionally and immunologically to the insulin-like growth factors. Proc. Nat. Acad. Sci. **84:** 2327–2330.

34. BASSAS, L., P. S. ZELENKA, J. SERRANO & F. DE PABLO. 1987. Insulin and IGF receptors are developmentally regulated in the chick embryo eye lens. Exp. Cell Res. **168:** 561–566.

35. BASSNETT, S. & D. C. BEEBE. 1990. Localisation of insulin-like growth factor-1 binding sites in the embryonic chicken eye. Invest. Ophthalmol. **8:** 1637–1643.

36. CHAMBERLAIN, C. G., J. W. McAVOY & N. A. RICHARDSON. 1991. The effects of insulin and basic fibroblast growth factor on fibre differentiation in rat lens epithelial explants. Growth Factors **4:** 183–188.

37. REDDAN, J. R. 1982. Control of cell division in the ocular lens, retina, and vitreous humor. *In* Cell Biology of the Eye. D. S. McDevitt, Ed.: 299–375. Academic Press. New York, N.Y.

The Role of Fibroblast Growth Factor in Early *Xenopus* Development

DAVID KIMELMAN

Department of Biochemistry SJ-70
University Of Washington
Seattle, Washington 98195

INTRODUCTION

Since the pioneering experiments of Mangold and Spemann in the 1920s, it has been clear that intercellular communication plays a major role in the development of the vertebrate embryo.[22] The attempt to identify the mechanism of these signaling events has both plagued and fascinated biologists over the last 70 years. Beginning 4 years ago with the discovery that a low molecular weight protein secreted by a *Xenopus* cell line (the XTC factor)[20] was competent to convert preectodermal cells to a mesodermal fate, a growing body of evidence has implicated members of the fibroblast growth factor (FGF) and transforming growth factor-β (TGF-β) families in these early signaling events. In this report I will focus mainly on the role of FGF in the early development of the *Xenopus* embryo.

EARLY *XENOPUS* DEVELOPMENT

The earliest signaling events in *Xenopus* development result in the formation of mesoderm at the equator of the embryo in response to signals that pass from the vegetal hemisphere to the animal hemisphere during the early cleavage stages. These signals convert the bottom part of the animal hemisphere from an ectodermal fate to a mesodermal fate with a variety of tissue types specified along the dorsal-ventral axis (FIGURE 1; reviewed in References 3, 6, and 15). For example, the notochord will form from the most dorsal cells of the embryo, whereas the blood will form from the most ventral cells. Other tissues such as the muscle will be specified in a more complex manner with contributions from cells across the dorsal-ventral axis of the equator.

Further signaling from the most dorsal mesoderm (termed the Spemann organizer) will both generate and pattern the neural ectoderm from the cells above the equator on the dorsal side of the embryo. Additional signals from this region will also organize the lateral mesodermal cells. Thus without these early signaling events the egg would only be able to form undifferentiated ectoderm and endoderm.

ASSAY FOR MESODERM INDUCTION

An assay developed by Nieuwkoop has been used extensively to test for the formation of mesoderm by potential inducing factors.[12] The upper one-quarter of the embryo (the "animal cap") is removed when the embryo reaches the 4000-cell stage (late blastula) and cultured in a buffered saline solution. In the absence of exogenous

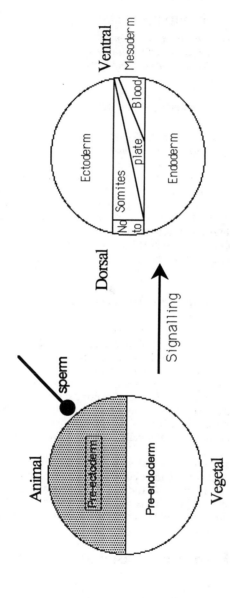

FIGURE 1. Patterning of the mesoderm in early development. At the time of fertilization (left diagram) the embryo is divided into animal (preectodermal) and vegetal (preendodermal) hemispheres. After a period of signaling, a fate map can be drawn for the pregastrula embryo (right diagram). The mesoderm occupies the equatorial region of the embryo and is divided into many cell types including notochord (noto), somites, lateral plate (plate), and blood.

factors, the explant will form into a sphere of undifferentiated ectoderm. The same result is found in the presence of most added factors. However if acidic or basic FGF or supernatant from the XTC cell line is added to the explant, mesoderm will form.[8,18,20] This can be observed directly as the explant loses its spherical shape and appears quite distorted. If the explants are examined histologically, a variety of mesodermal cell types are found, including notochord, pronephros, mesenchyme, and muscle.[18,20] Alternatively, the formation of muscle-specific products can be measured by immunohistochemistry or RNAse protection.[8,18]

PRESENCE OF bFGF RNA IN THE *XENOPUS* EMBRYO

The original discovery that bovine basic FGF (bFGF) could stimulate mesoderm formation led to a search for an FGF-like molecule in the early *Xenopus* embryo. Using a cDNA encoding bovine basic FGF, several isolates were obtained from *Xenopus* cDNA libraries produced from oocyte and neurula stage RNA.[7,8] All of these clones hybridize to three transcripts in the early embryo. The most abundant is a 1.5-kilobase (kb) transcript which is present in the oocyte and throughout the early embryonic period. The other two transcripts are present in the oocyte at 5% of the level of the 1.5 kb mRNA. Of these, a 4.5-kb transcript disappears when the oocyte matures and then is resynthesized at a much higher level during the neurula stages. The other transcript is 2.5 kb and present at a constant level from oogenesis to gastrulation.

The 4.5-kb transcript encodes the *Xenopus* bFGF protein. This protein is the same length as the mammalian bFGFs, and is identical at 83% of the amino acids in the peptide sequence. When *Xenopus* bFGF is synthesized in a bacterial expression system using a cDNA clone of the 4.5-kb transcript, the protein is able to convert ectoderm to mesoderm in the animal cap assay system.[7] Interestingly, whereas the animal cap must be exposed to bovine brain bFGF for 90 minutes to maximize mesoderm induction,[17] the recombinant *Xenopus* bFGF can cause the same effect after only 10 minutes.[5] Therefore the small number of amino acid differences between *Xenopus* and bovine bFGF may have large effects on the binding of bFGF to the *Xenopus* receptor.

The 1.5-kb mRNA is transcribed in the opposite direction of the 4.5-kb bFGF-encoding transcript and is thus referred to as an antisense transcript.[9] As shown in FIGURE 2, the sense and antisense transcripts share a 900-base-pair region of overlap. Nucleotide sequence analysis of the overlap region in cDNA copies of the two RNAs demonstrates that they must be transcribed from the same gene since there are no detectable sequence differences. The antisense RNA has an open reading frame that could encode a 25-kDa protein (FIGURE 2).[9] Although the existence of this protein has not yet been demonstrated, the amino acid sequence encoded by the open reading frame is highly conserved in mammalian species.[9]

The presence of both the sense and antisense RNAs in the oocyte, with the antisense RNA in approximately 20-fold excess, suggested that both RNAs might be hybridized in the oocyte. Typically it would be very difficult to determine conclusively if the two RNAs were present as a hybrid, particularly as there is a very small amount of the bFGF 4.5-kb transcript in the oocyte. However, a double-strand dependent RNA modifying enzyme has been shown to be released from the oocyte nucleus (germinal vesicle) at time that the oocyte matures.[1] Using a polymerase chain reaction (PCR) based assay it was found that the two RNAs were not modified before oocyte maturation, but were extensively modified after the oocyte matured, and only in the region that is shared between the two transcripts.[9] This demonstrated

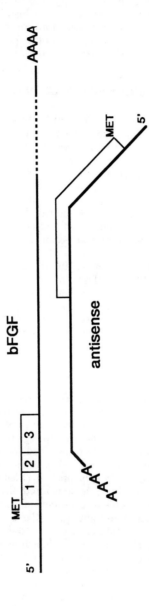

FIGURE 2. Schematic representation of the sense and antisense bFGF transcripts. The 4.5-kb sense transcript is drawn at top and the 1.5 kb antisense transcript is at the bottom, with the 5′ end and poly A tail indicated for each transcript. The parallel region indicates the 900 base pairs of complementarity between the two transcripts. The open reading frames on the two transcripts are represented by open boxes beginning with the initiating methionine (MET). The three exons comprising the bFGF peptide are shown.

that the sense and antisense RNAs must be in the cytoplasm prior to maturation such that they are separated from the modifying enzyme. Furthermore, they must be in the partial double strand configuration shown in FIGURE 2 for the modifying enzyme to modify the RNAs only in the overlap region. Since the sense transcript disappears soon after the modification reaction occurs, one possibility is that the antisense transcript targets the sense transcript for degradation when the oocyte matures. Experiments to test this hypothesis are under way.

PRESENCE OF THE bFGF PROTEIN AND THE FGF RECEPTOR IN THE *XENOPUS* EMBRYO

The bFGF protein has been detected in the embryo during the early cleavage stages,[7,19] which is the period during which mesoderm induction must occur. Surprisingly, the protein is also present at the same levels in the oocyte, which is essentially one large cell, as in the early embryo. Since only a portion of the preectodermal cells are normally converted to mesoderm, bFGF must only interact with its receptor in a limited region of the embryo. Whether this occurs by prelocalization of the protein in the oocyte or by selective release of the protein during the early cleavage stages is not yet known.

The bFGF receptor and its RNA have been identified in the *Xenopus* embryo.[4,11] The RNA encoding the receptor is maternal, but it is not translated until fertilization.[11] The receptor protein is maximally present during the early cleavage stages, but its levels decrease toward the end of the period during which the animal cells are competent to respond to mesoderm.[4] The loss of competence may therefore be due to disappearance of the receptor. Interestingly, bFGF or XTC factor treated animal cap cells retain the FGF receptor RNA, whereas untreated animal cap cells do not.[11] This suggests that once mesoderm is formed in response to inducers, it retains the capacity to respond to bFGF, either as an inducer or as a mitogen.

The receptor does not appear to be localized within the embryo, and hence the specification of mesoderm is likely to be due to the presence of bFGF.[4,11] This result is in agreement with previous studies demonstrating that the animal cap, which would normally become ectoderm, can be converted to mesoderm by the addition of bFGF.[8,18]

OTHER MESODERM-INDUCING FACTORS

Based on the quantitative assays for mesoderm induction using muscle-specific markers, it was clear that bFGF was a weak inducer of muscle.[8] This led to a search for other inducing factors. One of these, TGF-β1, has the property of a synergistic inducer.[8] By itself TGF-β1 is not able to induce mesoderm, but it greatly potentiates the formation of muscle produced by bFGF. In contrast, TGF-β2 is by itself a potent inducer of mesoderm.[13] Recently the XTC factor has been identified as activin A, a member of the TGF-β family.[21] One reason that other members of the TGF-β family may work in mesoderm induction is that they mimic the action of activin.

Two other TGF-β family members have been found to be present in the early embryo. One of these, Vgl, is present in the vegetal pole of the embryo due to localization of its RNA.[23,25] It is a novel member of the TGF-β family, although it has not yet been shown to have a function. Activin B has recently been found to be transcribed in the embryo beginning in the late cleavage stages.[24] Neither RNA encoding activin A or B is present in the oocyte although it is possible that one of the activin proteins is present at that time.

REGULATION OF MESODERM BY INDUCING FACTORS

As mentioned above, mesoderm is comprised of a large number of different cell types, each with its own unique distribution within the embryo. The two known embryonic factors, bFGF and XTC factor (activin A), have been used to investigate the precise role of these factors in mesoderm induction. Using a careful histological analysis, XTC factor has been shown to induce more dorsal-type mesoderm such as notochord and muscle, whereas bFGF induces more ventral-type mesoderm such as mesenchyme and mesothelium.[5] Additionally, an analysis for genes containing homeodomain sequences that are expressed at either the anterior or posterior ends of the embryo has shown that XTC factor appears to be an inducer of anterior mesoderm, whereas bFGF seems to induce posterior mesoderm.[2,14] Taken together these results suggest that the inducing factors are used both to specify cell type and position within the cleavage stage embryo.

MODEL FOR MESODERM INDUCTION

Although there are not yet enough data to present a convincing model for mesoderm induction, one possibility is suggested by the current data. As shown in FIGURE 3, three signaling events may be utilized to initially pattern the mesoderm. bFGF may induce the formation of mesodermal cells on the ventral side of the embryo, and perhaps also induce the lateral mesoderm. Another as of yet undetermined factor, here called V, would induce the formation of the dorsal and perhaps lateral mesoderm, including the Spemann organizer. The Spemann organizer would produce activin B, which then organizes the dorsal and lateral mesoderm. V could be an activin, although no activin A or B RNA was detected in the early cleavage stage embryo[24] which is when this factor would have to act.

This model does not account for the formation of anterior and posterior mesoderm, which has been shown to be separately regulated by XTC (activin) and bFGF. This model also does not include Vgl, the TGF-β analogue which has not yet been shown to have a role in the early embryo. Finally, it should be stated that these three factors would only determine the initial state of the mesoderm. Other factors such as the product of the Int-1 gene and retinoic acid would direct the further development of the embryo.[10,16]

CONCLUSION

bFGF is shown to be a potent inducer of mesoderm during early amphibian development. RNA-encoding bFGF is found in the oocyte, and the bFGF protein is found in the early embryo. Together with evidence demonstrating the presence of the FGF receptor in the early embryo, bFGF is suggested to be a natural mesodermal inducer. Formation of the entire mesoderm must be due to the action of multiple factors including activin B. The presence of an antisense bFGF RNA in the oocyte which causes the modification of the RNA encoding bFGF suggests that bFGF may be under complex regulation during the development of the embryo.

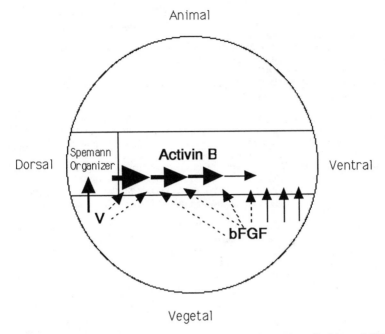

FIGURE 3. A possible three-signal model for the induction of mesoderm. In this model bFGF induces ventral (solid arrows) and perhaps lateral mesoderm (dashed arrows). The V factor induces dorsal mesoderm including the Spemann organizer and perhaps lateral mesoderm. Activin B is then produced by the organizer and patterns the lateral mesoderm, perhaps by a morphogenetic gradient.

ACKNOWLEDGMENTS

I wish to thank M. Kirschner for his advice throughout the course of this work and R. Cornell, A. Maas, and J. Northrop for critically reading this manuscript.

REFERENCES

1. BASS, B. L. & H. WEINTRAUB. 1988. An unwinding activity that covalently modifies its double-stranded RNA substrate. Cell **55:** 1089–1098.
2. CHO, K. W. Y. & E. M. DE ROBERTIS. 1990. Differential activation of *Xenopus* homeobox genes by mesoderm-inducing growth factors and retinoic acid. Genes Dev. **4:** 1910–1916.
3. DAWID, I. B., T. D. SARGENT & F. ROSA. 1989. The role of growth factors in embryonic induction in amphibians. Cell **59:** 767–770.
4. GILLESPIE, L. L., G. D. PATERNO & J. M. W. SLACK. 1989. Analysis of competence: receptors for fibroblast growth factor in early *Xenopus* embryos. Development **106:** 203–208.
5. GREEN, J. B. A., G. HOWES, K. SYMES, J. COOKE & J. C. SMITH. 1990. The biological effects of XTC-MIF: quantative comparison with *Xenopus* bFGF. Development **108:** 173–183.
6. GURDON, J. B. 1987. Embryonic induction—molecular prospects. Development **99:** 285–306.

7. KIMELMAN, D., J. A. ABRAHAM, T. HAAPARANTA, T. M. PALISI & M. W. KIRSCHNER. 1988. The presence of fibroblast growth factor in the frog egg: its role as a natural mesoderm inducer. Science 242: 1053–1056.

8. KIMELMAN, D. & M. KIRSCHNER. 1987. Synergistic induction of mesoderm by FGF and TGF-β and the indentification of an mRNA coding for FGF in the early Xenopus embryo. Cell 51: 869–877.

9. KIMELMAN, D. & M. W. KIRSCHNER. 1989. An antisense mRNA directs the covalent modification of the transcript encoding fibroblast growth factor in Xenopus oocytes. Cell 59: 687–696.

10. MCMAHON, A. P. & R. T. MOON. 1989. Ectopic expression of the proto-oncogene Int-1 in Xenopus embryos leads to duplication of the embryonic axis. Cell 58: 1075–1084.

11. MUSCI, T. J., E. AMAYA & M. W. KIRSCHNER. 1990. Regulation of the fibroblast growth factor receptor in early Xenopus embryos. Proc. Nat. Acad. Sci. USA 87: 8365–8369.

12. NIEUWKOOP, P. D. 1969. The formation of the mesoderm in urodelean amphibians I. The induction by the endoderm. Roux's Arch. Entwickl Mech. Org. 162: 341–373.

13. ROSA, F., A. B. ROBERTS, D. DANIELPOUR, L. L. DART, M. B. SPORN & I. B. DAWID. 1988. Mesoderm induction in amphibians: the role of TGF-β2-like factors. Science 239: 783–785.

14. RUIZ I ALTABA, A. & D. A. MELTON. 1989. Involvement of the Xenopus homeobox gene Xhox3 in patterning the anterior-posterior axis. Cell 57: 317–326.

15. RUIZ I ALTALBA, A. & D. A. MELTON. 1990. Axial patterning and establishment of polarity in the frog embryo. Trends Genet. 6: 57–64.

16. SIVE, H. L., B. W. DRAPER, R. M. HARLAND & H. WEINTRAUB. 1990. Identification of a retinoic acid–sensitive period during primary axis formation in Xenopus laevis. Genes Dev. 4: 932–942.

17. SLACK, J. M., H. V. ISAACS & B. G. DARLINGTON. 1988. Inductive effects of fibroblast growth factor and lithium ion on blastula ectoderm. Development 103: 581–590.

18. SLACK, J. M. W., B. G. DARLINGTON, J. K. HEATH & S. F. GODSAVE. 1987. Mesoderm induction in early Xenopus embryos by heparin-binding growth factors. Nature London 326: 197–200.

19. SLACK, J. M. W. & H. V. ISAACS. 1989. Presence of basic fibroblast growth factor in the early Xenopus embryo. Development 105: 147–153.

20. SMITH, J. C. 1987. A mesoderm-inducing factor is produced by a Xenopus cell line. Development 99: 3–14.

21. SMITH, J. C., B. M. J. PRICE, K. VAN NIMMEN & D. HUYLEBROECK. 1990. Identification of a potent Xenopus mesoderm-inducing factor as a homologue of activin A. Nature 345: 729–731.

22. SPEMANN, H. & H. MANGOLD. 1924. Uber induction von embryonalagen durch implantation artfremder organis atoren. Roux's Arch. Dev. Biol. 100: 599–638.

23. TANAHILL, D. & D. A. MELTON. 1989. Localized synthesis of the Vg1 protein during early Xenopus development. Development 106: 775–785.

24. THOMSEN, G., T. WOOLF, M. WHITMAN, S. SOKOL, J. VAUGHAN, W. VALE & D. A. MELTON. 1990. Activins are expressed in Xenopus embryogenesis and can induce axial mesoderm and anterior structures. Cell 63: 485–493.

25. WEEKS, D. L. & D. A. MELTON. 1987. A maternal mRNA localized to the vegetal hemisphere in Xenopus eggs codes for a growth factor related to TGF-β. Cell 51: 861–867.

The Role of Basic Fibroblast Growth Factor in Prenatal Development in the Rat[a]

CHARLES S. NICOLL,[b] LIMING LIU, ELAINE ALARID,
AND SHARON M. RUSSELL

*Department of Integrative Biology and
the Cancer Research Laboratory
University of California
Berkeley, California 94720*

INTRODUCTION

A growing body of evidence indicates that the fibroblast growth factors (FGFs) play important roles in developmental processes. Extracts of rat embryos stimulate fibroblast proliferation *in vitro* in a manner similar to that of purified FGF,[1] and a protein related to the FGFs is transiently expressed in the kidney of mouse embryos.[2] FGF or FGF-like proteins have been detected in the developing chick brain,[3] and FGF causes mesodermal induction in frog embryos.[4,5] Furthermore, FGF is found in the human placenta[6,7] and it is a potent angiogenic factor.[8–10] Thus, it could affect embryonic and or fetal development in various ways, including regulating the development of the vascular system in the conceptus. However, our knowledge of the role of the FGFs in developmental processes is still superficial.

Studies on the regulation of the development of mammalian embryos and fetuses are complicated by their inaccessability. Procedures that allow the direct application of treatments to the conceptus are highly invasive and frequently traumatic.[11] For example, "hypophysectomy" of the fetus of some species actually involves decapitation.[12] Such invasive procedures may affect growth of the conceptus indirectly by altering maternal physiology and/or placental functions. In addition, studies on the development of embryos and fetuses in species with short gestational period are also complicated by the fact the internal milieu to which the conceptus is exposed is continually changing, particularly during the latter part of pregnancy. For example, in rodent species, several types of placental hormones are secreted in different amounts during the second half of gestation[13] and fetal endocrine glands differentiate and become functional during this period. Although *in vitro* procedures obviate some of these problems, cell, tissue, and organ culture methods are also problematic.

EMBRYO AND FETAL TRANSPLANTATION

In order to avoid these problems we have used transplantation procedures to study the development of rat embryos and fetal tissues *in vivo*. Our systems allow the

[a] This research was supported by National Institutes of Health Grant HD14661.
[b] Address correspondence to Professor Nicoll at the Department of Integrative Biology, LSA 281.

internal environment in which the transplants are developing to be readily controlled and experimentally modified. When placed under the kidney capsule of syngeneic hosts, several fetal structures grow rapidly and tissue differentiation occurs in an essentially normal manner.[14–18] In fact, over periods ranging from 11–14 days fetal reproductive tracts, intestines, and paws attain a degree of growth that ranges from 60%–85% of that which they would achieve if they were left to grow *in situ* (TABLE 1).

Transplanted 10-day embryos also grow rapidly and attain a size and relative growth increment that are greater than those of any of the transplanted fetal structures.[19] However, in comparison to the degree of growth they would have achieved if they had developed *in utero,* growth of the embryo transplants is severely retarded (TABLE 1). They show less than 4% as much growth as their counterparts grown *in situ.*

Fetal paws grow equally well in male and female hosts and in animals of widely divergent age and growth rates.[11] Thus, while the internal environment of adult rats is highly favorable for supporting growth of fetal structures, it is much less salubrious for the growth of embryos. Accordingly, the internal milieu of the hosts may be lacking in a factor or factors that are needed specifically for embryonic growth.

TABLE 1. Absolute and Relative Growth of Transplanted Whole 10-Day Rat Embryos and of Structures from 15- to 16-Day Fetuses[a]

Type of Transplant	Days of Incubation	Wet Weight (mg)		Fold Increase	Percent of *In Situ* Growth
		Initial	Increase		
Fetal paws	11	2.0	22	11	85
Fetal intestines	11	1.2	68	57	80
Fetal male reproductive tract	14	—	—	—	60
10-day embryos	12	0.6	194	324	3.6

[a]Data from Cooke et al.,[15,17] Liu et al.,[19] and Alarid et al.[22]

Despite the severe growth retardation of the embryo transplants, however, tissue differentiation in them was only slightly retarded relative to that which occurred *in utero.*[19]

INTRARENAL INFUSION

We have developed infusion methods that allow us to test for direct effects of hormones, growth factors, and antisera to them on the growth and development of transplanted embryos and fetal structures.[18,20] One of these methods involves inserting a catheter into the inferior suprarenal artery, which branches from the right renal artery.[20] The external end of the catheter is attached to an osmotic minipump, which, when placed in the peritoneal cavity of the hosts, serves to infuse substances directly into the right kidney. Thus, the transplants growing on that kidney are exposed directly to the infused materials and the grafts on the contralateral kidney serve as internal controls. Other control procedures include infusion of normal rabbit serum or the solvent used for the hormones and growth factors.

When an antiserum to basic fibroblast growth factor was infused into the right kidney bearing 10-day embryo transplants,[20] it inhibited their growth by about 50%

TABLE 2. Effects of an Antiserum to Basic FGF on Growth of Rat Embryos and Fetal Structures[a]

Type of Transplant	Treatment Period (days)	Growth as Percent of Controls	$p <$
10-day embryos	10	−50	0.004
16-day fetal paws	10	−15	NS[b]
16-day fetal intestines	11	−29	0.05

[a]Data from Liu et al.[20,21] All transplants were grown in intact hosts.
[b]NS = not significant.

(TABLE 2). By contrast, the same antiserum had a much less striking effect on the growth of fetal paw and intestine transplants.[21] In fact, only the intestines showed a significant degree of growth retardation (TABLE 2). Infusion of a low dose of recombinant basic FGF (bFGF) increased growth of embryo transplants by about 75% in intact hosts[20] (TABLE 3) but it had no effect on growth of fetal paws in such hosts.[21] In hypophysectomized hosts in which growth of the fetal paw and intestinal transplants was impaired by about 65%, the infused bFGF did stimulate growth of the paws by about 45%[21] but it had no effect on the intestines (TABLE 3). Thus, growth of the embryo transplants appears to be more dependent upon and responsive to bFGF than is that of the fetal structures.[20,21]

A similar disparity was seen in the effects of these treatments on tissue differentiation. Infusion of the antiserum to bFGF had no evident effect on bone formation in the paws but it did impede villus formation in the intestines.[21] When grown on the kidney of hypophysectomized hosts, the paw and intestine transplants showed severe retardation of tissue differentiation in addition to impaired growth.[15,17] Infusion of bFGF into the kidney of such hosts promoted some bone formation in the paws and caused rudimentary development of the villi in the intestines, but it did not restore their development to normal.[21] However, much more striking effects were obtained with the embryo transplants that were grown in intact hosts.[20] Infusion of the antiserum to the bFGF completely suppressed the development of endodermal derivatives and of many of the tissues derived from the mesoderm. Differentiation of other mesodermal structures was partially suppressed (TABLE 4). The antiserum had no apparent effect on the differentiation of ectodermally derived tissues.[20]

These results indicate that FGF plays a significant role in the growth of rat embryos and it is evidently of critical importance for the differentiation of endodermal and some mesodermal structures. Growth and differentiation of ectodermal

TABLE 3. Effects of Infusing Recombinant Basic FGF (5 μg/day) on Growth of Transplanted Rat Embryos and Fetal Structures[a]

Type of Transplant	Treatment Period (days)	Growth as Percent of Controls	$p <$
10-day embryos	6	+74	0.005
16-day fetal paws	9	+46	0.05
16-day fetal intestines	11	+13	NS[b]

[a]Data from Liu et al.[20,21] The embryos were grown in intact hosts, whereas the fetal structures were placed on the kidneys of hypophysectomized hosts.
[b]NS = not significant.

TABLE 4. Effects of Anti-FGF on Tissue Differentiation in Transplanted Rat Embryos[a]

Type of Tissue	Control	Treated
Ectodermal		
Skin, hair follicles, sweat glands	+ +	+ +
Mesodermal		
Skeletal muscle, cartilage, ossification		
centers, blood vessels	+ +	+
Adipose tissue, kidney, transitional		
epithelium	+ +	−
Endodermal		
Intestinal villi, salivary gland, pancreas,		
bronchus	+ +	−

[a]The antiserum was infused for 10 days (from days 3–13 posttransplantation). At the end of the experiment, the transplants were fixed in neutral buffered formalin, imbedded in paraffin, and sectioned serially. Every 10th section was mounted on a slide and stained with hematoxylin and eosin before microscopic examination. All of the mounted sections were examined carefully for the presence of the various tissue types. Each group contained six transplants. The control transplants were on the left (noninfused) kidney, and treated ones were on the right, infused kidney of the host rats. + +, present and abundant; + present but sparse; −, not present. Results from Liu and Nicoll.[20]

structures appear to be relatively independent of bFGF. Differentiation of tissues in fetal paws and intestines may also be influenced by basic FGF but its role in these structures is less striking than it is in the embryos.

DEVELOPMENT OF REPRODUCTIVE TRACTS

The role of bFGF in the development of rat reproductive tracts was also investigated using our transplantation and infusion procedures.[22] The tracts are mesodermally derived, and several observations suggest that bFGF could be involved in their development and/or functions.[23–31] The urogenital (UG) sinuses or genital ridges of 16-day male and female rat fetuses were grown on the kidneys of adult hosts for 14 days. The sex of the hosts was matched to that of the transplants. Infusion of the antiserum to bFGF into the kidney of hosts bearing UG sinuses of male or female fetuses did not affect transplant growth or tissue differentiation (TABLE 5). A similar lack of effect of the antiserum was seen with the transplants of the genital ridge of

TABLE 5. Effects of Antiserum to Basic FGF on Growth of Transplanted Fetal Rat Reproductive Tracts[a]

Fetal Structure	Sex	Growth as Percent of Controls	$p <$
Urogenital sinus	Female	−25	NS[b]
	Male	−23	NS
Genital ridge	Female	−14	NS
	Male	−36	0.005

[a]The treatment period was 14 days. Data from Alarid et al.[22]
[b]NS = not significant.

female fetuses (TABLE 5). However, the antiserum caused a significant inhibition of the growth of the male genital ridge transplants (TABLE 5).

Histological examination of the transplants showed that the urogenital sinuses of both sexes had developed normally. The male genital ridges that were exposed to normal rabbit serum and those that served as internal controls in the hosts given the antiserum to bFGF showed normal development of the epididymis, the vas deferens, and testis. However, the male genital ridge transplants that were exposed to the antiserum to bFGF showed severe impairment of development of the epididymis by 7 days, and it was completely absent by 14 days (TABLE 6). Thus, most of the reduction in the size of the male genital ridge transplants caused by the antiserum to bFGF may be due to suppression of epididymal development.

These results indicate that of the various organs that develop in the reproductive tracts of male and female fetuses, the epididymis is specifically dependent on bFGF for its growth and differentiation. Preliminary studies indicate that infusion of bFGF can restore the growth and development of male genital ridge transplants to normal when grown in castrated host rats in which their development is severely impaired because of the lack of androgens (Alarid and Nicoll, unpublished). Thus, bFGF may be an autocrine or paracrine mediator of the actions of androgens in this structure.

TABLE 6. Effects of Antiserum to Basic FGF on the Number of Epididymal Profiles in Transplanted Male Genital Ridges[a]

Days of Incubation	Number of Profiles per Tract		$p <$
	Controls[b]	Anti-bFGF	
7	3.1 ± 0.6	0.6 ± 0.4	0.01
14	17.3 ± 3.8	0	0.001

[a]Data from Alarid et al.[22]
[b]The controls were exposed to normal rabbit serum.

DISCUSSION

Our results indicate that bFGF plays a major role in the development of rats during the embryonic period. Growth of embryos was strikingly influenced by blocking endogenous bFGF with the antibody, and by infusing the growth factor into the kidneys bearing the transplants (TABLES 2 and 3). Our findings also indicate that differentiation of endodermal derivatives is completely suppressed by the antiserum (TABLE 4). This is a novel and important discovery. Furthermore, mesodermally derived tissues show at least a quantitative difference in their degree of dependence on bFGF for their differentiation. The antibody to FGF completely suppressed differentiation of some mesodermal structures but only partially suppressed the differentiation of others (TABLE 4).

Development of fetal paws and intestines showed some evidence of a dependence on and responsiveness to bFGF, but the role of that growth factor in the development in these structures is much less striking than it is for development during the embryonic period. Accordingly, these results suggest that fetal structures are less dependent on FGF than are these tissues during the embryonic period. However, the results with the transplanted reproductive tracts show that this generalization is not entirely correct. All of the various structures that develop from the UG sinus and the genital ridge transplants are of mesodermal origin, yet only the epididymis showed evidence of a dependence on bFGF for its development. Hence, the finding that differentiation of the various mesodermal tissues in the embryo was variably dependent on FGF[20] is corroborated by our findings with the reproductive tract structures.[22]

Evidence from various sources indicates that growth factors besides bFGF may play a role in embryonic and fetal development in mammalian species.[11,32-35] Our transplantation and intrarenal infusion procedures will allow the role of these growth factors in developmental processes to be studied *in vivo,* and interactions among them can be investigated profitably.[36]

ACKNOWLEDGMENTS

We are indebted to Dr. Andrew Baird of the Whittier Institute, La Jolla, California, for providing us with the antibody to the basic FGF.

REFERENCES

1. HOFFMAN, R. S. 1940. The growth-activating effects of extracts of adult and embryonic tissues of rat on fibroblast colonies *in vitro.* Growth **4:** 361–372.
2. RISAU, W. & P. EKBLOM. 1986. Production of a heparin binding angiogenesis factor by the embryonic kidney. J. Cell Biol **103:** 1101–1108.
3. RISAU, W. 1986. Developing brain produces an angiogenesis factor. Proc. Nat. Acad. Sci. USA **83:** 3855–3860.
4. SLACK, J. M. W., B. F. DARLINGTON, H. K. HEATH & S. F. GODSAVE. 1988. Heparin binding growth factors as agents of mesoderm induction in early *Xenopus* embryo. Nature **326:** 297–299.
5. KIMELMAN, D. & M. KIRSCHNER. 1987. Synergistic induction of mesoderm by FGF and TGF-β and the identification of an mRNA coding for FGF in the early *Xenopus* embryo. Cell **51:** 869–873.
6. GOSPODAROWICZ, D., J. CHENG, G. M. LUI, D. K. FUJII, A. BAIRD & P. BOHLEN. 1985. Fibroblast growth factor in the human placenta. Biochem. Biophys. Res. Commun. **128:** 554–559.
7. MOSCATELLI, D., M. PRESTA & D. B. RIFKIN. 1986. Purification of a factor from human placenta that stumulates capillary endothelial cell protease production, DNA synthesis, and migration. Proc. Nat. Acad. Sci. USA **83:** 2091–2097.
8. GOSPODAROWICZ, D., N. FERRARA, L. SCHWEIGERER & G. NEUFELD. 1987. Structural characterization and biological functions of fibroblast growth factor. Endocr. Rev. **8:** 95–112.
9. THOMAS, K. A. 1987. Fibroblast growth factors. FASEB J. **1:** 434.
10. BAIRD, A., F. ESCH, P. MORMEDE, N. UENO, N. LING, P. BOHLEN, S. Y. YING, W. B. WEHRENBERG & R. GUILLEMIN. 1986. Molecular characterization of fibroblast growth factor: distribution and biological activities in various tissues. Recent Prog. Horm. Res. **42:** 143–158.
11. COOKE, P. S. & C. S. NICOLL. 1989. Hormonal regulation of fetal growth. *In* Handbook of Human Growth and Developmental Biology. E. Meisami & P. S. Timeras, Eds. 2(Part B): 3–22. CRC Press. Boca Raton, Fla.
12. JOST, A. 1979. Fetal hormones and fetal growth. Contrib. Gynecol. Obstet. **5:** 1–20.
13. TALAMANTES, F. & L. OGREN. 1988. The placenta as an endocrine organ: polypeptides. *In* The Physiology of Reproduction. E. Knobil & J. Neill, Eds.: 2093–2128. Raven Press Ltd. New York, N.Y.
14. NICOLL, C. S. 1991. Use of transplanted mammalian embryos and fetal structures to analyze the role of growth factors in the regulation of growth and differentiation. Zool. Sci. **8:** 215–223.
15. COOKE, P. S., S. M. RUSSELL & C. S. NICOLL. 1983. A transplant system for studying

hormonal control of growth of fetal tissues: effects of hypophysectomy, growth hormone, prolactin, and thyroxine. Endocrinology **112:** 806–812.

16. COOKE, P. S., C. U. YONEMURA & C. S. NICOLL. 1984. Development of thyroid hormone dependence for growth in the rat: a study involving transplanted fetal, neonatal, and juvenile tissues. Endocrinology **115:** 2059–2064.

17. COOKE, P. S., C. U. YONEMURA, S. M. RUSSELL & C. S. NICOLL. 1986. Growth and differentiation of fetal rat intestine transplants: dependence on insulin and growth hormone. Biol. Neonate **49:** 211–218.

18. COOKE, P. S., L. HIGA & C. S. NICOLL. 1986. Insulin but not GH directly stimulates growth of transplanted fetal rat paws. Am. J. Physiol. **251:** Endocrinol. Metab. **14:** E624–629.

19. LIU, L., S. M. RUSSELL & C. S. NICOLL. 1987. Growth and differentiation of transplanted rat embryos in intact, diabetic and hypophysectomized hosts: comparison with their growth *in situ*. Biol. Neonate **52:** 307–316.

20. LIU, L. & C. S. NICOLL. 1988. Evidence for a role of basic fibroblast growth factor in rat embryonic growth and differentiation. Endocrinology **123:** 2027–2031.

21. LIU, L., S. M. RUSSELL & C. S. NICOLL. 1990. Analysis of the role of basic fibroblast growth factor in growth and differentiation of transplanted fetal rat paws and intestines. Endocrinology **126:** 1764–1770.

22. ALARID, E., G. CUNHA, P. YOUNG & C. S. NICOLL. Evidence for an organ- and sex-specific role of bFGF in the development of the mammalian reproductive tract. Endocrinology **129.** (In press.)

23. STORY, M. T., J. SASSE, D. KAKUSKA, S. C. JACOBS & R. K. LAWSON. 1988. A growth factor in bovine and human testes structurally related to basic fibroblast growth factor. J. Urol. **140:** 422–427.

24. JACOBS, S., M. STORY, J. SASSE & R. LAWSON. 1988. Characterization of growth factors derived from the rat ventral prostate. J. Urol. **139:** 1106–1110.

25. SMITH, E. P., W. E. RUSSELL, F. S. FRENCH & E. M. WILSON. 1989. A form of basic fibroblast growth factor is secreted into the adluminal fluid of the rat coagulating gland. Prostate **14:** 353–365.

26. FAUSER, B. C. J., A. BAIRD & A. J. W. HSUEH. 1988. Fibroblast growth factor inhibits luteinizing hormone-stimulated androgen production in cultured rat testicular cells. Endocrinology **123:** 2935–2941.

27. KOOS, R. D. & C. E. OLSON. 1989. Expression of basic fibroblast growth factor in the rat ovary: detection of mRNA using reverse transcription polymerase chain reaction amplification. Mol. Endocrinol. **3:** 2041–2048.

28. NEUFELD, G., N. FERRARA, L. SCHWEIGERER, R. MITCHELL & D. GOSPODAROWICZ. 1987. Bovine granulosa cells produce basic fibroblast growth factor. Endocrinology **121:** 597–603.

29. BRIGSTOCK, D. R., R. B. HEAP, P. J. BARKER & K. D. BROWN. 1990. Purification and characterization of heparin-binding growth factors from porcine uterus. Biochem. J. **266:** 273–282.

30. GOSPODAROWICZ, D. & N. FERRARA. 1989. Fibroblast growth factor and the control of pituitary and gonad development and function. J. Steroid Biochem. **23:** 183–191.

31. GONZALEZ, A. M., M. BUSCAGLIA, M. ONG & A. BAIRD. 1990. Distribution of basic fibroblast growth factor in the 18 day rat fetus: localization in the basement membranes of diverse tissues. J. Cell Biol. **110:** 753–766.

32. THORBURN, G. D., I. R. YOUNG, M. DOLLING, D. W. WALKER, C. A. BROWNE & C. C. CARMICHAEL. 1984. Growth factors in fetal development. *In* Growth and Maturation Factors. G. Guroff, Ed. **3:** 175–201. Wiley-Interscience. New York, N.Y.

33. MERCOLA, M. & C. D. STILES. 1988. Growth factor superfamilies and mammalian embryogenesis. Development **102:** 451.

34. MUMMERY, C. L., A. J. M. VAN DEN EIJNDEN-VAN RAAIJ. 1990. Growth factors and their receptors in differentiation and early murine development. Cell Differ. Dev. **30:** 1–18.

35. DE PABLO, F., L. A. SCOTT & J. ROTH. 1990. Insulin and insulin-like growth factor I in early development: peptides, receptors and biological events. Endocr. Rev. **11:** 558–577.

36. NICOLL, C. S. & S. M. RUSSELL. 1991. Regulation of embryonic and fetal development: IGF-FGF interactions. *In* Modern Concepts of Insulin-like Growth Factors. E. M. Spencer, Ed.: 25–35. Elsevier. New York, N.Y.

Basic FGF and
Growth of Arterial Cells[a]

MICHAEL A. REIDY AND VOLKHARD LINDNER

Department of Pathology, SJ-60
University of Washington
Seattle, Washington 98195

The cells of the arterial wall are capable of both synthesizing and responding to basic fibroblast growth factor (bFGF),[1-4] and recent studies have suggested that FGF might be stored in an "inactive pool" in the vessel wall. The extracellular matrix is one such potential site,[5] although several studies also show that bFGF is stored in the nucleus.[6,7] Indeed, our own immunolocalization shows that the nuclei of arterial endothelial and smooth muscle cells (SMC) are invariably stained with antibodies to bFGF with little or no staining of the extracellular matrix.[8] The presence of these mitogens in the arterial wall has no obvious effect since *in vivo* the proliferation of endothelial and smooth muscle cells is extremely low. One interesting feature of bFGF is that it does not possess a signal sequence,[9] and therefore the mechanism by which it is secreted is unclear. Following injury and cell death, however, bFGF is thought to be released[10,11] and to interact with the surrounding cells. Thus when arteries are subjected to mechanical injuries or possibly focal cell death induced by an atherogenic agent,[12] the resultant cell death with release of bFGF may be the stimulus for subsequent cell proliferation.

In this brief paper we examine the importance of bFGF as a mitogen for the arterial wall and discuss whether it plays a significant role in cellular proliferation following arterial injury.

ENDOTHELIAL CELL GROWTH AND bFGF

In vivo, endothelial cells have a limited capacity for regrowth,[13,14] which is a major concern in ensuring the patency of vascular grafts. The lack of endothelial regrowth may also play a role in the high rate of restenosis seen after angioplasty.

It is possible to study this inability of endothelial cells to regrow by removing the endothelial cells from large arteries of experimental animals. Thus when rat carotid arteries are denuded of endothelium with a balloon catheter, the endothelial cells cease to replicate approximately 6 weeks later. At this time the artery is still only partially repopulated with endothelium.[15] In a series of experiments, we have tried to understand why endothelial cell regrowth stops and have attempted to modulate this response. The details of these studies have been recently reviewed[16] and therefore will not be discussed here, but one significant finding is that in the first few weeks following denudation the proliferating endothelial cells stain strongly with an antibody raised against human recombinant bFGF (FIGURE 1).[15] Several weeks later, when endothelial replication has stopped, the cells show markedly less staining with the same bFGF antibody. Normal uninjured endothelial cells also show little staining

[a] These studies were supported by grants from the National Institutes of Health (P50HL42270) and a grant from the Council for Tobacco Research.

for bFGF. Thus there seemed to be an association of endothelial cell replication with the presence of cellular bFGF. To some extent this concept is supported by an experiment in which endothelial regrowth was shown to be significantly modified by simply changing the procedure of denudation.[15,17] In this experiment, endothelial cells were removed from the arterial wall in a manner that caused little or no trauma to the underlying medial SMCs of the arterial wall (filament loop denudation), and endothelial replication remained high until total regrowth was achieved (FIGURE 2). At present we are still unsure how a variation in the severity of a mechanical injury can directly influence the regrowth of endothelium, but interestingly, the endothelial cells of these arteries always stained strongly for bFGF. This finding is in contrast to

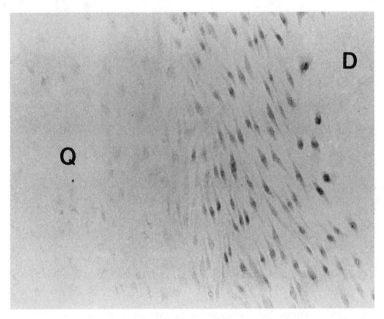

FIGURE 1. *En face* preparation of rat aortic endothelium stained with an antibody against bFGF 48 hours after partial denudation. Endothelial cells adjacent to the still denuded area stain strongly for bFGF. Quiescent endothelium distant from wounded area shows little staining for bFGF. D = denuded area; Q = quiescent endothelium. (200×).

the staining seen in endothelial cells from balloon catheter denuded arteries which show minimal staining for bFGF; these are the cells that do not continue to proliferate.

The above data are at best suggestive that endothelial cell growth is dependent upon bFGF, and therefore an experiment was undertaken to determine if endothelial cell regrowth could be directly influenced by the administration of bFGF. In these studies,[1] rat carotid arteries were denuded of endothelium with a balloon catheter, and as previously shown, endothelial regeneration stopped after approximately 6 weeks. Basic FGF was then administered to these animals and the replication of endothelium determined by [³H]thymidine autoradiography. A dra-

Total Endothelial Regrowth

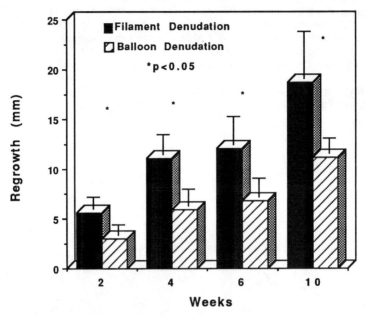

FIGURE 2. Endothelial regrowth in the rat carotid artery after denudation with a balloon catheter and filament loop. At all times after injury, significantly more regrowth was observed after denudation with the filament loop. [Means ± standard errors of the mean (SEM).] (From Reference 15.© by U.S. and Canadian Academy of Pathology, Inc.)

matic increase in the replication was observed in endothelial cells of the injured arteries after the administration of bFGF, but cells distant from the site of injury or from uninjured arteries showed only a small increase in their replication (FIGURE 3). Since the mitogen was administered intravenously over a period of several hours, it seems likely that all endothelial cells would be exposed to the same concentration of bFGF. One possible explanation for the different responses is that quiescent endothelial cells do not bind bFGF. It is difficult to directly measure bFGF binding to endothelial cells *in vivo*, so we utilized bFGF conjugated to the ribosomal toxin saporin (kindly provided by D. Lappi and A. Baird).[3,18] Saporin alone is not toxic to cells since it is not internalized, but when conjugated with bFGF, this "mitotoxin" will be bound and internalized by cells expressing the FGF receptor. In uninjured arteries this mitotoxin caused no loss of endothelium or any subsequent increase in their replication which is invariably observed after injury. One explanation for this finding is that these cells do not bind bFGF since they are neither stimulated by the mitogen nor killed by the mitotoxin. This result emphasizes that expression of the appropriate receptor is equally important as the presence of bFGF.

An increase in cell replication is not necessarily followed by an increase in cell growth, so we next determined whether bFGF would stimulate an increase in endothelial cell outgrowth. As mentioned above, endothelial regrowth slowed down markedly 6 weeks after denudation and in these arteries total endothelial regrowth does not occur. Using this model of denudation of the rat carotid artery, bFGF was injected intravenously into the animals every 4 days for a period of 8 weeks. FIGURE 4 shows that bFGF caused a dramatic increase in the extent of endothelial cell regrowth as compared to the controls and, in a few animals, total endothelial cell regrowth was achieved.[1] To our knowledge, this is the first example in which endothelial growth has been significantly increased *in vivo*, and this finding suggests that total endothelial growth can be achieved despite the size of the arterial injury. This result might also suggest that bFGF is normally required for the stimulation of endothelial cells, and that its absence may lead to an inhibition of further growth. Until more is known about the synthesis and secretion of bFGF, this latter statement is at best speculative.

The above studies show that endothelial cell regrowth can be stimulated by bFGF but no data are available to suggest that bFGF is important or even necessary for normal endothelial cell replication *in vivo*. To answer this question, we asked whether endothelial cell regeneration could be inhibited by administration of an anti-bFGF antibody. In this *in vivo* study, arteries were partially denuded of endothelium immediately following intravenous administration of a neutralizing antibody raised against human recombinant bFGF. This protocol did not inhibit endothelial cell replication (FIGURE 5) despite the fact that *in vitro*, this antibody inhibits 3T3 cell replication induced by exogenous bFGF. In the same animals, however, the identical concentration of antibody was capable of inhibiting SMC proliferation (see below). Again this result is not conclusive since not all the released bFGF may be accessible to the antibody prior to its binding to the endothelial cells. However, the observations that bFGF can only be detected inside endothelial cells

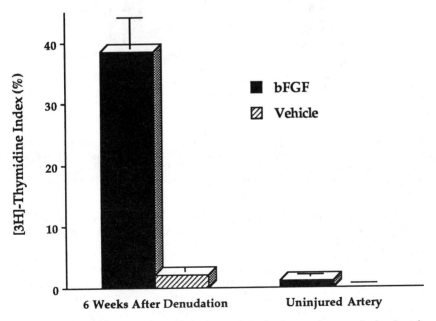

FIGURE 3. Endothelial cell replication in the rat carotid artery at the wound edge 6 weeks after balloon catheterization. Infusion of bFGF (120 μg) caused a dramatic increase in replication of endothelial cells near the wound edge. Uninjured endothelium of the contralateral carotid artery showed a small increase in proliferation. (Means ± SEM.)

and is not thought to be secreted suggest the existence of an intracrine pathway for bFGF.

SMOOTH MUSCLE CELLS AND bFGF

Acidic and basic FGF are potent mitogens for vascular smooth muscle cells, which themselves can synthesize aFGF and bFGF.[4,19] Our interest in the effects of bFGF on SMC was heightened when we found that this mitogen was highly effective in promoting endothelial cell growth since addition of exogenous bFGF was equally likely to target SMCs as well as endothelial cells. This would be especially true in a

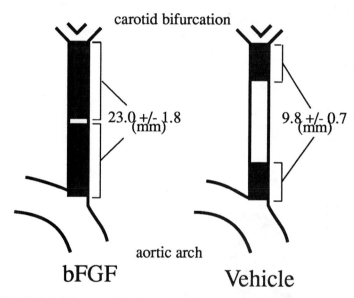

FIGURE 4. Endothelial regrowth in the rat carotid artery after injecting bFGF intravenously for 8 weeks (12 μg every 4 days starting 2 weeks after balloon catheter denudation). Treatment with bFGF significantly increased endothelial regrowth. (Means ± SEM.) (Reproduced from Reference 1 by copyright permission of the American Society of Clinical Investigation.)

damaged artery where the loss of endothelium would enhance the permeability of molecules into the arterial wall. Administration of bFGF (120 μg/rat over 8 hours) to animals immediately following endothelial denudation causes a 33-fold increase in the replication rate of SMCs as compared to controls (FIGURE 6).[3] The SMCs of uninjured arteries in the same animals showed no increase in replication. Evidence of bFGF binding to these cells was obtained using the FGF-saporin conjugate as described above. Administration of the mitotoxin caused widespread SMC loss from the media of injured arteries, but no detectable changes were observed in the SMCs of control, uninjured arteries. These data suggest that SMCs have the ability to bind and internalize bFGF presumably via a high-affinity receptor, although more direct evidence will be required to validate this issue. The data from uninjured vessels imply

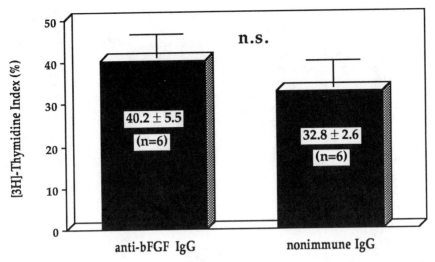

FIGURE 5. Effect of anti-bFGF antibody on endothelial cell replication at the wound edge in the rat aorta 42 hours after partial denudation with a balloon catheter. Anti-bFGF IgG or nonimmune IgG was injected prior to denudation. No significant difference in replication was observed between treatment groups. (Means ± SEM.)

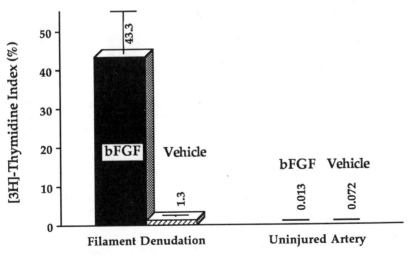

FIGURE 6. Smooth muscle replication in the rat carotid artery 41 hours after denudation with the filament loop. Infusion of bFGF (120 μg over 8 hours) caused a significant increase (33-fold) in replication. Infusion of bFGF had no effect on SMC replication in the uninjured carotid artery. (Means ± SEM.)

that the SMCs of normal vessels are either not exposed to the mitotoxin due to the presence of an intact endothelium, or that these quiescent cells do not bind bFGF. We favor the former possibility since in the above studies, bFGF was administered immediately after removal of the endothelium and it is unlikely that high-affinity receptors could be induced and expressed in this very brief time to bind the injected bFGF. However, bFGF might bind to the extracellular matrix and be released over a period of time. In this case, if FGF receptors were rapidly expressed by these cells, then subsequent proliferation could occur.

Clear evidence of the effect of bFGF on SMCs was obtained when bFGF was given daily for 14 days to animals whose carotid artery had been subjected to balloon catheter injury.[3] The addition of bFGF in this manner caused an approximate doubling in the size of the intimal lesion (FIGURE 7). An important consequence of

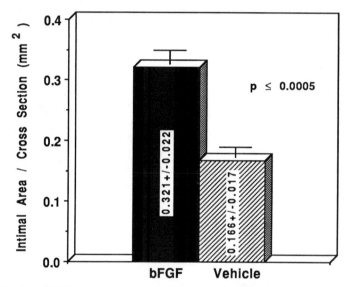

FIGURE 7. Intimal thickening of the rat carotid artery 2 weeks after balloon catheterization and injection of bFGF (12 μg/day intravenously). (Means $\pm$ SEM.) (Reference 3 by permission of the American Heart Association, Inc.)

this result is that it illustrates that bFGF would be an unsuitable choice as an endothelial mitogen *in vivo* since its presence would lead to a very significant increase in the size of arterial lesions. However, if there are differing sensitivities for the effect of bFGF between endothelial cells and SMCs then this might possibly allow for the selective stimulation of only endothelial cells.

Our previous studies have shown that endothelial cell loss and platelet adherence do not play a major role in stimulating smooth muscle cell replication,[20,21] and yet the degree of injury inflicted on the arterial wall has been found to significantly influence this event. One possibility is that injury causes cell death and release of intracellular bFGF which subsequently initiates cell proliferation. This concept would agree with the finding that when the arteries were denuded of endothelium with minimal injury (filament loop), as compared to the widespread injury caused by a balloon catheter,[22]

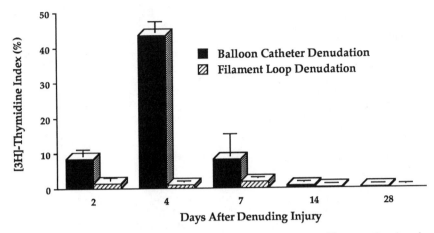

FIGURE 8. Replication of medial smooth muscle cells in the rat carotid artery after denudation with a balloon catheter and a filament loop. [Means ± standard deviations (SD).]

the subsequent SMC proliferation was diminished by approximately 10-fold (FIGURE 8).[17] The hypothesis that injury-induced release of FGF is an important factor for SMC growth was tested by using an antibody raised against bFGF. This antibody recognizes an 18 kDa band in extracts of rat artery, and *in vitro* blocks bFGF in a mitogenesis assay. Immediately prior to balloon catheter injury the antibody was administered intravenously and SMC replication was determined by injecting three doses of [³H]thymidine (24–40 hours after denudation). Control animals were also subjected to balloon denudation but received an injection of nonimmune immunoglobulin G (IgG).[8] As seen in FIGURE 9, the antibody to bFGF significantly reduced

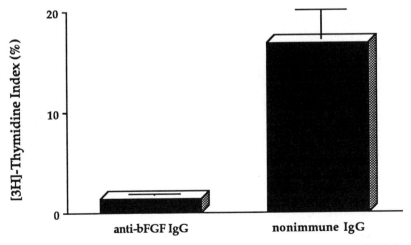

FIGURE 9. Effect of anti-bFGF antibody on smooth muscle cell replication induced by balloon catheter injury. A single intravenous injection of 10 mg anti-bFGF IgG prior to denudation caused a significant reduction in SMC proliferation 41 hours later. (Means ± SEM.)

the replication of the smooth muscle cells by $> 80\%$. This result strongly implies that bFGF is a major factor in controlling the growth of SMCs following vascular injury. This fact may have importance in certain clinical situations such as angioplasty where cell injury is prevalent and restenosis by cellular growth is an unfortunate occurrence.

Medial SMC proliferation is usually followed by migration of cells into the intima. This process takes place 3 to 4 days after surgery and a combination of cell migration and proliferation leads to the formation of an intimal lesion.[22] It is unclear if those cells that have been stimulated to divide are also the cells that migrate, and there are limited data to suggest that SMCs can migrate to the intima without entering the cell cycle.[23] Since the bFGF antibody caused a $> 80\%$ inhibition of the entry of cells into the cell cycle, we next asked whether the antibody would also cause an equal reduction in the size of the intimal lesion. The antibody against bFGF was given to animals immediately prior to balloon catheter denudation, and 8 days later the size of the intimal lesion was determined. Interestingly, this treatment did not lead to any significant decrease in the intimal lesion.[8] One explanation for this result is that bFGF bound to the matrix may be released gradually[24] when an effective concentration of the antibody is not present. Therefore, subsequent bFGF-driven replication might occur at later times. Alternatively, migration of SMCs into the intima and proliferation of intimal SMCs might be controlled by other, as yet undetermined factors. In this regard it is worth remembering that intimal SMCs behave differently from medial SMCs and at least *in vitro* they synthesize PDGF-B chain which is capable of initiating replication.[25]

In summary, our studies show that basic FGF appears to play a major role in controlling the growth of both endothelial and smooth muscle cells. It appears that endothelial cells may require bFGF for their proliferation and exogenous addition of bFGF can reinitiate endothelial cell replication in situations when these cells appeared to have senesced. Basic FGF is also a mitogen for SMCs, and in our animal models bFGF is capable of increasing replication to a much greater extent than other mitogens. Of greater importance is the fact that the response of SMCs to vascular injury can be inhibited by action of an anti-bFGF antibody. An attractive hypothesis is that bFGF may play a role in atherosclerosis where focal cell death followed by release of bFGF from either endothelium, SMCs, or macrophages may be responsible for the development of intimal lesions.

ACKNOWLEDGMENTS

We wish to thank Drs. Richard A. Majack, Douglas A. Lappi, and Andrew Baird who were involved in these experiments. The technical assistance of Eric Olson, Stella Chao, Colleen Irvin, and Abel Eng was greatly appreciated.

REFERENCES

1. LINDNER, V., R. A. MAJACK & M. A. REIDY. 1990. J. Clin. Invest. **85:** 2004–2008.
2. VLODAVSKY, I., R. FRIDMAN, R. SULLIVAN, J. SASSE & M. KLAGSBRUN. 1987. J. Cell. Physiol. **131:** 402–408.
3. LINDNER, V., D. A. LAPPI, A. BAIRD, R. A. MAJACK & M. A. REIDY. 1991. Circ. Res. **68:** 106–113.
4. GOSPODAROWICZ, D., N. FERRARA, T. HAAPARANTA & G. NEUFELD. 1988. Eur. J. Cell Biol. **46:** 144–151.

5. VLODAVSKY, I., J. FOLKMAN & R. SULLIVAN. 1987. Proc. Nat. Acad. Sci. USA **84**: 2292–2296.
6. BOUCHE, G., N. GAS, H. PRATS, V. BALDIN, J.-P. TAUBER, J. TEISSIÉ & F. AMALRIC. 1990. Proc. Nat. Acad. Sci. USA **84**: 6770–6774.
7. RENKO, M., N. QUARTO, T. MORIMOTO & D. B. RIFKIN. 1990. J. Cell. Physiol. **144**: 108–114.
8. LINDNER, V. & M. A. REIDY. 1991. Proc. Nat. Acad. Sci. USA **88**: 3739–3743.
9. ABRAHAM, J. A., A. MERGIA, J. L. WHANG, A. TUMULO, J. FRIEDMAN, K. A. HJERRILD, D. GOSPODAROWICZ & J. C. FIDDES. 1986. Science **233**: 545–548.
10. GAJDUSEK, C. M. & S. CARBON. 1989. J. Cell. Physiol. **139**: 570–579.
11. MCNEIL, P. L., L. MUTHUKRISHNAN, E. WARDER & P. A. D'AMORE. 1989. J. Cell Biol. **109**: 811–822.
12. ROSS, R., T. N. WIGHT, E. STRANDNESS & B. THIELE. 1984. Am. J. Pathol. **114**: 79–93.
13. REIDY, M. A., A. W. CLOWES & S. M. SCHWARTZ. 1983. Lab. Invest. **49**: 569–575.
14. REIDY, M. A., D. STANDAERT & S. M. SCHWARTZ. 1982. Arteriosclerosis **2**: 216–220.
15. LINDNER, V., M. A. REIDY & J. FINGERLE. 1989. Lab. Invest. **61**: 556–563.
16. REIDY, M. A. 1985. Lab. Invest. **53**: 513–520.
17. FINGERLE, J., Y. P. T. AU, A. W. CLOWES & M. A. REIDY. 1990. Arteriosclerosis **10**: 1082–1087.
18. LAPPI, D. A., D. MARTINEAU & A. BAIRD. 1989. Biochem. Biophys. Res. Commun. **160**: 917–923.
19. WINKELS, J. A., R. FRIESEL, W. H. BURGESS, R. HOWK, T. MEHLMAN, R. WEINSTEIN & T. MACIAG. 1987. Proc. Nat. Acad. Sci. USA **84**: 7124–7128.
20. FINGERLE, J., R. JOHNSON, A. W. CLOWES, M. W. MAJESKY & M. A. REIDY. 1989. Proc. Nat. Acad. Sci. USA **86**: 8412–8416.
21. REIDY, M. A. & S. M. SCHWARTZ. 1981. Lab. Invest. **44**: 301–308.
22. CLOWES, A. W., M. M. CLOWES & M. A. REIDY. 1983. Lab. Invest. **49**: 327–333.
23. CLOWES, A. W. & S. M. SCHWARTZ. 1985. Circ. Res. **56**: 139–145.
24. FLAUMENHAFT, R., D. MOSCATELLI, O. SAKSELA & D. B. RIFKIN. 1990. J. Cell. Physiol. **140**: 75–81.
25. WALKER, L. N., D. F. BOWEN-POPE & M. A. REIDY. 1986. Proc. Nat. Acad. Sci. USA **83**: 7311–7315.

Basic Fibroblast Growth Factor in Neurons and Its Putative Functions[a]

K. UNSICKER, C. GROTHE, D. OTTO,
AND R. WESTERMANN

Department of Anatomy and Cell Biology
University of Marburg
D-3550 Marburg, Germany

INTRODUCTION

Fibroblast growth factors (FGFs), initially identified as mitogens with prominent angiogenic properties, are now recognized as multifunctional growth factors with notable actions on neural cells. Although their physiological roles in the nervous system are still far from being fully understood the distribution of FGF proteins, mRNAs, and receptors for FGFs in the central nervous system (CNS) along with the established trophic, differentiation-inducing, and proliferative effects of FGFs on neurons and glial cells *in vitro* suggest important functions of FGFs in the developing, mature, and lesioned nervous system *in vivo*. A crucial handicap for assigning *in vivo* functions to FGFs is the fact that most molecular forms of the FGF family lack conventional signal peptides and that modalities of their release from intact cells are not known. It is therefore important to bear in mind that assumptions about putative roles of the FGFs in the nervous system are extrapolations from data obtained by *in vitro* and *in vivo* administration of the factors, which has demonstrated the efficacy of the factors as pharmacological tools.

This article is intended to provide a brief survey of the distribution of FGFs and their presumed roles in the nervous system. More extensive reviews on this topic are available including very recent ones.[1-5]

LOCALIZATION OF MEMBERS OF THE FGF GENE FAMILY AND THEIR RECEPTORS IN THE NERVOUS SYSTEM

Acidic and basic FGFs including several molecular mass forms of basic FGF greater than 18 kDa occur in the embryonic and adult CNS.[6-8] By immunocytochemistry and *in situ* hybridization, acidic and basic FGF and their mRNAs have been localized to neurons.[9-14] There is no substantial evidence that glial cells synthesize and store FGFs *in situ,* although cultured astroglial cells can express basic FGF.[15] Localization of FGFs in distinct sets of neurons, e.g., in the brain stem,[16] and absence from others suggests neuron-type-specific functions.

Acidic FGF and its mRNA increase in several CNS regions during development suggesting putative roles as a differentiation and maintenance factor for neurons or other cells during later stages of development.[17,14] With regard to basic FGF, Kalcheim and Neufeld reported a peak of basic FGF/immunoreactive material in

[a]Research from our laboratory described in this article was supported by grants from Deutsche Forschungsgemeinschaft, BMFT, and Fonds der Chemischen Industrie.

the embryonic chick spinal cord at day 10,[18] suggesting a transient role. A complex picture concerning the temporal and spatial pattern of basic FGF/immunoreactivity emerged from a study by Grothe *et al.* demonstrating both transient expression during development and postnatal plateauing of the protein in certain populations of the rat brain stem.[16] These results indicate that basic FGF may have a broader spectrum of functions than acidic FGF both in the developing and in the mature brain.

Acidic and basic FGFs follow distinct regulatory patterns subsequent to a brain lesion. While acidic FGF appears in the wound fluid without any up regulation in the adjacent tissue,[19,20] basic FGF shows a marked increase in reactive astrocytes bordering the lesion.[21]

Specific receptors for FGF on neural cells comprise both low and high affinity systems, the latter with a K_d of approximately o.2 nM.[22] In the developing chick nervous system FGF receptor mRNA initially appears in the germinal neuroepithelium,[23] a localization that would be in accord with the demonstrated *in vitro* role of FGF to affect proliferation and differentiation of neuronal precursor cells.[24–26] FGF receptor mRNA seems to be absent from migrating cells, but returns in several neuron populations during the period of differentiation and in their mature state.[23,27] There is as yet no convincing demonstration of FGF receptor mRNA in glial cells *in situ*.

EFFECTS OF FGFs ON NEURONS AND GLIAL CELLS *IN VITRO*

Promotion of neuron survival, neurite growth, and differentiation (including transmitter synthesis) are key functions that FGFs exert on cultured neurons.[28–31] Neuron populations addressed include a large number from the CNS and one peripheral population, chick ciliary ganglionic neurons. Although the presence of receptors for FGF on neurons and its survival-supporting effect for neurons in single cell cultures[2] suggest that FGF can act directly on neurons, multiple effects on glial cells make it conceivable that promotion of neuronal survival in mixed neuron–glial cell cultures may be the result of an indirect action. On most neurons basic FGF is approximately 200-fold more potent than acidic FGF, whose potency can be dramatically enhanced by heparin.[31]

On astrocytes, oligodendrocytes, and Schwann cells FGF shows a broad repertoire of effects. These include stimulation of proliferation, expression of several proteins associated with differentiation, and alterations in membrane structure.[32–38]

EFFECTS OF FGFs ON NEURONS AND GLIAL CELLS *IN VIVO*

The pronounced trophic effect of FGF for cultured neurons prompted experiments aiming to establish FGFs as neurotrophic factors of physiological relevance. However, administration to chick embryos has failed to produce consistent results with regard to neuronal rescue in the ciliary ganglion during the period of ontogenetic neuron death.[39–41] In contrast, FGF consistently induces the acetylcholine-synthesizing enzyme choline acetyltransferase in the embryonic chick spinal cord *in vivo* (Grothe and Wewetzer, unpublished) indicating that it may act as a transmitter-promoting rather than a survival-promoting factor for neurons *in vivo*. Whether antibodies to FGFs administered during crucial phases of neuronal ontogeny may cause developmental deficits has not been assessed as yet. Several groups have applied FGFs in brain and nerve lesions and neurological disorders. Thus, FGF has been shown to maintain dorsal root ganglionic and cholinergic medial septal neurons after sciatic nerve and fimbria-fornix transections, respectively, in a manner similar to nerve growth factor.[42-45] Interestingly, neither in the sciatic nerve nor in the septohippocampal pathways have receptors for FGF and retrograde axonal transport been demonstrated,[46,47] in contrast to several other pathways in the CNS[47] (Grothe, unpublished). It is conceivable, therefore, that the beneficial effects of FGF in these systems are mediated via glial or vascular cells. Likewise, indirect modes of action cannot be ruled out in three other lesion paradigms, where FGF has been shown to promote neuronal survival, axonal regeneration, and transmitter synthesis. In the visual system, FGF prevents death of axotomized retinal ganglion cells after an optic nerve cut.[48] In a mouse model of Parkinson's disease FGF induces reappearance of nigrostriatal dopaminergic nerve fibers and up regulation of dopamine and tyrosine hydroxylase.[49] Finally, FGF maintains sympathetic preganglionic neurons in the intermediolateral column of the spinal cord after destruction of the target organ, the adrenal medulla.[50,51] Whether, in all these instances, the actions of exogenously administered FGF mimic an endogenous FGF or a member of the FGF family, or are totally unrelated to the physiological significance of FGF, remains to be elucidated.

PUTATIVE *IN VIVO* ROLES OF FGF IN THE NERVOUS SYSTEM

The presence of specific receptors for FGF on neural cells and lack of a leader sequence for most FGF family members create a substantial dilemma in all speculations on a physiological role of FGFs in the nervous system. Given most FGFs were not released from intact cells, their extracellular actions would require permeabilized or dying cells, both during development and in the adult. Cell death, a phenomenon intrinsic to most morphogenetic events and crucial to the development of the nervous system,[52,53] may provide a physiological source for FGFs. Liberated FGF might then act as a neurotrophic molecule or affect glial cells, a hypothesis that can be tested by local administration of anti-FGF antibodies to CNS areas with massive ontogenetic cell death and subsequent analysis of neuron populations projecting into those areas and of local glial cell proliferation and differentiation. In analogy, application of anti-FGF antibodies to focal brain lesions might provide an insight into the modes of action and target cells of FGF in the adult lesioned nervous system.

The conceptual restrictions of FGFs acting solely as "lesion factors" would be immediately abolished with the discovery of a cellular export mechanism. Research on proteins, as, e.g., the multidrug resistant protein, serving as carriers for other proteins that lack a leader sequence is in full swing[54] and will possibly reveal a mechanism that permits the release of FGFs from intact cells.

The knowledge concerning *in vivo* and *in vitro* effects of exogenously applied FGF accumulated so far suggests that endogenous FGFs may be multifunctionally affecting neurons, glial cells, and vasculature. Effects on neurons may be trophic or imply modulation of neurotransmission. With regard to the latter, it has recently been shown that basic FGF enhances long-term potentiation in hippocampal slices, an effect that it shares with EGF.[55] Thus, FGFs may belong into a category of factors, together with several neuromodulatory peptides, that combine mitogenic, trophic, and neuromodulatory properties to ensure plasticity both in the developing and adult nervous system.

REFERENCES

1. UNSICKER, K., D. BLOTTNER, D. DREYER, C. GROTHE, A. LAGRANGE, D. OTTO & R. WESTERMANN. 1990. Basic FGF and its implications for physiological and lesion-induced neuron death. *In* Regulation of Gene Expression in the Nervous System. J. de Vellis, R. Perez-Polo, A. M. Giuffrida Stella, Eds.: 31–47. Allan R. Liss. New York, N.Y.

2. UNSICKER, K., C. GROTHE, G. LÜDECKE, D. OTTO & R. WESTERMAN. Fibroblast growth factors: their roles in the central and peripheral nervous system. Neurotrophic Factors. S. E. Loughlin & J. H. Fallon, Eds. Academic Press. New York, N.Y. (In press).

3. WALICKE, P. 1989. Novel neurotrophic factors, receptors and oncogenes. Annu. Rev. Neurosci. **12:** 103–126.

4. WALICKE, P. 1990. Fibroblast growth factor (FGF): a multifunctional growth factor in the CNS. Adv. Neural Regeneration Res.: 103–114.

5. WESTERMANN, R., C. GROTHE & K. UNSICKER. 1990. Basic fibroblast growth factor (bFGF), a multifunctional growth factor for neuroectodermal cells. J. Cell Sci. Suppl. **13:** 97–117.

6. GOSPODAROWICZ, D., G. NEUFELD & L. SCHWEIGERER. 1986. Molecular and biological characterization of fibroblast growth factor, an angiogenic factor which also controls the proliferation and differentiation of mesoderm and neuroectoderm derived cells. Cell Differ. **19:** 1–17.

7. RISAU, W., P. GAUTSCHI-SOVA & P. BÖHLEN. 1988. Endothelial cell growth factors in embryonic and adult chick brain are related to human acidic fibroblast growth factor. EMBO J. **7:** 959–926.

8. PRESTA, M., M. RUSNATI, J. A. M. MAIER & G. RAGNOTTI. 1988. Purification of basic fibroblast growth factor from rat brain: identification of a Mr 22,000 immunoreactive form. Biochem. Biophys. Res. Commun. **155:** 1161–1172.

9. PETTMANN, B., G. LABOURDETTE, M. WEIBEL & M. SENSENBRENNER. 1986. The brain fibroblast growth factor (FGF) is localized in neurons. Neurosci. Lett. **68:** 175–180.

10. JANET, T., M. MIEHE, B. PETTMANN, G. LABOURDETTE & M. SENSENBRENNER. 1987. Ultrastructural localization of fibroblast growth factor in neurons of rat brain. Neurosci. Lett. **80:** 153–157.

11. GONZALES, A. M., M. BUSCAGLIA, M. ONG & A. BAIRD. 1990. Distribution of basic fibroblast growth factor in the 18-day rat fetus: localization in the basement membranes of diverse tissues. J. Cell Biol. **110:** 753–765.

12. GROTHE, C. & K. UNSICKER. 1990. Immunocytochemical mapping of basic fibroblast growth factor in the developing and adult rat adrenal gland. Histochemistry **94:** 141–147.

13. EMOTO, N., A. M. GONZALES, P. A. WALICKE, E. WADA, D. M. SIMMONS, S. SHIMASAKI & A. BAIRD. 1989. Basic fibroblast growth factor (FGF) in the central nervous system: identification of specific loci of basic FGF expression in the rat brain. Growth Factors **2:** 21–29.

14. SCHNÜRCH, H. & W. RISAU. Differentiating and mature neurons express the acidic fibroblast growth factor gene during chick neural development. EMBO J. (In press.)

15. FERRARA, N., F. OUSLEY & D. GOSPODAROWICZ. 1988. Bovine brain astrocytes express basic fibroblast growth factor, a neurotrophic and angiogenic mitogen. Brain Res. **462:** 223–323.

16. GROTHE, C., K. ZACHMANN & K. UNSICKER. 1991. Basic FGF-like immunoreactivity in the developing and adult rat brain stem. J. Comp. Neurol. **305:** 328–336.

17. ISHIKAWA, R., K. NISHIKORI & S. FURUKAWA. Developmental changes in distribution of acidic fibroblast growth factor in rat brain evaluated by a sensitive twosite enzyme immunoassay. J. Neurochem. (In press.)

18. KALCHEIM, C. & G. NEUFELD. 1990. Expression of basic fibroblast growth factor in the nervous system of early avian embryos. Development **109:** 203–215.

19. NIETO-SAMPEDRO, M., R. LIM, D. J. HICKLIN & C. W. COTMAN. 1988. Early release of glial maturation factor and acidic fibroblast growth factor after rat brain injury. Neurosci. Lett. **86:** 361–365.

20. ISHIKAWA, R., K. NISHIKORI & S. FURUKAWA. Appearance of nerve growth factor and

acidic fibroblast growth factor with different time courses in the cavity-lesioned cortex of the rat brain. Neurosci. Lett. (In press.)

21. FINKLESTEIN, S. P., P. J. APOSTOLIDES, C. G. CADAY, J. PROSSER, M. F. PHILIPS & M. KLAGSBRUN. 1988. Increased basic fibroblast growth factor (bFGF) immunoreactivity at the site of focal brain wounds. Brain Res. **460**: 253–259.

22. WALICKE, P. 1989. Novel neurotrophic factors, receptors, and oncogenes. Annu. Rev. Neurosci. **12**: 103–126.

23. HEUER, J. G., C. S. VON BARTHELD, Y. KINOSHITA, P. C. EVERS & M. BOTHWELL. 1990. Alternating phases of FGF receptor and NGF receptor expression in the developing chicken nervous system. Neuron **5**: 283–296.

24. BARTLETT, P. F., H. H. REID, K. A. BAILEY & O. BERNARD. 1988. Immortalization of mouse neural precursor cells by the c-myc oncogene. Proc. Nat. Acad. Sci. USA **85**: 3255–3259.

25. GENSBURGER, C., G. LABOURDETTE & M. SENSENBRENNER. 1987. Brain basic fibroblast growth factor stimulates the proliferation of rat neuronal precursor cells in vivo. FEBS Lett. **217**: 1–5.

26. CATTANEO, E. & R. McKAY. 1990. Proliferation and differentiation of neuronal stem cells regulated by nerve growth factor. Nature **347**: 762–765.

27. WANAKA, A., M. JOHNSON JR. & J. MILBRANDT. 1990. Localization of FGF receptor mRNA in the adult rat central nervous system by in situ hybridization. Neuron **5**: 267–281.

28. WALICKE, P., W. M. COWAN, N. UENO, A. BAIRD & R. GUILLEMIN. 1986. Fibroblast growth factor promotes survival of dissociated hippocampal neurons and enhances neurite extension. Proc. Nat. Acad. Sci. USA **83**: 3012–3016.

29. MORRISON, R. S., A. SHARMA, J. DE VELLIS & R. A. BRADSHAW. Basic fibroblast growth factor supports the survival of cerebral cortical neurons in primary culture. Proc. Nat. Acad. Sci. USA **83**: 7537–7541.

30. WALICKE, P. 1988. Basic and acidic fibroblast growth factors have trophic effects on neurons from multiple CNS regions. J. Neurosci. **8**: 2618–2627.

31. UNSICKER, K., H. REICHERT-PREIBSCH, R. SCHMIDT, B. PETTMANN, G. LABOURDETTE & M. SENSENBRENNER. 1987. Astroglial and fibroblast growth factors have neurotrophic functions for cultured peripheral and central nervous system neurons. Proc. Nat. Acad. Sci. USA **84**: 5459–5463.

32. PETTMANN, B., M. WEIBEL, M. SENSENBRENNER & G. LABOURDETTE. 1985. Purification of two astroglial growth factors from bovine brain. FEBS Lett. **189**: 102–108.

33. WEIBEL, M., M. PETTMANN, G. LABOURDETTE, M. MIEHE, E. BOCK & M. SENSENBRENNER. 1985. Morphological and biochemical maturation of rat astroglial cells grown in a chemically defined medium: influence of an astroglial growth factor. Int. J. Dev. Neurosci. **3**: 617–630.

34. ROGISTER, B., P. LEPRINCE, B. PETTMANN, G. LABOURDETTE, M. SENSENBRENNER & G. MOONEN. 1988. Brain basic fibroblast growth factor stimulates the release of plasminogen activators by newborn rat cultured astroglial cells. Neurosci. Lett. **91**: 321–326.

35. SANETO, R. P. & J. DE VELLIS. 1985. Characterization of cultured rat oligodendrocytes proliferating in a serum-free, chemically defined medium. Proc. Nat. Acad. Sci. USA **82**: 3509–3513.

36. ECCLESTON, P. A. & D. H. SILBERBERG. 1985. Fibroblast growth factor is a mitogen for oligodendrocytes in vitro. Dev. Brain Res. **21**: 315–318.

37. BESNARD, F., F. PERRAUD, M. SENSENBRENNER & G. LABOURDETTE. 1989. Effects of acidic and basic fibroblast growth factors on proliferation and maturation of cultured rat oligodendrocytes. Int. J. Dev. Neurosci. **7**: 401–409.

38. PRUSS, R. M., P. F. BARTLETT, J. GAVRILOVIC, R. P. LISAK & S. RATTRAY. 1981. Mitogens for glial cells: a comparison of the response of cultured astrocytes, oligodendrocytes and Schwann cells. Brain Res. **254**: 19–35.

39. DREYER, D., A. LAGRANGE, C. GROTHE & K. UNSICKER. 1989. Basic fibroblast growth factor prevents ontogenetic neuron death in vivo. Neurosci. Lett. **99**: 35–38.

40. HENDRY, I. A., D. A. BELFORD, M. F. CROUCH & C. E. HILL. 1990. Parasympathetic

neurotrophic factors. Int. J. Dev. Neurosci. Abstr. 8th Biennial Meeting, Int. Soc. Devl. Neuroscience, June 16–22, Bal Harbor, Florida.

41. OPPENHEIM, R. W., D. PREVETTE & F. H. FULLER. 1990. In vivo treatment with basic fibroblast growth factor during development does not alter naturally occuring neuronal death. Soc. Neurosci. Abstr. **16:** 1135.

42. OTTO, D., K. UNSICKER & C. GROTHE. 1987. Pharmacological effects of nerve growth factor and fibroblast factor applied to the transectioned sciatic nerve on neuron death in adult rat dorsal root ganglia. Neurosci. Lett. **83:** 156–160.

43. OTTO, D., D. FROTSCHER & K. UNSICKER. 1989. Basic fibroblast growth factor administered in gel foam rescue medial septal neurons after fimbria fornix transections. J. Neurosci. Res. **22:** 83–91.

44. ANDERSON, K. J., D. DAM, S. LEE & C. W. COTMAN. 1988. Basic fibroblast growth factor prevents death of lesioned cholinergic neurons in vivo. Nature **332:** 360–362.

45. WILLIAMS, L. R., S. VARON, G. M. PETERSON, K. WICTORIN, W. FISCHER, A. BJÖRKLUND & F. H. GAGE. 1986. Continuous infusion of nerve growth factor prevents basal forebrain neuronal death after fimbria fornix transection. Proc. Nat. Acad. Sci. USA **83:** 9231–9235.

46. FERGUSON, I. A., J. B. SCHWEITZER & E. M. JOHNSON, JR. 1990. Basic fibroblast growth factor: receptor-mediated internalization, metabolism, and anterograde axonal transport in retinal ganglion cells. J. Neurosci. **10:** 2176–2189.

47. FERGUSON, I. A., A. WANAKA & E. M. JOHNSON, JR. 1990. bFGF undergoes receptor-mediated retrograde transport in CNS neurons. Soc. Neurosci. Abstr.: 824.

48. SIEVERS, J., B. HAUSMANN, K. UNSICKER & M. BERRY. 1987. Fibroblast growth factors promote the survival of adult rat retinal ganglion cells after transection of the optic nerve. Neurosci. Lett. **76:** 157–162.

49. OTTO, D. & K. UNSICKER. 1990. Basic FGF reverses chemical and morphological deficits in the nigrostriatal system of MPTP-treated mice. J. Neurosci. **10:** 1912–1921.

50. BLOTTNER, D., R. WESTERMANN, C. GROTHE, P. BÖHLEN & K. UNSICKER. 1989. Basic fibroblast growth factor in the adrenal gland. Eur. J. Neurosci. **1:** 417–478.

51. BLOTTNER, D. & K. UNSICKER. 1990. Maintenance of intermediolateral spinal cord neurons by fibroblast growth factor administered to the medullectomized rat adrenal gland: dependence on intact adrenal innervation and cellular organization of implants. Eur. J. Neurosci. **2:** 378–382.

52. BARDE, Y. A. 1989. Trophic factors and neuronal survival. Neuron **2:** 1525–1534.

53. OPPENHEIM, R. W. 1989. The neurotrophic theory and naturally occuring motoneuron death. TINS **12:** 252–255.

54. GUESCH, A., E. HARTMANN, K. ROHDE, A. RUBARTELLI, R. SITIA & T. A. RAPOPORT. 1990. A novel pathway for secretory proteins? TIBS **15:** 86–88.

55. TERLAU, H. & W. SEIFERT. 1990. Fibroblast growth factor enhances long-term potentiation in the hippocampal slice. Eur. J. Neurosci. **2:** 973–977.

Manipulation of the Wound-Healing Process with Basic Fibroblast Growth Factor[a]

JEFFREY M. DAVIDSON[b,c] AND
KENNETH N. BROADLEY[b]

[b]Department of Pathology
Vanderbilt University School of Medicine
Nashville, Tennessee 37232-2561

[c]Department of Veterans Affairs Medical Center
1310 24th Avenue South
Nashville, Tennessee 37212-2602

INTRODUCTION

Traumatic injury to the skin or other organs initiates a cascade of events that, in normal individuals, leads to the repair of the injury site with restoration of many of the functional properties including blood flow, tissue strength, and a barrier to infection. Beginning with the formation of a fibrin clot and the discharge of platelet contents, a set of signals is sent forth from the wound site to recruit first leukocytes and then connective tissue cells to participate in the various stages of wound repair. Although the concept of growth factors, or cytokines, as mediators of this process has been espoused for many years, only in the last decade has there been sufficient biochemical characterization of these activities to permit their application, as recombinant gene products, to the process of wound repair.

Basic fibroblast growth factor (bFGF) was among the first cellular growth factors to be evaluated in the context of wound repair.[1,2] Although not initially known to be present at the wound site, the mitogenic and angiogenic properties of this cytokine[3,4] suggested that it could have properties consistent with the augmentation of formation of granulation tissue, the organ of connective tissue repair. Thus, early studies with material purified from either pituitary or cartilage sources were able to demonstrate *in vivo* activity of partially purified bFGF in stimulating new blood vessel formation and accumulation of fibroblasts at wound sites.

With the large-scale production advent of recombinant bFGF, it became possible to design large-scale *in vivo* experiments to examine the biological effects of fibroblast growth factor.

This article will review some of our recent findings made with respect to biological properties of bFGF. First, we show that, while single doses of bFGF have distinct and positive effects on the repair process, those effects are considerably modified when the cytokine is continuously present at the site of injury. Secondly, we cite results that examine the interactions between bFGF and TGF-β, another cytokine known to be present at the wound site.

[a]Supported by National Institutes of Health grant AG06528, the Department of Veterans Affairs, and Synergen, Inc.

EXPERIMENTAL APPROACH

Animal Models

The principal animal species used for bFGF studies has been the rat. Although our initial studies used normal, juvenile animals, we have since established that growth factors show greater efficacy in full-grown animals (4 months in the Fischer 344 or Sprague-Dawley strains). Three distinct types of injury models have been used in this species: (a) experimental granulation tissue, produced by the subcutaneous ventral implantation of 2 mm × 10 mm polyvinyl alcohol sponges; (b) the transcutaneous, longitudinal, dorsal, incisional wounds closed by wound clips; and (c) ligament repair, modeled by a transecting wound of the median collateral ligament of the hind limb.

Quantitative estimates of healing are based on different parameters, depending on the wound model. In granulation tissue, the normal parameters include the amount and relative proportion of collagen present. This is calculated from the concentration of hydroxyproline relative to other amino acids in a hydrolysate of the total tissue homogenate. DNA content is used as an indication of cellularity, while total protein content, measured by a colorimetric assay, is used as an independent indicator of noncollagen proteins.[1] Principal parameters used in the incisional model are the tensile properties of the wound. These include breaking energy, breaking strength, and tensile strength. Similar parameters can be measured in the ligament model. Because it is important to distinguish between overriding inflammatory processes and the specific stimulation of fibroplasia or angiogenesis, each experimental wound is examined, at least in part, by conventional histology.

In some experimental circumstances the conventional histologic observations have been extended by use of ultrastructural analysis, by evaluation of the expression of the variety of matrix genes, or by *in situ* hybridization.

Healing-Impaired Models

Although normal, adult rats can display the stimulatory activities of cytokines, healing-impaired models have several advantages. First normal wound healing is unlikely to be significantly improved by the addition of exogeneous stimuli; at best, one can only obtain modest stimulation of a variety of wound-repair parameters when using any one of a large spectrum of cytokines in experimental animals or man. Secondly, the key clinical problems of wound repair involve individuals whose healing capacity has been reduced by one or more complicating conditions which may include the diabetic state, administration of high levels of glucocorticoids, administration of chemotherapeutic substances, or simply the onset of advanced age. Two models have been used extensively in this laboratory: the diabetic rat, chemically induced by streptozotocin,[5,6] and the aging Fischer 344 rat (> 26 months). Both of these models show substantial retardation in both incisional strengthening and granulation tissue formation relative to 4-month-old control animals.

Bioavailability of cytokines is a crucial issue in the development of stimulated wound-healing models. Our studies have made extensive use of a simple, commercially available slow-release system (Innovative Research of America, Toledo). Small pellets (2.25 mm in diameter) can be implanted subcutaneously, within implanted sponges, or over the site of ligament transection. These pellets are calibrated to release continuously the desired amount of cytokine for a number of days or weeks, thus acting as a local delivery site.

EXPERIMENTAL FINDINGS

The panels in FIGURE 1 illustrate data obtained in normal rats upon administration of recombinant bFGF (Synergen) locally in the sponge-implantation model. In the granulation tissue experiments discussed, acute FGF administration was always made on day 3 after implantation to allow the endogenous inflammatory events to

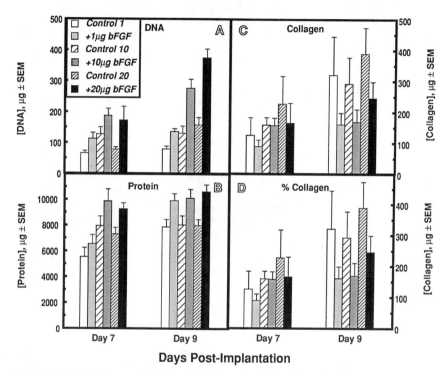

FIGURE 1. Effect of recombinant basic FGF on granulation tissue formation in the Sprague Dawley rat. 400 g male Sprague Dawley rats were each implanted with 4 polyvinyl alcohol sponge discs on day 0. On day 3, two sponges in each animal ($n = 5$) received the indicated treatment. Control sponges received an injection of the vehicle, 1 mg/ml bovine serum albumin in phosphate-buffered saline. Animals were sacrificed on days 7 and 9 after implantation, sponges were removed, and the indicated parameters were determined in tissue homogenates. All data represent means plus or minus the standard error of the mean for a total of 10 sponges in 5 animals. **Panel A:** Total DNA, mg per sponge. **Panel B:** Total protein, mg per sponge. **Panel C:** Total collagen, mg per sponge. **Panel D:** Relative collagen content, percent.

subside. Panel A shows the classic effect of bFGF on the accumulation of cells in experimental granulation tissue. Significant stimulation of DNA accumulation was present with as little as 1 μg of the cytokine, but greater doses of 10 or 20 μg showed an additional effect at day 7. By day 9, a dose-dependent relationship between the amount of bFGF administered to the wound site and the amount of DNA emerged. Panel B shows that bFGF was also able to stimulate protein accumulation in experimental sponges, although there was not a clear dose relationship.

In contrast to the mitogenic activity of bFGF, the treatment of sponges with this cytokine causes a characteristic *decrease* in the amount of collagen present. This is true both on an absolute (FIGURE 1, panel C) and relative (FIGURE 1, panel D) basis. In this experiment, the effect on collagen accumulation did not appear to have a dose-dependent relationship, and it was much more evident on day 9 after implantation, at which time collagen accumulation has become a significant component of the repair process. These findings echo data reported in earlier publications[7] that have shown the mitogenic, chemotactic, and collagenolytic activity of cells isolated from wounds and subsequently treated with bFGF.

Continuous exposure of connective tissue or wounds to bFGF produces striking results. FIGURE 2 (right panel) illustrates the gross morphology present after an experiment in which a slow-release pellet releasing 1.0 µg per day of bFGF was placed in a subcutaneous sponge. The continuous release of the cytokine induced an enormous angiogenic response which was far in excess of that seen with any single-dose treatment, both in terms of magnitude and of area affected. FIGURE 3, an experiment done in collaboration with Barry Oakes at Monash University, illustrates a similar type of hyperplasia which was obtained when slow release bFGF pellets were implanted over the site of ligament transection. Interestingly, although there was an enormous amount of connective tissue formation produced by this treatment, the tensiometry of bFGF-stimulated ligament repair indicated that this massive granulation tissue had relatively low tensile strength, at least up to six weeks after injury (data not shown).

The histological effects of continuous bFGF treatment were also striking (data not shown). There was an enormous proliferation of blood vessels in fibroblasts surrounding the implantation site, and consistent with the single-injection experiments, there was somewhat less obvious fibroplasia. The biochemical effects of continuous release contrasted sharply with those of the acute dose. FIGURE 4 illustrates the biochemical changes that occurred when experimental granulation tissue was exposed to pellets releasing either 1 or 2.5 µg per day of heparin plus bFGF. In these experiments, biochemical parameters were determined at days 4, 7, and 10 after implantation. Although there was a negative effect on collagen accumulation at 7 days, by 10 days after implantation there was a clear increase in total collagen content of wounds exposed to either 1 or 2.5 µg per day of the cytokine. The chemical analysis of these wounds was confined to that granulation tissue contained within the sponge. As could be seen from FIGURE 2, there was an obvious stimulus of new granulation tissue formation beyond the limits of the sponge which was not included in this biochemical analysis and which undoubtedly would have added to the total mass connective tissue estimated. Unlike the acute dose, continuous exposure to bFGF appeared to increase collagen content over control levels. However, continuous release bFGF still brought about somewhat diminished or unchanged relative collagen content in this experimental system. Thus, continuous release bFGF had a rather different effect on fibroplasia than did the single injection. The lower panels of FIGURE 4 illustrate the massive increases in protein content and cellularity that were brought about by continuous exposure of wounds to bFGF. These experimental findings suggest that wound tissues react in a qualitatively and quantitatively different fashion to cytokines depending on their mode of presentation.

DISCUSSION

The data in this paper illustrate several aspects of bFGF biology that are important to understanding its role in the wound-healing process. That bFGF is

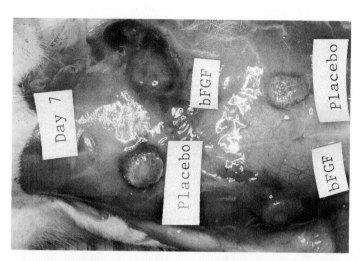

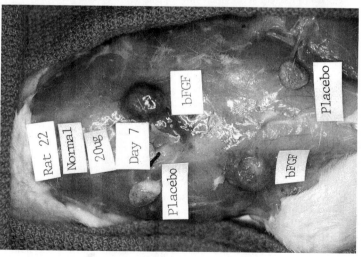

FIGURE 2. Effect of slow-release bFGF on experimental granulation tissue. This photograph illustrates views of sponges in the implantation site showing the distinct and striking effect of acute dosing (left panel) and continuous-release bFGF (right panel) on granulation tissue formation. While the single dose of 20 μg had a significant impact on granulation tissue formation, the cytokine effect was amplified by the exposure of wound tissue to this stimulus over a 7-day period.

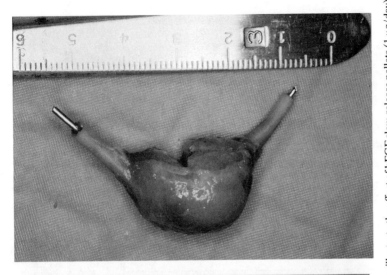

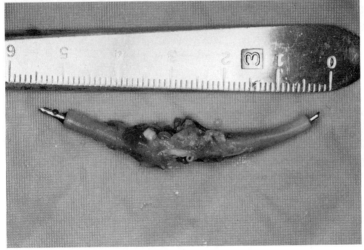

FIGURE 3. Effect of slow-release bFGF on soft tissue repair. These figures illustrate the effect of bFGF slow-release pellets (1 µg/day) when implanted over the transected median collateral ligament of the rat that links the femur and the tibia. At six weeks, there remained an enormous mass of granulation tissue, which on histological examination was loosely organized. **Left panel:** Effect of placebo pellet. **Right panel:** Slow-release bFGF effect. Tensiometric data indicated that, despite the large amount of tissue present, tensile strength was not significantly different from the placebo-treated ligaments (data not shown).

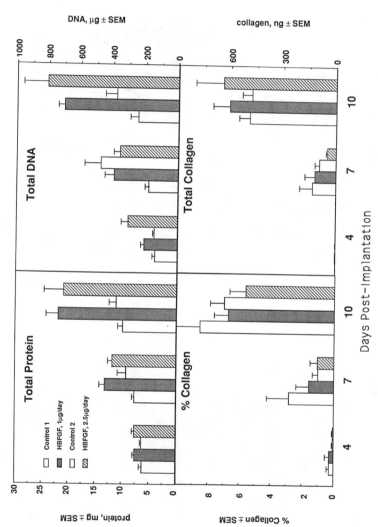

FIGURE 4. Biochemical effects of slow-release recombinant FGF on granulation tissue formation. The general outline of this experiment is similar to that indicated in FIGURE 1, with the exception that, instead of injecting a solution of bFGF into the sponge tissue on day 3, slow-release pellets were calibrated to release either 1 or 2.5 μg per day of human recombinant bFGF complexed with heparin (provided by Synergen, Inc.). Animals (400 g Sprague Dawley rats, *n* = 5) were sacrificed on the indicated days after sponge implantation. The most significant difference between the slow release and acute injection was observed in the parameter of total collagen, where treated sponges showed increased collagen content at the end of 10 days. Differences between the three sets of control values may represent a systemic effect.

intrinsically involved in wound repair is evidenced by at least one set of studies in which we have shown that neutralizing antibody, if present continuously at the wound site, can reduce the rate of wound healing.[8] From a wide variety of *in vitro* and *in vivo* studies, it can be concluded that bFGF is a highly desirable stimulus in cases where increased cell numbers and increased vascularity are desired. With conventional (single-dose) delivery schemes, the net result of bFGF treatment at the mid-stage of wound repair is to reduce the amount of matrix (collagen) present relative to cells and blood vessels.

As shown by a number of other investigators, bFGF is capable of strongly inducing neutral proteases such as plasminogen activator and collagenase in both epithelial cells and fibroblasts.[9] We have shown that this growth factor acts selectively on cells at different stages in the healing process, and the collagenase-stimulatory activity of bFGF seems to be reduced in cells derived from a mature wound.[7] It is highly likely that other matrix-degrading enzymes such as stromelysin and the two forms of gelatinase are stimulated by treatment with the cytokine. Indeed, our preliminary data using *in situ* hybridization on FGF-treated wounds has indicated that stromelysin activity was somewhat increased in FGF-treated wounds.[10]

Cytokine Interactions

In wounds, the stimulation by exogenous bFGF acts in the context of an unknown number of endogenous cytokines. The markedly different, dramatic actions of TGF-β on both wound tissue and cultured cells suggested that bFGF and TGF-β can interact in both *in vivo* and *in vitro* model wound systems. A clear example of the interaction of these cytokines came from a recent study in which it was shown that the prior and continuous presence of bFGF could either abrogate or sharply reduce the contraction of collagen gels by skin fibroblasts and granulation tissue fibroblasts.[11] Under these experimental conditions, bFGF was able to sharply reduce the stimulated contraction that was evoked by TGF-β. This result suggests that bFGF could be an important adjoint in situations where wound contracture (e.g., burns, scars) produces serious complications. The interaction of bFGF and TGF-β is further underscored by a more recent series of experiments in which we have compared the effects of varied proportions of the two cytokines on the production of collagen and elastin by a variety of connective tissue cells.[12] These studies revealed a biphasic effect of bFGF on TGF-β action. At low doses, bFGF augmented the TGF-β-stimulated production of both collagen and elastin by fibroblasts and smooth muscle cells, while at high doses (10–100 ng/ml), TGF-β-stimulated production of collagen and elastin was significantly reduced to or below control levels. The inhibition was also evident at the mRNA level by *in situ* hybridization. Thus, synergism and antagonism between these two cytokines can be evoked by adjusting their relative proportions.

bFGF action is also modified by the presence of TGF-β *in vivo*. By placing slow-release pellets releasing each of these cytokines a few millimeters apart, there is a dramatic stimulation of local granulation tissue formation in the zone between the two growth factor sources that is indicative of cytokine synergism. Furthermore, studies with sponge implants in both normal and especially diabetic and aging animals have revealed that the combination of the two cytokines could fully restore both the mass of granulation tissue and its collagen content, as well as restoring the tensile properties of incisional wounds to control values. Other combinations with bFGF may also be effective, although not as synergistic.[13,14]

The wound-healing system has evolved in vertebrate animals and man as a highly redundant process in which multiple factors are called into play to bring about regrowth of damaged tissue, a functional vascular bed, and a barrier to infection. Since, negative mutations of FGF are not known to occur, it is difficult to demonstrate directly that this or any single cytokine is absolutely essential to the wound-repair process. The same cytokines may be intimately involved in early, critical developmental events.[15] It is very likely that connective tissue cells and other cells that participate in the repair process are primed to react with a wide variety of signals, and it may be the relative proportion or sequence of these signals that determines the fate and progression of the wound. In addition, bFGF is only one member of a very large family of heparin-binding cytokines present in tissues and cells.[3] Some of these other FGFs may be more critical to the developmental process, while others may be most evident during abnormal growth. It is not understood which of the forms of FGF is most crucial to the process of wound repair.

One approach to the understanding of the role of FGF in the process may come about through the introduction of local negative mutations. It has recently been shown in our laboratory that experimental wounds are very active in the uptake of foreign DNA. Thus, it is not unreasonable to begin experiments in which antisense molecules or plasmids expressing the antisense sequences are used to ask whether direct reduction in the amount of bFGF mRNA activity at the wound has any effect on repair. Conversely, the transient introduction of extra copies of the bFGF gene under regulation of vigorous promoters may be a productive means of augmenting the repair process when drug delivery by topical or other means is problematic.

As the sophistication of molecular biology generates new forms of FGF, we have no doubt that the process of wound healing will continue to be a vital paradigm for understanding growth factor action. At the chemical and cellular level, wound repair is an enormously complicated process, representing the attempt by the adult organism to regenerate lost function. Through exploration of the reaction of cells and tissues to bFGF, we can begin to appreciate the role of the factor in the process and the means by which we may use this potent molecule to correct a variety of defects in wound healing.

REFERENCES

1. DAVIDSON, J. M., M. KLAGSBRUN, K. E. HILL, A. BUCKLEY, R. SULLIVAN, P. S. BREWER & S. C. WOODWARD. 1985. Accelerated wound repair, cell proliferation, and collagen accumulation are produced by a cartilage-derived growth factor. J. Cell Biol. 100: 1219–1227.
2. BUNTROCK, P. M., K. D. JENTZSCH & G. HEDER. 1982. Stimulation of wound healing using brain extract with fibroblast growth factor (FGF) activity. I. Quantitative and biochemical studies into formation of granulation tissue. Exp. Pathol. 21: 46–53.
3. BAIRD, A. & P. A. WALICKE. 1989. Fibroblast growth factors. Br. Med. Bull. 45(2): 438–452.
4. GOSPODAROWICZ, D. 1990. Fibroblast growth factor. Chemical structure and biologic function. Clin. Orthop. (257): 231–248.
5. BROADLEY, K. N., A. M. AQUINO, B. HICKS, J. A. DITESHEIM, G. S. McGEE, A. DEMETRIOU, S. C. WOODWARD & J. M. DAVIDSON. 1989–90. The diabetic rat as an impaired wound healing model: stimulatory effects of transforming growth factor-beta and basic fibroblast growth factor. Biotechnol. Ther. 1(1): 55–68.
6. TSUBOI, R. & D. B. RIFKIN. 1990. Recombinant basic fibroblast growth factor stimulates wound healing in healing-impaired db/db mice. J. Exp. Med. 172(1): 245–251.
7. BUCKLEY-STURROCK, A., S. C. WOODWARD, R. M. SENIOR, G. L. GRIFFIN, M. KLAGSBRUN & J. M. DAVIDSON. 1989. Differential stimulation of collagenase and chemotactic

activity in fibroblasts derived from rat wound repair tissue and human skin by growth factors. J. Cell. Physiol. **138:** 70–78.

8. BROADLEY, K. N., A. M. AQUINO, S. C. WOODWARD, A. STURROCK, Y. SATO, D. B. RIFKIN & J. M. DAVIDSON. 1989. Monospecific antibodies indicate that basic fibroblast growth factor is intrinsically involved in wound repair. Lab. Invest. **61:** 571–575.

9. RIFKIN, D. B. & D. MOSCATELLI. 1989. Recent developments in the cell biology of basic fibroblast growth factor. J. Cell Biol. **109:** 1–6.

10. QUAGLINO, D., JR., K. N. BROADLEY & J. M. DAVIDSON. 1989. Role of basic fibroblast growth factor on extracellular gene matrix expression during wound healing in normal and diabetic rats. ATTI X Riunone Naz. Soc. It. Studio Connettivo: 129.

11. FINESMITH, T. H., K. N. BROADLEY & J. M. DAVIDSON. 1990. Fibroblasts from wounds of different stages of repair vary in their ability to contract a collagen gel in response to growth factors. J. Cell. Physiol. **144**(1): 99–107.

12. DAVIDSON, J. M. & O. ZOIA. 1990. Basic fibroblast growth factor and transforming growth factor-α antagonize the stimulation of elastin production by transforming growth factor-β1 in fibroblasts and smooth muscle cells. Matrix **10:** 254–255.

13. GREENHALGH, D. G., K. H. SPRUGEL, M. J. MURRAY & R. ROSS. 1990. PDGF and FGF stimulate wound healing in the genetically diabetic mouse. Am. J. Pathol. **136**(6): 1235–1246.

14. LYNCH, S. E., R. B. COLVIN & H. N. ANTONIADES. 1989. Growth factors in wound healing. Single and synergistic effects on partial thickness porcine skin wounds. J. Clin. Invest. **84**(2): 640–646.

15. HEATH, J. K. & A. G. SMITH. 1989. Growth factors in embryogenesis. Br. Med. Bull. **45**(2): 319–336.

Preclinical Wound-Healing Studies with Recombinant Human Basic Fibroblast Growth Factor

JOHN C. FIDDES,[a,d] PATRICIA A. HEBDA,[b]
PETER HAYWARD,[c] MARTIN C. ROBSON,[c]
JUDITH A. ABRAHAM,[a] AND CORINE K. KLINGBEIL[a]

[a]California Biotechnology, Inc.
2450 Bayshore Parkway
Mountain View, California 94043

[b]Department of Dermatology
University of Pittsburgh
School of Medicine
Pittsburgh, Pennsylvania 15261

[c]Division of Plastic Surgery and
Wound Healing Laboratory
University of Texas Medical Branch and
Shriners Burns Institute
Galveston, Texas 77550

Basic fibroblast growth factor (bFGF)[1-3] is a polypeptide that exhibits a wide range of *in vitro* biological activities such as the stimulation of cell mitogenesis and chemotaxis. The mitogenic effects of bFGF are directed primarily towards cells of mesodermal or neuroectodermal origin such as fibroblasts, endothelial cells, smooth muscle cells, chondrocytes, osteoblasts, preadipocytes, and melanocytes. Additionally, it has been found recently that bFGF is also a mitogen for certain ectodermal cell types such as skin keratinocytes.[4,5] As would be predicted from the observation that bFGF is a potent mitogen and chemoattractant for capillary endothelial cells, bFGF has been shown *in vivo* to be an effective stimulator of neovascularization in systems such as the chicken chorioallantoic membrane and the rabbit eye angiogenesis assays.

While much is known about the *in vitro* and angiogenic activities of bFGF, little is known about its natural *in vivo* role. It has been proposed, however, based on the wide range of cell types that are known to be bFGF responsive, that exogenously administered bFGF should accelerate the rate at which wounds heal. For example, the bFGF-stimulated process of angiogenesis is considered key to the early stages in the healing of injured tissue, and the proliferation of bFGF-responsive cells such as fibroblasts and keratinocytes is an essential part of healing.

In normal, healthy individuals, the majority of wounds heal adequately without the need for outside intervention. In many cases, however, underlying clinical problems result in poorly or nonhealing wounds. Examples include pressure sores (decubitus ulcers), venous ulcers, and diabetic ulcers, as well as traumatic wounds in individuals with compromised healing abilities such as diabetics or patients receiving chemotherapy.

[d]Present affiliation: ImmuLogic Pharmaceutical Corporation, 855 California Avenue, Palo Alto, California 94304.

Prior to initiating the clinical evaluation of growth factors such as bFGF, it was necessary to demonstrate efficacy in animal models of healing. Given that the targeted clinical indications include impaired healing situations, the efficacy of bFGF was examined in healthy and healing-impaired models. The purpose of this article is to review our experience with the topical administration to a variety of animal wounds of recombinant human bFGF made in *Escherichia coli.*[6]

SUBCUTANEOUS SPONGE IMPLANTS

The subcutaneous implantation of polyvinyl alcohol (PVA) sponges has been used to generate an artificial wound space. The healing process can be followed over time by a combination of histological and biochemical analyses of the new tissue that grows into the interstices of the sponge material. Using the sponge implant model with rats, Davidson *et al.* showed that local injection of a single dose of bovine-cartilage-derived growth factor, now known to be identical to bFGF, caused a substantial increase in the rate of deposition of new granulation tissue.[7]

Klingbeil and coworkers took a similar approach as a first step towards studying the *in vivo* effects of recombinant human bFGF. The inert PVA sponge material was implanted subcutaneously in rats, and injected once with bFGF or control saline 4 days later. By this time much of the acute inflammatory response associated with the surgical procedure had subsided. A subset of the implanted sponges was then harvested every 2 to 3 days for 17 days. The tissue within sections of each harvested sponge was examined histologically for evidence of an inflammatory response, granulation tissue development, and angiogenesis.

At the time of injection 4 days after implantation, histological analysis of the sections showed them to contain primarily fibrin and some polymorphonuclear cells as well as lymphocytes and plasma cells. By 6 and 8 days after injection, the bFGF-treated sponges showed greater numbers of fibroblasts and new capillaries than did contralateral control sponges taken from the same animal. Representative sponges taken 8 days posttreatment and stained with hematoxylin and eosin are shown in FIGURE 1. In the case of the bFGF-treated implant (FIGURE 1A), the entire diameter of the sponge was filled with granulation tissue while the control sponge taken from the same animal almost completely lacked granulation tissue. By days 11 to 14, both the bFGF-injected and saline-injected sponges were filled with fibroblasts and collagen, but the bFGF sponges were more densely filled and contained more organized tissue. Another appreciable difference was the significantly greater number of new capillaries in the bFGF sponges than in the controls taken from the same animals.

Representative sponges taken 8 and 11 days posttreatment and stained with Mason's trichrome to identify collagen are shown in FIGURE 2. Eight days after injection with bFGF (FIGURE 2A), there was a dramatic increase in the appearance of cellular material and collagen compared to the control sponges (FIGURE 2B). An additional 3 days later, there was a substantial increase in the amount of collagen and tissue organization. By day 17, the bFGF-treated and control sponges were filled with granulation tissue to a similar extent (data not shown).

The results of these experiments indicate that recombinant human bFGF acts to increase neovascularization, the number of fibroblasts, and the deposition of collagen in the wound space. To quantify this effect, a granulation score was assigned based on the extent of granulation tissue development and angiogenesis as judged on

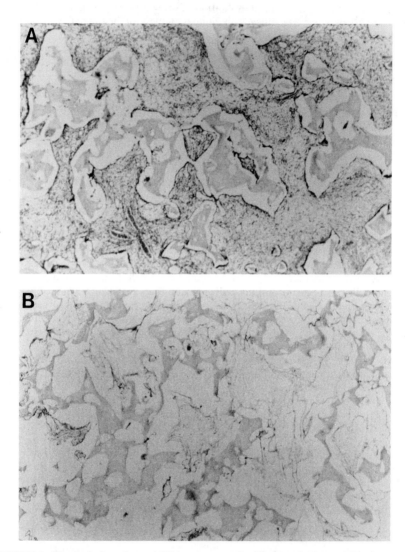

FIGURE 1. Histological sections of PVA sponges excised from male Sprague Dawley rats 8 days after a single injection of either (A) 10 μg recombinant bFGF or (B) phosphate-buffered saline (PBS). The animals were anesthetized, a 2-cm incision was made on the ventral side, and 4 sterile sponges (1-cm diameter, 2-mm thick) were implanted beneath the subcutaneous fascia against the muscle in 4 separate pockets. On day 4 after sponge implantation, 2 of the 4 sponges per animal were injected with recombinant bFGF and 2 with PBS, each in a volume of 25 μl. After removal, sponges were sectioned and stained with hematoxylin and eosin and magnified 40×. Both the bFGF-treated (A) and control (B) sponges were taken from the same animal. (Klingbeil *et al.,* unpublished data.)

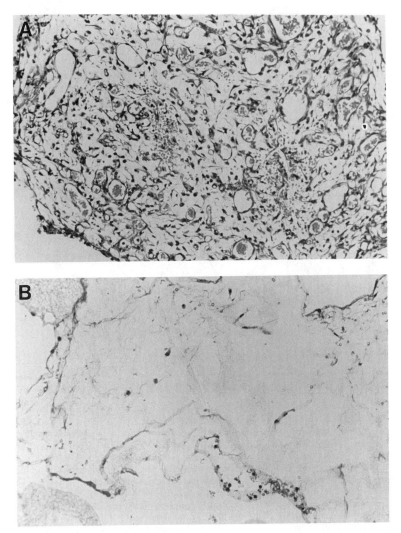

FIGURE 2. Histology of granulation tissue deposited in PVA sponges treated as described in FIGURE 1 either with 10-μg bFGF or PBS. **A:** Sponge excised 8 days after bFGF treatment. **B:** Sponge excised 8 days after control treatment. The histological sections were stained with Mason's trichrome to visualize collagen, and magnified 200×. Both of the 8-day sponges were taken from the same animal. (Klingbeil *et al.*, unpublished data.)

a blind basis by two independent investigators. The granulation score for sponges injected with doses of bFGF ranging from 0.66 to 10.0 μg/sponge showed significant increases between days 8 and 14 when compared to sponges in the same animals injected with buffer control (FIGURE 3). A dose-dependent effect was not discernable.

PARTIAL-THICKNESS PIGSKIN WOUNDS

Partial-thickness excisional wounds in porcine skin have been used previously as a model for human epidermal healing since pigskin is considered to be similar to human skin with respect to the relative thickness of the epidermis and dermis, the sparsity of hair, and the presence of an underlying adipose rather than muscle layer.[8] Repair of a partial-thickness wound is dependent on the migration and proliferation of keratinocytes both from the edges of the wound and from remaining epidermal appendages.

In this model, the partial-thickness wounds were made on the back of the pig using an electrodermatome. Each rectangular wound was approximately 1 cm^2 in

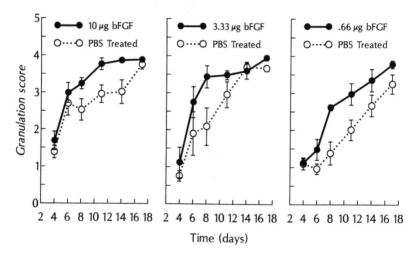

FIGURE 3. Stimulation of granulation tissue deposition in the sponge model. Four days after implantation, sponges were injected with bFGF (0.66, 3.33, or 10.0 µg per sponge) or PBS buffer ($n = 60$), and then a subset of each treatment group was removed on that day and 4, 6, 8, 11, 14, and 17 days later. Following staining with either hematoxylin and eosin or Mason's trichrome for light microscopy, sections of the sponges were evaluated blind on a score of 0–4 by two investigators to indicate the degree of proliferation of granulation tissue. A score of 4 indicates that 100% of the sectioned field of the sponge, which includes the entire diameter cross section, was filled with well-organized granulation tissue (fibroblasts, collagen, and developing vascularization). Significance ($P \leq 0.05$) as determined by a paired t-test was seen on days 8, 11, and 14 (10-µg dose); days 6, 8, and 11 (3.33-µg dose); and days 6, 8, 11, and 15 (0.66-µg dose). (Klingbeil *et al.*, unpublished data.)

surface area and 0.3 mm deep, and up to 175 such wounds were made per animal. The wounds were divided into treatment groups that received either topical bFGF in saline, saline, or no treatment. Healing of the epidermal defect was evaluated daily on a macroscopic level following excision of a subset of the wounds to a sufficient breadth and depth to include the entire wound bed of the original partial-thickness injury. After excision, the tissue samples were treated to separate the epidermis from the dermis, and the epidermal specimens were evaluated for the presence or absence of a defect. Only wounds with no visible epidermal defect were scored as healed; wounds with even a slight remaining defect were still scored as unhealed.

In one experiment, Hebda *et al.* compared a single 10-μg dose of bFGF given on day 0 to multiple 10-μg doses given on days 0, 1, and 2 (FIGURE 4).[9] A total of five animals were used in this experiment, and the treatment groups (30–35 wounds/group per animal) were randomized over the surfaces of the animals (FIGURE 4A). The healing curves (FIGURE 4B and C) indicate that the single and multiple applications of bFGF accelerated epidermal healing when compared to both the saline and untreated controls. The single application of 10 μg bFGF on day 0 resulted in an approximately 21% acceleration in the rate of healing (FIGURE 4B); the calculated time in which 50% of the wounds were completely healed (HT_{50}) was 4.5 days for the single-dose bFGF-treated wounds and 6.1 days for the comparable saline-treated wounds. Histologic analysis of samples taken from the healing wounds confirmed that bFGF treatment enhanced the rate of epithelialization, and also showed a bFGF-mediated stimulation of dermal healing as evidenced by accelerated neovascularization and collagen deposition. In another experiment, single applications on day 0 of 0.1, 1.0, and 10 μg bFGF were compared. Both the 1.0- and 10-μg bFGF doses gave an acceleration in healing of about 25%, but the 0.1-μg dose was found to be ineffective (data not shown).

FULL-THICKNESS WOUNDS IN DIABETIC MICE

The animals used in both the sponge-implant and partial-thickness wound-healing models are healthy and do not have any healing impairments. To assess the ability of bFGF to treat poorly healing wounds, several animal models of healing impairment have been developed. One of these makes use of the genetically diabetic, db/db mouse which develops many symptoms of diabetes such as hyperglycemia, glycosurea, polyurea, and polydipsea. In addition, the db/db mouse is hyperphagic and becomes obese. Healing rates in db/db mice are impaired by about 35–50% as compared to their phenotypically normal, heterozygous controls, db/+m mice.

In the db/db wound healing model, a single full-thickness excisional wound, 1.5 cm in diameter, was made on the flank of each mouse. Excised tissue included the epidermis, dermis, and the panniculus carnosus muscle. The bFGF or control solutions (vehicle) were administered topically in a 20-μl volume and the wounds were covered with a transparent occlusive dressing. The change in wound area was measured by planimetry, and the rate of wound closure for each individual wound was evaluated as the percent decrease in total wound area that resulted from a combination of contraction, granulation, and epithelialization.

The healing rates of bFGF-treated and control wounds in the db/db mice and in their heterozygous controls were compared by Klingbeil *et al.*[10] (FIGURE 5). In the case of the vehicle-treated, db/+m control mice, healing was rapid and 80% closure was achieved in 12 days. A slight acceleration, only statistically significant at day 6, was seen when the wounds were treated with a single dose of bFGF applied at 1 μg/cm². In contrast, vehicle-treated wounds in the db/db mice healed much more slowly. The wounds initially expanded in area and did not return to their initial size until 6 days postinjury. This may have been the result of physical forces generated by the obesity of the animal. Reduction in wound area then proceeded at a rate such that approximately 75% closure was not reached until day 30. This impairment was substantially overcome by a single 1 μg/cm² treatment with bFGF on day 0. A statistically significant difference in percentage wound closure was seen from days 6–30. The time required for a 50% reduction in wound area improved from 18 to 8 days.

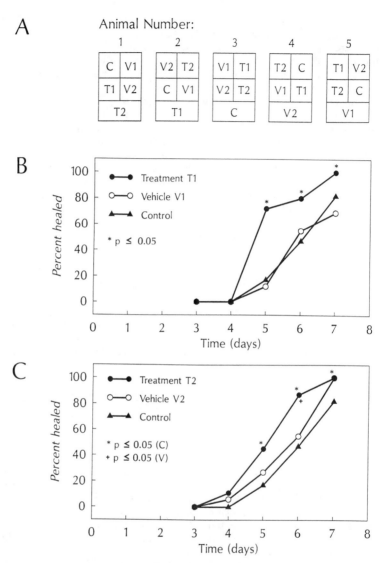

FIGURE 4. Stimulation of the rate of epidermal wound healing by recombinant bFGF. **A:** Experimental design. Partial-thickness injuries on each of 5 female Yorkshire pigs were divided into 5 treatment groups each containing about 30 individual wounds. The treatment groups (T1 = 10-μg bFGF on day 0; V1 = vehicle control on day 0; T2 = 10-μg bFGF on days 0, 1, and 2; V2 = vehicle control on days 0, 1, and 2; C = untreated control) were distributed among the 5 animals as shown. **B:** Treatment with a single dose of bFGF. The percentage of fully healed epidermal specimens ($n = 20$–25) is shown on days 3, 4, 5, 6, and 7 postwounding. The T1 group showed a greater percentage of healed wounds on days 5, 6, and 7 compared to either V1 or C ($p \leq 0.05$). **C:** Treatment with multiple doses of bFGF. The T2 group showed a greater percentage of healed wounds on days 5, 6, and 7 compared to C, and on day 6 compared with V2 ($p \leq 0.05$). (From Hebda *et al.*[9] with permission from the *Journal of Investigative Dermatology.*)

The macroscopic appearance on days 0 and 12 of db/db wounds that were either treated with bFGF or treated with the vehicle is shown in FIGURE 6. The acceleration in healing caused by the bFGF treatment is clearly seen by the reduction in the size of the wound area by day 12.

The diabetic model was also used to investigate the dose-response relationship of single bFGF treatments (FIGURE 7). Maximal healing acceleration was seen with both the 1.0- and 10-μg/cm^2 treatments, and no effect was seen with the 0.01-μg/cm^2 treatment. The apparently intermediate effect seen with the 0.1-μg/cm^2 treatment was not statistically significant.

BACTERIALLY INFECTED, CHRONIC GRANULATING WOUNDS

Clinically, the bacterial or fungal infection of wounds is a frequent cause of retarded healing. In many cases, infected wounds are treated by the topical administration of antimicrobial agents, but evidence is now accumulating that these agents

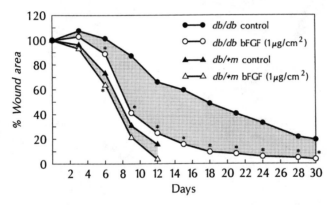

FIGURE 5. Effect of bFGF on the closure of full-thickness excisional wounds in diabetic (db/db) and heterozygous control (db/+m) mice. A single, 1-μg/cm^2 dose of bFGF was used. Total wound areas were measured every 3 days from days 0–28 and then on day 30 in the db/db, and every 3 days from days 0–12 in the db/+m. The mean starting wound areas were 2.3 ± 0.2 cm^2 (db/db) and 2.0 ± 0.3 cm^2 (db/+m). The number of animals in each treatment group ranged from 7–10. (From Klingbeil *et al.*[10] with permission from Alan R. Liss, New York.)

are cytotoxic and may, therefore, be inhibitory to the healing process.[11] To investigate whether a growth factor such as bFGF could be an alternative therapeutic for infected wounds, Hayward *et al.* used a chronic granulating wound model generated by infecting full-thickness rat burns with *E. coli.*[12]

In this model, full-thickness burns (approximately 10 × 3 cm) were made on the back of rats by immersion in 100°C water, and were then either inoculated with 10^8 *E. coli* or left uninfected. Five days after burning (day 5), the eschar from the wounds was excised and the wounds were either treated with varying concentrations of topically applied bFGF five times over a two-week period, or were left untreated. The wounds were monitored by quantitative bacteriology to confirm bacterial levels, and the rate of healing was assessed by planimetry to measure the surface area of the contracting wounds.

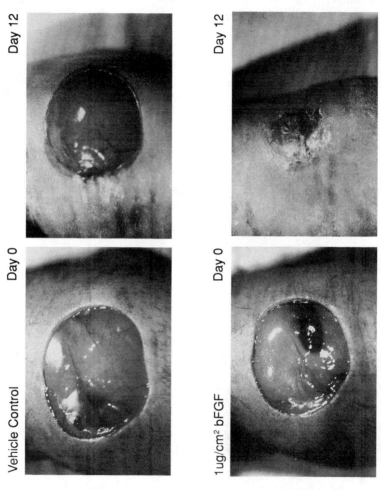

FIGURE 6. Full-thickness wounds in db/db mice on days 0 and 12. Wounds were treated on day 0 with either the vehicle control or with bFGF at 1 $\mu g/cm^2$.

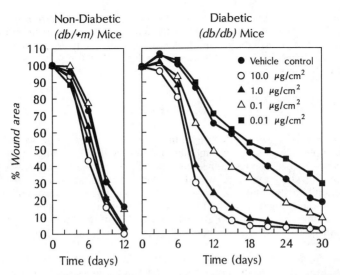

FIGURE 7. Effect of varying doses of bFGF on the closure of full-thickness excisional wounds in diabetic (db/db) and heterozygous control (db/+m) mice. Single bFGF doses of 10, 1.0, 0.1, and 0.01 μg/cm² were used. The number of animals in each treatment group ranged from 7–10. (From Klingbeil *et al.*[10] with permission from Alan R. Liss, New York.)

In the experiment shown in FIGURE 8, the wounds were either infected but not treated with bFGF, or infected and treated with bFGF at concentrations of 1, 10, or 100 μg/cm². A significant healing retardation was seen with the bacterial infection, and this was not overcome by either the 1- or 10-μg/cm² bFGF doses. The 100-μg/cm² dose did, however, result in a substantial acceleration in the rate of

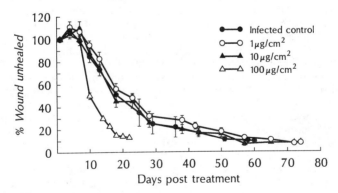

FIGURE 8. Acceleration in the rate of closure of chronic granulating burn wounds by bFGF. Wounds generated in male Sprague Dawley rats were divided into 5 groups ($n = 20$) that were either noninfected; infected but not treated; or infected and treated with bFGF at 1, 10, or 100 μg/cm² on days 5, 9, 12, 15, and 18. The percentage of wound remaining open is plotted against time in days after the start of treatment. The acceleration in the rate of closure with the 100-μg/cm² treatment is statistically significant ($p \leq 0.01$).

healing of the infected wound such that the healing rate of the group treated with this bFGF dose was indistinguishable from that of the noninfected controls. In another experiment, the administration of a single 100-μg/cm^2 dose of bFGF did not give the accelerated healing seen with the five doses of 100 μg/cm^2 given over two weeks (data not shown).

DISCUSSION

We have used several different animal models to assess the wound-healing potential of topically applied, recombinant human bFGF. The results obtained from these models—subcutaneous sponge implants in rats, partial-thickness skin wounds in pigs, full-thickness excisional wounds in diabetic mice, and bacterially infected, chronic granulating wounds in rats—show that exogenously administered bFGF can accelerate the healing process, and that bFGF exhibits this activity towards several different wound types. Basic FGF has been shown to stimulate angiogenesis, granulation, contraction, and epithelialization, all key steps in the healing of injured tissue.

In the sponge model, a single injection of between 0.66 to 10.0 μg bFGF accelerated the processes of angiogenesis and granulation tissue deposition over a period of 6–14 days postinjection. The results presented here are similar to those reported previously using bovine bFGF purified from cartilage (cartilage-derived growth factor or CDGF),[7] and are in contrast to the results obtained in the same model with epidermal growth factor (EGF), which required either repeated administration or slow release to obtain a stimulatory effect on granulation tissue accumulation.[13,14]

Basic FGF is found ubiquitously in tissue but is not believed to be actively synthesized under normal conditions[15]; rather the bFGF has been found to be stored in tissue bound to the extracellular matrix component, heparan sulfate.[16] It is assumed that the stored bFGF is released in an active form by the action of proteases and endoglycosidases. Thus, while exogenous bFGF can stimulate wound healing in the sponge model, it is also possible that endogenous bFGF plays a role in this process. Recently, Broadley et al. showed that when an anti-bFGF antibody was implanted into PVA sponges to which no exogenous bFGF had been added, there was a substantial reduction in the rate at which granulation tissue developed,[17] thus supporting the belief that bFGF has a natural role in this healing process.

The PVA-sponge-implant model assesses dermal healing only. Results from the partial-thickness pigskin wounds show, however, that bFGF can also accelerate epidermal healing, a result consistent with the recent observation that bFGF is a mitogen for human skin keratinocytes.[4,5] In this model, a single application of bFGF was found to be as efficacious as repeat applications, and activity was seen in the 1.0–10.0 μg/cm^2 range. Again, the results with bFGF are to be contrasted with the observation that EGF can also promote healing in this model, but that repeat doses of EGF are required to accelerate healing.[18,19]

The stimulatory effect of bFGF has also been demonstrated in models of healing impairment such as full-thickness excisional wounds in the diabetic mouse.[10] In this case, a single treatment with topical bFGF accelerated wound closure to a rate that was comparable to that seen in the normal, heterozygous control mice. We have obtained similar results with full-thickness wounds in two other situations of impaired healing.[10] These are the genetically obese (ob/ob) mouse, a strain that grows to twice the size of its heterozygous controls and that exhibits increased lipogenesis, decreased lipolysis, and transient hyperglycemia; and prednisolone-treated hairless (hr/hr) mice in which the glucocorticoid treatment slows down healing possibly through inhibiting the inflammatory response. The db/db mouse has also been used by Greenhalgh *et al.*[20] and Tsuboi and Rifkin[21] to demonstrate the healing effect of bFGF on full-thickness wounds. In addition, Phillips *et al.*[22] and Broadley *et al.*[23,24] have administered streptozotocin to rats to generate a diabetic healing impairment in both incisional and sponge wounds that can be overcome with bFGF.

Treatment with bFGF has also been shown to overcome the healing impairment caused by chronic bacterial infection.[12] In this case, however, multiple treatments (five doses over a two-week period) of bFGF at 100 μg/cm^2 were needed for efficacy. Neither a single 100-μg/cm^2 dose nor multiple lower doses resulted in a healing acceleration. Quantitative bacteriology carried out on the healing wounds showed that the bFGF effect was not antibacterial, and that the healing acceleration took place in the presence of an elevated bacterial count ($> 10^5$ organism/g tissue). In

| | bFGF (μg/cm^2) | | | | |
	0.01	0.1	1.0	10	100
Mouse full-thickness:					
hr/hr			+		
hr/hr with steroids		+	+	+	
ob/+		+	+	+	
ob/ob	−	+	+	+	
db/m+			+		
db/db	−	−	+	+	
Rat full-thickness:					
Acute contamination			+	+	+
Chronic contamination			−	−	+
Pig partial-thickness:					
Normal			−	+	+

FIGURE 9. Dose-response ranges for topical bFGF in several different wound-healing models. A plus indicates efficacy at a given concentration, a minus indicates no effect, and a blank indicates that the bFGF concentration has not been tested in that model. All results were with a single administration of bFGF with the exception of the chronically contaminated rat wound which required 5 doses for efficacy.

another study, Stenberg *et al.* showed that topical bFGF could also overcome the inhibition of wound contraction in rats caused by an acute bacterial infection.[25]

A summary of our dose-response data obtained with bFGF in a variety of healing situations is given in FIGURE 9. In all cases, the bFGF concentration is expressed as amount of bFGF applied per wound surface area (μg/cm^2). A striking feature of the combined dose-response data is that bFGF appears to act in an all-or-none fashion in wound-healing models. For the most part, statistically significant, intermediate levels of efficacy are not observed. We interpret this to mean that the topical administration of bFGF initiates a cascade of events that probably involve cell recruitment and proliferation, followed by the release of additional growth factors and inhibitors by the cells that have newly populated the healing wound. Also, there is a reasonable consistency in that efficacy is normally found with a single bFGF dose in the 1–10 μg/cm^2 range. The one exception is the chronically infected, granulating rat wound which requires multiple bFGF doses of 100 μg/cm^2 for efficacy. These dose-response data have formed the basis of dosing regimens currently being evaluated in the clinic for the treatment of pressure sores, venous ulcers, diabetic ulcers, and donor-site skin grafts.

REFERENCES

1. GOSPODAROWICZ, D., N. FERRARA, L. SCHWEIGERER & G. NEUFELD. 1987. Endocr. Rev. **8:** 95–114.
2. RIFKIN, D. B. & D. MOSCATELLI. 1989. J. Cell Biol. **109:** 1–6.
3. BAIRD, A. & P. A. WALICKE. 1989. Br. Med. Bull. **45:** 438–452.
4. O'KEEFE, E. J., M. L. CHIU & R. E. PAYNE. 1988. J. Invest. Dermatol. **90:** 767–769.
5. SHIPLEY, G. D., W. W. KEEBLE, J. E. HENDRICKSON, R. J. COFFEY & M. R. PITTELKOW. 1989. J. Cell. Physiol. **138:** 511–518.
6. THOMPSON, S. A., A. A. PROTTER, L. BITTING, J. C. FIDDES & J. A. ABRAHAM. 1990. Methods Enzymol. **198:** 96–116.
7. DAVIDSON, J. M., M. KLAGSBRUN, K. E. HILL, A. BUCKLEY, R. SULLIVAN, P. S. BREWER & S. C. WOODWARD. 1985. J. Cell Biol. **100:** 1219–1227.
8. WINTER, G. D. 1971. *In* Epidermal Wound Healing. H. Maibach & D. Rovee, Eds. Year Book Medical Publishers, Inc. Chicago, Ill.
9. HEBDA, P. A., C. K. KLINGBEIL, J. A. ABRAHAM & J. C. FIDDES. 1990. J. Invest. Dermatol. **95:** 626–631.
10. KLINGBEIL, C. K., L. B. CESAR & J. C. FIDDES. 1991. *In* Clinical and Experimental Approaches to Dermal and Epidermal Repair: Normal and Chronic Wounds. A. Barbul, M. Caldwell, W. Eaglstein, T. Hunt, D. Marshall, E. Pines & G. Skover, Eds: 443–458. Alan R. Liss. New York, N.Y.
11. MCCAULEY, R. L., H. A. LINARES, V. PELLIGRINI, D. N. HERNDON, M. C. ROBSON & J. P. HEGGERS. 1989. J. Surg. Res. **46:** 267–274.
12. HAYWARD, P., J. HOKANSON, J. HEGGERS, J. FIDDES, C. KLINGBEIL, M. GOEGER & M. ROBSON. 1990. Paper presented at the 69th Annual Meeting of American Association of Plastic Surgeons, Hot Springs, Va., May 8.
13. BUCKLEY, A., J. M. DAVIDSON, C. D. KAMERATH, T. B. WOLT & S. C. WOODWARD. 1985. Proc. Nat. Acad. Sci. USA **82:** 7340–7344.
14. BUCKLEY, A., J. M. DAVIDSON, C. D. KAMERATH & S. C. WOODWARD. 1987. J. Surg. Res. **43:** 322–328.
15. ABRAHAM, J. A., J. L. WHANG, A. TUMOLO, A. MERGIA, J. FRIEDMAN, D. GOSPODAROWICZ & J. C. FIDDES. 1986. EMBO J. **5:** 2523–2528.
16. VLODAVSKY, I., J. FOLKMAN, R. SULLIVAN, R. FRIDMAN, R. ISHAI-MICHAELI, J. SASSE & M. KLAGSBRUN. 1987. Proc. Nat. Acad. Sci. USA **84:** 2292–2296.
17. BROADLEY, K. N., A. M. AQUINO, S. C. WOODWARD, A. BUCKLEY-STURROCK, Y. SATO, D. B. RIFKIN & J. M. DAVIDSON. 1989. Lab Invest. **61:** 571–575.
18. LAATO, M., J. NIINIKOSKI, B. GERDIN & L. LEBEL. 1986. Ann. Surg. **203:** 379–381.
19. BROWN, G. L., L. CURTSINGER, J. R. BRIGHTWELL, D. M. ACKERMAN, G. R. TOBIN, H. C. POLK, C. GEORGE-NASCIMENTO, P. VALENZUELA & G. S. SCHULTZ. 1986. J. Exp. Med. **163:** 1319–1324.
20. GREENHALGH, D. G., K. H. SPRUGEL, M. J. MURRAY & R. ROSS. 1990. Am. J. Pathol. **136:** 1235–1246.
21. TSUBOI, R. & D. B. RIFKIN. 1990. J. Exp. Med. **172:** 245–251.
22. PHILLIPS, L. G., P. GELDNER, J. BROU, S. DOBBINS, J. HOKANSON & M. C. ROBSON. 1990. Surg. Forum **41:** 602–603.
23. BROADLEY, K. N., A. M. AQUINO, B. HICKS, J. A. DITESHEIM, G. S. MCGEE, A. A. DEMETRIOU, S. C. WOODWARD & J. M. DAVIDSON. 1988. Int. J. Tissue React. **10**(6): 345–353.
24. BROADLEY, K. N., A. M. AQUINO, B. HICKS, J. A. DITESHEIM, G. S. MCGEE, A. A. DEMETRIOU, S. C. WOODWARD & J. M. DAVIDSON. 1989. Biotechnol. Ther. **1:** 55–68.
25. STENBERG, B. D., L. G. PHILLIPS, J. A. HOKANSON, J. P. HEGGERS & M. C. ROBSON. 1991. J. Surg. Res. **50:** 47–50.

Preclinical and Clinical Studies with Recombinant Human Basic Fibroblast Growth Factor

G. MAZUÉ, F. BERTOLERO, C. JACOB, P. SARMIENTOS,
AND R. RONCUCCI

Farmitalia Carlo Erba
Erbamont Group
Via Carlo Imbonati 24
20159 Milan, Italy

INTRODUCTION

In the last decade, many recombinant proteins were shown to have unique and very significant biological properties. Their number is continuously increasing, and recombinant proteins are effectively developed into pharmaceutical products as documented by the amount of IND (investigational new drug) (approximately 100 up to 1990) delivered by the U.S. Food and Drug Administration.[1] This is true also in Europe where, e.g., France granted 26 AMM (Autorisations De Mise sur le Marché).[2]

It means that the goal to successfully approach new therapeutical targets with recombinant biotechnological products has already been met and will be further expanded in the future.

But to undertake the development of a new protein, as will be described hereafter, requires the solution of unique problems of production, formulation, and finally of clinical development that result in a real and exciting adventure. This is particularly true with basic fibroblast growth factor (bFGF), a polypeptide belonging to a family of proteins also known as heparin-binding growth factors.[3,4] bFGF has a pleiotropism of biological activities, the most striking of which are associated with the growth of blood vessels *in vivo* and which can be related to wound healing and tissue regeneration.[5]

Despite the complexity of the problems encountered, a recombinant DNA process was developed to produce large quantities of recombinant human bFGF (rh-bFGF) which enabled us to undertake an adequate preclinical development and to start clinical studies in the wound healing of ocular and cutaneous lesions. Our preclinical and clinical studies with rh-bFGF (Lab. code FCE 26184) are here described.

PRODUCTION AND CHEMICAL CHARACTERISTICS

For the relatively low molecular weight of bFGF and for the lack of glycosilation in the natural molecule, we decided to produce bFGF in recombinant *Escherichia coli* cells.

A synthetic gene encoding the 155 amino acids of human bFGF was designed on the basis of the known amino acid sequence and constructed by assembling 12 synthetic oligonucleotides. The resulting DNA sequence was used to construct different expression plasmids for the production of rh-bFGF in *E. coli.* Parameters to be considered in order to optimize expression of recombinant proteins in bacteria include promoters, operators, ribosome binding sites (rbs), transcription terminators, plasmid copy number, and physiological growth conditions depending on the host strain. The majority of foreign genes expressed at high levels in *E. coli* are usually transcribed from promoters p_L, P_{lac}, P_{trp}, and their derivatives.

In initial attempts to express bFGF under the control of the promoters P_{trp} or p_L, we did not observe levels of recombinant product high enough to allow the development of a large-scale process. Thus, in a second series of experiments, we modified the 5' end of the bFGF gene taking into account the use of statistically "preferred" codons among highly expressed *E. coli* genes and the need to avoid the presence of excessively stable secondary structures in the encoded mRNA.

The "optimized" sequence was subcloned into an expression plasmid under the transcriptional control of the *E. coli* tryptophan promoter, while translation initiation was assured by the lambda cII ribosome binding site sequence. The resulting plasmid, pFC81, is schematically illustrated in FIGURE 1.

To comply with international guidelines for the production of recombinant proteins to be used as pharmaceuticals, this plasmid carries the tetracycline-resistance gene for selection instead of the more commonly used ampicilline-resistance gene.

One of the most important features of our production process is the use of an *E. coli* strain of the type B as host for pFC81. Indeed, previous experiments in our laboratory have indicated that the host type may drastically affect expression levels of several mammalian proteins.

Purification was performed using both ion exchange and heparin Sepharose chromatography followed by a gel filtration step for buffer exchange and freeze-drying.

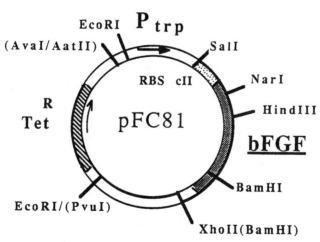

FIGURE 1. The plasmid used for expression of bFGF in *E. coli* is pFC81. It carries the bFGF gene under transcriptional and translational control respectively of the promoter P_{trp} and of the Shine-Dalgarno sequence CII. It also carries the gene coding for the resistance to the antibiotic tetracycline.

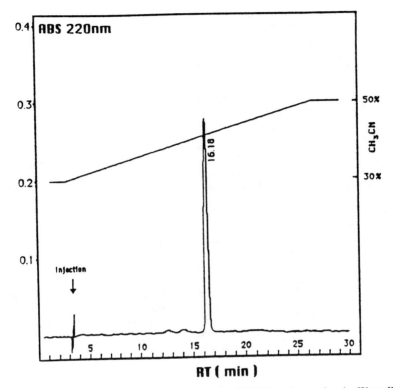

FIGURE 2. Reverse-phase HPLC of bFGF: a sample of bFGF was injected under Water Wisp autosampler control into a reverse phase HPLC column (Vydac CTR 214 C4; 0.46 × 25 cm). Proteins were eluted with a linear gradient of 30–50% acetonitrile in 0.1% trifluoroacetic acid (TFA) at a flow rate of 1 ml/minute and detected at 220 nm.

The purification procedure was facilitated by the fact that after cell breakage, most of the rh-bFGF was found in the soluble fraction. A medium scale purification yielded about 100 mg of pure rh-bFGF from 400 ml of cell broth. The final product was assayed by sodium dodecyl sulfate-polyacrilamide gel electrophoresis (SDS-PAGE) in reducing and nonreducing conditions, amino acid analysis, and reverse-phase high-performance liquid chromatography (HPLC), and found to be "apparently" homogeneous. A typical HPLC profile of the purified rh-bFGF is represented in FIGURE 2.

NEW TECHNOLOGIES

Two structural characteristics, worth considering, are associated with our product. First, N-terminal sequence analysis revealed the presence of double sequences. Indeed, the product consists of a constant equal mixture of the 154- and the 153-amino-acid forms starting respectively with NH$_2$-Ala-Ala-Gly-Ser-etc. and NH$_2$-Ala-Gly-Ser-etc.

The initial methionine was always removed, and about 50% of the isolated bFGF lost also the subsequent alanine residue. The resulting rh-bFGF was isolated in a completely reduced form, with the four cysteine residues all in the free thiol state. It was found to be fully active and biologically undistinguishable from other nonrecombinant as well as partially oxidized forms.

The above structural features did not hamper the pharmaceutical development of such rh-bFGF since formulations were defined and allowed us to generate pharmaceutical preparations stable enough for clinical evaluation and product development. On the contrary, in the absence of stabilizing agents, we did observe a quite high degree of instability of the fully reduced bFGF form.

The NH_2-end microheterogeneity was also acceptable due to its high batch-to-batch reproducibility. In addition, it is already known from the literature that the first 10–15 amino acids of bFGF are not essential for biological activity. We have now pursued further technological improvements in order to produce forms of recombinant bFGF that are highly homogeneous and stable even in pure preparations.

In fact we have developed an efficient enzymatic process able to generate a homogeneous 146-amino-acid form of bFGF. This process exploits a controlled proteolysis of the mixture of 154- and 153-amino-acid forms (UK Patent Application N. X; submitted for publication). Through such technology we are today producing large quantities of this bFGF for possible pharmaceutical development. In addition, the resulting homogeneous form was further processed with iodoacetic acid to isolate a partially carboxymethylated form which we have found to be more active and stable as compared to more "traditional bFGFs" (Caccia et al., The FGF Family Meeting, La Jolla, California, January 16–18, 1991). Besides obvious improvements in the pharmacy of the molecule, the "carboxymethylated bFGF" due to its prolonged stability at 37°C may open new therapeutic approaches such as those requiring a slow release in systemic applications.

BIOLOGICAL ACTIVITY

Biological and Pharmacological Properties

Basic fibroblast growth factor is a potent mitogen for cells of mesodermal and neurodermal origin. Specifically, bFGF can stimulate the proliferation of fibroblasts and of vascular derived cells in vitro.[6] It induces migration of endothelial cells as well as the release of matrix-degrading proteases.[7] Exposure to bFGF in vitro resulted also in the rearrangement of endothelial cells into tubular structures resembling blood capillaries.[8] Together these activities are directly associated with the growth of new blood vessels in vivo and to wound healing and tissue regeneration.

In the process of preclinical development, bioassays were selected to confirm relevant in vitro and in vivo biological activities attributed to bFGF.

In vitro the mitogenic response was studied in a variety of target cells mainly of vascular origin. The release of a matrix-degrading enzyme, i.e., tissue plasminogen activator, was investigated in bovine aortic endothelial cells.

An FGF receptor associated tyrosine kinase activated by the growth factor was reported in different cell systems to result in a unique 90–92 kDa phosphotyrosine-bearing protein.[9] The tyrosine kinase activity was studied in BALB/3T3 cells in response to rh-bFGF to confirm the specificity of the stimulation also at the molecular level.

In vivo the effect of rh-bFGF was evaluated on corneal lesions in the rabbit following several wounding procedures. This species was chosen since it is a time-tested animal model for ophthalmology.[10] Further *in vivo* studies were designed to evaluate the activity of rh-bFGF on the healing rate of the tympanic membrane perforations in rats.

In Vitro *Biological Activity*

rh-bFGF was shown to stimulate growth of human and bovine endothelial cells and of several cells of mesodermal origin. Depending on the different cell type tested, the drug presented in the different assays a range of concentrations that yield half-maximal growth (EC_{50}) spanning from 70 pg/ml in 3T3 cells to 100 ng/ml in human vascular smooth muscle cells (hVSM) (FIGURE 3). Capillary endothelial cells (ACE) were more sensitive to bFGF (EC_{50} = 0.3–1.0 ng/ml) as compared to both human and bovine large-vessel-derived endothelium.

The range of EC_{50} for the mitogenic activity of bFGF reported in the literature[11-13] is very broad and thus difficult to compare. It should be noted that considerable differences in sensitivity may be due to differences in the assay protocols, e.g., serum-free or serum-supplemented media during the assays, cell density, cell culture passage, etc. Two other major *in vitro* cellular activities, the induction of plasminogen activator synthesis in endothelial cells and the stimulation of a tyrosine kinase activity resulting in the specific phosphorylation of a 90–92 kDa protein in 3T3 cells, induced by the recombinant drug were studied during the preclinical development (data not shown).

In vivo *Biological Activity on the Cornea*

The wounding model of the rabbit cornea by iodine vapors described by Moses *et al.*[14] was applied to study the efficacy of rh-bFGF on the epithelial regeneration in the eye and showed a significantly accelerated epithelial healing rate in the drug-treated (200 ng/50 µl) as compared to control-treated eyes (TABLE 1).[15] Briefly, a further study on corneal healing was performed to determine the efficacy and the safety of rh-bFGF on a long-term corneal injury obtained by an initial anterior keratectomy followed by repeated epithelial swabbings (every 3 days) for 24 days, in order to prolong the injured status.[16]

Lesioned eyes were treated twice daily with 50 ng of rh-bFGF for a total of 27 days at which time histologic examinations of the corneals excluded the presence of neovascularization, inflammation, or abnormal giant cells. Qualitatively, parts of the substantia propria underlying the anterior limiting membrane of the cornea appeared to be more organized and showed a higher density of cells and collagen in rh-bFGF-treated eyes as compared to control eyes.

In vivo *Biological Activity on the Tympanic Membrane*

Preliminary data are available on the activity of FGF in the healing of tympanic perforations in rats.[17] Calibrated tympanic perforations (0.9 mm) were performed on both tympanic membranes of 30 rats. At day 0 each animal received 40 µl of placebo in the right ear and in the left ear 40 µl of a solution of rh-bFGF at the doses of 200, 400, and 2000 ng.

The quality of healing was assessed by otomicroscopy and by photonic histology. The mean time to closure of 5 and 6 days respectively in the two higher-dose groups

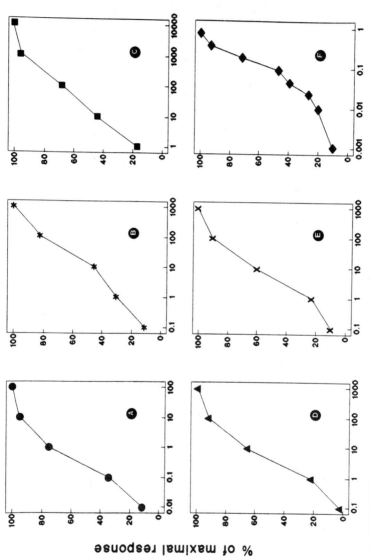

FIGURE 3. Effects of FCE 26184 on various human, bovine, and rodent mesoderm-derived cell types *in vitro*. **A:** Bovine adrenal capillary endothelial. **B:** Bovine aortic endothelial. **C:** Human vascular smooth muscle. **D:** Human umbilical vein endothelial. **E:** Baby hamster kidney (BHK cells); **F:** BALB/C mouse derived 3T3 cells. The concentration of FCE 26184 was plotted against the percent of maximal stimulation. Values were the mean of 4 determinations and standard deviations (SD) were always less than 20%.

was significantly reduced as compared to the 8 days required for the placebo-treated control ears to heal. No difference vs. placebo was found at the low dose. Myringitis was noticed mainly at the dose of 2000 ng. Results were promising but need to be complemented, before beginning in human beings, with a long-term follow-up of at least 3 months after tympanic regeneration, verifying the integrity of the membrane as well as the status of the middle ear.

DRUG ABSORPTION

Until now we have not succeeded in measuring any amount of circulating intact rh-bFGF after administration of therapeutic doses to experimental animals. The overall metabolic fate of the growth factor is thus unknown possibly due to a unique behavior of the molecule but also due to limits in the sensitivity of the quantitative analytical methods presently available to detect rh-bFGF in biological samples, which is a commonly encountered problem in the evaluation of biologic products.[18]

TABLE 1. Healing of Corneal Burns

Treatment[a] (50 μl × 2/day)	Healing Rate[b] (mm²/hour)
Vehicle only	1.12 ± 0.28 (n = 9)
FCE 26184 20 ng	1.33 ± 0.28 (n = 9)
FCE 26184 200 ng	1.57[c] ± 0.21 (n = 9)
EGF 100 ng	1.27 ± 0.31 (n = 11)

[a]Each animal was submitted to 2 daily topical applications until epithelial healing was completed (3–4 days). One eye was treated with the active drug substance, while the contralateral eye was used as control and treated with vehicle only.

[b]Photographs of fluorescein-stained wounds were scored by computerized image analysis. The wound surface was plotted against time, and the linear regression of the healing rate was calculated between 13 and 50 hours after injury. Data are the mean ± standard deviation.

[c]$p < 0.005$ (unpaired Student-Fisher t-test vs. control).

Nevertheless, in the process of preclinical development of rh-bFGF for ophthalmological applications in ocular wound healing, studies were designed to verify the possible drug absorption when given topically in three different situations that may arise in the course of clinical trials such as anterior keratectomy, corneal epithelial lesions (burns induced by iodine vapors), and penetrating autokeratoplasty that stimulates corneal graft in humans.

For this purpose, the distribution of ^{125}I-labeled rh-bFGF after topical application to the eye of rabbit was investigated in the three ocular wounding models reported above. These studies showed that ^{125}I-rh-bFGF did not penetrate into the inner structures of the lesioned eye, that it was not absorbed by the vessels of the bulbar conjunctiva or by the ocular annexes of the lacrymal apparatus. On the contrary, labeled rh-bFGF bound and was retained by the denuded corneal anterior limiting membrane after epithelial burns as demonstrated by autoradiography (FIGURE 4). In fact the labeled drug accumulated in the anterior limiting membrane up to the completed healing phase which occurred between 3 and 4 days. These data were in agreement with what had already been observed by others under similar experimental conditions.[19]

Finally it should be mentioned that radioactivity was found in blood and urine, but after applying a specific extraction procedure for bFGF on a heparin Sepharose column, it was impossible to recover any amount of intact bFGF, indicating that the radioactivity was due to breakdown products.

ANIMAL SAFETY

The safety evaluation of rh-bFGF was assessed for ophthalmological and skin topical application,[20] following classical procedures accepted for topical preparation and taking into account nonclinical regulatory considerations.[21]

As already mentioned when applied on the eye, intact or wounded, the rh-bFGF was not recovered systemically. Similarly preliminary results on bFGF applied on the

FIGURE 4. Autotoradiography of the cornea showing the fixation of [125] I-labeled rh-bFGF to the anterior limiting membrane 96 hours after iodine vapor lesions.

skin seem to confirm these observations. A systemic acute toxicity study was performed in 3 species: mouse, rat, and cynomolgus monkey at the doses of 5, 1, and 0.1 mg/kg by intravenous (iv) route. These dose levels were considered as the maximal expected ones that could be administered. They did not induce any mortality or noticeable clinical signs, including at the site of injection. No postmortem lesions were observed. The pharmaceutical formulation was used for all local tolerance studies. Rabbits (New Zealand white) were used to assess acute local eye irritation. Following the topical ocular administration of formulated FCE 26184, 0.1 ml of a solution containing 200 µg/ml, the ophthalmological examination revealed no changes in the conjunctiva, the cornea, and the iris at any time. No abnormal clinical signs were noted during the study.

Acute dermal irritation studies were also performed in rabbits. Formulated FCE 26184 was administered percutaneously in a volume of 0.25 ml per site (4 cm²) on scarified or intact skin. No deaths, or treatment related clinical signs, or local irritation was observed.

General toxicity and subchronic local studies were undertaken in rabbits (New Zealand white). Three concentrations, 20, 100, and 200 μg/ml, were daily instilled for 4 consecutive weeks on the eye of the animals. No mortality, no treatment-related clinical signs, no body weight or ophthalmological changes were observed. Laboratory investigations, organ weights, and necropsy did not reveal any drug-related changes. The application site (eye) as well as the route of elimination (lacrymal duct, nasal mucosa) were examined with particular attention.

In addition, 3 concentrations of 20, 100, and 200 μg/ml were daily administered percutaneously in a volume of 0.25 ml per application as already described for the skin. The same negative results were obtained as in the above-reported study. In conclusion, administration of a pharmaceutical preparation of rh-bFGF on normal eyes or on normal or scarified skin did not induce any sign of irritation, or signs of general toxicity. Further subacute iv toxicity studies are in progress in two species, rat and cynomolgus monkey, injected at the dose levels of 1, 10, and 100 μg/kg, but these data are not yet available.

In addition to the formal studies recommended by regulatory agencies, some others need to be developed ad hoc to address safety problems that may be encountered in particular applications, such as in the regeneration of the tympanic membrane, which we intend to explore.

CLINICAL STUDIES

Eye

A tolerance study in healthy volunteers was recently started with single, three times a day (TID) daily doses ranging from 500 to 1000 ng. The study is ongoing, and until now no side effects have been reported in this trial.

A second tolerance study is also in progress. It is a double-blind study randomized versus placebo in parallel groups performed on patients undergoing surgery for penetrating keratoplasty. This study, besides providing data on tolerance, can be expected also to provide data on a possible biological activity on the lesioned eye.

Patients undergoing penetrating keratoplasty were chosen since they present a standardized lesion obtained by trephination allowing a rigorous evaluation of the corneal reepithelization process by photography after fluorescein staining. A further advantage of this study consists in a timely standardized drug administration (3 daily topical treatments with 500 ng each of rh-bFGF) that can be rigorously performed in hospitalized patients. Till now 26 patients have been recruited, and although the code has not yet been broken, no side effect like pain or neovascularization of the cornea has been reported for any of the subjects treated with either active drug or placebo.

Skin

Due to poor availability of representative animal models and, on the other hand, having preclinical safety data already available, in 1990 we decided to verify the pharmacological activity of rh-bFGF in human beings. Two different human models were chosen based on the type of skin lesion. The first study, performed in healthy volunteers, was designed to evaluate the degree of epidermal regeneration following the induction of suction blisters[22] in the skin and will be presented in more detail hereafter.

The second study, here only mentioned, is ongoing and will consider the healing of the epidermis as well as of the dermis as it occurs in the wounds procured during split-thickness skin grafting at the donor site.

Skin Blister Study

Four epidermal suction blisters with a diameter of 8 mm each were made according to the method described by Kiistala[23] on both forearms of healthy volunteers.

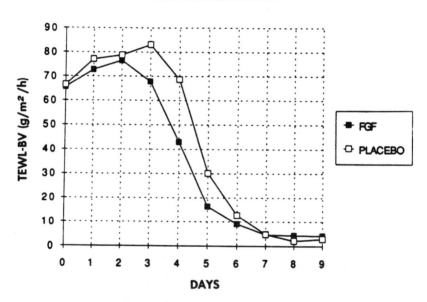

TEWL = Trans epidermal water loss
BV = Basal value

FIGURE 5. Transepidermal water loss (TEWL) during the reepithelization of suction blisters in a representative subject treated for 6 days with 500 ng of rh-bFGF/blister or with placebo. TEWL was measured daily for 9 consecutive days by evaporimetry. Data represent the mean values of TEWL calculated on four blisters per experimental point.

A standardized procedure that measures the transepidermal water loss (TEWL) by evaporimetry[24] was used to evaluate the degree of reepithelization of the lesioned skin, administering rh-bFGF on one and placebo on the controlateral forearm.

The TEWL of a constant skin surface was inversely proportional to the thickness of the epithelium covering the area, showing highest values after wounding which returned within 8 to 9 days to baseline values when reepithelization of the blister with all the epidermal layers was completed (FIGURE 5).

A double-blind study with this wounding model was designed for 24 healthy volunteers. Blisters on one arm were treated daily with 500 ng of rh-bFGF/blister, while the wounds on the controlateral arm were treated with placebo. TEWL of the wounded areas was measured for 9 days and in addition, 3 days after completed healing, skin biopsies were taken for histological examinations. Treatment with rh-bFGF resulted in a significantly increased healing rate evaluated by TEWL as described above. Histological examination showed an increase in cellularity, including melanocytes, and of neovascularization and of collagen maturation in rh-bFGF- vs. placebo-treated blisters. A clinical and histological follow-up study of the subjects at 6 months is now in progress.

CONCLUSIONS

The preclinical studies conducted until now have allowed us to start with few and well-defined clinical investigations on rh-bFGF. The initial results of the clinical studies are expected in the near future and will be very important for evaluating the further clinical development of this recombinant protein. Nevertheless, due to its manyfold activities, it is possible to imagine more and new areas of therapeutic interest for rh-bFGF such as in hard-tissue repair and in the nervous system.

SUMMARY

Basic fibroblast growth factor is a polypeptide belonging to a family of natural proteins also known as heparin-binding growth factors endowed with a pleiotropism of biological activities, the most striking of which are related to wound healing. Large quantities of recombinant human basic fibroblast growth factor (rh-bFGF) of a clinical grade were obtained and used to undertake preclinical and clinical studies. *In vivo* the wound healing effect of rh-bFGF was evaluated in experimental targets such as the cornea and the tympanic membrane, showing a significantly increased epithelial healing rate in drug-treated animals. The deposition of labeled rh-bFGF after topical applications in ocular wounding models did not result in a systemic absorption of the intact rh-bFGF molecule. The acute and the subchronic toxicity studies undertaken after iv and topical administration of a stable pharmaceutical formulation of rh-bFGF did not result in irritation, and no signs of general toxicity were observed. Altogether these data permitted us to start recently with human studies, which are still ongoing, aimed to evaluate the tolerability and the activity of rh-bFGF on tegumental targets such as the cornea and the skin.

REFERENCES

1. 1990. *In* Development Biotechnology Medicines. Annual Survey. Pharmaceutical Manufacturers Association. Washington, D.C.
2. 1990. Séance Extraordinaire du 26 Septembre 1990. Académie de Pharmacie. Paris, France.
3. THOMAS, K. 1987. Fibroblast Growth Factors. FASEB J. **6:** 434–440.
4. GOSPODAROWICZ, D., G. NEUFELD & L. SCHWEIGERER. 1986. Fibroblast growth factor. Mol. Cell. Endocrinol. **46:** 187–204.
5. BAIRD, A., F. ESCH, P. MORMEDE, *et al.* 1986. Molecular characterization of fibroblast growth factor: distribution and biological activities in various tissues. Recent Prog. Horm. Res. **42:** 143–205.

6. ESCH, F., A. BAIRD, N. LING, N. UENO, F. HILL, L. DENOROY, R. KLEPPER, D. GOSPO-DAROWICZ, P. BOHLEN & R. GUILLEMIN. 1985. Primary structure of bovine basic fibroblast growth factor (FGF) and comparison with the amino-terminal sequence of bovine brain acidic FGF. Proc. Nat. Acad. Sci. USA. **82:** 6507–6511.

7. PRESTA, M., D. MOSCATELLI, J. J. SILVERSTEIN & D. B. RIFKIN. 1986. Purification from a human hepatoma cell line of a basic fibroblast growth factor–like molecule that stimulates capillary endothelial cell plasminogen activator production, DNA synthesis, and migration. Mol. Cell. Biol. **6:** 4060–4066.

8. MONTESANO, R., J. D. VASSALLI, A. BAIRD, R. GUILLEMIN & L. ORCI. 1986. Basic fibroblast growth factor induces angiogenesis in vitro. Proc. Nat. Acad. Sci. USA. **83:** 7297–7301.

9. COUGHLIN, S. R., J. P. BARR, L. S. COUSENS, J.-L. FRETTO & L. T. WILLIAMS. 1988. Acidic and basic fibroblast growth factors stimulate tyrosine kinase activity in vivo. J. Biol. Chem. **263:** 988–993.

10. CINTRON, C., C. L. KUBLIN & H. COVINGTON. 1982. Quantitative studies on corneal epithelial wound healing in rabbits. Curr. Eye Res. **9:** 507–516.

11. BARR, J. P., L. S. COUSENS, C. T. LEE-NG, A. MEDINA-SELBY, F. R. MASIARZ, R. A. HALLEWELL, S. CHAMBERLAIN, J. BRADLEY, D. LEE, K. S. STEIMER, L. POULTER, A. L. BURLINGAME, F. ESCH & A. BAIRD. 1988. Expression and processing of biologically active fibroblast growth factors in the yeast *Saccaromyces cerevisiae*. J. Biol. Chem. **263:** 16471–16478.

12. SQUIRES, C. H., J. CHILDS, P. EISENBERG, P. J. POLVERINI & A. SOMMER. 1988. Production and characterization of human basic fibroblast growth factor from *Escherichia coli*. J. Biol. Chem. **263:** 16297–16302.

13. FOX, G. M., S. G. SCHIFFER, M. F. ROHDE, L. B. TSAI, A. R. BANKS & T. ARAKAWA. 1988. Production, biological activity and structure of recombinant basic fibroblast growth factor and an analog with cysteine replaced by serine. J. Biol. Chem. **263:** 18452–18458.

14. MOSES, R. A., G. PARKINSON & R. SCHUCHARDT. 1979. A standard large wound of the corneal epithelium in the rabbit. Invest. Ophthalmol. Vis. Sci. **18:** 103–106.

15. ASSOULINE, M., G. MONTEFIORE, G. PETROUTSOS, D. RAULAIS, L. LEVY-HUET, Y. POULIQUEN & Y. COURTOIS. 1989. The effect of recombinant human basic fibroblast growth factor (FCE 26184) on corneal epithelial wound healing in rabbits. Farmitalia Carlo Erba Int. Rep. 201i.

16. RIECK, P., M. ASSOULINE, Y. POULIQUEN & Y. COURTOIS. 1990. Effect of recombinant human basic fibroblast growth factor (FCE 26184) on corneal wound healing after keratectomy and repeated swabbing in rabbits. Farmitalia Carlo Erba Int. Rep. 204i.

17. MONDAIN, M., S. SAFFIEDINE & A. UZIEL. 1991. Fibroblast growth factor improves the healing of experimental tympanic membrane perforations. Acta Otolaryngol. Stockh. **111:** 337–341.

18. WEISSINGER, J. 1989. Nonclinical pharmacologic and toxicologic considerations for evaluating biologic products. Reg. Toxicol. Pharmacol. **10:** 255–263.

19. ASSOULINE, M., C. HUTCHINSON, K. MORTON, F. MASCARELLI, J. C. JEANNY, N. FAYEIN, Y. POULIQUEN & Y. COURTOIS. 1989. In vivo binding of topically applied human bFGF on rabbit corneal epithelial wound. Growth Factor **1:** 251–261.

20. 1989, 1990. Farmitalia C. Erba internal files. Milan, Italy.

21. WEISSINGER, J. 1990. Nonclinical regulatory considerations for wound-healing growth factors. Drug Dev. Res. **19:** 275–284.

22. LYONNET, S. Une nouvelle méthode d'étude de la cicatrisation cutanée: application à l'étude du bFGF. Mémoire Technique. Université Claude Bernard, Lyon I, Année Universitaire 1989/1990.

23. KIISTALA, U. & K. K. MUSTAKALLIO. 1967. Dermoepidermal separation with suction: electron microscopy and histochemical study of initial events of blistering on human skin. J. Invest. Dermatol. **48:** 466–477.

24. PINNAGODA, I., R. A. TUPKER, T. AGNER & J. SERUP. 1990. Guidelines for transepidermal water loss (TEWL) measurement. A report from the Standardization Group of the European Society of Contact Dermatitis. Contact Dermatitis **22:** 164–178.

Basic Fibroblast Growth Factor Protects Photoreceptors from Light-Induced Degeneration in Albino Rats[a]

MATTHEW M. LaVAIL,[b,d] ELLA G. FAKTOROVICH,[d]
JEANNE M. HEPLER,[d] KATHRYN L. PEARSON,[d]
DOUGLAS YASUMURA,[b,d] MICHAEL T. MATTHES,[d]
AND ROY H. STEINBERG[c,d]

[b]Department of Anatomy
[c]Department of Physiology
[d]Beckman Vision Center
University of California, San Francisco
San Francisco, California 94143-0730

One of the diverse roles of the fibroblast growth factors is to serve as survival-promoting factors in the central nervous system (CNS). In several regions of the CNS, for example, the application of basic fibroblast growth factor (bFGF) reduces the degree of neuronal death following axonal injury.[1-3] We considered the possibility that bFGF might play a similar role in certain forms of retinal degeneration, because the peptide (as well as acidic FGF) is found in the retina[4-12] and in the closely associated retinal pigment epithelium (RPE).[13,14] Moreover, bFGF induces retinal regeneration from the developing chick RPE.[15]

We recently found that in Royal College of Surgeons (RCS) rats with inherited retinal dystrophy, the rate of photoreceptor degeneration can be slowed dramatically by an intraocular injection of bFGF.[16] When an injection of bFGF was given subretinally (between the neural retina and RPE), extensive photoreceptor rescue occurred, in some cases across half of the retina. When the injection was made into the vitreous, rescue was even more widespread, covering almost the entire retina.[16] This represented the first time that the pace of an inherited retinal degeneration had been slowed significantly by pharmacological means.

In the RCS rat, the mutant gene is expressed in the RPE cell, and this secondarily leads to photoreceptor cell loss.[17] We have now examined the action of bFGF on a noninherited form of retinal degeneration that is thought to damage photoreceptor cells directly, constant light-induced photoreceptor degeneration in albino rats.[18] Moreover, we sought a more flexible model with which to study the action of bFGF than the slowly evolving retinal dystrophy of the RCS rat.[19]

[a]This work was supported in part by National Institutes of Health Research Grants EY01919, EY06842, EY01429, and Core Grant EY02162 and funds from the Retinitis Pigmentosa Foundation Fighting Blindness, Research to Prevent Blindness, and That Man May See, Inc. Dr. LaVail is the recipient of a Research to Prevent Blindness Senior Scientific Investigators Award.

341

MATERIALS AND METHODS

The methods are described elsewhere in detail.[16] Briefly, experiments were carried out on 2- to 4-month-old F344 or Sprague-Dawley albino rats obtained from Simonsen Laboratories and maintained in our cyclic light environment 7 or more days before intraocular injections. bFGF (1 μl) dissolved in phosphate-buffered saline (PBS) (1150 ng/μl) was injected transsclerally into the vitreous of anesthetized rats. For buffer control experiments, 1 μl of PBS alone was injected in some animals, and each of these was compared to uninjected animals. In other rats a dry needle was inserted into the subretinal space with no injection. Injections were made 2 days before the rats were placed into a constant light environment (115–130 ft-c) provided by fluorescent illumination as described elsewhere.[20] Eyes were taken after 1 week of light exposure. In some cases after constant light exposure, the animals were returned to cyclic light for 10 days before eye removal to allow for possible regeneration of photoreceptor inner and outer segments. In some experiments, the interval between injection and constant light exposure was varied. Eyes were prepared for histological analysis using 1 μm thick plastic sections cut along the vertical meridian of the eye.[16,19] All procedures involving the rats adhered to the ARVO Resolution on the Use of Animals in Research and the guidelines of the UCSF Committee on Animal Research.

RESULTS

The retinas of untreated albino rats reared in cyclic light, or those with PBS injected intravitreally, showed the expected normal histological structure, with the outer nuclear layer (ONL) consisting of 8–10 rows of photoreceptor nuclei (FIGURE 1a). After 1 week of constant light, these retinas showed the degeneration and loss of numerous photoreceptors demonstrated previously in albino rats.[21–23] The most severely damaged region of the retina was located in the superior hemisphere,[22] where the ONL was reduced to about 2 rows of nuclei, and almost all photoreceptor inner and outer segments were missing (FIGURE 1b). In more peripheral regions of the superior hemisphere and in the inferior hemisphere of the eye, the retina was somewhat less severely damaged. In most cases, few, if any, outer segments remained in any part of the retina.

When bFGF was injected intravitreally 2 days before the light exposure, far less photoreceptor degeneration occurred. For example, the most degenerated region of a Sprague-Dawley rat is shown in FIGURE 1c. In this case, the ONL consisted of 7–8 rows of photoreceptor nuclei, and inner and outer segments were still present, albeit fairly disorganized. Although some retinas showed a somewhat greater degree of degeneration than that shown in FIGURE 1c, particularly those of F344 rats, at least 4 or 5 rows of nuclei and some inner and outer segments were always present in the most degenerated retinal region. Thus, bFGF afforded a remarkable degree of protection from the damaging effect of constant light.

It is known that moderately light-damaged photoreceptors, where inner segments are still present, can regenerate their outer segments if the rats are kept in cyclic light after the damaging light exposure.[24] When uninjected or PBS-injected F344 rats were exposed to 1 week of constant light and then 10 days of cyclic light (one rod outer segment renewal time in the rat),[25] a small degree of outer segment renewal was seen. However, when bFGF-injected rats were exposed to the same light regimen, there was much more extensive regeneration of outer segments (data not

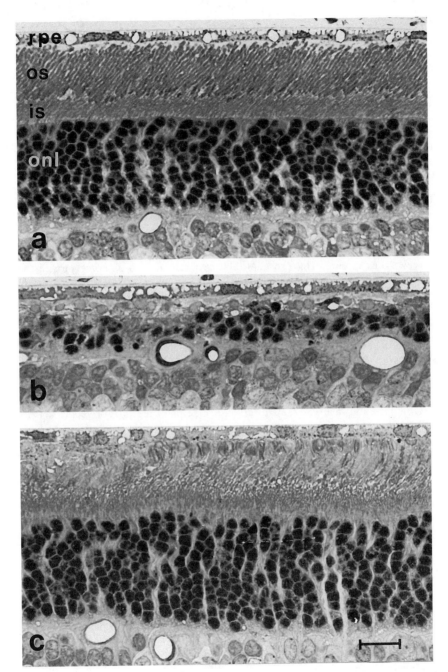

FIGURE 1. Plastic-embedded sections of 3-month-old Sprague-Dawley albino rats. **a:** Normal retina from a rat reared in cyclic light. **b:** Retina from an uninjected rat exposed to constant light for 7 days in which the outer nuclear layer is reduced to about 2 rows of nuclei, and no normal appearing photoreceptor inner or outer segments are present. **c:** Retina from a rat that received an intravitreal injection of bFGF 2 days before a 7-day constant light exposure. The outer nuclear layer is 7–8 rows in thickness, and many discrete photoreceptor inner and outer segments are present, albeit somewhat disorganized. is, inner segments; onl, outer nuclear layer; os, outer segments; rpe, retinal pigment epithelium. Toluidine blue stain. Scale bar, 15 μm.

shown). Quantitative assessment of the above findings are in preparation and will be presented elsewhere.

We also have obtained a number of other preliminary observations. For example, in the outer segment regeneration experiments described above, we asked if the enhanced regeneration was a direct effect of bFGF, or whether it was due simply to the retina being less severely damaged. To distinguish between these two possibilities, we injected bFGF or PBS *after* a 1-week constant light exposure, at the start of 10 days of cyclic light, a period in which outer segment regeneration would occur. We compared these rats with a group having the same light regimen, but without receiving the growth factor. In the few eyes examined, the degree of outer segment regeneration was the same in bFGF-injected, PBS-injected, and uninjected eyes. Thus, bFGF does not appear to enhance outer segment regeneration directly.

In RCS rats, the insertion of a dry needle into the subretinal space rescued photoreceptors in an area restricted to the site of injection and along the needle track.[16] To test whether this phenomenon occurs in light damage, we did comparable experiments by inserting a dry needle into the subretinal space 2 days before a 1-week light exposure. This procedure did, in fact, induce a photoreceptor rescue that was almost as great in magnitude as with bFGF. The area of rescue, however, was clearly not as broad as an intravitreal bFGF injection, but it extended beyond the boundaries of the immediate injection site and needle track.

Finally, we studied how long bFGF retains survival-promoting activity in the eye by varying the interval between bFGF injection and the beginning of a 1-week constant light exposure. Although we have not yet quantified the data, it is clear that postinjection times of 0 hours (into constant light immediately after injection), 3 and 6 hours, and 1 and 2 days all gave a maximal protective effect. At 4 and 6 days, the rescue effect appeared somewhat less than maximal, and it was minimal at 8 days. At postinjection intervals of 16 and 40 days, no photoreceptor rescue was seen. Thus, bFGF injected intravitreally retains at least some photoreceptor protective activity for up to 8 days.

DISCUSSION

We have demonstrated that bFGF protects retinal photoreceptor cells from light-induced degeneration. In most instances where bFGF has shown a survival-promoting activity in neuronal degeneration experiments,[1-3] it has been suggested that bFGF is a naturally occurring neurotrophic molecule. The implication is that axotomy interrupts retrograde transport of bFGF and prevents it from reaching the cell body. According to this idea, experimental application of exogenous bFGF would replace the missing growth factor. In the retina, many interactions occur between the photoreceptors and the RPE that are essential for photoreceptor function and viability.[26,27] Since bFGF is synthesized by the RPE,[13,14] bFGF is found in the interphotoreceptor matrix,[6] and rod outer segments have bFGF receptors,[10,28] it may be that bFGF acts normally as a neurotrophic factor from the RPE to photoreceptor cells, although this remains to be shown experimentally.

The idea that bFGF is a neurotrophic factor for photoreceptors has implications for both the mechanism of cell death and photoreceptor rescue in RCS rats and light damage. In RCS rats, the cascade of cytopathological events that stems from mutant gene expression may result in abnormal RPE synthesis or release of bFGF, or abnormal photoreceptor uptake or processing of the growth factor. It is also possible that the unusual accumulation of outer segment debris membranes and/or excessive accumulation of interphotoreceptor matrix in RCS rats[29] may bind bFGF, preventing it from reaching the photoreceptor cells. Damage due to constant light is considered to involve lipid peroxidation by free radicals at some stage in the mechanism of cell death.[18,30,31] However, the effects of light damage might also include a disruption of the normal photoreceptor-RPE interactions involving bFGF. While retinas undergoing light-induced degeneration do not show the accumulation of outer segment debris membranes seen in RCS rats, we have found recently that they do have an accumulation of interphotoreceptor matrix similar to that in RCS rats (unpublished observations). Thus, the two degenerative disorders both affect the architecture and, most probably, the chemical composition of the interphotoreceptor space. It is intriguing to speculate that in both conditions this might interrupt the putative neurotrophic action of bFGF. Obviously, more experimentation is needed on this question.

Alternatively, bFGF may *not* have a normal role in photoreceptor-RPE interactions and may *not* be a naturally occurring neurotrophic factor. Indeed, when it is considered that bFGF also rescues diverse CNS neuronal cell types, such as retinal photoreceptors,[16] retinal ganglion cells,[1] and neurons that project to the hippocampal formation,[2,3] a case can be made for bFGF acting as a nonspecific survival-promoting factor, exclusive of a normally occurring neurotrophic function. Furthermore, in the axotomy experiments discussed above, consideration that bFGF is a neurotrophic factor implies that bFGF is normally carried to neuronal cell bodies from the periphery by retrograde transport mechanisms. In fact, bFGF has recently been found to be transported in an anterograde direction by rat retinal ganglion cells, but not in a retrograde direction.[32]

Many questions remain about the specificity of neuronal rescue by bFGF, the molecular mechanisms of its survival-promoting action, and its normal role in neuronal metabolism. Light damage in the albino rat offers an excellent model for *in vivo* studies of these phenomena. The bFGF can be applied in a relatively noninvasive manner; photoreceptor damage is induced in a noninvasive manner; the degree of photoreceptor degeneration can be regulated by varying the intensity or duration of light; the concentration of injected peptide can be relatively well controlled within the eyeball; the period of degeneration is short; and the degeneration can be produced in commercially available animals independent of specific ages, compared to the age restrictions of an inherited disorder such as that found in RCS rats.

ACKNOWLEDGMENTS

We thank D. Gospodarowicz for helpful discussions and the gift of bFGF, and N. Lawson and G. Riggs for technical and secretarial assistance.

REFERENCES

1. SIEVERS, J., B. HAUSMANN, K. UNSICKER & M. BERRY. 1987. Fibroblast growth factors promote the survival of adult rat retinal ganglion cells after transection of the optic nerve. Neurosci. Lett. **76:** 157–162.
2. ANDERSON, K. J., D. DAM, S. LEE & C. W. COTMAN. 1988. Basic fibroblast growth factor prevents death of lesioned cholinergic neurons *in vivo*. Nature **332:** 360–361.
3. OTTO, D., M. FROTSCHER & K. UNSICKER. 1989. Basic fibroblast growth factor and nerve growth factor administered in gel foam rescue medial septal neurons after fimbria fornix transection. J. Neurosci. Res. **22:** 83–91.
4. D'AMORE, P. A. & M. KLAGSBRUN. 1984. Endothelial cell mitogens derived from retina and hypothalamus: biochemical and biological similarities. J. Cell Biol. **99:** 1545–1549.

5. BAIRD, A., F. ESCH, D. GOSPODAROWICZ & R. GUILLEMIN. 1985. Retina- and eye-derived endothelial cell growth factors: partial molecular characterization and identity with acidic and basic fibroblast growth factors. Biochemistry 24: 7855–7860.

6. PLOUET, J., F. MASCARELLI, O. LAGENTE, C. DOREY, G. LORANS, J. P. FAURE & Y. COURTOIS. 1986. Eye derived growth factor: a component of rod outer segment implicated in phototransduction. *In* Retinal Signal Systems, Degenerations and Transplants. E. Agardh and B. Ehinger, Eds.: 311–320. Elsevier Science Publishers. Amsterdam, the Netherlands.

7. JEANNY, J.-C., N. FAYEIN, M. MOENNER, B. CHEVALLIER, D. BARRITAULT & Y. COURTOIS. 1987. Specific fixation of bovine brain and retinal acidic and basic fibroblast growth factors to mouse embryonic eye basement membranes. Exp. Cell Res. 171: 63–75.

8. MASCARELLI, F., D. RAULAIS, M. F. COUNIS & Y. COURTOIS. 1987. Characterization of acidic and basic fibroblast growth factors in brain, retina and vitreous chick embryo. Biochem. Biophys. Res. Commun. 146: 478–486.

9. PLOUET, J., F. MASCARELLI, M. D. LORET, J. P. FAURE & Y. COURTOIS. 1988. Regulation of eye derived growth factor binding to membranes by light, ATP or GTP in photoreceptor outer segments. EMBO J. 7: 373–376.

10. PLOUET, J. 1988. Molecular interaction of fibroblast growth factor, light-activated rhodopsin and s-antigen. *In* Molecular Biology of the Eye: Genes, Vision and Ocular Disease. J. Piatigorsky, S. Toshimichi & P. S. Zelenka, Eds.: 83–92. Alan R. Liss, Inc. New York, N.Y.

11. HANNEKEN, A., G. A. LUTTY, D. S. McLEOD, F. ROBEY, A. K. HARVEY & L. M. HJELMELAND. 1989. Localization of basic fibroblast growth factor to the developing capillaries of the bovine retina. J. Cell. Physiol. 138: 115–120.

12. NOJI, S., T. MATSUO, E. KOYAMA, T. YAMAAI, T. NOHNO, N. MATSUO & S. TANIGUCHI. 1990. Expression pattern of acidic and basic fibroblast growth factor genes in adult rat eyes. Biochem. Biophys. Res. Commun. 168: 343–349.

13. SCHWEIGERER, L., B. MALERSTEIN, G. NEUFELD & D. GOSPODAROWICZ. 1987. Basic fibroblast growth factor is synthesized in cultured retinal pigment epithelial cells. Biochem. Biophys. Res. Commun. 143: 934–940.

14. STERNFELD, M. D., J. E. ROBERTSON, G. D. SHIPLEY, J. TSAI & J. T. ROSENBAUM. 1989. Cultured human retinal pigment epithelial cells express basic fibroblast growth factor and its receptor. Curr. Eye Res. 8: 1029–1037.

15. PARK, C. M. & M. J. HOLLENBERG. 1989. Basic fibroblast growth factor induces retinal regeneration *in vivo*. Dev. Biol. 134: 201–205.

16. FAKTOROVICH, E. G., R. H. STEINBERG, D. YASUMURA, M. T. MATTHES & M. M. LAVAIL. 1990. Photoreceptor degeneration in inherited retinal dystrophy delayed by basic fibroblast growth factor. Nature 347: 83–86.

17. MULLEN, R. J. & M. M. LAVAIL. 1976. Inherited retinal dystrophy: primary defect in pigment epithelium determined with experimental rat chimeras. Science 192: 799–801.

18. NOELL, W. K. 1980. Possible mechanisms of photoreceptor damage by light in mammalian eyes. Vision Res. 20: 1163–1171.

19. LAVAIL, M. M. & B. A. BATTELLE. 1975. Influence of eye pigmentation and light deprivation on inherited retinal dystrophy in the rat. Exp. Eye Res. 21: 167–192.

20. LAVAIL, M. M., G. M. GORRIN, M. A. REPACI, L. A. THOMAS & H. M. GINSBERG. 1987. Genetic regulation of light damage to photoreceptors. Invest. Ophthalmol. Visual Sci. 28: 1043–1048.

21. NOELL, W. K., V. S. WALKER, B. S. KANG & S. BERMAN. 1966. Retinal damage by light in rats. Invest. Ophthalmol. 5: 450–473.

22. RAPP, L. M. & T. P. WILLIAMS. 1980. A parametric study of retinal light damage in albino and pigmented rats. *In* The Effects of Constant Light on Visual Processes. T. P. Williams & B. N. Baker, Eds.: 135–159. Plenum Press. New York, N.Y.

23. LAVAIL, M. M., G. M. GORRIN, M. A. REPACI & D. YASUMURA. 1987. Light-induced retinal degeneration in albino mice and rats: strain and species differences. *In* Degenerative Retinal Disorders: Clinical and Laboratory Investigations. J. G. Hollyfield, R. E. Anderson & M. M. LaVail, Eds.: 439–454. Alan R. Liss, Inc. New York, N.Y.

24. WYSE, J. P. H. 1980. Renewal of rod outer segments following light-induced damage of the retina. Can. J. Ophthalmol. **15:** 15–19.
25. LAVAIL, M. M. 1981. Photoreceptor characteristics in congenic strains of RCS rats. Invest. Ophthalmol. Visual Sci. **20:** 671–675.
26. ZINN, K. M. & M. F. MARMOR. 1979. The Retinal Pigment Epithelium. Harvard University Press. Cambridge, Mass.
27. STEINBERG, R. H. 1986. Research update: report from a workshop on cell biology of retinal detachment. Exp. Eye Res. **43:** 695–706.
28. MASCARELLI, F., D. RAULAIS & Y. COURTOIS. 1989. Fibroblast growth factor phosphorylation and receptors in rod outer segments. EMBO J. **8:** 2265–2273.
29. LAVAIL, M. M., L. H. PINTO & D. YASUMURA. 1981. The interphotoreceptor matrix in rats with inherited retinal dystrophy. Invest. Ophthalmol. Visual Sci. **21:** 658–668.
30. ANDERSON, R. E., L. M. RAPP & R. D. WIEGAND. 1984. Lipid peroxidation and retinal degeneration. Curr. Eye Res. **3:** 223–227.
31. WIEGAND, R. D., J. G. JOSE, L. M. RAPP AND R. E. ANDERSON. 1984. Free radicals and damage to ocular tissues. *In* Free Radicals in Molecular Biology, Aging, and Disease. D. Armstrong, R. S. Sohal, R. G. Cutler & T. F. Slater, Eds.: 317–353. Raven Press. New York, N.Y.
32. FERGUSON, I. A., J. B. SCHWEITZER & E. M. JOHNSON, JR. 1990. Basic fibroblast growth factor: receptor-mediated internalization, metabolism, and anterograde axonal transport in retinal ganglion cells. J. Neurosci. **10:** 2176–2189.

Differential Localization and Possible Functions of aFGF and bFGF in the Central and Peripheral Nervous Systems[a]

F. ECKENSTEIN, W. R. WOODWARD,[b] AND R. NISHI

Department of Cell Biology and Anatomy
[b]Department of Neurology
Oregon Health Sciences University
3181 Southwest Sam Jackson Park Road
Portland, Oregon 97201

INTRODUCTION

Acidic and basic fibroblast growth factors (aFGF and bFGF, respectively) are two members of the family of heparin-binding growth factors (HBGFs) that are reproducibly found in adult nervous tissue. *In vitro,* aFGF and bFGF have been shown to have a wide variety of biological effects on cells from the nervous system. For example, both factors are potent mitogens for fibroblasts,[1] astrocytes,[2] oligodendrocytes,[3] Schwann cells,[4] and endothelial cells,[5,6] and can also serve as chemotactic signals for many of these cell types.[7] In addition, aFGF and bFGF can promote the survival in culture of many types of peripheral and central neurons.[8–12] However, the physiological relevance of these *in vitro* observations to the functions of aFGF and bFGF *in vivo* are obscure for the following reasons: (1) some of the *in vitro* effects of FGFs, such as the stimulation of mitogenesis of astrocytes, are not observed in the normal central nervous system (CNS) in spite of the fact that the CNS contains substantial levels of FGFs; (2) both aFGF and bFGF appear to lack signal peptide sequences, a feature that appears to be necessary for the efficient secretion of proteins from cells; (3) the pleiotrophic effects of the FGFs observed *in vitro* appear nonspecific considering the very specific cellular interactions that must guide CNS development and maintenance.

This apparent discrepancy between the abundant FGF effects observed *in vitro* and the possibly more restricted role these factors may play *in vivo* could be explained by the hypothesis that FGFs become available to responsive cells *in vivo* only at restricted locations and after specific events that cause the release of normal intracellular stores of FGFs. A first step towards testing this hypothesis is to determine the distributions of aFGF and bFGF in nervous tissue in order to determine whether cell lysis after injury or during developmental cell death may represent a plausible mechanism for the release of FGFs.

We have previously shown that aFGF and bFGF are differentially distributed in the nervous system with rat sciatic nerve containing very high levels of only aFGF, spinal cord containing both aFGF and bFGF, and primarily bFGF in rat cerebral cortex.[13] Here we report that bFGF is largely localized to nuclei and cytoplasm of astrocytes and of CA2 hippocampal neurons in the CNS, whereas aFGF is localized

[a]This work was supported by National Institutes of Health grants AG07424 (FE), NS17493 (FE), NS25767 (RN), a grant from the Amyotrophic Lateral Sclerosis Association (RN), and by a March of Dimes Basil O'Connor grant (FE).

in peripheral neurons. This differential distribution of aFGF and bFGF suggests differences in the functions of these two molecules in the development and maintenance of the central and peripheral nervous system.

METHODS

Preparation of Extracts and Supernatants

Animals were sacrificed by asphyxiation with carbon dioxide, tissues were dissected, frozen immediately, and stored at $-70°C$ for no longer than 14 days. Tissues were thawed and quickly homogenized in 10 ml/g of ice-cold 20 mM Tris, pH 8.2, the homogenates were centrifuged for 10 minutes at $15,000 \times g$, and supernatants were collected, and mitogenic activity present in the supernatants were determined as described below. Protein concentrations in supernatants was determined using a Coomassie blue binding assay. Tissues from at least three different animals were assayed for all data presented in this study.

Lesions

All operations were performed on deeply anesthetized animals. Sciatic nerves were transsected about 7 mm above the entry into the gastrocnemius muscle. Care was taken to clearly separate the two resulting nerve stumps in order to prevent possible regeneration. Optic nerves were lesioned by enucleation of the eyes. Animals were allowed to recover and were sacrificed after varying survival times, ranging from 1 to 45 days, and nerves were collected and processed as described above. Groups of at least three animals were employed for each postlesion time point.

Assay for Mitogenic Activity

The mitogenic effect of extracts and human recombinant aFGF (a gift from Dr. K. Thomas, Merck) or human recombinant bFGF (a gift from Dr. J. Abraham, California Biotechnology) was tested using a serum-free [^{3}H]thymidine incorporation assay as previously described.[13] Briefly, AKR-2B cells were transferred at a density of 10,000 cells per well into 24-well culture plates in McCoy's 5A medium supplemented with 5% (v/v) fetal bovine serum. Cultures were then incubated for 5 days until the cells formed a confluent monolayer. The medium was then replaced with serum-free medium (MCDB 402) and the cells were incubated for an additional 2 days at 37°C. Fresh MCDB 402, containing FGFs or diluted extracts, was then added and 22 hours later the cultures were pulsed with 1.0 μCi [^{3}H]thymidine. Cells were incubated in isotope for 1 hour after which the relative incorporation of [^{3}H]thymidine into cold 10% trichloracetic acid insoluble material was determined as previously described.[14] Total mitogenic activity present in extracts was determined by obtaining dose-response curves for the extracts, followed by calculating the concentration of extract necessary to induce a half-maximal effect (estimated dose for 50% stimulation, or ED_{50}), and one unit of mitogenic activity was defined as giving a half-maximal stimulation per milliliter of assay medium.

Neuronal and Astrocyte Cultures

Cultures from newborn rat cerebral cortex were established using a method similar to those described earlier.[15,16] Cultures were grown in serum-free medium with N-2 supplements and the percentage of specific cell types in the cultures was determined by immunohistochemical localization of GFAP and neurofilaments. Sensory neuron cultures were prepared from E10 chickens as described earlier.[12] Cultures were harvested using a small amount of 20 mM Tris, 5 mM EDTA, 0.25% Chaps, pH 7.8. Protein was determined in the lysate, bovine serum albumin was then added, the lysate was dialyzed against phosphate-buffered saline (PBS), followed by determination of mitogenic activity in the lysate.

Immunoprecipitation of bFGF

Immunoprecipitation was achieved by incubating tissue extracts or cell lysates with a monoclonal antibody specific for bFGF (a gift from Dr. C. Hart, Zymogenetics, Seattle, Wash), followed by precipitation of the immune complex with secondary antibodies and *Staphylococcus aureus* cell walls.

Immunohistochemical Localization of bFGF

The same monoclonal antibody to bFGF that monospecifically recognizes bFGF in Western blot experiments was used for immunohistochemical localization of bFGF. Mature female Sprague-Dawley rats were deeply anesthetized and perfused through the heart, first with 100 ml saline, then with fixative (300 ml 5% formalin in 100 mM sodium phosphate pH 7.2). The brains were rapidly dissected, blocked into coronal slices of approximately 4 mm, and postfixed for another 30 minutes. Frozen sections were cut, incubated in the monoclonal antibody at a dilution of 1:10,000, followed by a peroxidase-linked staining method. Specificity controls were performed by preincubation of the antibody with 1 μg/ml of bFGF.

RESULTS

Effects of Transection on FGF Levels in Central versus Peripheral Nerve

In an earlier study we had established that FGFs in extracts of nervous tissue could be quantified by measuring the stimulation of incorporation of [³H]thymidine into DNA in cultured mouse AKR-2B cells. In the present study we confirmed that transection of optic and sciatic nerves had the opposite effect on FGF levels in these two tissues: seven days posttransection mitogenic activity was induced about twofold in the distal stump of optic nerve, whereas mitogenic activity decreased to very low levels in the distal stump of sciatic nerve (FIGURE 1). The rapid decrease of aFGF in the distal stump of the sciatic nerve suggests that aFGF is localized to axons in the sciatic nerve, which degenerate when the nerve is transected. In contrast, the slight increase in mitogenic activity in the optic nerve suggests that the FGF is localized to elements other than axons such as nonneuronal cells.

Localization of bFGF in the Nucleus and Cytoplasm of Rat Astrocytes and CA2 Hippocampal Neurons

A monoclonal antibody to bFGF found in an earlier study to specifically recognize bFGF in Western blots of tissue extracts[13] was used here for immunohistochemical localization of bFGF formalin-fixed sections of adult rat CNS. Strong staining was observed in the nuclei of all astrocytes, and in CA2 hippocampal pyramidal cells, whereas the cytoplasm of the same cells was stained more weakly (FIGURE 2A and B). Double immunofluorescence staining of these sections with

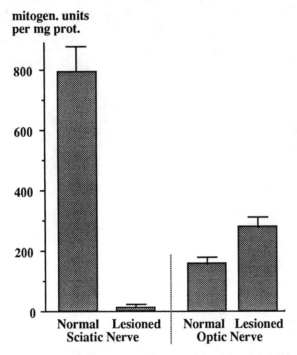

FIGURE 1. Effect of transection on mitogenic activity in peripheral (sciatic) versus central (optic) nerve. Mitogenic activity in extracts prepared from the distal stumps of the respective nerves was quantified 7 days after transection and compared to extracts of nontransected nerves. Note the virtual disappearance of activity from lesioned sciatic nerve in comparison to an increase in activity in optic nerve.

antibodies to the astrocytic marker GFAP and to bFGF unambiguously confirmed the characterization of bFGF-containing cells as astrocytes, and immunoelectron micrographic analysis confirmed that a large percentage of the bFGF immunoreactivity was nuclear in localization (not shown). Similarly, bFGF immunoreactivity was mostly observed in the nucleus of astrocytes in culture (FIGURE 2C and D). All staining reported here was abolished by preincubation of the monoclonal antibody with 1 µg/ml of bFGF. Western blots of cytoplasmic and nuclear fractions prepared by sucrose-density centrifugation confirmed that about two-thirds of bFGF immuno-

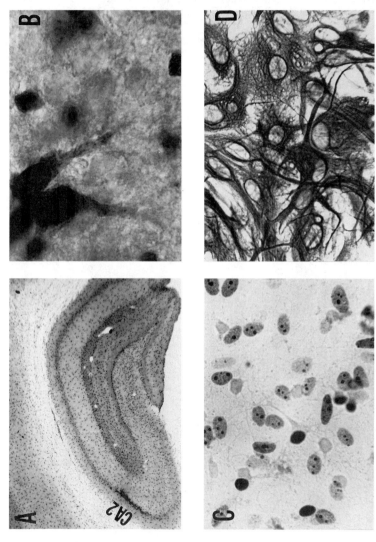

FIGURE 2. Immunohistochemical localization of bFGF. **A:** Low-power (18-fold final magnification) view of a coronal section through rat hippocampus. Nuclear staining of astrocytes is seen throughout the section, but note the strong staining of CA2 hippocampal neurons. **B:** Higher-power (500-fold final magnification) view of labeled neurons in CA2. Both nuclear and cytoplasmic staining is visible. **C:** Basic FGF immunoreactivity in cultured astrocytes. Note the punctate nuclear staining. **D:** Sister culture stained for the astrocyte marker GFAP. All immunohistochemical stains were performed using the avidin-biotin-peroxidase technique. (Reduced to 65%).

reactivity was associated with the nuclear fraction, and one-third with the cytoplasmic fraction (not shown).

Using the AKR-2B cell assay, the levels of bFGF were quantified in extracts from cultures of purified cortical astrocytes (>90% GFAP positive), and of cultures enriched for cortical neurons (>85% neurofilament positive). The specific activity of mitogen from astrocytes was found to be significantly higher than in crude extracts from rat cerebral cortex, and no measurable mitogenic activity was present in neuronal cultures. Using an indirect immunoprecipitation protocol, the monoclonal antibody to bFGF precipitated more than 90% of the mitogenic activity from the extract of astrocyte cultures (FIGURE 3).

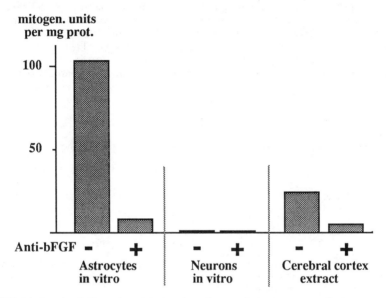

FIGURE 3. Levels of mitogenic activity in cultured rat cortical astrocytes, cortical neurons, and cerebral cortex. Mitogenic activity in extracts from cultures enriched for astrocytes or neurons was compared to mitogenic activity from extracts of cerebral cortex. The majority of the mitogenic activity is immunoprecipitated by a monoclonal antibody to bFGF.

Localization of aFGF in the Chicken Peripheral Nervous System

Extracts prepared from chicken sciatic nerve also contained large amounts of mitogenic activity, 90% of which was retained on a heparin-affigel column, even at NaCl concentrations as high as 0.75 M (TABLE 1). Most of the activity was recovered by the elution of the column with 1.5 M NaCl (FIGURE 4), strongly suggesting that chicken sciatic nerve contains large amounts of aFGF, similar to what has been shown in rat.

The Effect of Transection on Chicken Sciatic FGF

Sciatic nerves of three adult chickens were transected unilaterally, and the resulting distal and proximal stumps, as well as the normal contralateral nerves, were collected seven days later. Transection led to a more than 90% decrease in mitogenic activity in the distal stump, and an approximately 50% decrease in the proximal stump (FIGURE 5).

Presence of FGF in Sensory Neurons and in the Developing Sciatic Nerve

The observation that transection caused a dramatic reduction of mitogenic activity in the distal stump, but only a moderate reduction in the proximal stump, suggested that sciatic aFGF may be present mostly in axons running within the nerve. We investigated this possibility by growing sensory neurons from E10 chick embryos (which represent one of the neuronal populations projecting through the nerve) in culture in the presence of the antimitotic agent fluorodeoxyuridine. After one week *in vitro* the resulting cultures were virtually devoid of nonneuronal cells (less than 5% of the cells had a nonneuronal morphology). Extracts prepared from these cultures contained moderately high levels of mitogenic activity (FIGURE 6).

Mitogenic activity was similarly determined in extracts prepared from sciatic nerves dissected from chicken embryos at different developmental stages. Sciatic

TABLE 1[a]

	Mitogenic Activity (units)	Protein (mg)	Percent Recovery of Activity
Sciatic extract	5236	11.45	100
Flowthrough from heparin column	418	11.0	8
1.5 M NaCl eluate from column	4440	0.02	85

[a] A quantitative analysis of recoveries of mitogenic activity and of protein after heparin chromatography of chicken sciatic nerve extract is shown. Note that heparin chromatography allows near quantitative recovery of activity.

nerves from embryonic day 11 (the earliest time point assayed) already contained significant levels of mitogenic activity that were comparable to the mitogenic activity observed in extracts of E10 sensory neuron cultures. The levels of mitogenic activity in embryonic sciatic nerves increased to adult levels by embryonic day 16 (FIGURE 6).

DISCUSSION

The data presented here demonstrate that aFGF and bFGF are localized differentially in the nervous system. Using immunohistochemical methods, bFGF was found mainly in astrocytes, with the exception of a very few neuronal populations, such as the CA2 hippocampal pyramidal neurons. These observations disagree with earlier studies showing bFGF immunoreactivity in most cortical neurons.[17,18] Our observation, however, that biological bFGF activity is found only in cultured astrocytes and not in cortical neurons provides strong support for the specificity of the staining observed here.

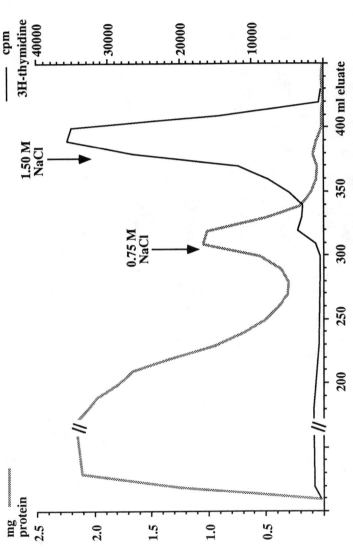

FIGURE 4. Heparin-affinity chromatography of mitogenic activity from chicken sciatic nerve. Soluble material from an extract of adult chicken sciatic nerve was loaded onto heparin-affigel at 0.4 M NaCl, washed, and the bound material eluted with buffer steps of 0.75 M NaCl and 1.5 M NaCL. Profile of protein (broken line) from the column is compared to mitogenic activity (solid line). The vast majority of mitogenic activity elutes at 1.5 M NaCl.

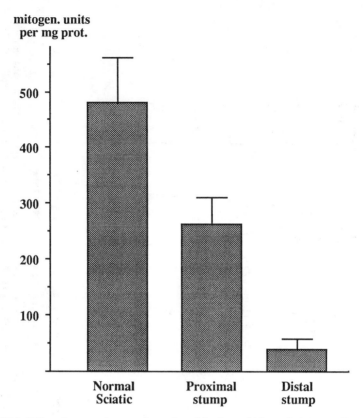

mitogen. units per mg prot.

FIGURE 5. Effect of transection on mitogenic activity from chicken sciatic nerve. Extracts of the proximal and distal stumps of the chicken sciatic nerve were prepared 7 days after transection and compared to extracts of normal nerve.

A significant amount of bFGF immunoreactivity was localized in the nucleus of stained cells. This observation appears less surprising considering reports that different cell types transfected with FGF-expression vectors appear to accumulate FGF in their nuclei.[19,20] In addition, deletion of a nuclear translocation sequence in FGF by site-directed mutagenesis abolishes the mitogenic activity of FGF, apparently without affecting the binding of FGF to its cell surface receptor.[21] It is thus likely that at least part of the physiological mechanism by which FGFs act requires direct interaction with nuclear components.

We observed a moderate increase of bFGF after transection of the optic nerve, suggesting that bFGF might be involved in the response of the adult CNS to injury, possibly by stimulating injury-induced gliosis. It is not clear, however, whether bFGF can be released efficiently from astrocytes, as bFGF lacks the signal peptide sequence common for secreted growth factors. On the other hand, Hatten and coworkers observed that the interaction of granule cells with astrocytes *in vitro* could be blocked by the addition of an antibody to bFGF.[22] This observation might suggest that a limited amount of bFGF might be present on the surface of astrocytes, possibly bound to heparan-proteoglycans.[23,24]

Chicken sciatic nerve contains high levels of a heparin-binding mitogen with characteristics similar to aFGF, which we have shown previously to be abundant in rat sciatic nerve.[13] Our observations that cultured sensory neurons contain significant levels of FGF-like mitogenic activity, and that FGF-like activity is lost from the distal stump of transected sciatic nerve, strongly suggest that aFGF in sciatic nerve is present mainly in axons. This conclusion is in good agreement with observations demonstrating FGF immunoreactivity in cultured sensory neurons.[25]

The function of neuronal aFGF in the adult peripheral nervous system (PNS) is currently unclear, especially as the absence of a signal peptide sequence from aFGF suggests that aFGF might not be released efficiently from healthy cells. It is, however, of interest to speculate that axonal degeneration may lead to the release of intracellular stores of aFGF. Such release could represent the first transient molecular signal indicating that axonal injury has occurred. Possible effects of aFGF under these circumstances might be the promotion of Schwann cell mitogenesis[4] and possibly also the chemotactic attraction and stimulation of macrophages.

The presence of high levels of aFGF in developing sciatic nerve observed here raises the possibility that aFGF might be involved in regulating some aspects of

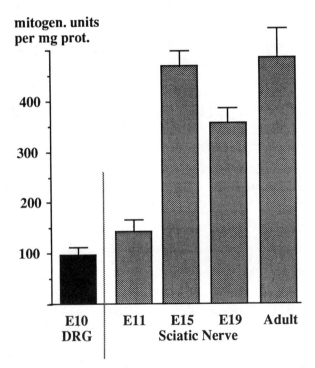

FIGURE 6. Mitogenic activity in cultures of pure E10 chicken sensory neurons (DRG) compared to developing chicken sciatic nerve. Extracts from E10 DRG cultures with <5% contamination from nonneuronal cells were prepared 7 days after plating and compared to extracts of sciatic nerve from the indicated days of embryonic development. Significant levels of mitogenic activity in the nerve can be detected as early as E11, and the possible neuronal origin of this activity is confirmed by observing detectable mitogenic activity in the culture of DRG neurons.

development in the PNS. Observations of interest in this respect are that axonal membranes have been reported to be mitogenic for Schwann cells, that this axonal mitogen binds to heparin, but differs from bFGF.[26] It is thus possible to speculate that axonal aFGF, presumably bound to heparan proteoglycans on the axonal surface, may stimulate the mitosis of Schwann cell precursors.

In addition, it is intriguing to hypothesize that neuronal aFGF may be involved in limiting cell death during development. We earlier have shown that FGFs can promote the survival in culture of peripheral neurons that *in vivo* depend on target-derived nerve growth factor[12] (NGF). It is generally accepted that these neurons compete for a limited supply of NGF after their axons reach their target, and that those neurons will die that do not gain access to sufficient amounts of NGF.

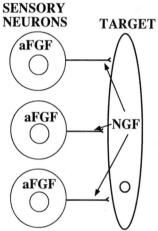
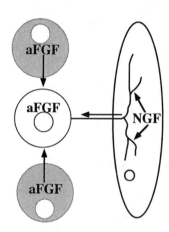

A: Pools of aFGF are intracellular at the start of
competition for NGF.

B: Cell death leads to release of
aFGF, resulting in transient
support of surviving neurons.

FIGURE 7. Hypothesis of aFGF function during PNS development. The figure shows a hypothetical model of how release of aFGF from dying sensory neurons during developmental cell death might be involved in stabilizing transiently the number of surviving neighboring neurons. During the time of stabilization, the surviving neurons might increase their field of innervation in the target, thus gaining increased access to NGF.

The data presented here raise the possibility that NGF-dependent neurons, such as sensory neurons, themselves contain an intracellular store of aFGF, which if released into the extracellular space may promote sensory neuron survival. Developmental cell death might represent the mechanism for such release, and this transient burst of aFGF then would likely result in promoting the survival of nearby vulnerable neurons for a limited time, allowing these neurons to establish denser innervation of the target, resulting in increased supply of NGF to these neurons. FIGURE 7 depicts this model of the role of aFGF during developmental cell death.

In conclusion, our observation that the mature and developing PNS contains abundant neuronal aFGF-like activity has led us to hypothesize that aFGF might have specific physiological functions. Clearly, this hypothesis need to be tested experimentally, preferentially by studying whether inhibition of aFGF activity by antibodies *in vivo* will affect regeneration or neuronal survival during development.

SUMMARY

We investigated the relative distribution of acidic and basic FGF (aFGF and bFGF) in the nervous system of the rat, using a combination of biological, biochemical, immunochemical, and immunohistochemical methods that can differentiate unambiguously between aFGF and bFGF. We found that different regions of the nervous system contained varying levels of aFGF and bFGF. In the central nervous system, bFGF was present nearly exclusively in astrocytes. Most neurons did not contain detectable amounts of bFGF immunoreactivity, with the notable exception of pyramidal cells in hippocampal area CA2. Interestingly, bFGF immunoreactivity was localized to the nucleus of both CA2 neurons and astrocytes. Astrocytes *in vitro* were also found to express bFGF, whereas cortical neurons in culture did not contain detectable amounts of bFGF. Transection of the optic nerve led to an approximately twofold increase of bFGF in the distal stump, which is consistent with the observation that bFGF is expressed by astrocytes.

Transection of rat and chicken sciatic nerve resulted in a rapid and complete disappearance of aFGF from the distal nerve stump, suggesting that aFGF is present in axons projecting through the sciatic nerve. We observed, in agreement with this notion, that cultured sensory neurons contain reasonably high levels of FGF-like bioactivity. Similar levels of activity were found in developing sciatic nerve, suggesting that neuronal aFGF might be involved in regulating the development of the peripheral nervous system.

ACKNOWLEDGMENTS

The enthusiastic technical help of Margaret Coulombe, Brain Boucher and Tracy Williams is acknowledged gratefully. We thank Dr. Gary Shipley for helpful discussions.

REFERENCES

1. GOSPODAROWICZ, D. 1974. Localization of a fibroblast growth factor and its effect alone and with hydrocortisone on 3T3 cell growth. Nature **249:** 123–127.
2. PETTMANN, B., M. WEIBEL, M. SENSENBRENNER & G. LABOURDETTE. 1985. Purification of two astroglial growth factors from bovine brain. FEBS **189:** 102–108.
3. ECCLESTON, P. A. & D. H. SILBERBERG. 1985. Fibroblast growth factor is a mitogen for oligodendrocytes in vitro. Brain Res. **353:** 315–318.
4. DAVIS, J. B. & P. STROOBANT. 1990. Platelet-derived growth factors and fibroblast growth factors are mitogens for rat Schwann cells. J. Cell Biol. **110:** 1353–1360.
5. GOSPODAROWICZ, D., S. MASSOGLIA, J. CHENG & D. K. FUJII. 1986. Effect of fibroblast growth factor and lipoproteins on the proliferation of endthelial cells derived from bovine adrenal cortex, brain cortex, and corpus luteum capillaries. J. Cell Physiol. **127:** 121–136.
6. SCHREIBER, A. B. 1985. A unique family of endothelial cell polypeptide mitogens: the antigenic and receptor cross-reactivity of bovine endothelial cell growth factor, brain-derived acidic fibroblast growth factor, and eye-derived growth factor-II. J. Cell Biol. **101:** 1623–1626.

7. BURGESS, W. H. & T. MACIAG. 1989. The heparin-binding (fibroblast) growth factor family of proteins. Annu. Rev. Biochem. **58:** 575–606.
8. MORRISON, R. S., A. SHARMA, J. DE VELLIS & R. A. BRADSHAW. 1986. Basic fibroblast growth factor supports the survival of cerebral cortical neurons in primary culture. Proc. Nat. Acad. Sci. USA **83:** 7537–7541.
9. UNSICKER, K., H. REICHERT, R. SCHMIDT, B. PETTMANN, G. LABOURDETTE & M. SENSEN-BRENNER. 1987. Astroglial and fibroblast growth factors have neurotrophic functions for cultured peripheral and central nervous system neurons. Proc. Nat. Acad. Sci. USA **84:** 5459–5463.
10. WALICKE, P., W. M. COWAN, N. UENO, A. BAIRD & R. GUILLEMIN. 1986. Fibroblast growth factor promotes survival of dissociated hippocampal neurons and enhances neurite extension. Proc. Nat. Acad. Sci. USA **83:** 3012–3016.
11. SCHUBERT, D., N. LING & A. BAIRD. 1987. Multiple influences of a heparin-binding growth factor on neuronal development. J. Cell Biol. **104:** 635–643.
12. ECKENSTEIN, F. P., F. ESCH, T. HOLBERT, R. W. BLACHER & R. NISHI. 1990. Purification and characterization of a trophic factor for embryonic peripheral neurons: comparison with fibroblast growth factors. Neuron **4:** 623–631.
13. ECKENSTEIN, F. P., G. D. SHIPLEY, & R. NISHI. 1991. Acidic and basic fibroblast growth factors in the nervous system: distribution and differential alteration of levels after injury of central versus peripheral nerve. J. Neurosci. **11:** 412–419.
14. SHIPLEY, G. D. (1986). A serum-free [^{3}H]thymidine incorporation assay for the detection of transforming growth factors. J. Tissue Culture Methods **10:** 117–123.
15. HUETTNER, J. E. & R. W. BAUGHMAN. 1986. Primary culture of identified neurons from the visual cortex of postnatal rats. J. Neurosci. **6:** 3044–3060.
16. MCCARTHY, K. D. & J. DE VELLIS 1980. Preparation of separate astroglial and oligoden-droglial cell cultures from rat cerebral tissue. J. Cell Biol. **85:** 890–902.
17. FINKLESTEIN, S. P., P. J. APOSTOLIDES, C. G. CADAY, J. PROSSER, M. F. PHILIPS & M. KLAGSBRUN. 1988. Increased basic fibroblast growth factor (bFGF) immunoreactivity at the site of focal brain wounds. Brain Res. **460:** 253–259.
18. PETTMANN, B., G. LABOURDETTE, M. WEIBEL & M. SENSENBRENNER. 1986. The brain fibroblast growth factor (FGF) is localized in neurons. Neurosci. Lett. **68:** 175–180.
19. RENKO, M., N. QURTO, T. MORIMOTO & D. B. RIFKIN. 1990. Nuclear and cytoplasmic localization of different basic fibroblast growth factor species. J. Cell Physiol. **44:** 108–114.
20. BALDIN, V., A. ROMAN, I. BOSC-BIERNE, F. AMALRIC & G. BOUCHE. 1990. Translocation of bFGF to the nucleus is G1 phase cell cycle specific in bovine aortic endothelial cells. EMBO J. **9:** 1511–1517.
21. IMAMURA, T., et al. 1990. Recovery of mitogenic activity of a growth factor mutant with a nuclear translocation sequence. Science **249:** 1567–1570.
22. HATTEN, M. E., M. LYNCH, R. E. RYDEL, J. SANCHEZ, J. JOSEPH-SILVERSTEIN, D. MOSCATELLI & D. B. RIFKIN. 1988. In vitro neurite extension by granule neurons is dependent upon astroglial-derived fibroblast growth factor. Dev. Biol. **125:** 280–289.
23. ROGELJ, S., M. KLAGSBRUN, R. ATZMON, M. KUROKAWA, A. HAIMOWITZ, Z. FUKS & I. VLODAVSKY. 1989. Basic fibroblast growth factor is an extracellular matrix component required for supporting the proliferation of vascular endothelial cells and the differen-tiation of PC12 cells. J. Cell Biol. **109:** 823–831.
24. GORDON, P. B., H. U. CHOI, G. CONN, A. AHMED, B. EHRMAN, L. ROSENBERG & V. B. HATCHER. 1989. Extracellular matrix heparan sulfate proteoglycans modulate the mitogenic capacity of acidic fibroblast growth factor. J. Cell. Physiol. **140:** 584–592.
25. JANET, T., C. GROTHE, B. PETTMANN, K. UNSICKER & M. SENSENBRENNER. 1988. Immuno-cytochemical demonstration of fibroblast growth factor in cultured chick and rat neurons. J. Neurosci. Res. **19:** 195–201.
26. RATNER, N., D. M. HONG, M. A. LIEBERMAN, R. P. BUNGE & L. GLASER. 1988. The neuronal cell-surface molecule mitogenic for Schwann cells is a heparin-binding protein. Proc. Nat. Acad. Sci. USA **85:** 6992–6996.

Functional Domains of Basic Fibroblast Growth Factor[a]

Possible Role of Asp-Gly-Arg Sequences in the Mitogenic Activity of bFGF

M. PRESTA, M. RUSNATI, C. URBINATI, M. STATUTO,
AND G. RAGNOTTI

Unit of General Pathology and Immunology
Department of Biomedical Sciences and Biotechnology
School of Medicine
University of Brescia
via Valsabbina 19
I-25123 Brescia, Italy

Examination of the primary sequence of basic fibroblast growth factor (bFGF) shows the presence of the amino acid sequence Asp-Gly-Arg (DGR), which is repeated twice at positions bFGF (46–48) and bFGF (88–90) of the 155 amino acid form of human bFGF.[1] Both DGR sequences are present in hydrophilic regions of bFGF (FIGURE 1), suggesting that they are localized on the exposed surface of the molecule. DGR sequence represents the inverse of the RGD sequence of the cell-binding domain of fibronectin and of various other cell-adhesive proteins.[2] The RGD sequence is recognized by integrin receptors, and RGD- and DGR-containing peptides can compete with adhesive proteins for integrin interaction.[2–4]

Here we have addressed the question of whether the DGR sequences of bFGF are involved in the biological activity exerted by this growth factor in endothelial cells. To this purpose we have investigated the capacity of two synthetic fragments bFGF (38–61) and bFGF (82–101), both containing the DGR sequence, and of synthetic RGD-containing tetra- and heptapeptides to exert an agonist and/or antagonist bFGF activity in transformed fetal bovine aortic endothelial GM 7373 cells.

EXPERIMENTAL DATA

Agonist Activity

Recombinant bFGF induces cell proliferation and plasminogen activator (PA) production in GM 7373 cells.[5] Peptides bFGF (38–61) and bFGF (82–101) (indicated as peptides A and C in FIGURE 1) induce a mitogenic response in these cells, without affecting significantly the levels of cell-associated PA activity (FIGURE 2A). At 3 µg/ml the two peptides cause a maximal increase in cell number, which corresponds to 30% and 75% respectively of the mitogenic activity exerted by 3 ng/ml bFGF. The

[a]This work was supported by a grant from Associazione Italiana per la Ricerca sul Cancro and a grant from CNR (Progetto Finalizzato Biotechnologie e Biostrumentazioni) to MP, and by grant n. 89.02759.04 from CNR to GR.

two peptides share with bFGF the same protein kinase C (PKC) dependent
intracellular signaling pathway.[5] Indeed, the mitogenic activity of bFGF and of the
two peptides is specifically inhibited by the PKC inhibitor H-7 (data not shown).

No mitogenic or PA-inducing activity is exerted by other bFGF fragments
(indicated in FIGURE 1) or by the RGD-containing tetra- and heptapeptides listed in
TABLE 1 (data not shown).

```
1                    20                   40    ***
MAAGSITTLPALPEDGGSG  AFPPGHFKDPKRLYCKNGGF  FLRIHPDGRVDGVREKSDPH

60                   80        ***         100
IKLQLQAEERGVVSIKGVCA NRYLAMKEDGRLLASKCVTD  ECFFFERLESNNYNTYRSRK

120                  140
YTSWYVALKRTGQYKLGSKT GPGQKAILFLPMSAKS
```

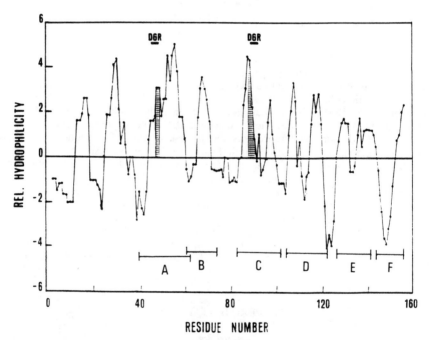

RESIDUE NUMBER

FIGURE 1. Amino acid sequence and hydrophilicity analysis of human bFGF. Amino acid
sequences corresponding to bFGF (38–61) and bFGF (82–101) (peptides A and C) are
underlined and DGR sequences are evidenced by asterisks and shaded areas.

Antagonist Activity

Peptides bFGF (38–61) and bFGF (82–101) partially inhibit the mitogenic
activity exerted by bFGF in GM 7373 cells, without affecting its PA-inducing capacity
(FIGURE 2B). The two peptides are specific inhibitors of bFGF mitogenic activity
since they do not influence cell proliferation induced by fetal calf serum (FCS),
epidermal growth factor (EGF), phorbol ester (TPA), or 1,2 diacylglycerol (diC8)

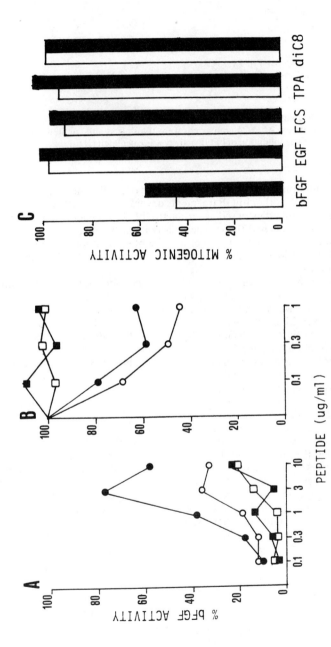

FIGURE 2. Effect of bFGF fragments on cell proliferation and PA production in GM 7373 cells. Cultures were incubated in 0.4% FCS with different amounts of bFGF (38–61) (open symbols) or of bFGF (82–101) (filled symbols) in the absence (A) or in the presence (B) of 3 ng/ml bFGF. After 24 hours, cells were trypsinized and counted (circles) or evaluated for cell-associated PA activity (squares). For further details on cell proliferation and PA activity assays see Reference 5. In panel C, cultures were incubated with 10% FCS or with 0.4% FCS containing 3 ng/ml bFGF, 30 ng/ml EGF, 30 ng/ml TPA, or 5 µg/ml diC8. Cultures received also 0.3 µg/ml bFGF (38–61) (open bar) or 0.3 µg/ml bFGF (82–101) (black bar). After 24 hours, cells were trypsinized and counted. Cell proliferation is expressed as percent of the proliferation exerted by each mitogen in the absence of the peptide.

TABLE 1. bFGF Antagonist Activity of RGD-Containing Peptides

Peptide Sequence	ID_{50} $(nM)^a$
Arg-Gly-Asp-Ser	5
Gly-Arg-Gly-Asp	2
Ser-Asp-Gly-Arg	5
Gly-Arg-Gly-Asp-Ser-Pro-Lys	6
Gly-Arg-Ala-Asp-Ser-Pro-Lys	No inhibition
Arg-Phe-Asp-Ser	> 60
Arg-Gly-Glu-Ser	> 60

^{a}GM 7373 cells were incubated in 0.4% FCS with 3 ng/ml bFGF in the presence of increasing concentrations of the peptides. After 24 hours, cells were trypsinized and counted. Then the dose of peptide causing a 50% inhibition of bFGF activity was calculated (ID_{50}).

(FIGURE 2C). No inhibition of the mitogenic activity and of the PA-inducing capacity of bFGF was exerted by the other bFGF fragments devoid of an agonist activity.

Also, monospecific antibodies raised against active bFGF fragments (anti-A and anti-C antibodies) fully quenched the mitogenic activity of the whole bFGF molecule, while antibodies raised against the other inactive bFGF peptides were inaffective. However, all the antibodies caused a 40–60% inhibition of the PA-inducing activity of bFGF (FIGURE 3).

To assess whether the biological activity exerted by peptides bFGF (38–61) and bFGF (82–101) is related to their DGR sequences, we have evaluated the effect of the tetrapeptide RGDS, which contains the inverse of the DGR sequence, on the mitogenic activity exerted by the whole bFGF molecule or by the two bFGF fragments. As shown in FIGURE 4, RGDS fully inhibits cell proliferation induced by these molecules, without affecting cell proliferation induced by acidic FGF (aFGF), FCS, TPA, diC8, or EGF. Moreover, the nature of the inhibition exerted by RGDS is competitive for bFGF. The ID_{50} of RGDS is in fact increased from 1 to 60 nM when bFGF concentration varies from 1 to 10 ng/ml.

To investigate the structural requirements for the inhibition of bFGF activity, a library of RGD-containing peptides was tested for bFGF-inhibitory activity. As shown in TABLE 1, a similar inhibitory activity is exerted by the tetrapeptides SDGR, RGDS, and GRGD and by the heptapeptide GRGDSPK, indicating that the orientation of the DGR sequence and the nature of the flanking amino acids do not affect the inhibitory activity of the peptide. However, the specificity of the inhibition of bFGF mitogenic activity is apparent from the lack of inhibitory activity of the heptapeptide containing a conservative substitution of the glycine residue with an alanine residue (GRADSPK) and from the highly reduced effectiveness (ID_{50} > 60 nM) of the tetrapeptides containing either a substitution of the glycine residue with a phenylalanine residue (RFDS) or of the aspartic acid residue with a glutamic acid residue (RGES).

Receptor Binding

The data indicate that the two DGR sequences present in the bFGF molecule are involved in the mitogenic activity exerted by this growth factor in cultured endothelial cells. These findings prompted us to evaluate the effect of bFGF fragments, of their corresponding antibodies, and of RGD-containing tetra- and heptapeptides on the binding of ^{125}I-bFGF to GM 7373 cells.

No inhibition of the binding of ^{125}I-bFGF to low- and high-affinity binding sites[6] was observed when cells were incubated for 2 hours at 4°C with 3 ng/ml of ^{125}I-bFGF in the presence of 30 μg/ml of biologically active or inactive RGD-containing peptides. Also, no modification of the amount of ^{125}I-bFGF bound to low- and high-affinity bFGF receptors was observed in cells incubated in the presence of active bFGF fragments at doses up to 500 μg/ml. Accordingly, preincubation of ^{125}I-bFGF with anti-bFGF (38–61) and/or anti-bFGF (82–101) antibodies performed in the same experimental conditions that fully quench the mitogenic activity of bFGF does not affect the binding of ^{125}I-bFGF to high- and low-affinity receptors. The binding was instead fully inhibited by preincubation of ^{125}I-bFGF with polyclonal anti-bFGF antibodies.

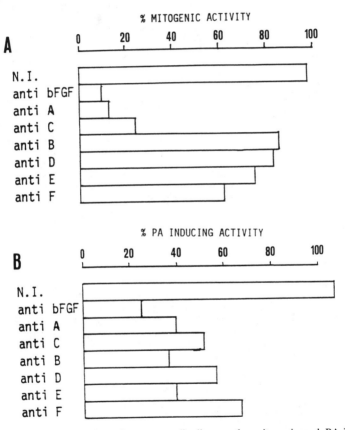

FIGURE 3. Effect of anti-bFGF fragment antibodies on the mitogenic and PA-inducing activity of bFGF. bFGF was incubated at 400 ng/ml with nonimmune rabbit immunoglobulin G (N.I.), polyclonal affinity-purified antihuman placental bFGF antibodies (anti bFGF), or with monospecific antibodies raised against bFGF (38–61) (anti A), bFGF (82–101) (anti C), bFGF (61–73) (anti B), bFGF (105–119) (anti D), bFGF (126–138) (anti E), or bFGF (141–155) (anti F), all at 15 μg/ml. Then, samples were assayed for mitogenic activity (A) and PA-inducing activity (B) on GM 7373 cells at the final concentration of 3 ng/ml bFGF. Data are expressed as percent of the activity exerted by 3 ng/ml of bFGF incubated in the absence of antibodies.

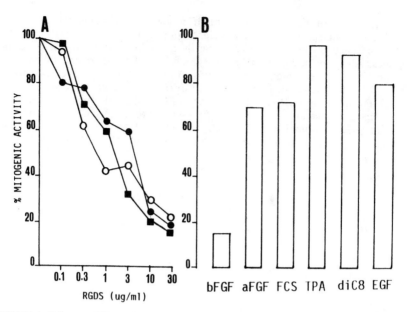

FIGURE 4. Effect of RGDS on cell proliferation in GM 7373 cells. **A:** Cells were incubated with 3 ng/ml bFGF (squares), 3 μg/ml bFGF (38–61) (open circles), or 3 μg/ml bFGF (82–101) (filled circles) in the presence of increasing concentrations of RGDS. **B:** Cells were incubated with 3 ng/ml bFGF, 3 ng/ml aFGF plus 10 μg/ml heparin, 10% FCS, 30 ng/ml TPA, 5 μg/ml diC8, or 100 ng/ml EGF in the presence of 30 μg/ml RGDS. Cells were trypsinized and counted after 24 hours of incubation. Cell proliferation is expressed as percent of the proliferation exerted by each mitogen in the absence of the peptide.

CONCLUSIONS

The results indicate that amino acid residues 38–61 and 82–101 of the 155 amino acid form of bFGF are involved in the mitogenic activity exerted by this growth factor in endothelial cells. The experimental evidence is twofold: synthetic peptides bFGF (38–61) and bFGF (82–101) partially agonize and antagonize the mitogenic activity of bFGF in GM 7373 cells; monospecific antibodies raised against these peptides completely quench the mitogenic activity of recombinant bFGF.

Peptides bFGF (38–61) and bFGF (82–101) contain a DGR sequence. This sequence represents the inverted sequence of the widespread recognition signal RGD. The RGD sequence is specifically recognized by the family of the integrin receptors, molecules that bind a variety of adhesive proteins like fibronectin, fibrinogen, laminin, vitronectin, and collagen.[2] Short peptides containing the RGD or the DGR sequence have been shown to inhibit the interaction of integrins with their ligands.[3,4] Here we have demonstrated that short RGD-containing peptides inhibit the mitogenic activity of bFGF, as well as that exerted by the two active bFGF fragments. Experimental evidence demonstrates that the inhibition is specific: (a) no effect on cell morphology was exerted by the peptides at concentrations that fully inhibit the mitogenic activity of bFGF; (b) the nature of the inhibition seems to be competitive for bFGF; (c) RGD-containing peptides inhibit the mitogenic activity of bFGF and of active bFGF fragments with the same potency. They are instead ineffective as inhibitors of aFGF, a heparin-binding mitogen which shares with bFGF the same cell membrane receptor,[7,8] shows a 55% amino acid sequence homology with bFGF, but does not contain DGR sequences[9]; and (d) RGD-containing peptides do not affect cell proliferation induced by serum, TPA, diC8, and EGF. These data strongly suggest that the DGR sequence, repeated twice in hydrophilic regions of the bFGF molecule at positions bFGF (46–48) and bFGF (88–90), and present both in peptide bFGF (38–61) and in peptide bFGF (82–101), may play a role in mediating the mitogenic activity of bFGF.

It should be pointed out that this role is not exerted through an interaction of DGR-containing bFGF sequences with bFGF cell membrane receptor. Indeed, active bFGF fragments, their corresponding antibodies, and RGD-containing peptides do not affect the interaction of ^{125}I-bFGF with its low- and high-affinity binding sites. Moreover, as stated above, RGD-containing peptides inhibit very poorly the mitogenic activity of aFGF, a molecule that shares with bFGF the same plasma membrane receptor.[7,8] Also, the putative receptor-binding domain of bFGF has been tentatively identified as the bFGF (102–129) sequence.[10]

On this basis, it is tempting to hypothesize that bFGF may interact via its DGR

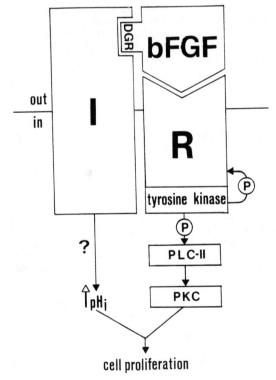

FIGURE 5. Postulated mechanism of interaction of bFGF with endothelial cell membrane. bFGF interacts with its cell membrane receptor (R) endowed with tyrosine kinase activity.[8] This leads to tyrosine phosphorylation of phospholipase C-II (PLC-II).[18] Activation of the enzyme causes degradation of phosphatidyl inositols and liberation of diacylglycerol, which activates PKC.[5] Moreover, DGR sequences of bFGF interact with an integrinlike molecule (I), possibly causing an increase in intracellular pH.[19] Both PKC activation and cytoplasmic alkalinization are necessary to induce endothelial cell proliferation.

sequences with an integrinlike molecule, and that this interaction may affect the mitogenic activity exerted by this growth factor in endothelial cells, as proposed in FIGURE 5. The structural requirements for the inhibition of bFGF activity by RGD-containing peptides mimics that observed for ligand/integrin interaction.[2-4] Also, bFGF promotes substrate adhesion of PC12 cells and of endothelial cells.[10,11] Preliminary experiments have shown that peptides bFGF (38–61) and bFGF (82–101), as well as bFGF, promote the adhesion of GM 7373 cells to nonadhesive plastic and that this capacity can be inhibited by RGD-containing tetra- and heptapeptides (M. Rusnati, unpublished observations). Moreover, extracellular matrix components have been shown to affect the mitogenic activity of bFGF in endothelial cells by modulating cell shape and intracellular pH.[12-15] It is interesting to note that experiments performed with a synthetic fragment of EGF have also revealed a possible functional interrelationship between this growth factor and the cell-adhesive protein laminin.[16] These findings bring new insights on the mechanism(s) that are responsible for the well-known modulation of the activity of soluble growth factors by extracellular matrix.[12,13,17]

DGR sequences do not appear to be involved in the PA-inducing activity of bFGF. Peptides bFGF (38–61) and bFGF (82–101) do not stimulate significantly PA production in GM 7373 cells. Also, bFGF fragments and RGD-containing tetra- and heptapeptides do not inhibit the PA-inducing capacity of recombinant bFGF. These findings suggest the possibility of dissociating at the structural level the mitogenic activity of bFGF from its PA-inducing capacity. The biological properties of a bFGF mutant recently characterized in our laboratory are in keeping with this hypothesis (see A. Isacchi et al., this volume).

REFERENCES

1. SOMMER, A., M. T. BREWER, R. C. THOMPSON, D. MOSCATELLI, M. PRESTA & D. B. RIFKIN. 1987. Biochem. Biophys. Res. Commun. 144: 543–550.
2. RUOSLATHI, E. & M. D. PIERSHBACHER. 1986. Cell 44: 517–518.
3. RUOSLATHI, E. & M. D. PIERSHBACHER. 1987. Science 238: 493–497.
4. HYNES, R. O. 1987. Cell 48: 549–554.
5. PRESTA, M., J. A. M. MAIER & G. RAGNOTTI. 1989. J. Cell Biol. 109: 1877–1884.
6. MOSCATELLI, D. 1987. J. Cell. Physiol. 131: 123–130.
7. NEUFELD, G. & D. GOSPODAROWICZ. 1986. J. Biol. Chem. 261: 5631–5637.
8. RUTA, M., W. BURGESS, D. GIVOL, J. EPSTEIN, N. NEIGER, J. KAPLOW, G. CRUMLEY, C. DIONNE, M. JAYE & J. SCHLESSINGER. 1989. Proc. Nat. Acad. Sci. USA 86: 8722–8726.
9. THOMAS, K. A. & G. GIMENEZ-GALLEGO. 1986. Trends Biochem. Sci. 11: 81–84.
10. BAIRD, A., D. SCHUBERT, N. LING & R. GUILLEMIN. 1988. Proc. Nat. Acad. Sci. USA 85: 2324–2328.
11. SCHUBERT, D., N. LING & A. BAIRD. 1987. J. Cell Biol. 104: 635–643.
12. INGBER, D. E., J. A. MADRI & J. FOLKMAN. 1987. In Vitro Cell. Dev. Biol. 23: 387–394.
13. INGBER, D. E. & J. FOLKMAN. 1989. J. Cell Biol. 109: 317–330.
14. INGBER, D. E. 1990. Proc. Nat. Acad. Sci. USA 87: 3579–3583.
15. INGBER, D. E., D. PRUSTY, J. V. FRANGIONI, E. J. CRAGOE, JR., C. LECHENE & M. A. SCHWARTZ. 1990. J. Cell Biol. 110: 1803–1811.
16. EPSTEIN, D. A., Y. V. MARSH, B. B. SCHRYVER & P. J. BERTICS. 1989. J. Cell. Physiol. 141: 420–430.
17. GOSPODAROWICZ, D., G. GREENBURG & R. C. BIRDWELL. 1978. Cancer Res. 38: 4155–4171.
18. BURGESS, W. H., C. A. DIONNE, J. KAPLOW, R. MUDD, R. FRIESEL, A. ZILBERSTEIN, J. SCHLESSINGER & M. JAYE. 1990. Mol. Cell. Biol. 10: 4770–4777.
19. SCHWARTZ, M. A., G. BOTH & C. LECHENE. 1989. Proc. Nat. Acad. Sci. USA 86: 4525–4529.

A Mutant of Basic Fibroblast Growth Factor that Has Lost the Ability to Stimulate Plasminogen Activator Synthesis in Endothelial Cells

A. ISACCHI,[a] L. BERGONZONI,[a] M. STATUTO,[b]
M. RUSNATI,[b] R. CHIESA,[b] P. CACCIA,[a] P. SARMIENTOS,[a]
M. PRESTA,[b] AND G. RAGNOTTI[b]

[a]Department of Biotechnology
Farmitalia Carlo Erba
Viale Bezzi, 24
20146 Milano, Italy

[b]Department of Biomedical Sciences and Biotechnology
School of Medicine
University of Brescia
25124 Brescia, Italy

INTRODUCTION

Basic fibroblast growth factor (bFGF) is a well-characterized polypeptide with mitogenic activity in several cell types. In endothelial cells, bFGF stimulates proliferation, migration, and synthesis of plasminogen activators (PA).[1,2]

bFGF has been demonstrated to interact with cell-associated tyrosine kinase receptors,[3] but little is known on the mechanism that mediates its intracellular action. Two regions, corresponding to residues 33–77 and 102–129 in the primary structure of the 155 amino acid bFGF molecule, have been proposed to be involved in the receptor-binding and mitogenic activity of this growth factor.[4] In addition, we have recently observed that two synthetic peptides, overlapping amino acid residues 38–61 and 82–101, exert a partial agonist and antagonist activity in mitogenic assays on endothelial cells.[5] However, these peptides do not affect PA production in bFGF-treated and untreated endothelial cells, suggesting that the PA-inducing activity of bFGF may depend upon a functional domain that differs from those involved in the mitogenic activity of this growth factor.

To investigate the structure-function relationship of bFGF, we have designed a number of mutants on the basis of homology to related growth factors belonging to the FGF family. In these new molecules, highly conserved regions, which could possibly be endowed with functional importance, were deleted or modified in single amino acid residues.

Here we describe a recombinant deletion mutant of the 155 amino acid form of human bFGF (M1-bFGF) lacking amino acid residues 27–32 (Lys-Asp-Pro-Lys-Arg-Leu). This mutant interacts with endothelial cell membrane bFGF receptor in a manner undistinguishable from the wild-type bFGF; it also appears to have the same capacity of bFGF to stimulate early signal transduction events. Moreover, the deletion mutant retains the capacity to induce cell proliferation when added to cultured endothelial cells and to NIH 3T3 fibroblasts. However, M1-bFGF has lost the ability to stimulate PA synthesis in endothelial cells, suggesting the possibility that the PA-induction activity follows a pathway independent from the receptor-mediated proliferation activity.

RESULTS

Biochemical Characterization of the M1-bFGF Mutant

M1-bFGF, a deletion mutant of the 155 amino acid form of human bFGF lacking amino acid residues 27–32, was constructed by site-directed mutagenesis. A schematic representation of the mutant is shown in FIGURE 1. The recombinant protein was efficiently expressed in *Escherichia coli* and purified from the soluble fraction of cell extracts by heparin Sepharose chromatography. The purified M1-bFGF molecule was more than 95% pure, as evaluated by sodium dodecyl sulfate-polyacrylamide gel electrophoresis (SDS-PAGE) and high-performance liquid chromatography (HPLC) analysis (data not shown). Protein concentration of M1-bFGF and of the recombinant wild-type bFGF used in all experiments was determined by amino acid analysis.

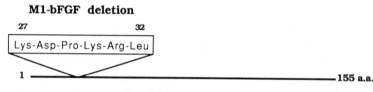

bFGF sequence

FIGURE 1. Schematic representation of the M1-bFGF mutant. Amino acid residues 27–32, which were deleted in the M1-bFGF mutant, are indicated in the box.

A detailed analysis of the heparin-binding capacity of the recombinant protein on heparin affinity HPLC showed that M1-bFGF has the same affinity for heparin as the wild-type molecule (data not shown).

To prevent possible degradation of the bFGF and M1-bFGF molecules during incubation with cells, all the following biological and receptor-binding assays were performed in the presence of soluble heparin, which is known to stabilize the tertiary structure of bFGF.[6]

Interaction with High-Affinity Receptors

We compared the ability of recombinant bFGF and M1-bFGF to bind to high-affinity binding sites in GM7373 cells. Cells were incubated for 2 hours at 4°C with a constant amount of ^{125}I-bFGF in the presence of increasing concentrations of either unlabeled bFGF or unlabeled M1-bFGF, both maintained in the presence of an equimolar concentration of soluble heparin. Under these conditions, M1-bFGF was as effective as bFGF in competing with ^{125}I-bFGF for its high-affinity binding sites (FIGURE 2A). Moreover, chemical cross-linking studies confirmed that ^{125}I-bFGF and ^{125}I-M1-bFGF bind to the same M_r 140,000 plasma membrane receptor (data not shown).

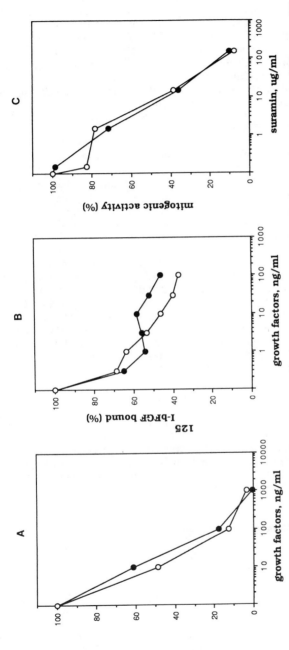

FIGURE 2. A: Competition for binding of ^{125}I-bFGF to GM7373 cells by bFGF and M1-bFGF. GM7373 cells were incubated for 2 hours at 4°C with ^{125}I-bFGF (8 ng/ml) and increasing concentrations of unlabeled bFGF (filled circles) or M1-bFGF (open circles). The incubation was performed in the presence of 10 μg/ml heparin. Then binding of ^{125}I-bFGF to bFGF plasma membrane receptor was quantitated and expressed as percent of the binding in the absence of unlabeled molecules. **B:** Down regulation of bFGF plasma membrane receptor. GM7373 cells were incubated at 37°C with increasing concentrations of bFGF (filled circles) or of M1-bFGF (open circles). After 2 hours, cells were washed with 2 M NaCl buffered at pH 4.0, to remove bound bFGF, incubated for 2 hours at 4°C with ^{125}I-bFGF (8 ng/ml), and assayed for bFGF high-affinity binding sites. **C:** Dose-response curve to suramin. GM7373 cells were incubated in 0.4% FCS with 30 ng/ml of bFGF (filled circles) or M1-bFGF (open circles) in the presence of increasing concentrations of suramin. After 24 hours, cells were trypsinized and counted. Data are expressed as percent of the mitogenic activity observed in the absence of suramin.

Binding of bFGF to its membrane receptors causes a down regulation of the number of high-affinity binding sites.[7] Accordingly, a 2-hour incubation of GM7373 cells causes a dose-dependent disappearance of ^{125}I-bFGF high-affinity binding sites, with similar dose-response curves for bFGF and for M1-bFGF (FIGURE 2B). Moreover, the kinetics of down regulation of ^{125}I-bFGF high-affinity binding sites after bFGF or M1-bFGF treatment were found to be identical (data not shown).

Finally, suramin, a molecule known to dissociate bFGF from its high-affinity binding sites,[8] inhibits the mitogenic activity of bFGF and M1-bFGF in a similar manner (FIGURE 2C). The effect was dose dependent, with an ED_{50} equal to 5 $\mu g/ml$. These results are consistent with a similar interaction of bFGF and M1-bFGF with cell surface receptors.

Signal Transduction

The mechanisms of bFGF signal transduction are not fully understood but may include activation of protein tyrosine kinase activity, phosphorylation of phospholipase C-γ with activation of protein kinase C, and stimulation of early gene transcription. We have compared the two recombinant proteins bFGF and M1-bFGF for their capacity to elicit the early responses that follow their interaction with cell surface receptors.

The same pattern of tyrosine phosphorylation is observed when bFGF or M1-bFGF is added for 15 minutes at 37°C to NIH 3T3 cells in serum-free conditions (FIGURE 3A, lanes 2 and 3). Tyrosine-phosphorylated proteins include the typical M_r 90,000 substratum and an M_r 140,000 molecule, possibly corresponding to the activated bFGF receptor or to phosphorylated protein phospholipase C-γ.

Evidence is increasing that activation of protein kinase C is part of the mitogenic response to bFGF.[9-11] Indeed, the protein kinase inhibitor 1-(5 isoquinolynsulfonyl)-2-methylpiperazine (H-7) inhibits the mitogenic response to bFGF, with an ID_{50} equal to 6 μM. Accordingly, an equal response to H-7 was found for M1-bFGF (FIGURE 3B).

To determine the capacity of bFGF and M1-bFGF to stimulate protooncogene transcription, we analyzed expression of the *c-fos* gene. Quiescent GM7373 cells were treated with 30 ng/ml of bFGF or M1-bFGF for different times. Total RNA was prepared and analyzed by Northern blot hybridization. As shown in FIGURE 3C, M1-bFGF and bFGF increased *c-fos* mRNA levels to a similar degree, the maximal stimulation being attained after 1 hour of induction (lanes 3 and 5). Dehydropholate reductase (DHFR) mRNA levels were also determined by subsequent hybridization to verify the relative amount of RNA per lane (data not shown).

Mitogenic and Chemiotactic Activity

The mitogenic activity of bFGF and M1-bFGF was evaluated as their capacity to stimulate DNA synthesis on 3T3 cells and to support the proliferation of ABAE cells.

As shown in FIGURE 4A, M1-bFGF exerts a mitogenic activity in NIH 3T3 cells. The effect is dose dependent, with an ED_{50} equal to 0.3–1 ng/ml, very similar to that observed for bFGF.

In the second mitogenesis assay, the two molecules were compared for their ability to support the proliferation of ABAE cells. The results, shown in FIGURE 4B, are consistent with those obtained on 3T3 cells and suggest that bFGF and M1-bFGF have a similar mitogenic effect on endothelial cells.

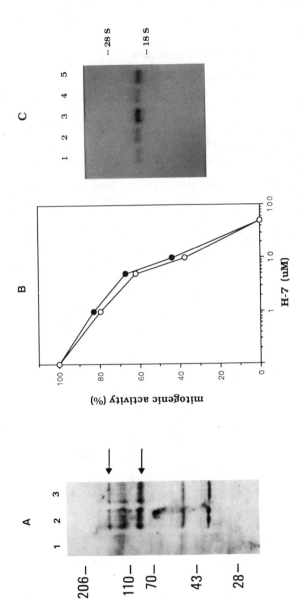

FIGURE 3. A: Tyrosine phosphorylation in intact NIH 3T3 cells. NIH 3T3 cells were incubated for 15 minutes in serum-free medium with no addition (lane 1), M1-bFGF 20 ng/ml (lane 2), or bFGF 20 ng/ml (lane 3). Then cells were lysed in reducing sample buffer, run on a 5–12% SDS-PAGE, immunoblotted with antiphosphotyrosine antibodies, and immunocomplexes were visualized by incubation with ^{125}I-protein A. The arrowheads mark the bands induced by both bFGF and M1-bFGF. **B:** Effect of H-7 on GM7373 cell proliferation. Cultures were incubated in 0.4% FCS with 30 ng/ml bFGF (filled circles) or M1-bFGF (open circles) in the presence of increasing concentrations of H-7. After 24 hours, cells were trypsinized and counted. Data are expressed as percent of the mitogenic activity exerted in the absence of H-7. **C:** Induction of c-fos protooncogene expression. Quiescent GM7373 cells were stimulated with 30 ng/ml of M1-bFGF (lanes 2 and 3) or bFGF (lanes 4 and 5). After 30 minutes (lanes 2 and 4) or 1 hour (lanes 3 and 5), RNA was isolated and Northern blot analysis was performed according to standard procedures, using a ScaI-NcoI fragment from pc-fos as a probe. Lane 1 represents RNA from unstimulated cells.

Finally, we have compared bFGF and M1-bFGF for their ability to stimulate migration of ABAE cells in a Boyden chamber assay. As reported in FIGURE 5, no difference could be evidenced in the chemiotactic activity exerted by bFGF or M1-bFGF.

Plasminogen-Activator Induction

GM7373 cells respond to a 24-hour treatment with bFGF with a 10- to 20-fold increase in PA activity,[10] which can be identified as urokinase-type PA by immunological methods (data not shown). To better characterize the biological activity of M1-bFGF, we have evaluated its capacity to stimulate PA production in different endothelial cell types. As shown in FIGURE 6A, in all cell types examined, M1-bFGF was dramatically less efficient than bFGF in stimulating PA activity. Dose-response curves on GM7373 cells have shown that M1-bFGF is about 100 times less active than bFGF (data not shown).

To assess the involvement of changes in mRNA expression in the different behavior of bFGF and M1-bFGF, the steady-state levels of urokinase mRNA were analyzed by Northern blot. FIGURE 6B illustrates a time course of urokinase mRNA induction in GM7373 cells after the addition of 30 ng/ml of bFGF or M1-bFGF. Urokinase mRNA was increased as early as 2 hours after the addition of bFGF and reached a peak at 6 hours, when a 10-fold increase over control levels was observed. This stimulation was not observed after M1-bFGF addition. These data are consis-

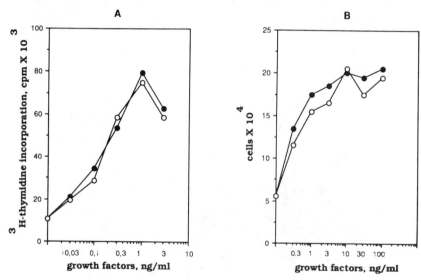

FIGURE 4. A: DNA synthesis in NIH 3T3 cells. Quiescent cultures of NIH 3T3 cells were incubated in fresh medium containing ^{3}H-thymidine and increasing concentrations of bFGF (filled circles) or M1-bFGF (open circles). After 24 hours, the amount of radioactivity incorporated into the trichloroacetic acid–precipitable material was measured. **B:** Proliferation activity on ABAE cells. ABAE cells were seeded in 35-mm dishes at 2000 cells/dish. After 16 hours, cells were added with bFGF (filled circles) or M1-bFGF (open circles) in the presence of 10% FCS. Then medium was changed every other day. Cells were trypsinized and counted after 6 days.

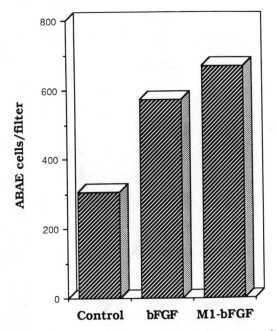

FIGURE 5. Chemiotactic activity on ABAE cells. The experiment was performed in Boyden blind well chambers. Medium containing 1 ng/ml bFGF or M1-bFGF was placed in the lower chamber. 5×10^4 ABAE cells were placed in the upper chamber. After 4 hours incubation, the cells on the lower surface of the collagen-coated filter separating the two chambers were counted.

tent with those obtained at the protein level and indicate that the deletion mutant M1-bFGF has a dramatically reduced capacity to stimulate PA production in endothelial cells.

DISCUSSION

In order to understand the action of bFGF, effort has been focused on the elucidation of the signal transduction pathways underlying its activity. Our approach is based on the construction of recombinant mutants of the wild-type molecule, which can be used to identify functional domains in the structure of bFGF involved in specific biological functions.

Here we describe the biological features of M1-bFGF, a deletion mutant of bFGF lacking amino acid residues 27–32 of the 155 amino acid primary structure. This mutant, which can be recovered as a soluble protein from *E. coli* cell extracts, retains unaltered affinity for heparin, suggesting that the deleted region, although representing a cluster of positively charged residues, is not involved in bFGF interaction with this polysaccharide.

Our results indicate that M1-bFGF interacts with bFGF receptors on endothelial cells in a manner indistinguishable from that of the wild-type protein, as judged by its ability to bind to and to down regulate the high-affinity binding sites for bFGF.

Early events in signal transduction of bFGF include stimulation of tyrosine

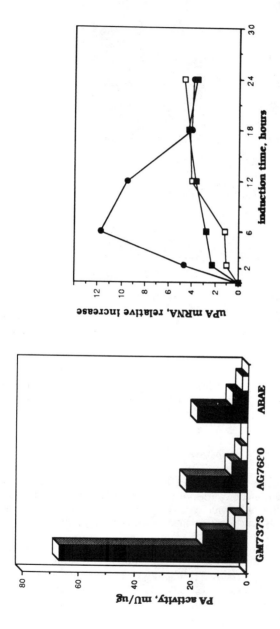

FIGURE 6. A: PA-inducing activity in different endothelial cell lines. FBAE GM7373 cells, FBAE AG 7680 cells, and ABAE cells were incubated in the presence of 30 ng/ml of bFGF (black), or of M1-bFGF (hatched). White columns represent unstimulated cells. After 24 hours, cells were extracted and assayed for cell-associated PA activity. B: Time course of induction of urokinase mRNA in GM7373 cells. Twenty-five micrograms of total RNA from cultures of quiescent GM7373 cells untreated (circles) or stimulated with 30 ng/ml of bFGF (filled squares) or M1-bFGF (open squares) were analyzed by Northern blot hybridization with a urokinase probe. The relative amount of RNA loaded per lane was verified by subsequent hybridization to a dihydrofolate reductase probe. The blots were scanned, and the intensities of the bands are presented relative to the lowest value of mRNA from untreated cells (time 2 hours).

kinase activity and phosphorylation of specific proteins, including phospholipase C-γ.[13] The same phosphorylation pattern was observed after the addition of bFGF or M1-bFGF to NIH 3T3 cells. In addition, the two molecules have an equal capacity to stimulate c-fos expression in endothelial cells.

When bFGF and M1-bFGF were compared for their mitogenic activity, no difference could be evidenced in their ability to stimulate DNA synthesis on 3T3 cells and to sustain proliferation of ABAE endothelial cells. However, M1-bFGF is at least 100 times less potent than bFGF in stimulating plasminogen activator production in endothelial cells, as shown both at protein and at mRNA levels.

bFGF has to remain active in the extracellular environment for several hours in order to induce endothelial cells to proliferate and to produce PA.[12] Thus, DNA synthesis and PA production represent two late events that have been dissociated at the structural level in the M1-bFGF molecule. These data suggest that two different pathways mediate these two responses to bFGF. Indeed, we have shown that the mitogenic and the PA-inducing activity of bFGF in endothelial cells can be pharmacologically dissociated by inhibitors of protein kinase C which block the mitogenic response to bFGF but do not alter its induction of PA activity.[10]

Several hypotheses can be made to explain the inability of M1-bFGF to stimulate PA production. The amino acid sequence deleted in M1-bFGF (Lys-Asp-Pro-Lys-Arg-Leu) is similar to the aFGF Lys-Lys-Pro-Lys-Leu-Leu sequence, corresponding to amino acids 23–28, which has been proposed to be involved in the nuclear targeting and biological activity of aFGF.[14] Thus, it is tempting to hypothesize that amino acid residues 27–32 represent a functional domain involved in the interaction of bFGF with the cell nucleus and that this interaction is required for the induction of the expression of the uPA gene in endothelial cells. Experiments are in progress to assess this hypothesis.

REFERENCES

1. MOSCATELLI, D., M. PRESTA & D. B. RIFKIN. 1986. Proc. Nat. Acad. Sci. USA **83:** 2091–2095.
2. PRESTA, M., D. MOSCATELLI, J. JOSEPH-SILVERSTEIN & D. B. RIFKIN. 1986. Mol. Cell. Biol. **6:** 4060–4066.
3. LEE, P. L., D. E. JOHNSON, L. S. COUSENS, V. A. FRIED & L. T. WILLIAMS. 1989. Science **245:** 57–60.
4. BAIRD, A., D. SCHUBERT, N. LING & R. GUILLEMIN. 1988. Proc. Nat. Acad. Sci. USA **85:** 2324–2328.
5. PRESTA, M., M. RUSNATI, C. URBINATI, M. STATUTO & G. RAGNOTTI. Ann. N.Y. Acad. Sci. (This volume.)
6. BURGESS, W. H. & T. MACIAG. 1989. Annu. Rev. Biochem. **58:** 575–606.
7. MOSCATELLI, D. 1988. J. Cell Biol. **107:** 753–759.
8. MOSCATELLI, D. & N. QUARTO. 1989. J. Cell Biol. **109:** 2519–2527.
9. KAIBUCHI, K., T. TSUDA, A. KIKUCHI, T. TANIMOTO, T. YAMASHITA & Y. TAKAI. 1986. J. Biol. Chem. **261:** 1187–1192.
10. PRESTA, M., J. A. MAIER & G. RAGNOTTI. 1989. J. Cell Biol. **109:** 1877–1884.
11. TSUDA, T., K. KAIBUCHI, Y. KAWAHARA, H. FUKUZAKI & Y. TAKAI. 1985. FEBS Lett. **191:** 205–210.
12. PRESTA, M., L. TIBERIO, M. RUSNATI, P. DELL'ERA & G. RAGNOTTI. Cell Regul. (In press.)
13. BURGESS, W. H., C. DIONNE, J. KAPLOW, R. MUDD, R. FRIESEL, A. ZILBERSTEIN, J. SCHLESSINGER & M. JAYE. 1990. Mol. Cell. Biol. **10:** 4770–4777.
14. IMAMURA, T., K. ENGLEKA, X. ZHAN, Y. TOKITA, R. FOROUGH, D. ROEDER, A. JACKSON, J. MAIER, T. HLA & T. MACIAG. 1990. Science **249:** 1567–1570.

A Role for Fibroblast Growth Factor in Oligodendrocyte Development

RANDALL D. McKINNON, TOSHIMITSU MATSUI,[a]

MARIANO ARANDA,[b] AND MONIQUE DUBOIS-DALCQ

Laboratory of Viral and Molecular Pathogenesis
National Institute of Neurological Disorders and Stroke
Building 36, Room 5D04
National Institutes of Health
Bethesda, Maryland 20892

INTRODUCTION

Myelin is a membranous sheath surrounding axons, formed by oligodendrocytes in the central nervous system (CNS) and by Schwann cells in the peripheral nervous system (PNS). Its function is to increase the speed of electrical conduction and thus the efficiency of information transfer in the nervous system. Myelin assembly is a late event in nervous tissue development, occurring shortly after birth in rodents, and thus is not essential for much of prenatal CNS development in these species. However the destruction of myelin in the adult, as seen in human diseases such as multiple sclerosis, can compromise CNS functions and can sometimes be fatal. Since the repair of damage incurred after a demyelinating lesion involves processes that recapitulate the events of early development,[1] an understanding of the processes controlling normal development of myelin-forming cells may lead to the design of strategies for the treatment of this human disease. We are therefore investigating the factors involved in regulating the growth and differentiation of the myelin-forming cells in the developing rodent CNS, and describe a potential role for the fibroblast growth factors in this cell lineage.

THE O-2A LINEAGE

In the rodent CNS, oligodendrocyte precursor cells originate in the germinal neuroepithelium surrounding the ventricles during late embryonic development, migrate into regions destined to become white matter, then differentiate into myelin-forming oligodendrocytes shortly after birth.[2,3] While the timing of these events varies slightly in different brain regions, the sequence of events is similar. The cells involved in CNS myelin formation have been best studied in the rat optic nerve, a CNS structure that contains no neuronal cell bodies and just three identifiable differentiated macroglial cell types, oligodendrocytes, and two types of astrocytes (reviewed in Reference 4). Type 1 astrocytes appear first in the nerve during embryonic development, followed by oligodendrocytes starting from birth then type

---

[a]Present affiliation: Kobe University School of Medicine, Chuo-ku, Kusunoki-cho, Kobe 650 Japan.

[b]Present affiliation: Department of Physiology and Biophysics, University of California, Irvine, California 92717.

2 astrocytes at two weeks postnatally.[4] A progenitor cell that is thought to migrate into the nerve in late embryonic development[5] can form either oligodendrocytes or type 2 astrocytes under different conditions in culture, and has therefore been termed the O-2A progenitor cell.[6] Collectively these three cell types make up the O-2A lineage, while type 1 astrocytes represent a separate lineage of macroglial cells.

We have examined the factors controlling the growth and differentiation of O-2A progenitor cells isolated from the perinatal rat cerebral hemispheres. These cells behave much like their counterparts from the optic nerve,[7] although they can be isolated in sufficient quantities to allow an examination at the molecular level.[8] When isolated from other glial cells they quickly differentiate into either oligodendrocytes or type 2 astrocytes depending on the culture conditions. This differentiation can be followed by monitoring the emergence of either oligodendrocyte-specific or astrocyte-specific transcripts, as shown in FIGURE 1. When cultured in media containing a low level (0.5%) of fetal bovine serum (FBS), the majority of progenitor cells differentiate into oligodendrocytes as illustrated by the emergence of transcripts encoding one of the major myelin proteins, myelin basic protein (MBP) (FIGURE 1, lanes 3–7). When cultured in the presence of 10% FBS they differentiate along the type 2 astrocyte pathway, as determined by the emergence of transcripts encoding glial fibrillary acidic protein (GFAP) (FIGURE 1, lanes 8–10).

Factors produced by other neural cells in the developing CNS are likely to regulate the differentiation of O-2A progenitor cells. For example, when O-2A progenitor cells are cultured in the presence of type 1 astrocytes, they differentiate into oligodendrocytes on the same time scale *in vitro* as seen *in vivo*.[9-11] The examination of factors produced by mixed glial cultures that affect the differentiation of isolated O-2A progenitor cells has led to an understanding of the role of polypeptide growth factors in this lineage (see Reference 12 for a review). We will focus on the biological effects of basic fibroblast growth factor (bFGF), a ligand that is present in the CNS during development,[13] is produced by cultured glial cells,[14,15] and as reported in other manuscripts in this volume has a remarkable range of biological activities in a variety of mesoderm- and neuroectoderm-derived cells.

FGF BLOCKS O-2A PROGENITOR CELL DIFFERENTIATION

Both basic and acidic FGFs are mitogens for cultured O-2A progenitor cells.[8,16,17] After 9 days in the continued presence of bFGF under low serum culture conditions, these cells continued to proliferate as progenitors and did not differentiate into oligodendrocytes as indicated by the absence of transcripts encoding MBP.[8] This effect requires the constant presence of active ligand, since on removal of bFGF the cells quickly differentiate.[18] This effect is seen with both basic and acidic FGF, but not with KGF (R. McKinnon and J. Rubin, unpublished data), suggesting a specificity in the response of O-2A progenitor cells to various members of the FGF family of related ligands.

We have extended these studies using the polymerase chain reaction (PCR) technique as a sensitive measure for the expression of myelin gene transcripts. RNA isolated from cultured cells was reverse transcribed into cDNA then amplified by PCR, using synthetic primers specific for either MBP or a control transcript cyclophilin (CYC). The MBP primers were designed to amplify a region of cDNA representing alternative spliced forms of the MBP transcript,[19] generating a range of PCR products including the mature spliced forms of 383 to 584 base pairs (bp), while the CYC primers gave a single PCR product of 669 bp (FIGURE 2). The PCR

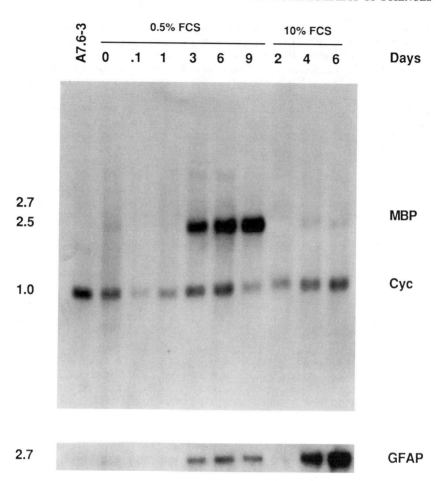

FIGURE 1. Bipotential differentiation of cultured O-2A progenitor cells. Northern blot analysis of cytoplasmic RNA (0.1 μg/lane) isolated from O-2A progenitor cells grown for the indicated times in media supplemented with either 0.5% or 10% fetal calf serum. The blot was probed with radiolabeled cDNAs for oligodendrocyte-specific (MBP) and astrocyte-specific (GFAP) transcripts as described.[8] RNA sizes are in kilobases. The cyclophilin transcript (CYC) was used to normalize for the amount of RNA loaded.

products generated in this analysis were validated by their predicted size, restriction endonuclease digestion patterns, and by hybridization to the appropriate RNA targets by Northern blot analysis.

As was seen by Northern blot analysis, progenitor cells isolated from mixed glial cultures did not express detectable levels of MBP (FIGURE 2, lane 1). When cultured in the continued presence of bFGF these progenitors remained undifferentiated as indicated by the absence of the predicted PCR products from the MBP primers (FIGURE 2, lane 2). Other myelin-specific (proteolipid protein, myelin-associated glycoprotein) and astrocyte-specific gene products (GFAP) were also not detected under these conditions, indicating that this represented a general block on progenitor cell differentiation rather than a specific effect on MBP gene expression (R. McKinnon, data not shown). After removing bFGF from the culture medium, the emergence of MBP transcripts was clearly detected by 72 hours (FIGURE 2, lanes 4 and 5), demonstrating that the block in differentiation induced by bFGF was reversible.

This effect on O-2A progenitor cell differentiation appears to be unique among factors demonstrated to act on these cells in culture; platelet-derived growth factor (PDGF) is mitogenic but promotes only a limited number of cell divisions before differentiation (see below), while insulinlike growth factor (IGF-1) and possibly transforming growth factor (TGF-β) may act at later stages of progenitor cell

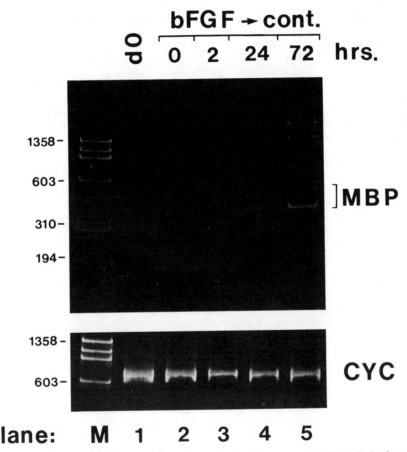

FIGURE 2. FGF blocks differentiation of O-2A glial progenitor cells. RT-PCR analysis of total cell RNA isolated from purified O-2A progenitor cells (lane 1) and progenitors cultured for 5 days in the presence of 10 ng/ml bFGF (lane 2) then for an additional 2, 24, or 72 hours in medium without bFGF (lanes 3–5). For each lane, 10 μg total cellular RNA was converted to first strand cDNA, then 50 ng was subjected to PCR amplification with myelin basic protein (MBP) and cyclophilin (CYC) specific primers. DNA sizes are in kilobases. (lane M) marker: ø×174 DNA digested with *Hae* III endonuclease.

maturation to promote oligodendrocyte differentiation.[12] Epidermal growth factor (EGF) has been reported to block the expression of MBP transcripts,[20] although this has not yet been shown in purified cultures and could be an indirect effect. The response of O-2A progenitor cells to bFGF is comparable to that of cultured myoblasts,[21] suggesting that ability of FGF to promote the proliferation of undifferentiated progenitor cells could be a general function of FGF signaling during development.

FGF AND PDGF HAVE DISTINCT EFFECTS ON CULTURED O-2A PROGENITOR CELLS

O-2A progenitor cells have distinct morphologies when cultured in the presence of different polypeptide growth factors. In the presence of bFGF, O-2A progenitors are stellate in appearance with short processes (FIGURE 3A), while in the presence of PDGF they are bipolar with long processes (FIGURE 3B). When cultured in the absence of these growth factors, O-2A progenitor cells differentiate into oligodendrocytes with a complex branching morphology (FIGURE 3C). When progenitor cells cultured in the presence of bFGF are exposed to PDGF their stellate morphology is converted to bipolar, and upon removal of PDGF from these cultures they revert back to stellate (R. McKinnon, C. Smith, T. Behar, T. Smith, and M. Dubois-Dalcq, submitted). Thus the morphology of O-2A progenitor cells exhibits a plasticity in the presence of these polypeptide growth factors. Furthermore, the bipolar shape that has up to now been considered characteristic of O-2A progenitor cells[16] is actually diagnostic of the migratory state and is an indication of the presence of PDGF in the culture medium.

PDGF has two important biological effects on cultured O-2A progenitor cells that are distinct from the effects of bFGF. First, while PDGF is mitogenic for these cells,[16,22,23] it promotes proliferation for only a limited number of cell divisions prior to their differentiation into oligodendrocytes.[16,22] PDGF is thus said to drive a molecular clock that controls the timing of progenitor cell differentiation.[22] Second, human PDGF (but not bFGF) induces migration[16] and is chemotactic[24] for cultured O-2A progenitors. PDGF is produced by astrocytes in culture[23] and by neurons in vivo,[25,26] and is present from embryonic day 16 during CNS development.[23] Thus PDGF may be responsible for the migration of O-2A progenitor cells from the subependymal zone into white matter regions of the CNS during myelination.

FGF INCREASES THE SENSITIVITY OF O-2A PROGENITORS TO PDGF

One of the principal effects of bFGF on cultured O-2A progenitor cells, in addition to preventing their terminal differentiation, is to maintain a high level of PDGF receptors.[8] These cells express PDGF α but not PDGF β receptors in the undifferentiated state, and the levels of PDGF α receptors decrease as the cells differentiate.[12] In the presence of bFGF, in contrast, the expression of PDGF α receptors is maintained at a high level.[8] The relative increase in PDGF α-receptor levels can be seen as early as 24 hours after bFGF treatment (FIGURE 4, lane 4), and in the continued presence of bFGF these receptors continue to accumulate (FIGURE 4, lane 6). This effect is specific for the PDGF α-receptor, since there is no detectable increase in expression of the PDGF β receptor under these conditions.[8]

The increase in PDGF α receptors represents an increase in functional respon-

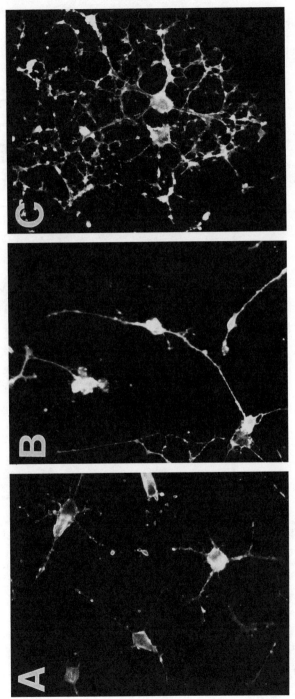

FIGURE 3. Morphology of O-2A glial progenitor cells. Progenitors were purified from mixed glial cell cultures and grown for 3 days in media containing (A) 10 ng/ml bFGF, (B) 10 ng/ml PDGF-AA, or (C) in control media. The cells were fixed and stained with monoclonal O4 (A and B) and anti-GC (panel C) antibodies then fluorescent conjugated second antibodies as previously described.[8]

siveness to PDGF, since bFGF-treated progenitor cells have an increased sensitivity to PDGF as assayed by their proliferative response to low doses of PDGF[8] and the ability of low concentrations of PDGF to induce the bipolar morphology (R. McKinnon, unpublished). Thus one of the principal effects of FGF in addition to promoting proliferation of undifferentiated O-2A progenitor cells may be to increase

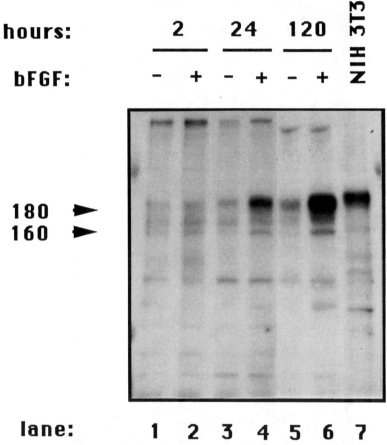

FIGURE 4. bFGF maintains a high level of PDGF α receptors in cultured O-2A progenitor cells. Immunoblot analysis of total cell lysates (100 μg/lane) probed with anti-PDGF α-receptor antiserum. Progenitor cells were cultured for 2 hours (lanes 1 and 2), 24 hours (lanes 3 and 4), or 5 days (lanes 5 and 6) in the absence (lanes 1, 3, and 5) or presence (lanes 2, 4, and 6) of 10 ng/ml bFGF. Lane 7: NIH 3T3 fibroblasts. Protein sizes are in kilodaltons.

the sensitivity of these cells to PDGF. FGF has also been shown to induce nerve growth factor (NGF) dependence[27] and expression of the NGF receptor[28] in sympathoadrenal progenitor cells. Thus the ability of FGF to promote responsiveness to other growth and differentiation factors may be a common theme during development.

A PROPOSED ROLE FOR FGF IN OLIGODENDROCYTE DEVELOPMENT

Our studies are consistent with the proposal that bFGF, or some related ligand localized to the subependymal zone in the developing CNS, acts to promote proliferation and prevent the differentiation of O-2A progenitor cells and to increase their sensitivity to PDGF. PDGF produced by neurons[25,26] and/or astrocytes,[23,25] in turn, would act as a chemoattractant to direct the migration of these progenitors into regions of the CNS destined to become white matter. Movement of the progenitor cell away from its source of FGF would release the block to differentiation, placing the cell under the PDGF-driven mitotic clock which would control the extent of proliferation during migration out to the axonal tracts.

Other growth factors that are present in the CNS during development may also have roles in this lineage. For example, TGF-β may have a role in down regulating the PDGF response through receptor transmodulation, as has been shown in other systems.[29] In addition, IGF-1 is found in white matter tracts during oligodendrocyte differentiation[30] and could have a role in stimulating the production of myelin in differentiated oligodendrocytes. An analysis of the effects of these factors on purified O-2A lineage cells in culture should further our understanding of the roles of these polypeptide growth factors in O-2A progenitor cell maturation and differentiation, and direct studies on the potential therapeutic use of these growth factors during recovery from a demyelinating event.

SUMMARY

The differentiation of oligodendrocyte–type 2 astrocyte (O-2A) glial progenitor cells into myelin-forming oligodendrocytes is influenced by several polypeptide growth factors. Exposure of O-2A progenitors to basic fibroblast growth factor (bFGF) promotes their sustained proliferation, blocks their differentiation, and maintains both a high level of platelet-derived growth factor (PDGF) α-receptors and PDGF sensitivity. Exposure to PDGF, in contrast, promotes only a limited number of cell divisions prior to their differentiation and triggers progenitor cell migration. In the continued presence of bFGF the cells have a stellate morphology with short processes. Upon addition of PDGF these stellate cells become bipolar with long processes, and on removal of PDGF their morphology reverts back to stellate. Thus the phenotype of O-2A progenitor cells in response to these growth factors is plastic. Our studies suggest that bFGF (or a related ligand) in the CNS may sensitize O-2A progenitors to PDGF and thereby initiate their ability to migrate into white matter tracts prior to the onset of myelination.

ACKNOWLEDGMENTS

We are grateful to our colleagues at the National Institutes of Health involved with various stages of this work, including Jeff Rubin and Stu Aaronson (National Cancer Institute), and Carolyn Smith, Tom Smith, Toby Behar, and Regina Armstrong (National Institute of Neurological Disorders and Stroke). We would also like to thank Ray Rusten for photography.

REFERENCES

1. DUBOIS-DALCQ, M. & R. ARMSTRONG. 1990. Bioessays **12:** 569–576.
2. REYNOLDS, R. & G. P. WILKIN. 1988. Development **102:** 409–425.
3. LEVINE, S. M. & J. E. GOLDMAN. 1988. J. Comp. Neurol. **277:** 441–455.
4. RAFF, M. 1989. Science **243:** 1450–1455.
5. SMALL, R., P. RIDDLE & M. NOBLE. 1987. Nature **328:** 155–157.
6. RAFF, M. C., R. H. MILLER & M. NOBLE. 1983. Nature **303:** 390–396.
7. BEHAR, T., F. A. MCMORRIS, E. A. NOVOTNY, J. L. BARKER & M. DUBOIS-DALCQ. 1988. J. Neurosci. Res. **21:** 168–180.
8. MCKINNON, R. D., T. MATSUI, M. DUBOIS-DALCQ & S. A. AARONSON. 1990. Neuron **5:** 603–614.
9. ABNEY, E. R. , P. P. BARTLETT & M. C. RAFF. 1981. Dev. Biol. **83:** 301–310.
10. RAFF, M. C., E. R. ABNEY & J. FOK-SEANG. 1985. Cell **42:** 61–69.
11. DUBOIS-DALCQ, M., T. BEHAR, L. HUDSON & R. A. LAZZARINI. 1986. J. Cell. Biol. **102:** 384–392.
12. RICHARDSON, W. D., M. RAFF & M. NOBLE. 1991. Semin. Neurosci. **2:** 445–454.
13. GONZALEZ, A.-M., M. BUSCAGLIA, M. ONG & A. BAIRD. 1990. J. Cell Biol. **110:** 753–765.
14. HATTEN, M. E., M. LYNCH, R. E. RYDEL, J. SANCHEZ, J. JOSEPH-SILVERSTEIN, D. MOSCATELLI & D. B. RIFKIN. 1988. Dev. Biol. **125:** 280–289.
15. EMOTO, N., A.-M. GONZALEZ, P. A. WALICKE, E. WADA, D. M. SIMMONS, S. SHIMASAKI & A. BAIRD. 1989. Growth Factors **2:** 21–29.
16. NOBLE, M., K. MURRAY, P. STROOBANT, M. D. WATERFIELD & P. RIDDLE. 1988. Nature **333:** 560–562.
17. BÖGLER, O., D. WREN, S. C. BARNETT, H. LAND & M. NOBLE. 1990. Proc. Nat. Acad. Sci. USA **87:** 6368–6372.
18. MCKINNON, R. D. & M. DUBOIS-DALCQ. 1990. Ann. N.Y. Acad. Sci. **605:** 358–359.
19. ZELLER, N. K., M. J. HUNKELER, A. T. CAMPAGNONI, J. SPRAGUE & R. A. LAZZARINI. 1984. Proc. Nat. Acad. Sci. USA **81:** 18–22.
20. SHENG, H. Z., A. TURNLEY, M. MURPHY, C. C. A. BERNARD & P. F. BARTLETT. 1989. J. Neurosci. Res. **23:** 425–432.
21. CLEGG, C. H., T. A. LINKHART, B. B. OLWIN & S. D. HAUSCHKA. 1987. J. Cell Biol. **105:** 949–956.
22. RAFF, M. C., L. E. LILLIEN, W. D. RICHARDSON, J. F. BURNE & M. D. NOBLE. 1988. Nature **333:** 562–565.
23. RICHARDSON, W. D., N. PRINGLE, M. J. MOSLEY, B. WESTERMARK & M. DUBOIS-DALCQ. 1988. Cell **53:** 309–319.
24. ARMSTRONG, R. C., L. HARVATH & M. E. DUBOIS-DALCQ. 1990. J. Neurosci. Res. **27:** 400–407.
25. YEH, H.-J., K. G. RUIT, Y.-X. WANG, W. C. PARKS, W. D. SNIDER & T. F. DEUEL. 1991. Cell **64:** 209–216.
26. SASAHARA, M., J. W. U. FRIES, E. W. RAINES, A. M. GOWN, L. E. WESTRUM, M. P. PROSCH, D. T. BONTHRON, R. ROSS & T. COLLINS. 1991. Cell **64:** 217–227.
27. STEMPLE, D. L., N. K. MAHANTHAPPA & D. J. ANDERSON. 1988. Neuron **1:** 517–525.
28. BIRREN, S. J. & D. J. ANDERSON. 1990. Neuron **4:** 189–201.
29. BATTEGAY, E. J., E. W. RAINES, R. A. SEIFERT, D. F. BOWEN-POPE & R. ROSS. 1990. Cell **63:** 515–524.
30. ANDERSSON, I. K., D. EDWALL, G. NORSTEDT, B. ROZELL, A. SKOTTNER & H.-A. HANSSON. 1988. Acta. Physiol. Scand. **132:** 167–173.

afGF Content Increases with Malignancy in Human Chondrosarcoma and Bladder Cancer[a]

DENIS BARRITAULT, BÉATRICE GROUX-MUSCATELLI,
DANIÈLE CARUELLE, MARIE-CATHERINE VOISIN,[b]
DOMINIQUE CHOPIN,[c] SANDRINE PALCY,
AND JEAN-PIERRE CARUELLE

Department of Biotechnology
[b]*Department of Anatomopathology*
[c]*Department of Urology*
University of Paris Val de Marne
Avenue du Général de Gaulle
94010 Creteil Cedex, France

Acidic and basic fibroblast growth factors (a and bFGFs) are the most abundant and studied growth factors from the FGF family.

In vitro, both forms of FGFs have a wide spectrum of effects on cell proliferation and differentiation. *In vivo,* a growing body of evidence suggests that FGFs play a key role in homeostatic mechanisms such as embryogenesis cellular growth, vascularization, tissue repair, etc. (for review, see Reference 1).

bFGF, which was first purified to homogeneity from a rat chondrosarcoma tissue[2,3] taking advantage of its affinity to heparin, has also been found in a large number of tissues and appears to have a quasi-ubiquitous distribution (for review, see Reference 1).

aFGF was originally isolated from nervous tissue where it seemed to be almost exclusively located.[1] More recently, several reports have indicated that aFGF was more widely distributed and that large amounts of aFGF could also be localized and purified from other sources such as kidney[4] or heart.[5]

Clearly, the quasi-ubiquitous distribution of a and bFGFs suggests the presence of several regulatory mechanisms controlling the tissue repartition, local bioavailability, and multiple biological activities of these factors and those of their receptors.

In order to increase our understanding of some of these mechanisms, more is to be known on the tissue distribution of these molecules. We have therefore developed specific and sensitive enzyme immuno assays for FGFs[6,7] and studied the distribution of FGF immunoreactive materials in several developing, adult, and pathological tissues and fluids.[8,9]

Among several screened cancers, chondrosarcoma and bladder carcinoma were more specifically studied for their content in aFGF. Indeed, the presence of aFGF had not yet been reported either in cartilage or in chondrosarcoma,[2] while *in vitro* studies have shown that in contrast to bFGF, aFGF was associated with the motility

[a]This work was supported by grants from the University of Paris Val de Marne, INSERM, Caisse Nationale d'Assurance Maladie des Travailleurs Salariés (CNAMST), Association de la Recherche Contre le Cancer (ARC), Ligue Nationale Française Contre le Cancer, and Fondation pour la Recherche Médicale.

387

in a rat-derived bladder carcinoma cell line,[10] suggesting the possible involvement of aFGF in the progression of bladder tumors.

We have therefore measured and localized aFGF in these two types of tumors as well as studied the variation of aFGF immune material content in tumor extracts and in urine as a function of the degree of evolution of the tumor.

MATERIALS AND METHODS

Patients

Chondrosarcoma Studies

Specimens of tumoral cartilage were obtained from several chondrosarcomas of G2 and G4 histological gradings according to Broders's classification.[11]

Specimens of control cartilage were obtained from femoral heads of patients treated for cervical fracture and from articular tibial plate.

The samples were frozen in liquid nitrogen and stored at $-70°C$ until use.

Bladder Cancer Studies

Bladder tumors were harvested from cystectomy specimens. Normal bladders were obtained during organ procurement for transplantation from cadaverous donors and stored. These tissue samples were treated as for chondrosarcoma.

Patients included in this study had a definitive pathological diagnosis and were staged according to the Union Internationale Contre le Cancer (UICC) classification (TABLE 2).

Twenty-four-hour urine samples were obtained prior to histological examinations. Grossly infected urines were excluded (bacteria $\geq 10^5$). Samples were centrifuged at $3000 \times g$, aliquoted (5×4 ml), and stored at $-80°C$ until use.

Tissue Extraction

Frozen tissues were powdered in liquid nitrogen and homogenized in a Turrax blender in 2 M NaCl 0.1 M phosphate buffer pH 7.4 (1 ml/100 mg tissue). Samples were then centrifuged at 4°C for 30 minutes at $40,000 \times g$. The ionic strength for the supernatant was adjusted with a conductimeter to 0.32 mΩ (corresponding approximatively to 0.60 M NaCl) with 100 mM phosphate-buffered saline (PBS) pH 7.4, 0.1% bovine serum albumin (BSA). For some of these crude extracts, further purification of aFGF was achieved as follows: heparine-Sepharose (50 mg of dry powder/g of tissue) was equilibrated in PBS, 0.65 M NaCl, 0.1% BSA and added to the crude extract solution. The mixture was then gently stirred overnight at 4°C and poured into a 10-ml disposable column. The gel was then washed three times in PBS 0.65 M NaCl, 0.1% BSA. The fractions eluted in the same buffer containing 1 M and 2 M NaCl were collected and further studied.

Enzyme Immunoassays for a and bFGFs

Quantification of a or bFGF was performed with a competitive enzyme immunoassay (EIA, acetyl choline esterase) as already described[6] with slight modifications.

Briefly mouse antirabbit monoclonal immunoglobulin G antibodies (IgG) were adsorbed onto 96 well plates, 10% BSA was used to saturate protein binding sites. After washing with 10^{-2} M phosphate buffer/0.05% tween 20, pH 7.4, each well was filled with 300 μl of EIA buffer and the plates were stored at 4°C until use. Assays were performed in a total volume of 150 μl with 50 μl of each of the following: diluted a or bFGF polyclonal antiserum, a or bFGF acetyl choline esterase conjugate (FGF-ACHE). Final ionic concentration for all theses assays was adjusted at 1 M NaCl and values compared to a standard displacement curve obtained under the same ionic strength conditions using a known concentration of highly purified (> 98%) a or bFGF from bovine brain, or human recombinant sources.

The characteristics of the anti-aFGF immune serum used have been studied elsewhere by means of Elisa, immunoblotting, and inhibition of biological activity.[6] It reacts with aFGF and does not cross-react with bFGF (< 0.1%).

Anti-bFGF immune serum was kindly given by Jean Plouet, and its characteristics are described elsewhere.[7]

In these conditions, the maximum sensitivity defined as B/B0 = 50% was 2.5 ng/ml (125 pg/tube) and the detection limit defined as B/B0 = 80% was 0.5 ng/ml ± 0.2 ng/ml (25 pg/tube).

Statistical Analysis

Data were analyzed using Yates modified χ^2 test. Sensitivity, specificity, and positive and negative predictive values were calculated with standard formulae. Fisher's exact test and Tukey's quick test were used for small samples and in chondrosarcoma studies.

RESULTS

Chondrosarcoma Studies

Measure of a and bFGF Immune Reactive Material in Normal Cartilage and Chondrosarcoma

Values obtained after the measure of FGF immune reactive material in various surgical specimens from cartilage or chondrosarcoma are presented in TABLE 1. The measures are performed in crude extracts and after elution of crude tissue extracts from heparin-Sepharose chromatography as described in Materials and Methods.

aFGF was measured in all the samples studied while bFGF was measured in a more limited number of cases.

Bladder Cancer Studies

Bladder Extracts

FIGURE 1 presents the values obtained in the measure of the aFGF immunoreactive material eluted at 1 M NaCl from heparin-Sepharose from normal urothelium and transitional cell carcinoma. The significance of the difference (1.77 ± 2 ng/g of tissue versus 20.36 ng/g in TCC) was assessed by Tukey's quick test ($p < 0.05$).

TABLE 1. aFGF and bFGF Content in Crude Extracts (CE) and after 1 M Elution on Heparin-Sepharose Chromatography in Normal Cartilage and in Grade II and IV Chondrosarcoma

	Normal Cartilage		G 2 Chondrosarcoma		G 4 Chondrosarcoma	
	CE	1 M Na Cl	CE	1 M Na Cl	CE	1 M Na Cl
aFGF ng/g tissue						
Number (n) of cases	6	4	13	8	2	2
n of extractions	14	6	16	8	5	2
Mean Value	12.35	ND[b]	33.25	2.75	48.61	11.12
SD[a]	4.27	—	17.55	2.74	11.72	1.52
bFGF ng/g tissue						
Number of (n) of cases	1		5	1		
n of extractions	1		5	1		
Mean value	108.33		67.18		51.2	
SD	—		46.34		—	

[a]SD: standard deviation.
[b]ND: nondetectable.

Urine Samples

The EIA was considered positive for urine samples with a B/B0 ratio below 80%, and the distribution of the positive aFGF immunoreactivity in 469 urine samples is presented in TABLE 2.

There was a significant difference between the group of patients with invasive disease (T2-T3/M+) and the control group ($p < 0.001$), giving a sensitivity of 83% and a specificity of 92%. The difference was also significant between patients with invasive disease and those with prostatic disease ($p < 0.001$).

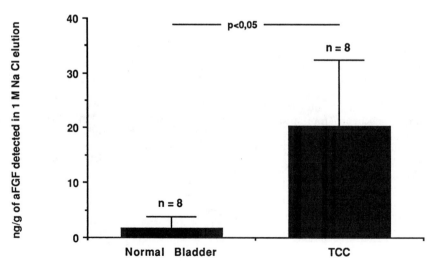

FIGURE 1. aFGF in bladder extracts. aFGF from 8 normal bladders and 8 TCC of bladder was measured after extraction and 1 M NaCl elution from heparin-Sepharose chromatography ($p < 0.05$ according to Tukey's quick test).

DISCUSSION

Chondrosarcoma and bladder cancer are the first two examples of cancers where a correlation can be made between the degree of dedifferentiation and evolution of the tumor and the content of aFGF immune reactive material.

The first part of the results relates the detection and measure of aFGF immune reactive material in chondrosarcoma. Previous reports have shown the presence of bFGF in chondrosarcoma while no aFGF was detected after purification on heparin-Sepharose.[3] Our immunoassay revealed small amounts of aFGF (compared to bFGF) in crude extracts of normal cartilage (12.35 ng/g) but failed to reveal material after heparin-Sepharose fractionation. However, in chondrosarcoma, we were able to detect a significant amount of aFGF immunoreactive material after heparin-Sepharose chromatography. The sensitivity of our immuno assay, 25 pg per sample (detected in a volume of 50 μl), is approximately tenfold greater than the sensitivity of biological assays.[5]

TABLE 2. Acidic Fibroblast Growth Factor Immunoreactivity in Urine

Control group (*n* = 114)		
Healthy volunteers	0/16	(0%)
Noninfectious, noninflammatory urological disease	2/24	(8%)
Infectious or inflammatory urinary tract conditions	8/74	(10%)
Transitional cell carcinoma of the bladder[a] (*n* = 150)		
To	10/70	(14%)
TIS	2/14	(14%)
TA	1/10	(10%)
T1	11/32	(34%)
T2-T3	13/15	(86%)
Metastatic	7/9	(77%)
Benign prostatic hypertrophy (*n* = 132)	14/132	(10%)
Carcinoma of the prostate (*n* = 73)		
Metastatic CaP	7/57	(12%)
Localized CaP	0/16	(0%)

[a]UICC classification: To tumor detected on specimen; TIS, carcinoma *in situ;* TA, papillary tumor without invasion of the lamina propria; T1, invasion of the lamina propria; T2, superficial muscle invasion; T3, deep muscle invasion or invasion to the perivesical fat.

Measures of bFGF in crude extracts of grade II chondrosarcoma indicate a content of bFGF approximatively twice greater than that of aFGF. Since the previously used bioassays for detecting bFGF are approximatively 10-fold more sensitive than for aFGF, a 20-fold difference occurs in the estimation of bFGF versus aFGF. The difference is even more amplified using collagenase digestion which increases the yield of bFGF extraction while aFGF yield is decreased (data not shown).

The aFGF immunoreactive material detected after heparin-Sepharose chromatography was further characterized as identical to aFGF on the basis of (a) electrophoretical comigration with native and recombinant highly purified a FGF, (b) Western blot characterization, and (c) by its ability to be potentiated by heparin in *in vitro* bioassays.[12] The increase of aFGF immunoreactive material with the degree of dedifferentiation of the tumor suggests that aFGF may be implicated in the tumor evolution. This implication is further supported by immunochemical analysis, indicating a distribution of intense immunolabeling at the cellular level, hardly detectable in normal cartilage (Groux *et al.,* in preparation). The specificity of aFGF in the evolution of chondrosarcoma can also be stressed by an absence of significant difference in bFGF content and also by the fact that epidermal growth factor could not be detected in cartilage or chondrosarcoma extracts (in preparation).

However, the real implication of an intracellular angiogenic growth factor like aFGF in the progression of an avascular tumor like chondrosarcoma remains to be clarified.

The second part of our results indicates that aFGF or aFGF-like molecules are also associated with bladder cancer. bFGF has already been found in extracts of urological tumors and in the urine of patients with urogenital disease.[13-15] Most of these experiments have involved biological assays, precluding their use in clinical practice. We have adapted an enzyme immunoassay previously described[6] for the aFGF-related molecules to the detection of such immunoreactivity in tissue extracts and in unconcentrated urine. This was achieved by using higher ionic strength conditions for immunodetection which probably prevents masking of epitopes by interacting molecules. This immunoreactivity was present at a significantly higher level in tumor biopsies and in urine of patients with bladder cancer than in a control group with inflammatory nonneoplastic conditions and in patients with benign prostatic hypertrophy or carcinoma of the prostate. There was a relationship with the stage of the tumor, the frequency of detection increasing with the progression of the disease (up to 85% in T2T3 categories).

The high sensitivity (83%) and specificity (92%) obtained with this original enzyme immunoassay applied to unconcentrated urine samples warrant a multicenter clinical evaluation of aFGF as a potential marker for bladder cancer, as well as a useful tool in the case of chondrosarcoma for supporting anatomopathologist gradings of these tumors.

ACKNOWLEDGMENTS

The authors wish to thank Dr. Grassi (Dr. Grassi is a fellow from the Association de la Recherche sur la Polyarthrite) and Dr. Pradelles (CEA, Saclay) for their gift of acetyl choline esterase.

REFERENCES

1. BURGESS, W. H. & T. MACIAG. 1989. The heparin-binding (fibroblast) growth factor family of proteins. Annu. Rev. Biochem. **58:** 575–606.
2. BEKOFF, M. C. & M. KLAGSBRUN. 1982. Characterization of growth factors in human cartilage. J. Cell. Biochem. **20:** 237–245.
3. SHING, Y., J. FOLKMAN, R. SULLIVAN, C. BUTTERFIELD, J. MURRAY & M. KLAGSBRUN. 1984. Heparin affinity, purification of a tumor-derived capillary endothelial cell growth factor. Science **223:** 1296–1299.
4. GAUTSCHI-SOVA, P., Z. P. JIANG, M. FRATER-SCHRODER, P. BOHLEN. 1987. Acidic fibroblast growth factor is present in neuronal tissue: isolation and chemical characterization from bovine kidney. Biochemistry **26:** 5844–5847.
5. SASAKI, H., H. HOSHI, Y. M. HONG, T. SUZUKI, T. KATA, M. SARTOT, H. YOUKI, K. KARUBE, S. KONNO, M. ONODERAT, T. SARTO & S. AOYAGI. 1989. Purification of acidic fibroblast growth factor from bovine heart and its localization in the cardiac myocytes. J. Biol. Chem. **264:** 17606–17612.
6. CARUELLE, D., J. GRASSI, J. COURTY, B. GROUX-MUSCATELLI, P. PRADELLES, D. BARRITAULT & J. P. CARUELLE. 1988. Development and testing of radio and enzyme

immunoassays for acidic fibroblast growth factor (aFGF). Anal. Biochem. **173:** 328–339.

7. GROUX-MUSCATELLI, B., Y. BASSAGLIA, J. P. CARUELLE, D. BARRITAULT & J. GAUTRON. 1990. Proliferating satellite cells express acidic fibroblast growth factor during *in vitro* myogenesis. Dev. Biol. **142:** 380–385.

8. CARUELLE, D., B. GROUX-MUSCATELLI, A. GAUDRIC, C. SESTIER, G. COSCAS, J. P. CARUELLE & D. BARRITAULT. 1989. Immunological study of acidic fibroblast growth factor (aFGF) distribution in the eye. J. Cell. Biochem. **29:** 117–128.

9. JOUANNEAU, J., J. GAVRILOVIC, D. CARUELLE, M. JAYE, G. MOENS, J. P. CARUELLE & J. P. THIERY. 1991. Secreted or non-secreted forms of aFGF produced by transfected epithelial cells influence cell morphology, motility and invasive potential. Proc. Nat. Acad. Sci. USA **194:** 252–259.

10. VALLES, A. M., B. BOYER, J. BADET, G. C. TUCKER, D. BARRITAULT & J. P. THIERY. 1990. Acidic fibroblast growth factor is a modulator of epithelial plasticity in a rat bladder carcinoma cell line. Proc. Nat. Acad. Sci. USA **87:** 1124–1128.

11. BRODERS, A. C., R. HARGRAVE & H. W. MEYERDIG. 1939. Pathologic features of soft tissue fibrosarcoma. Surg. Gynecol. Obstet. **69:** 267.

12. SCHREIBER, A. B., J. KENNEY, W. J. KOWALSKI, R. FRIESEL, T. MEHLMAN & T. MACIAG. 1985. Interaction of endothelial cell growth factor with heparin: characterization of receptor and antibody recognition. Proc. Nat. Acad. Sci. USA **82:** 6138–6142.

13. CHODAK, G. W., C. J. SCHEINER & B. R. ZETTER. 1981. Urine from patients with transitional-cell carcinoma stimulates migration of capillary endothelial cells. N. Engl. J. Med. **305:** 869–874.

14. CHODAK, G. W., Y. SHING, M. BORGE, S. M. JUDGE & M. KLAGSBRUN. 1986. Presence of heparin binding growth factor in mouse bladder tumors and urine from mice with bladder cancer. Cancer Res. **46:** 5507–5510.

15. CHODAK, G. W., V. HOSPELHORN, S. M. JUDGE, R. MAYFORTH, H. KOEPPEN & J. SASSE. 1988. Increased levels of fibroblast growth factor–like activity in urine from patients with bladder or kidney cancer. Cancer Res. **48:** 2083–2088.

Isolation of Basic FGF Receptors from Adult Bovine Brain Membranes[a]

AGNES MEREAU, MYLENE PERDERISET, DOMINIQUE
LEDOUX, ISABELLE PIERI, JOSE COURTY,
AND DENIS BARRITAULT

Biotechnology Laboratory of Eucaryotic Cells
University of Paris XII
94010 Creteil, France

INTRODUCTION

Biological activities of acidic and basic fibroblast growth factors (aFGF and bFGF) are mediated by their interactions with specific cell surface membrane receptors characterized as a 125–165 kDa glycoprotein.[1] As previously described, FGF receptors have been purified and identified as a tyrosine kinase protein from embryonic tissues.[2] From this structural domain, so called flg, bek, and cek cDNAs coding for various forms of FGF receptors have been identified[3,4] and expressed as functional receptors in transfected cells.[4] We have previously reported that adult brain contained a large amount of high-affinity receptors for bFGF.[5] In this report, we describe a purification procedure for FGF receptors from adult bovine brains, and studies of some properties of these highly purified proteins.

MATERIALS AND METHODS

Briefly, CHAPS solubilized membrane fractions were applied onto an anionic column and the elution was performed with buffer A (0.1 M NaCl, 20 mM HEPES pH = 7.4, 0.2% CHAPS, 0.1 mM PMSF, 1 μg/ml pepstatin, 1 μg/ml leupeptin, 5 U/ml aprotinin) containing 1 M NaCl. Partially purified proteins were dialysed against buffer A and applied successively onto a WGA-Sepharose and bFGF-Sepharose affinity columns. WGA-Sepharose elution was obtained with buffer A containing 0.3M N-acetyl glucosamine, and bFGF-Sepharose elution with buffer A containing 1 M NaCl. All collected fractions were tested for their ability to inhibit FGF binding to insolubilized membranes[6] and analyzed by cross-linking experiments.

RESULTS AND DISCUSSION

Starting from 25 adult bovine brains, we obtained 240 μg of highly purified proteins containing 110 μg of a major 140-kDa polypeptide as estimated by sodium dodecyl sulfate (SDS) gel scanning (FIGURE 1a). Cross-linking of iodinated aFGF or

[a]This work was supported by grants from the "Ministère de l'Éducation Nationale," the "Ligue Nationale Contre le Cancer," "l'Association pour la Recherche Contre le Cancer" and from INSERM C.E. n°872002 and n°892003.

a
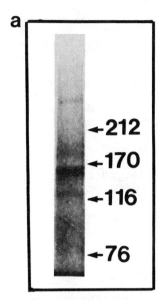

b
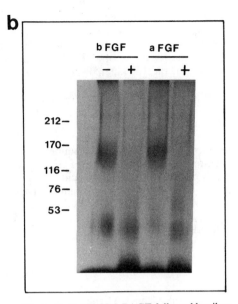

FIGURE 1. a: Highly purified fractions were subjected to 5–7% SDS-PAGE followed by silver staining. **b:** Cross-linking was performed by incubating highly purified FGF receptor fractions with 100 pM of ^{125}I-aFGF or ^{125}I-bFGF in the presence (+) or absence (−) of a 100-fold excess of unlabeled FGF for 1 hour at 4°C. Dissuccinimidyl suberate bifunctional reagent was added at a final concentration of 0.1 mM, and the reaction mixture was incubated for 15 minutes at room temperature. Samples were subjected to 5–7% SDS-PAGE followed by autoradiography.

TABLE 1[a]

		Eluted Iodinated Proteins	
Affinity Column	Heparinase Treatment	NaCl (2 M)	Heparin (500 μg/ml)
bFGF Sepharose	+	95%	90%
	−	95%	92%
Protamine-agarose	−	97%	95%
Lysozyme-agarose	−	7%	ND
Sepharose	−	4%	ND

[a]Highly purified fractions were iodinated as described in Reference 6. Radiolabeled fractions were incubated with (+) or without (−) heparinase (1 U/ml) in 20 mM Hepes pH = 7.4 containing 0.15 M NaCl, 0.6 mM $CaCl_2$, and 0.4 mM $MgSO_4$. The efficiency of heparinase treatment was checked in crude unsolubilized membrane preparations showing a reduction of 70% of ^{125}I-bFGF binding. Iodinated proteins were loaded onto different mini–affinity columns (200 μl). Columns were washed with buffer A and eluted with buffer A containing NaCl (2 M) or heparin (500 μg/ml) as noted above. Controls were performed using Sepharose alone or a lysozyme-Sepharose affinity column exhibiting the same amount of anionic charge as the protamine agarose affinity column.

bFGF to highly purified fractions yielded a single specific 160-kDa cross-linked complex highly correlated with the major presence of a 140-kDa polypeptide in purified fraction (FIGURE 1b), suggesting that this 140-kDa entity is a common binding site for aFGF and bFGF. Treatment with N-glycanase (25 U/ml) demonstrated that the 140-kDa polypeptide contained at least 20-kDa N-linked carbohydrates (result not shown). As shown in TABLE 1, heparinase treatment did not affect the affinity of purified proteins for immobilized bFGF. Treated compounds exhibited no shift in mobility by SDS-polyacrylamide gel electrophoresis (SDS-PAGE) (result not shown). These results indicated that purified entities did not contain detectable heparan sulfate side chains that could be involved in the ability of these entities to bind immobilized bFGF. Protamine, an inhibitor of FGF mitogenic activities, has been reported to displace ^{125}I-bFGF bound to bovine brain membranes.[7] Affinity of highly purified proteins for protamine agarose (TABLE 1) suggested a direct interaction of these proteins with protamine. These results suggest that protamine agarose columns could be used in the FGF receptor purification procedure. Achievement of a primary sequence of these bFGF binding proteins as well as production of antibodies is now in progress.

ACKNOWLEDGMENTS

We thank Jacques Avignan for his help during the redaction of this manuscript.

REFERENCES

1. BURGESS, W. H. & T. MACIAG. 1989. Annu. Rev. Biochem. **58:** 575–606.
2. LEE, P. L., D. E. JOHNSON, L. S. COUSENS, V. A. FRIED & L. T. WILLIAMS. 1989. Science **245:** 57–60.
3. PASQUALE, E. B. 1990. Proc. Nat. Acad. Sci. USA **87:** 5812–5816.
4. DIONNE, C. A., G. CRUMLEY, F. BELLOT, J. M. KAPLOW, G. SEARFOSS, M. RUTA, W. H. BURGESS, M. JAYE & J. SCHLESSINGER. 1990. EMBO J. **9:** 2685–2692.
5. LEDOUX, D., A. MEREAU, M. C. DAUCHEL, D. BARRITAULT & J. COURTY. 1989. Biochem. Biophys. Res. Commun. **159:** 290–296.
6. MEREAU, A., I. PIERI, C. GAMBY, J. COURTY & D. BARRITAULT. 1989. Biochimie **71:** 865–871.
7. DAUCHEL, M. C., J. COURTY, A. MEREAU & D. BARRITAULT. 1989. J. Cell. Biochem. **39:** 411–420.

Adult Brain but Not Kidney, Liver, Lung, Intestine, and Stomach Membrane Preparations Contain Detectable Amounts of High-Affinity Receptors to Acidic and Basic Growth Factors[a]

DOMINIQUE LEDOUX, AGNES MEREAU,
MICHAEL JAYE,[b] MARCELLE MISKULIN,
DENIS BARRITAULT, AND JOSE COURTY

Biotechnology Laboratory of Eucaryotic Cells
University of Paris XII
94010 Creteil, France

[b]*Rorer Biotechnology*
King of Prussia, Pennsylvania 19406

The fibroblast growth factor (FGF) family consists of polypeptide growth factors characterized by heparin-binding affinity, amino acid sequence homology, and the

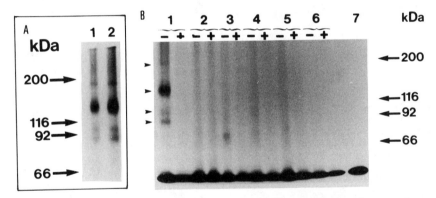

FIGURE 1. Cross-linking experiments of acidic and basic ^{125}I-FGF to various tissue membrane preparations. Four nanograms per milliliter (250 pM) of acidic (A, lane 1; B, lanes 1–6) or 1.6 ng/ml (100 pM) of basic (A, lane 2) ^{125}I-FGF were incubated for 60 minutes with 100 µg of membrane preparations of brain (A, lanes 1–2; B, lane 1), kidney, liver, lung, intestine, or stomach (B, lanes 2–6). At the end of the incubation, bound ^{125}I-FGFs were cross-linked with 0.1 mM disuccinimidyl suberate (DSS). Membranes were pelleted (5600 × *g*, 4°C, 15 minutes), washed with phosphate-buffered saline (PBS) 2.15 M NaCl, and solubilized by a buffer containing 2% sodium dodecyl sulfate (SDS), 10% glycerol, 70 mM Tris-HCl (pH = 6.8), and 5% β-mercaptoethanol. The solubilized material was applied to a 5–7% SDS-polyacrylamide gel electrophoresis (SDS-PAGE) according to Laemmli. Fixed and dried gel was subjected to autoradiography for 2-day exposure. In B, ^{125}I-aFGF was incubated in the presence (+) or not (−) of a 500-fold excess of native acidic FGF.

[a]This work was supported by grants from the "Ministère de l'Éducation Nationale," the "Ligue Nationale Contre le Cancer," "l'Association pour la Recherche Contre le Cancer," and from INSERM C.E. No. 872002 and No. 892003.

397

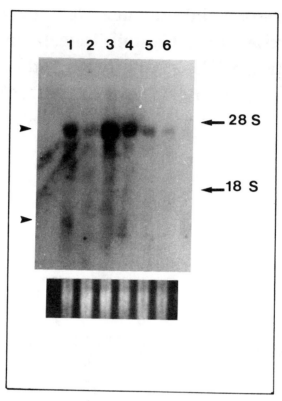

FIGURE 2. Northern blot analysis of FGF receptor mRNA. Ten micrograms of poly A$^+$ mRNA obtained from brain, kidney, lung, liver, intestine, or stomach (lanes 1–6) were electrophoresed on a 1% formaldehyde-agarose gel and transferred onto Hybond-N membrane. Hybridization was performed using ^{32}P-labeled *bek* as a probe (1.5 × 10^6 cpm/ml) under conditions of high stringency (5 × SSPE, 5× Denhardt's, 0.5% SDS, 100 μg/ml salmon sperm DNA at 65°C). Filter was then washed 2×with 2 × SSPE, 0.1% SDS followed by 1 × at 50°C with 0.1 × SSPE, 0.1% SDS for 15 minutes/wash.

ability to stimulate the mitogenesis of a wide variety of cells of epithelial, mesenchymal, and neuronal origin.[1] One of the first steps of induction of the mitogenic signal is a specific interaction with high-affinity cell surface receptors which have been characterized in cultured cells[2] and tissue membrane preparations.[3,4] In this report, we studied and compared the presence of high-affinity receptors for acidic and basic FGFs in various adult guinea pig tissues using Scatchard and cross-linking experiments. In addition, the expression of FGF receptor mRNA was determined in the same tissues by Northern blot analysis using human cDNA *bek* as a probe.[5]

Distribution of basic and acidic FGF receptors was therefore examined in adult brain, kidney, liver, lung, intestine, and stomach membrane preparations. Interestingly, Scatchard analyses showed that only brain membranes contained detectable amounts of high-affinity receptors for acidic (K_d = 180 pM) and basic (K_d = 15 pM) FGFs (results not shown). Cross-linking experiments revealed that this family of receptors was characterized by four molecular species of M_r 175, 125, 95, and 70 kDa which could easily bind acidic (FIGURE 1A and B; lane 1) as well as basic (FIGURE 1A, lane 2) FGFs specifically. However none of these molecular species was detected in kidney, liver, lung, intestine, or stomach membranes (FIGURE 1B, lanes 2–6). Moreover, cross competition experiments indicated that these two forms of FGF interacted with the same families of receptors (results not shown).

In order to expand these results, distribution of FGF receptor mRNA was investigated in the same tissues. Northern blot analysis on poly A$^+$ mRNA was then performed using a human cDNA probe that encodes for the first immunoglobulinlike domain of the FGF receptor type 2 (*bek*). As shown on FIGURE 2, a 4.5-kilobase (kb) transcript was revealed in all tissues with differential expression. An additional transcript of about 1.5 kb was detected only in brain tissue.

The presence of large amounts of FGF high-affinity receptors only in adult brain compared to other adult organs suggests specific functions of these molecules in the nervous system. Clearly, more information on the structure of these receptors is needed to study the relationship between these brain-derived receptors and the large distribution of *bek* mRNA and to further allow a better understanding of the physiological role of FGF molecules in the brain.

ACKNOWLEDGMENTS

We thank J. Avignan for his help during the redaction of this manuscript and Dr. P. J. Munson for giving the Ligand program.

REFERENCES

1. BURGESS, W. H. & T. MACIAG. 1989. Annu. Rev. Biochem. **58:** 575–606.
2. MOENNER, M., B. CHEVALLIER, J. BADET & D. BARRITAULT. 1986. Proc. Natl. Acad. Sci. USA **83:** 5024–5028.
3. COURTY, J., M. C. DAUCHEL, A. MEREAU, J. BADET & D. BARRITAULT. 1988. J. Biol. Chem. **263:** 11217–11220.
4. LEDOUX, D., A. MEREAU, M. C. DAUCHEL, D. BARRITAULT & J. COURTY. 1989. Biochem. Biophys. Res. Commun. **159:** 290–296.
5. DIONNE, C. A., G. CRUMLEY, F. BELLOT, J. M. KAPLOW, G. SEARFOSS, M. RUTA, W. H. BURGESS, M. JAYE & J. SCHLESSINGER. 1990. EMBO J. **9:** 2685–2692.

Characterization of the FGFR-3 Gene and Its Gene Product

KATHLEEN KEEGAN,[a] DANIEL E. JOHNSON,[b]
LEWIS T. WILLIAMS,[b] AND MICHAEL J. HAYMAN[a]

[a]Department of Microbiology
State University of New York at Stony Brook
Stony Brook, New York 11794

[b]Department of Medicine
Howard Hughes Medical Institute
Program of Excellence in Molecular Biology
Cardiovascular Research Institute
University of California
San Francisco, California 95143

The fibroblast growth factors (FGFs) are a family of polypeptide growth factors involved in numerous biological activities including mitogenesis, angiogenesis, and wound healing.[1] Fibroblast growth factor receptors (FGFRs) have been identified in chicken, mouse, and human.[2-4] The amino acid sequence of these receptors predicts a protein containing either two or three immunoglobulinlike domains, a transmembrane domain, and an intracellular tyrosine kinase domain.[2] We have isolated a cDNA, 17B, from the human chronic myelogenous leukemia cell line (CML), K562, with the same structural features as the three immunoglobulin-domain form of FGFRs. The sequence[5] predicts a protein containing a hydrophobic signal sequence, three immunoglobulinlike loops, a transmembrane domain, and tyrosine kinase domain (FIGURE 1). The tyrosine kinase domain of 17B, like other FGFRs, is split into two parts by a 14-amino-acid insert and the juxtamembrane domain is longer than other tyrosine kinase receptors. These features distinguish FGFRs as a distinct class of tyrosine kinase receptors. The kinase domain of this cDNA, called 17B, has 75% amino acid identity to human FGFR-1/flg,[4] 77% to FGFR-2/bek[4] and 94% to chicken Cek2.[6] That 17B is more homologous to Cek2 than to the other FGFRs suggests that it may be the human homologue of Cek2.

To confirm that clone 17B is encoded by a different gene than FGFR-1, we performed Southern blot analysis. DNA probed with either FGFR-1 or 17B shows a

HUMAN FGFR 3

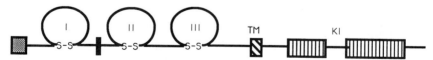

FIGURE 1. Schematic representation of the structural features of human FGFR-3. The FGFR-3 (17B) sequence predicts a protein with a hydrophobic leader sequence (shaded box), three extracellular immunoglobulinlike domains (I, II, and III) with disulfide linked cysteine residues (S), a short acidic sequence (black box), a transmembrane domain (TM, diagonal striped box), and an intracellular tyrosine kinase domain (vertical striped box). The tyrosine kinase domain is split by a 14-amino-acid insert sequence (KI).

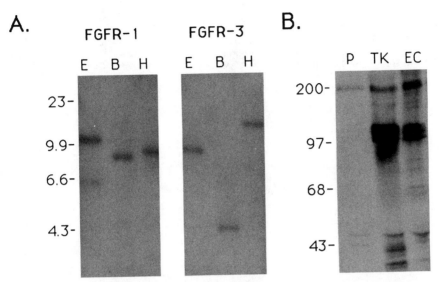

FIGURE 2. A: Southern blot analysis of FGFR-1 and FGFR-3. K-562 DNA was digested with EcoRI (E), BamHI (B), or HindIII (H), separated on a 1% agarose gel, transferred to nitrocellulose, and probed with $[\alpha^{32}P]$dCTP-labeled FGFR-1 or FGFR-3 (17B) probes. A distinct pattern of hybridization is seen with each probe. Sizes of molecular weight standard, in kilobases are indicated at the left. **B:** Expression of FGFR-3 in COS cells. 17B was electroporated into COS cells and, after 48 hours, cells were labeled with ^{35}S-methionine and immunoprecipitated with preimmune serum (P), antiserum against the kinase domain (TK) or against the extracellular domain (EC). Molecular mass standards are denoted at left in kDa.

different pattern of hybridization suggesting the genes are distinct (FIGURE 2A). Chromosomal mapping studies using rodent/human hybrid cell lines localize 17B to chromosome 4 (K. Keegan, M. J. Hayman, and N. Spurr, unpublished results).

To characterize the protein product of 17B, we assayed transient expression in COS cells. Cells were transfected with a plasmid containing 17B and immunoprecipitated with antiserum that recognizes either the kinase domain or the extracellular domain of the cDNA (FIGURE 2B). Both antisera specifically recognize a series of polypeptides with M_rs of 97–125 kDa that are not recognized by the preimmune serum (FIGURE 2B).

Functional analysis of the protein encoded by 17B has shown that it can be activated by both aFGF and bFGF. 17B RNA was transcribed *in vitro,* injected into *Xenopus* oocytes, and assayed in a $^{45}Ca^{2+}$ efflux assay.[5] Addition of either acidic or basic FGF to injected oocytes resulted in a rapid and large efflux of $^{45}Ca^{2+}$ indicating activation of the receptor. Control oocytes, injected with either water or antisense RNA, showed no such efflux. This establishes 17B as an additional member of the FGFR family; thus we have named it FGFR-3.

REFERENCES

1. BURGESS, W. H. & T. MACIAG. 1989. Annu. Rev. Biochem. **58:** 575–606.
2. LEE, P. L., D. E. JOHNSON, L. S. COUSSENS, V. A. FRIED & L. T. WILLIAMS. 1989. Science **245:** 57–60.

3. SAFRAN, A., A. AARON, A. ORR-URTEREGER, G. NEUFELD, P. LONAI, D. GIVOL & Y. YARDEN. 1990. Oncogene **5:** 635–643.
4. DIONNE, C. A., G. CRUMLEY, F. BELLOT, J. N. KAPLOW, G. SEARFOSS, M. REITA, W. H. BURGESS, M. JAYE, & J. SCHLESSINGER. 1990. EMBO J. **9:** 2685–2692.
5. KEEGAN K., D. E. JOHNSON, L. T. WILLIAMS, & M. J. HAYMAN. 1991. Proc. Nat. Acad. Sci. USA **88:** 1095–1099.
6. E. B. PASQUALE. 1990. Proc. Nat. Acad. Sci. USA **87:** 5812–5816.

Novel Human FGF Receptors with Distinct Expression Patterns

JAANA KORHONEN, JUHA PARTANEN, ELINA EEROLA,
SATU VAINIKKA, VESA ILVESMÄKI,
RAIMO VOUTILAINEN, MERVI JULKUNEN,[a]
TOMI MÄKELÄ , AND KARI ALITALO

Laboratories for Cancer Biology and Endocrinology
Departments of Virology and Pathology
University of Helsinki
and
[a] Department of Obstetrics and Gynecology
Helsinki University Central Hospital
00290 Helsinki 29, Finland

INTRODUCTION

PCR cloning of tyrosine kinases from K562 human erythroleukemia cells revealed two cDNAs with strong homology to members of the fibroblast growth factor receptor (FGFR) gene family.[1] Two full-length cDNAs for these genes were isolated, one of which probably represents FGFR-3, the human homologue of the chick Cek-2 gene.[2,3] The other cDNA also encodes a relatively aFGF[4] specific member of the FGFR family, which we have named FGFR-4.[5] The amino acid sequence of FGFR-4 is 56–57% identical with both *flg* and *bek,* the two previously characterized FGFRs, which are 71% identical with each other. FGFR-4 maps to chromosome 5q33-qter.[6] We have analyzed FGFR mRNA expression in various tissues and physiological conditions.

Here we show evidence of FGFR mRNA expression in fetal and adult endocrine tissues, active in metabolism of steroid hormones and in endometrial tissue, which undergoes periodic angiogenesis and regression in a steroid-regulated fashion. We have also looked at FGFR expression tumors and tumor cell lines derived from these tissues.

RESULTS AND DISCUSSION

In fetal tissues, the highest levels of FGFR-4 mRNA were found in the adrenals and testes (data not shown). In order to compare the levels of FGFR-4 expression in fetal and adult adrenals, total RNA was isolated and analyzed by Northern blotting and hybridization with cDNA fragments from the 5′ half of the four receptors. TABLE 1 shows that in the human fetal adrenal glands, FGFR-3 and FGFR-4 are expressed more abundantly than in adult adrenals. In contrast, an adrenal carcinoma was negative for all four FGFRs.

In human fetal and adult liver, all FGFR mRNAs except *flg* are found. Interestingly, a hepatocellular carcinoma expresses *flg* and besides *flg* the Hep G2 hepatocellular carcinoma cell line has also high levels of FGFR-4 and FGFR-3 mRNAs. In contrast, only a barely detectable amount of *bek* mRNA is seen in these cells.

The human endometrium undergoes repeated proliferation and differentiation phases during each menstrual cycle. During the proliferative phase, blood vessels grow into the endometrium, and during the luteal phase, the endometrial tissue is sloughed off. As the FGFs have been implicated in the formation of new blood vessels (angiogenesis), it was of interest to know the pattern of expression of the FGFRs in the endometrium sampled during different phases of the menstrual cycle. For this purpose, polyadenylated RNA isolated from the samples was analyzed. All four FGFR mRNAs were expressed in the endometrium but there was no distinct correlation between the mRNA expression and different phases of the menstrual cycle (TABLE 2). The *flg* mRNA was expressed at highest levels, and both FGFR-3 and FGFR-4 expression was relatively low compared to other FGFR mRNAs. There was some variation in *bek* mRNA in the individual samples, but again no consistent differences were seen.

Besides the endometrium, we analyzed different endometrial carcinoma cell lines and tumors. As shown in TABLE 1, three of five and four of five endometrial carcinoma cell lines express FGFR-4 and FGFR-3 mRNAs, respectively. The RL 95-2 endometrial carcinoma cell line lacks FGFR transcripts. The AN3 CA carcinoma cell line is metastatic and differs from the other endometrial carcinoma cell lines because of enhanced levels of all four mRNAs, especially *flg* mRNA.

We find that the novel FGFR-4 gene is highly expressed in human fetal testes and adrenals. The other FGFR mRNAs are also highly expressed in fetal, but less in adult adrenals. This is consistent with the reported growth regulatory role of FGF during the fetal development of the adrenal gland.[7,8] Human fetal liver has small amounts of FGFR-4 mRNA, but enhanced levels are seen in the Hep G2 hepatocellular carcinoma cell line.

TABLE 1. Expression of FGF Receptor mRNAs in Fetal and Adult Tissues and Tumor Cells[a]

Sample	FGFR-4	FGFR-3	*flg*	*bek*
HF adrenal	+ +	+ +	−	+ +
HA adrenal	−	−	−	+ +
Adrenal CA	−	−	−	−
HF liver	+	+	−	+
HA liver	+	+ +	−	(+)
HEP CA	−	+	+	(+)
Hep G2	+ + +	+ + +	+ + +	−
RL 95-2	−	−	−	−
KLE	−	+	−	+
HEC-1-B	+	+	−	−
HEC-1-A	+	+	−	−
AN3 CA	+ +	+ +	+ + +	+ +

[a]Total RNA was isolated according to Chirgwin[9] and analyzed by Northern blotting using 20–30 μg of total RNA per lane. Filters were hybridized with the FGFR-4, FGFR-3, flg, and bek cDNAs (inserts of plasmids pCD115 and pCD116, respectively; kind gifts from Dr. Graig Dionne and Michael Jaye (Rhone-Poulenc Rorer Central Research, King of Prussia, Pa.]. The samples are as follows: human fetal (HF) and adult (HA) adrenals, respectively; adrenal carcinoma (CA) from a patient suffering from Cushing's syndrome. Human fetal (HF) and adult (HA) liver, a hepatocellular carcinoma (HEP CA) and a hepatocellular carcinoma cell line, Hep G2. RL 95-2: endometrial carcinoma cell line; KLE, HEC-1-A, HEC-1-B: endometrial adenocarcinoma cell lines. AN3 CA: endometrial adenocarcinoma cell line from a metastatic tumor.

TABLE 2. Expression of FGF Receptor mRNAs in the Endometrium during Different Phases of the Menstrual Cycle[a]

Phase	FGFR-4	FGFR-3	*flg*	*bek*
3–4	+	+	+++	−
6	+	+	+++	+
7	+	+	+++	+
9	+	++	++	++
13	+	+++	+++	+++
13/3	+	−	−	++
20/6	−	+	+++	+
21/8	−	−	+	+
22/10	−	−	−	+
24/9	+	+	+++	++
29/12	+	+	++	++

[a]Endometrial samples were obtained from apparently healthy women undergoing hysterectomy for uterine fibroids. Polyadenylated RNA was isolated as described by Sambrook et al.[10] and 8-μg aliquots were subjected to Northern hybridization analysis. The autoradiographic hybridization signals were normalized relative to the β-actin (p41, a kind gift from Dr. M. Buckingham, Pasteur Institute, Paris, France) signal and are expressed as: (−) no signal and (+), (++), (+++) signals of increasing intensity. The day of the menstrual cycle is marked as days after menses/days after ovulation.

Endometrial samples obtained during different phases of the menstrual cycle express all four FGFR mRNAs, but there does not appear to exist any consistent cyclic variation of their levels despite considerable differences in the growth status of the endometrium. Interestingly, highest levels of all four FGFR mRNAs are found in an anaplastic endometrial adenocarcinoma cell line, AN3 CA. Overall, our results indicate that the four different FGFR mRNAs (*flg, bek,* FGFR-3, and FGFR-4) are all independently expressed in both cell culture and *in vivo*. The finding of enhanced FGFR levels in two cell lines from advanced carcinomas may reflect cellular selection during tumor growth.

REFERENCES

1. PARTANEN, J., T. P. MÄKELÄ, R. ALITALO, H. LEHVÄSLAIHO & K. ALITALO. 1990. Proc. Nat. Acad. Sci. USA **87:** 8913–8917.
2. KEEGAN, K., D. E. JOHNSON, L. T. WILLIAMS & M. J. HAYMAN. 1991. Proc. Nat. Acad. Sci. USA **88:** 1095–1099.
3. PASQUALE, E. B. 1990. Proc. Nat. Acad. Sci. USA **87:** 5812–5816.
4. JAYE, M., R. HOWK, W. BURGESS, G. A. RICCA, I. M. CHIU, M. W. RAVERA, S. J. O'BRIEN, W. S. MODI, T. MACIAG & W. N. DROHAN. 1986. Science **233:** 541–545.
5. PARTANEN, J., T. P. MÄKELÄ, E. EEROLA, J. KORHONEN, H. HIRVONEN, L. CLAESSON-WELSH, K. HUEBNER & K. ALITALO. 1991. EMBO J. **10:** 1347–1354.
6. EEROLA, E., J. PARTANEN, L. CANNIZZARO, K. HUEBNER & K. ALITALO. Genes, Chromosomes & Cancer. (In press.)
7. CLAUDE, P., I. M. PARADA, K. A. GORDON, P. A. D'AMORE & J. A. WAGNER. 1988. Neuron **1:** 783–790.
8. STEMPLE, D. J., N. K. MAHANTHAPPA & D. J. ANDERSON. 1988. Neuron **1:** 517–525.
9. CHIRGWIN, J. M., A. E. PRZYBYLA, R. J. MACDONALD & W. J. RUTTER. 1979. Biochemistry **18:** 5294–5299.
10. SAMBROOK, J., E. F. FRITSCH & T. MANIATIS. 1989. *In* Molecular Cloning: a Laboratory Manual. Cold Spring Harbor Laboratory. Cold Spring Harbor, N.Y.

Expression of Fibroblast Growth Factor Receptors during Chick Brain Development

HUBERT HONDERMARCK,[a] DIDIER THOMAS,[a]
JOSÉ COURTY,[b] DENIS BARRITAULT,[b]
AND BÉNONI BOILLY[a]

[a]Laboratory of Biology of Growth Factors
University of Lille I
59655 Villeneuve d'Ascq Cedex, France

[b]Biotechnology Laboratory of Eucaryotic Cells
University of Paris XII
94010 Creteil, France

INTRODUCTION

Fibroblast growth factors (FGFs) are present in embryonic brain[1,2] and may be involved in brain development since, *in vitro*, the proliferation of neuroblasts, astroglial, oligodendroglial, and endothelial cells as well as neurite outgrowth[3] is stimulated by these growth factors. The effects of FGFs are mediated by membrane receptors of M_r ranging from 85 K to 165 K; other membrane interaction sites related to heparinlike molecules were also described but their role is still unknown.[3] In order to understand the biological function of FGFs during the development of brain, we looked for FGF receptors in chick embryonic brain on cellular membrane preparations using radioreceptor assays and cross-linking experiments with ^{125}I-labeled acidic or basic FGF. FGFs were purified from bovine brain[4] and radioiodinated using the chloramine T method. Brain membrane preparations, radioreceptor assays, and cross-linking experiments were conducted as previously described.[5] Messengers of the Bek form of FGF receptor were detected on Northern blot by hybridization with a [32]P cDNA probe for human Bek receptor.[6]

RESULTS

Radioreceptor Assay

Competitive binding studies were conducted using an isotopic dilution of ^{125}I-FGFs with unlabeled FGFs. Scatchard analysis of the data shared two classes of binding sites. The first class presented a high affinity (K_d of 100 pM for aFGF and 25 pM for bFGF), the second a low affinity (K_d = 10 pM for aFGF and 3 nM for bFGF). Cross competition between aFGF and bFGF indicated that these two growth factors interacted with common high- and low-affinity binding sites. The membrane capacity of high-affinity binding sites decreased from E7 (1 ± 0.2 pmole/mg of protein) to E10 (0.5 ± 0.2 pmol/mg of protein) and remained constant until P1. The membrane capacity of low-affinity binding sites was constant from E7 to E15 (20 ± 5 pmole/mg of protein) but increased at P1 (75 pmole/mg of protein).

Cross-linking Experiments

Interaction sites for FGF were identified by cross-linking between FGF and brain membranes at various stages of development. Analysis of the cross-linked complexes by sodium dodecyl sulfate-polyacrylamide gel electrophoresis (SDS-PAGE) and radioautography yielded two major distinct labeled bands with an apparent molecular mass of 110 K and 145 K for each stage studied. Densitometric analysis of the radioautograms showed, for the same amount of radiolabeled FGF and membrane protein, a decrease of relative intensity of the cross-linking complexes during development.

Northern Blot Analysis

Hybridization of poly A^+ mRNAs with a ^{32}P cDNA probe for Bek revealed a 4.5-kilobase (kb) transcript at all stages studied (E7 to P1) but with a more intense signal at E7. A 1.5-kb transcript was also detected.

DISCUSSION

Two classes of FGF binding sites were detected on embryonic chick brain membranes. The values of K_d obtained for high- and low-affinity sites are similar to that described in adult brain.[5,7] Cross-linking experiments revealed two putative receptor forms of M_r 95 K and 130 K. These values correspond to those reported for the receptor purified from adult bovine brain membranes (85 K and 125 K),[8] or from the whole chick embryo (130 K),[9] as well as those obtained from fetal rat cultured hippocampal neurons (85 K and 135 K).[10] The number of high-affinity receptors decreased twofold from E7 to E10. Such a decrease of FGF receptor number during development was also reported for membrane preparations of whole chick embryo[11] and of mouse placenta.[12] In contrast, the number of low-affinity binding sites increased threefold from E15 to P1. These changes during brain development suggest that the temporal action of FGFs may be regulated by the expression of high- and low-affinity membrane binding sites.

REFERENCES

1. MASCARELLI, F., D. RAULAIS, M. E. COUNIS & Y. COURTOIS. 1987. Biochem. Biophys. Res. Commun. **146:** 478–486.
2. RISAU, W., P. GAUTSCHI-SOVA & P. BÖHLEN. 1988. EMBO J. **7:** 959–962.
3. BAIRD, A. & P. BÖHLEN. 1990. *In* Handbook of Experimental Pharmacology. Peptide Growth Factors and Their Receptors. M. B. Sporn & A. B. Roberts, Eds. **95:** 369–418. Springer-Verlag. Berlin, Heidelberg, New York.
4. MOENNER, M., B. CHEVALLIER, J. BADET & D. BARRITAULT. 1986. Proc. Nat. Acad. Sci. USA. **85:** 5024–5028.
5. COURTY, J., C. LORET, B. CHEVALLIER, M. MOENNER & D. BARRITAULT. 1987. Biochimie **69:** 511–516.
6. DIONNE, G. A., G. CRUMLEY, F. BELLOT, J. M. KAPLOW, G. SEARFOSS, M. RUTA, W. H. BURGESS, M. JAYE & J. SCHLESSINGER. 1990. EMBO J. **9:** 2685–2692.
7. LEDOUX, D., A. MEREAU, M. C. DAUCHEL, D. BARRITAULT & J. COURTY. 1989. Biochem. Biophys. Res. Commun. **159:** 590–296.

8. MEREAU, A., I. PIERI, C. GAMBY, J. COURTY & D. BARRITAULT. 1989. Biochimie 71: 865–871.
9. LEE, P. L., D. E. JOHNSON, L. S. COUSENS, V. A. FRIED & L. T. WILLIAMS. 1989. Science. 245: 57–60.
10. WALICKE, P. A., J. J. FEIGE & A. BAIRD. 1989. J. Biol. Chem. 264: 4120–4126.
11. OLWIN, B. B. & S. D. HAUSCHKA. 1990. J. Cell Biol. 110: 503–509.
12. HONDERMARCK, H., J. COURTY, D. LEDOUX, V. BLANCKAERT, D. BARRITAULT & B. BOILLY. 1990. Biochem. Biophys. Res. Commun. 169: 272–281.

Role of FGFs and FGF Receptors in Human Carcinogenesis

D. BIRNBAUM,[a] O. deLAPEYRIERE,[a] J. ADNANE,[b]
C. DIONNE,[c] G. CRUMLEY,[c] M. JAYE,[c]
J. SCHLESSINGER,[d] L. XERRI,[a] C. ESCOT,[e]
P. GAUDRAY,[b] AND C. THEILLET[e]

[a]INSERM U119
Institut Paoli-Calmettes
27, Boulevard Leî Roure
13009 Marseille, France

[b]LGMCH
Nice, France

[c]Rhone-Poulenc Rorer Central Research
King of Prussia, Pennsylvania

[d]New York University Medical Center
New York, New York

[e]INSERM U248 and
Centre P. Lamarque
Montpellier, France

Evidence for the participation of fibroblast growth factors (FGFs) in malignant cell proliferation is well documented in cellular or animal models.[1] The question is now whether FGFs play a role in human carcinogenesis and to what extent. In order to answer this question we have studied two models, representative of different abnormal proliferations.

We detected coamplification of HST/FGFK and INT2, both located on chromosomal band 11q13, in 16% of breast carcinomas[2] (FIGURE 1). Similarly, we observed amplification of one FGF receptor (FGFR) gene (either FLG or BEK) in 12% of tumors[3] (FIGURE 1). Taken together these results point to a role of FGFs (at least of two of them) in human epithelial neoplasias. However, as also observed by others,[4] expression of HST and INT2 is either low[2] or absent altogether in the amplified tumors. Moreover, in few cases, BCL1, a locus located about one megabase centromeric of HST/INT2,[5] is the only amplified marker.[6] BCL1 corresponds, on chromosome 11, to the break-point region of the translocation t(11;14) (q13;q32) observed in some B-cell malignancies. It is therefore probable that another gene, present in the neighborhood of BCL1, is the actual "driver" of the 11q13 amplicon. Still, the amplification of the FGFR genes remains an argument in favor of a participation of FGFs in human cancer. Unless FGFR genes themselves turn out not to be the "drivers" of their respective amplicons.

Due to their properties, FGFs may play an important role in proliferations that present a strong angiogenic component. Therefore we investigated the involvement of FGFs in Kaposi's sarcomas (KS). Indeed, a cascade of growth factors is thought to intervene in the pathogenesis of KS,[7] which are skin tumors mainly associated with Acquired Immune Deficiency Syndrome (AIDS) and characterized by complex vascular lesions showing spindle-shaped cells of undetermined origin. Using in situ

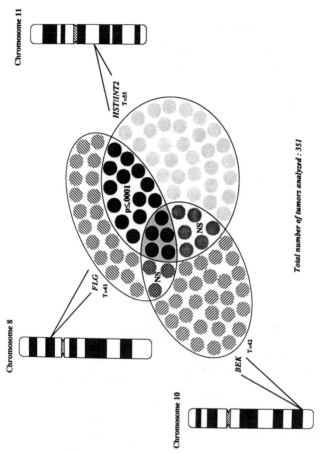

FIGURE 1. Amplification of *FGF* and *FGFR* genes in breast carcinomas. A series of mammary tumor DNAs were analyzed for amplification of *FGF* and *FGFR* genes. None out of 238 showed amplification of either *FGFB, FGF5,* or *FGF6*.[2] A total of 351 tumors were investigated for *HST/FGFK, INT2, BEK,* and *FLG. HST* and *INT2* were coamplified in 55 samples (15.7%). This systematic coamplification can be explained by the small distance separating the two genes on band q13 of chromosome 11. Two FGF receptor genes were also amplified in a significant number of breast carcinomas: 42 (12%) for *BEK* (present on band q26 of chromosome 10) and 41 (11.7%) for *FLG* (located on chromosome 8, band p12). The three amplified populations present overlaps (shaded areas). The only overlap statistically significant is between amplifications of *HST/INT2* and of *FLG*, since 44% of the *FLG*-amplified tumors were also amplified for *HST/INT2.* This result is of particular interest since *HST* was shown to activate the tyrosine kinase activity of *FLG*.[9] In addition, *HST/INT2* amplifications take place in similar subpopulations of relatively well-differentiated, low-grade tumors, in patients presenting metastatic invasion of axillary lymph nodes.[3]

TABLE 1. *In Situ* Hybridizations with *FGF* Probes on KS Lesions[a]

Number of Cases (Serology)	Histological Features	*In Situ* Hybridization with	
		FGFB	*FGF5*
2 (HIV⁺/HIV⁻)	Normal skin	−	−
2 (HIV⁺/HIV⁻)	Vasculitis	−	−
5 (HIV⁺)	KS	+	+
1 (HIV⁻)	KS	+	−

[a]Biopsy samples from five AIDS-KS and one non-AIDS-KS were assayed by *in situ* hybridization onto paraformaldehyde-fixed, paraffin-embedded skin sections for the presence of *FGF* gene transcripts, using labeled single-stranded RNA antisense probes. *FGF5* expression was detected in the characteristic KS spindle-shaped cells in the 5 samples from HIV⁺ patients. *FGFB* transcripts were detected in KS cells as well as in epidermis of both HIV⁺ and HIV⁻ patients. These results strengthen the observations done on KS cell lines[7] and the hypothesis that *FGF* genes are involved in the pathogenesis of Kaposi's sarcomas. The absence of expression of *FGFB* and *FGF5* in vasculitis suggests that abnormal angiogenesis alone does not require these two *FGF* genes' expression.

hybridization, we detected expression of *FGFB* and *FGF5* genes[8] in the spindle-shaped cells (TABLE 1).

Although these results speak in favor of FGFs participating in human carcinogenesis, more work is required to determine their real degree of involvement. On this issue depends the possibility of starting a specific therapy, targeted on the interactions between FGFs and FGFRs.

REFERENCES

1. GOLDFARB, M. 1990. Cell Growth Diff. **1:** 439–445.
2. THEILLET, C., X. LEROY, O. DELAPEYRIERE, J. GROSGEORGES, J. ADNANE, S. RAYNAUD, J. SIMONY-LAFONTAINE, M. GOLDFARB, C. ESCOT, D. BIRNBAUM & P. GAUDRAY. 1989. Oncogene **4:** 915–922.
3. ADNANE, J., P. GAUDRAY, C. DIONNE, G. CRUMLEY, M. JAYE, J. SCHLESSINGER, P. JEANTEUR, D. BIRNBAUM & C. THEILLET. 1991. Oncogene **6:** 659–663.
4. FANTL, V., M. RICHARDS, R. SMITH, G. LAMMIE, G. JOHNSTONE, D. ALLEN, W. GREGORY, G. PETERS, C. DICKSON & D. BARNES. 1990. Eur. J. Cancer **26:** 423–429.
5. HAGEMEIJER, A., M. LAFAGE, M-G. MATTEI, J. SIMONETTI, E. SMIT, O. DELAPEYRIERE & D. BIRNBAUM. 1991. Genes Chrom. Cancer **3:** 210–214.
6. THEILLET, C., J. ADNANE, P. SZEPETOWSKI, M-P. SIMON, P. JEANTEUR, D. BIRNBAUM & P. GAUDRAY. 1988. Oncogene **5:** 147–149.
7. ENSOLI, B., Z. SALAHUDDIN & R. GALLO. 1989. Cancer Cells **1:** 93–96.
8. XERRI, L., J. HASSOUN, J. PLANCHE, V. GUIGOU, J-J. GROB, P. PARC, D. BIRNBAUM & O. DELAPEYRIERE. 1991. Am. J. Pathol. **138:** 9–15.
9. MANSUKHANI, A., D. MOSCATELLI, D. TALARICO, V. LEVYTSKA & C. BASILICO. 1990. Proc. Nat. Acad. Sci. USA **87:** 4378–4382.

cAMP-Mediated Regulation of Adrenocortical Cell bFGF Receptors

C. SAVONA AND J. J. FEIGE

INSERM U244
DBMS, LBIO, BRCE
Grenoble Center for Nuclear Studies, 85X
38041 Grenoble Cedex, France

Mesoderm-derived steroidogenic adrenocortical cells are mitogenically stimulated by both basic and acidic fibroblast growth factors (bFGF and aFGF, respectively).[1] When grown in primary culture, these cells have been reported to express the bFGF gene and to synthesize the bFGF protein.[2] The aim of the present study was to characterize the receptors that mediate the mitogenic effects of bFGF on cultured bovine adrenocortical cells and to investigate their possible regulation by the trophic hormone ACTH. Besides its steroidogenic action, ACTH is also known to stimulate adrenal growth *in vivo*[3] and this has long been very difficult to reconcile with its antiproliferative action observed *in vitro*.[4]

METHODS

Bovine adrenocortical cells (zona fasciculata-reticularis) were prepared by tryptic dissociation of freshly collected glands and grown in primary culture in Ham's F12 medium supplemented with 12.5% horse serum and 2.5% fetal calf serum. Human recombinant basic FGF was a generous gift from Dr. L. Cousens (Chiron, Emeryville, Calif.) It was radioiodinated by the chloramine T method. Specific activities of 50–100 μCi/μg were routinely obtained. ^{125}I-bFGF binding to adrenocortical cells was measured using a radioreceptor assay performed essentially as described by Moscatelli.[5] Cross-linking of ^{125}I-bFGF to intact adrenocortical cells was performed in the presence of 0.25 mM disuccinimidyl suberate (DSS) as previously described,[6] and the cross-linked proteins were analyzed by electrophoresis on 7.5% polyacrylamide–0.1% sodium dodecyl sulfate (SDS) gels and autoradiography.

RESULTS

Scatchard analysis of ^{125}I-bFGF binding to bovine adrenocortical cells demonstrated the presence of two binding systems with affinities of 2 pM (2000 sites/cell) and 400 pM (120,000 sites/cell), presenting the same biochemical features (lectin affinity, heparinase sensitivity, pH, and salt sensitivity) as the high-affinity and low-affinity receptors from other cell types.[5-8] Cross-linking of ^{125}I-bFGF to adrenocortical cells with DSS revealed two radiolabeled complexes: a major band at 145 kDa and a minor band at 125 kDa. Treatment of the cells with heparitinase decreased the binding to the low-affinity receptor but did not modify the pattern of cross-linked species, indicating that these cross-linked proteins represent the high-affinity receptors (data not shown).

We investigated the regulation of bFGF receptor expression by ACTH. This was performed by cross-linking ^{125}I-bFGF to adrenocortical cells pretreated by the hormone for various periods of time at 37°C. ACTH appeared to increase the binding and cross-linking of bFGF to its receptors (FIGURE 1). This increase was

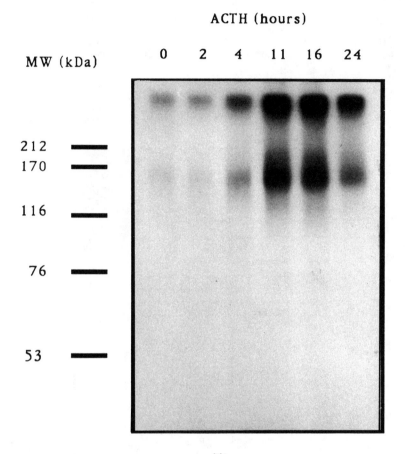

FIGURE 1. Time course of ACTH effect on the binding of ^{125}I-bFGF to its receptors. Bovine adrenocortical cells cultured under standard conditions (1.6×10^6 cells/8-cm^2 well) were incubated for 15 hours in Ham's F12 medium containing 0.5% fetal calf serum and 10^{-9} M ACTH for the indicated periods of time. At the end of the incubation, ^{125}I-bFGF (100 µCi/µg, 0.5×10^6 cpm/ml) was cross-linked to its receptors. The ligand-receptor complexes were analyzed by 7.5% SDS-polyacrylamide gel electrophoresis (SDS-PAGE) and visualized by autoradiography.

time dependent. It was barely detectable after 4 hours of ACTH treatment, reached a maximum (fold stimulation) after 11–16 hours, and decreased after 24 hours (FIGURE 1). Since the steroidogenic effects of ACTH are known to be mediated by cAMP, we investigated whether 8-bromo cAMP, a membrane-permeable analogue of cAMP and forskolin, a direct activator of adenylate cyclase, could mimic the effects of ACTH. Results in FIGURE 2 clearly indicate that such is the case and that stimulation of bFGF binding to its high-affinity receptors by ACTH is mediated by activation of adenylate cyclase.

DISCUSSION

We report here that ACTH, the major trophic hormone of the adrenal cortex, is a potent stimulator of the expression of bFGF receptors. The mechanism of this

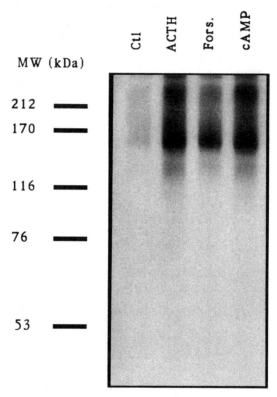

FIGURE 2. Effects of 8-bromo-cAMP and forskolin on the binding of ^{125}I-bFGF to its receptors. BAC cells cultured under standard conditions (1.6×10^6 cells/8-cm^2 well) were incubated for 15 hours in Ham's F12 medium containing 0.5% fetal calf serum, 1 mM isobutylmethylxanthine, and either no addition (Ctl), or 10^{-9} M adrenorticotropin (ACTH), or 25 μM forskolin (Fors.), or 1 mM 8-bromo-cAMP (cAMP). At the end of the incubation, ^{125}I-bFGF (100 μCi/μg, 0.5×10^6 cpm/ml) was cross-linked to its receptors. The ligand-receptor complexes were analyzed by 7.5% SDS-PAGE and visualized by autoradiography.

stimulation appears to be cyclic AMP dependent since it is mimicked by 8-bromo cAMP or forskolin. High-affinity bFGF receptors have been previously reported to be negatively regulated by the phorbol ester TPA, a well-known activator of protein kinase C,[9,10] and it has been suggested that this effect could be mediated by direct phosphorylation of the FGF receptor by protein kinase C.[10] Similarly, bFGF binding could be up regulated through direct phosphorylation of the high-affinity bFGF receptors by the cAMP-dependent protein kinase. The kinetics of this up regulation, however, do not favor this hypothesis. Another possibility is that ACTH stimulates bFGF receptor expression at the gene level. This hypothesis, however, has been rejected since the stimulation of bFGF binding by ACTH is still observed in the presence of RNA synthesis inhibitors (DRB, actinomycin D) or protein synthesis inhibitors (cycloheximide) (data not shown). More probably, ACTH is acting through the recycling to the cell surface of an intracellular pool of receptors, independently of bFGF receptor neosynthesis. Such an effect, independent of protein synthesis, has also been observed for adrenocortical cell TGFβ receptors.[11]

Although the mechanism of the cAMP-mediated regulation of bFGF receptors remains to be fully elucidated, this observation will bring new information for the understanding of the proliferative effects of ACTH on adrenocortical cells *in vivo.*[3,4] Also, recent data support the hypothesis that the regulation of human fetal adrenal growth by ACTH could be mediated by the stimulation of bFGF and insulin-like growth factor II expression.[12] Our data suggest that ACTH-induced increase in bFGF receptor expression may also participate in this process.

REFERENCES

1. ESCH, F., A. BAIRD, N. LING, N. UENO, F. HILL, L. DENOROY, R. KLEPPER, D. GOSPODAROWICZ, P. BÖHLEN & R. GUILLEMIN. 1985. Proc. Nat. Acad. Sci. USA **82:** 6507–6511.
2. SCHWEIGERER, L., G. NEUFELD, J. FRIEDMAN, J. A. ABRAHAM, J. C. FIDDES & D. GOSPODAROWICZ. 1987. Endocrinology **120:** 796–800.
3. DALLMAN, M. F. 1985. Endocr. Res. **10:** 213–242.
4. SCHIMMER, B. P. 1980. Adv. Cyclic Nucleotide Res. **13:** 181–207.
5. MOSCATELLI, D. 1987. J. Cell. Physiol. **131:** 123–130.
6. FEIGE, J. J. & A. BAIRD. 1988. J. Biol. Chem. **263:** 14023–14029.
7. WALICKE, P. A., J. J. FEIGE & A. BAIRD. 1989. J. Biol. Chem. **264:** 4120–4126.
8. OLWIN, B. B. & S. D. HAUSCHKA. 1989. J. Cell. Biochem. **39:** 443–454.
9. HOSHI, H., M. KAN, H. MIOH, J.-K. CHEN & W. L. MCKEEHAN. 1988. FASEB J. **2:** 2797–2800.
10. DOCTROW, S. R. 1989. J. Cell. Biochem. Suppl. 13B: 153.
11. COCHET, C., J. J. FEIGE & E. M. CHAMBAZ. 1988. J. Biol. Chem. **263:** 5707–5713.
12. MESIANO, S., S. H. MELLON, D. GOSPODAROWICZ, A. M. DIBLASIO & R. B. JAFFE. 1991. Proc. Nat. Acad. Sci. USA **88:** 5428–5432.

Local Fate and Distribution of Locally Infused Basic FGF

The Example of the Rat Brain and the *Xenopus* Tail Mesenchyme

ANA-MARIA GONZALEZ, MARINO L. BUSCAGLIA,
JOHN FULLER, RICHARD DAHL, LAURIE S. CARMAN,
AND ANDREW BAIRD

Department of Molecular and Cellular Growth Biology
The Whittier Institute for Diabetes and Endocrinology
at Scripps Memorial Hospital
9894 Genesee Avenue
La Jolla, California 92037

Basic fibroblast growth factor (FGF) is a multifunctional growth factor *in vitro* and *in vivo* with a wide range of activities in different vertebrates.[1] In the mammalian fetus, its synthesis and storage are widely distributed, but in adult life, despite its large distribution, the brain is the only tissue showing a detectable level of synthesis.[2] One of the hallmarks that characterizes basic FGF is its high affinity for heparin *in vitro* and for highly sulfated glycosaminoglycans *in vivo*. This feature can modify the availability of basic FGF stored in the extracellular matrix and has been used extensively for its purification from several diverse tissues.[1] Because the levels of basic FGF mRNA expression in normal adult peripheral tissues are undetectable and basic FGF is present in high amounts in the extracellular matrix, we[3,4] and others[5,6] have hypothesized that the extracellular matrix constitutes a mechanism by which the cell can store basic FGF outside the cell for long periods of time.

In this paper, we report the fate of basic FGF after local injection in the adult rat brain and in the *Xenopus* tadpole tail fin. Immunohistochemistry and histoautoradiography studies reveal that the fate of exogenous basic FGF follows a similar pattern in these two different systems. It is characterized by an exceptionally long local binding (FIGURE 1). At the time of injection, basic FGF is detected in the intercellular spaces and then progressively binds to the extracellular matrix (ECM). After 2 days in the brain and 20 days in the tadpole, the basic FGF is associated with specific target cells (i.e., the blood vessels and also ependymal cells in the brain). Basic FGF is thus not only stored locally in the ECM, but is progressively transferred to other structures. When analyzed by Western blotting (FIGURE 2), the 18-kDa basic FGF shows a rather slow catabolism into smaller sized immunoreactive basic FGFs. The 18-kDa basic FGF can still be detected after four and seven days in the rat brain and after 4, 7, 10, and 20 days in the tail mesenchyme of the tadpoles.

The results presented here suggest that basic FGF is not only stored in the ECM but can also be specifically transferred and then internalized by specific cell populations, after being actively concentrated over target tissues. The findings support the hypothesis that the matrix can act as a long-term storage site that sequesters a biologically active basic FGF.

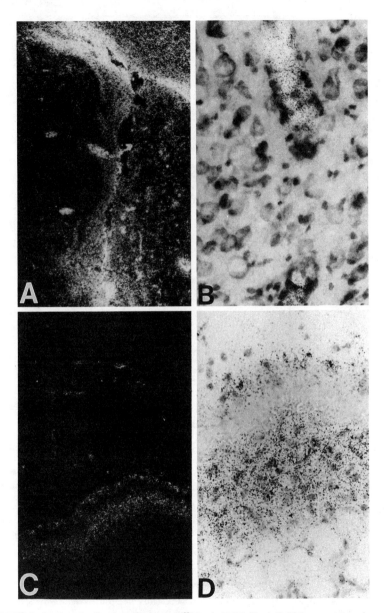

FIGURE 1. *In vivo* distribution of injected ^{125}I-basic FGF. Dark field image at two days after injection into the striatal parenchyma of the rat brain shows the radioactivity is localized in the corpus callosum, the parenchyma, and over the blood vessels (panel A). Bright field image shows the localization of silver grains over blood vessels seven days postinjection (panel B). Panel C is a dark field image of the tadpole tail fin 12 days postinjection showing localization over blood vessels, fibroblasts, and local skin epithelium. Panel D is a bright field view of panel C, showing silver grains in the skin epithelium.

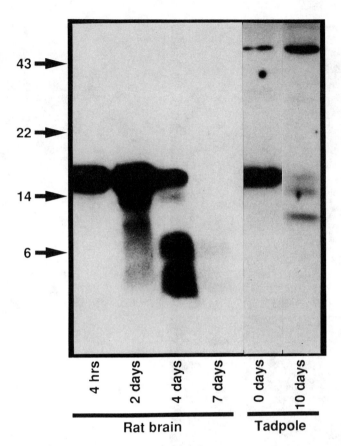

FIGURE 2. *In vivo* metabolism of injected basic FGF. The infused 18-kDa basic FGF form shows a slow catabolism into lower molecular weight forms after injection in the rat brain or in the tadpole tail fin. Injected tissue extracts were purified for basic FGF by absorption to heparin-Sepharose. ^{125}I-basic FGF was injected into the rat brain, and a sodium dodecyl sulfate-polyacrylamide gel electrophoresis was performed on extracts of tissues obtained from the injection site. The tadpole was injected with nonradiolabeled basic FGF; extracted FGF was detected by Western blotting using a guinea pig antirecombinant human basic FGF antibody followed by ^{125}I-labeled protein A.

REFERENCES

1. BAIRD, A. & P. BÖHLEN. 1990. Fibroblast growth factors. *In* Growth Factors. M. Sporn & A. Roberts, Eds. 369–417. Springer-Verlag. New York, N.Y.
2. EMOTO, N., A. M. GONZALEZ, P. A. WALICKE, E. WADA, D. M. SIMMONS, S. SHIMASAKI & A. BAIRD. 1989. Identification of specific loci of basic fibroblast growth factor synthesis in the rat brain. Growth Factors **2:** 21–29.
3. BAIRD, A. & N. LING. 1987. Fibroblast growth factors are present in the extracellular matrix produced by endothelial cells in vitro: implications for a role of heparinase-like enzymes in the neovascular response. Biochem. Biophys. Res. Commun. **142:** 428–435.
4. BAIRD, A. & P. WALICKE. 1989. Fibroblast growth factors. Br. Med. Bull. **45:** 438–452.

5. VLODAVSKY, I., J. FOLKMAN, R. SULLIVAN, R. FRIDMAN, R. ISHAI-MICHAELI, J. SASSE & M. KLAGSBRUN. 1987. Endothelial cell–derived basic fibroblast growth factor: synthesis and deposition into subendothelial extracellular matrix. Proc. Nat. Acad. Sci. USA **84:** 2292–2296.
6. FOLKMAN, J., M. KLAGSBRUN, J. SASSE, M. G. WADZINSKI, D. INGBER & I. VLODAVSKY. 1988. A heparin-binding angiogenic protein—basic fibroblast growth factor—is stored within basement membrane. Am. J. Pathol. **130:** 393–400.

Multiple Molecular Weight Forms of Basic Fibroblast Growth Factor Are Developmentally Regulated in the Rat Central Nervous System

SUZANNE GIORDANO, LARRY SHERMAN,[a] AND
RICHARD MORRISON[b]

*R.S. Dow Neurological Sciences Institute
& Comprehensive Cancer Center
1120 Northwest 20th Avenue
Portland, Oregon 97209*

*[a]Department of Cell Biology and Anatomy
Oregon Health Sciences University
3181 Southwest Sam Jackson Park Road
Portland, Oregon 97210*

Basic fibroblast growth factor (bFGF) is a heparin-binding protein originally identified in pituitary and brain extracts. Although bFGF has been associated with the nervous system since its initial identification, its function in neural tissue is not clearly understood.[1] All of the major cell types comprising the central nervous system (CNS) respond to bFGF. Thus, bFGF has been observed to stimulate the proliferation of astrocytes and oligodendrocytes[2] and to support the survival of postmitotic neurons.[3] bFGF has also been shown to enhance the survival of lesioned cholinergic neurons in the basal forebrain, suggesting that bFGF may influence developmental events in brain and may also play a role in wounding and regenerative responses in the CNS.[4] It is not clear how one protein could be responsible for this multitude of effects. However, recent investigations have documented the existence of multiple molecular weight forms of bFGF protein. *In vitro* experiments have shown that multiple molecular weight forms of approximately 18, 21, 22, and 24 kDa can be transcribed from a single mRNA species. These proteins represent amino terminal extensions of the 18 kDa bFGF and appear to be initiated at leucine codons upstream of the first methionine.[5,6] Furthermore, the different forms of bFGF appear to have distinct intracellular localizations, suggesting that they may have unique activities. The different molecular weight forms of bFGF may have unique cellular localizations and different patterns of temporal expression which could account for their diverse activities within the CNS.

Therefore, we investigated the expression of bFGF protein forms in the developing rat CNS in order to determine (1) if bFGF expression is developmentally regulated in the CNS and (2) if bFGF expression varies between different brain regions. Western blot analysis of proteins isolated from adult rat brain demonstrated the presence of four bFGF protein forms with approximate molecular weights of 18, 21, 22, and 34 kDa (FIGURE 1A). All four forms bound to heparin and their staining on blots was abolished in the presence of human recombinant bFGF (not shown). In

---

[b]To whom correspondence should be addressed.

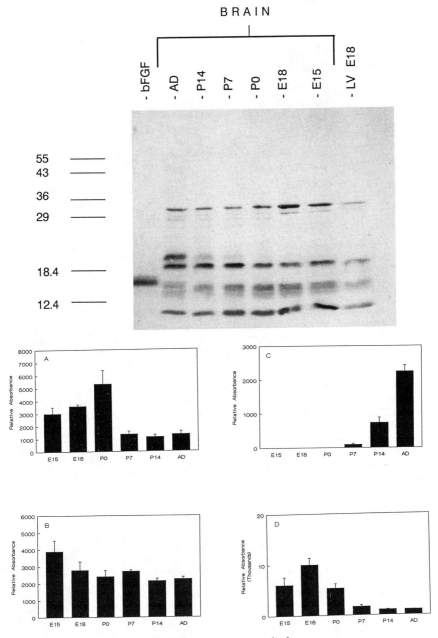

FIGURE 1. Legend on overleaf.

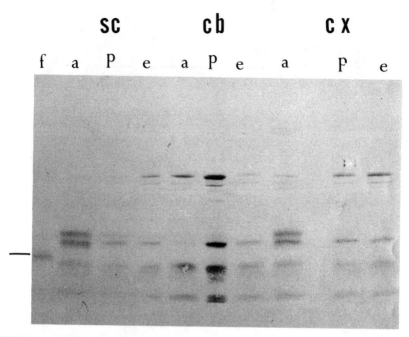

FIGURE 2. Expression of bFGF in specific regions of the developing rat central nervous system. Proteins were extracted, purified, and visualized as explained in FIGURE 1. Western blot of proteins extracted from rat spinal cord (SC), cerebellum (Cb), and cortex (Cx) at ages E18 (e), P0 (p), and adult (a); 600 μg of total protein was incubated with heparin Affigel.

contrast, the pattern of bFGF expression changed dramatically during development (FIGURE 1A and B). Embryonic brain contained only three bFGF forms: 18, 21, and 34 kDa. Expression of the 22-kDa form, which was so prominent in adult brain, was not observed until after birth (P0). The expression of this form steadily increased reaching maximum levels in the adult. The expression of the 18- and 21-kDa forms peaked at E18 and P0 respectively and were both significantly reduced in adult brain. A small reduction in the expression of the 34-kDa form was also observed during this time course (FIGURE 1B).

FIGURE 1. Developmental time course of bFGF expression in rat brain. **A:** Proteins were extracted from entire rat brain in buffer containing 2M NaCl. Equal amounts of total protein (600 μg) were subsequently incubated with Affigel-heparin (Biorad) to enrich for bFGF. Eluted proteins were loaded onto sodium dodecyl sulfate polyacrylamide gel and analyzed by western blotting using the DE6 monoclonal antibody specific for bFGF and a secondary antibody conjugated to alkaline phosphatase. Proteins were extracted from rat brain at ages E15, E18, P0, P7, P14, and adult and from rat liver at age E18 (labeled LV E18). In addition 20 ng of human recombinant bFGF (18kD) and molecular weight markers were run as standards. **B:** Data from three separate western blots (including blot shown above) were compiled and graphed to analyze changes in the expression of the four bFGF protein forms during the observed stages of development. The relative absorbance of each band is shown for all developmental time points examined. A, 18 kDa; B, 21 kDa; C, 22 kDa; D, 34 kDa.

bFGF expression was also examined in several subregions of the developing rat CNS. The spinal cord and cortex showed the same pattern of bFGF expression as observed in whole brain. However, in the cerebellum both the 21- and 22-kDa forms were absent in the adult (FIGURE 2).

Alterations in the expression of bFGF forms during brain development suggest that individual molecular weight forms may have discrete functions in the central nervous system.

REFERENCES

1. KALCHEIM, C. & G. NEUFELD. 1990. Development **109:** 203–215.
2. SILBERBERG, D. H. & A. P. A. ECCLESTON. 1985. Dev. Brain Res. **21:** 315–318.
3. MORRISON, R. S., A. SHARMA, J. DE VELLIS & R. BRADSHAW. 1986. Proc. Nat. Acad. Sci. USA **83:** 7537–7541.
4. ANDERSON, K. J., D. DAM, S. LEE & C. W. COTMAN. 1988. Nature 360–361.
5. FLORKIEWICZ, R. Z. & A. SOMMER. 1989. Proc. Nat. Acad. Sci. USA **86:** 3978.
6. PRATS, R., M. KAGHAD, C. PRATS, M. KLAGSBURN, J. M. LELIAS, P. LIAUZUN, P. CHALON, J. P. TAUBER, F. AMALRIC, J. SMITH & D. CAPUT. 1989. Proc. Nat. Acad. Sci. USA **86:** 1836.

Internalized bFGF is Translocated to the Nuclei of Venular Endothelial Cells and Established Fibroblast Cell Lines

JAMES R. HAWKER AND HARRIS J. GRANGER

Microcirculation Research Institute and
Department of Medical Physiology
College of Medicine
Texas A&M University
College Station, Texas 77843

Basic fibroblast growth factor (bFGF) is a potent mitogen for endothelial cells, fibroblasts, and other cell types of mesenchymal origin. The molecular mechanisms by which bFGF binding to its cellular receptor transduces signals to the nucleus to

TABLE 1. Comparison of Intracellular Distribution of ^{125}I-bFGF after Internalization (4 hours, 37°C) in Different Cell Lines

Cell Line[a]	Cell Fraction[b]	Femtomole Bound/ 10^5 Cells	Percent Intracellular Distribution	Stimulation of DNA Synthesis[c]	FGF Receptor MW (kDa)[d]
CVEC	Nuclei	0.278	28	Yes	110
	Cytoplasm	0.702	72		
	Receptor	0.203			
	Matrix	1.28			
BHK-21	Nuclei	0.194	14	Yes	135, 121
(CCL 10)	Cytoplasm	1.2	86		
	Receptor	0.125			
	Matrix	0.753			
Balb/c 3T3	Nuclei	0.43	17	Yes	160
(CCL 163)	Cytoplasm	2.1	83		
	Receptor	0.33			
	Matrix	1.72			
CHO-K1	Nuclei	0.0352	6.2	Yes	133, 121
(CCL 61)	Cytoplasm	0.534	93.8		
	Receptor	0.044			
	Matrix	0.324			

[a] CVEC were isolated from 15 μm beads as described.[1] The other cell lines were from the American Type Culture Collection.

[b] Cell cultures were washed with acidic (pH 4) and high salt (2 M) buffers to remove remaining high- and low-affinity binding, respectively, after internalization as described.[5] Cells were washed in this order since we have found that washing cells with 2 M NaCl does diminish high-affinity binding.[4] Cell monolayers were then trypsinized, homogenized, and nuclei isolated by microcentrifugation through a sucrose cushion as described,[6] before counting different cell fractions.

[c] Measured by ^{3}H-thymidine incorporation into quiescent cultures treated with bFGF under low serum conditions as described.[2]

[d] ^{125}I-bFGF (10 ng/ml) was chemically cross-linked to cell surface receptors and analyzed by sodium dodecyl sulfate-polyacrylamide gel electrophoresis as described.[3,4]

424

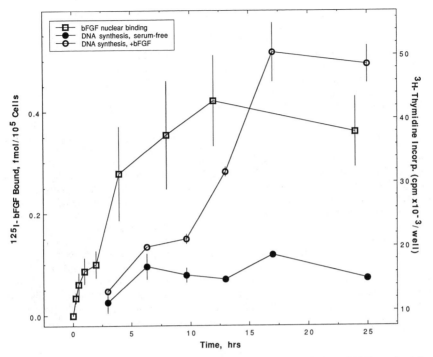

FIGURE 1. Kinetics of bFGF nuclear binding vs. the kinetics of initiation of DNA synthesis in CVEC. The data shown are from separate experiments. Nuclear binding of ^{125}I-bFGF ($n = 5$) at increasing time points after growth factor addition was measured as described in TABLE 1. The DNA synthesis assay with 5 ng/ml bFGF ($n = 3$) was also performed as described in the legend to TABLE 1. Error bars represent standard deviations.

initiate DNA replication and cell division are unknown. We have begun investigating the hypothesis that internalization and nuclear translocation of surface-bound growth factor may play a role in nuclear signal transduction. We have compared internalization and nuclear translocation, receptor cross-linking, and biological activity of bFGF in quiescent coronary venular endothelial cells (CVEC), baby hamster kidney (BHK-21) fibroblasts, Balb/c 3T3 fibroblasts, and chinese hamster embryo (CHO) cells.

All cell types exhibited dose-dependent stimulation of DNA synthesis by bFGF in low serum medium, cross-linking of ^{125}I-bFGF to one or more receptor species, and three cell types showed specific translocation of a significant proportion of internalized ^{125}I-bFGF (14–28%) to the nucleus (TABLE 1). Interestingly, CVEC accumulated the highest proportion (28%) of internalized ^{125}I-bFGF in the nucleus, while CHO cells accumulated relatively little in the nucleus (6.2%) but substantial amounts into the cytoplasm (TABLE 1).

Isolated nuclei from parallel control cultures were shown to be pure by microscopy, propidium iodide staining, and by alkaline phosphatase activity, a marker for endothelial cell plasma membranes. Specificity of nuclear ^{125}I-bFGF binding was shown in control experiments. ^{125}I-bFGF bound to cells at 4°C or heat-inactivated ^{125}I-bFGF (5 minutes, 90°C) displayed residual nuclear and cytoplasmic accumulation compared to control cultures treated with native ^{125}I-bFGF at 37°C. Moreover, ^{125}I-cytochrome c, an unrelated protein of similar size and charge as bFGF, also showed background nuclear and cytoplasmic binding, ruling out nonspecific uptake.

In all cell types the major nuclear species are the 18-kilodalton (kDa) species and small amounts of the 16-kDa fragment. In contrast, cytoplasmic bFGF displayed more extensive degradation to the 16-kDa and smaller fragments.

Nuclear binding of ^{125}I-bFGF is low but detectable within 15 minutes in CVEC, nearly reaching equilibrium at 4–8 hours after growth factor addition (FIGURE 1). Cytoplasm, on the other hand, continues to accumulate bFGF for up to 24 hours. Somewhat surprisingly, nuclear ^{125}I-bFGF accounts for up to ~40% of the total internalized growth factor at early time points (0–1 hours). The proportion of nuclear bound bFGF decreases or levels off at later time points as the cytoplasm accumulates more FGF. Nuclear bFGF binding has reached equilibrium by the time of initiation of DNA synthesis (FIGURE 1). These data suggest, but do not prove, that nuclear bFGF binding may play a role in the initiation of DNA synthesis. Further experiments are required to answer this question.

The reasons for differences in nuclear and cytoplasmic binding of bFGF between different cell lines are unknown. They may relate to the nature of the FGF receptor species and the signal transduction pathways operant in each cell type. It is well established that many cell lines, e.g., Balb/c 3T3 cells, require the action of more than one growth factor for maximal mitogenic response, whereas venular endothelial cells, because of their unique physiological role, may have evolved a mechanism to rapidly respond to a single growth factor, e.g., bFGF. In support of results presented here, two papers have been recently published documenting nuclear binding of acidic and basic FGF in target cells.[7,8]

REFERENCES

1. SCHELLING, M. E., C. J. MEININGER, J. R. HAWKER & H. J. GRANGER. 1988. Am. J. Physiol. **254:** H1211–H1217.
2. CHAROLLAIS, R. H. & J. MESTER. 1989. J. Cell Physiol. **137:** 559–564.
3. NEUFELD, G. & D. GOSPODAROWICZ. 1985. J. Biol. Chem. **260:** 13860–13868.
4. SCHELLING, M. E., J. R. HAWKER & H. J. GRANGER. 1989. FASEB J. **3:** A1317.
5. MOSCATELLI, D. 1988. J. Cell. Biol. **107:** 753–759.
6. BUNCE, C. M., J. A. THICK, J. M. LORD, D. MILLS & G. BROWN. 1988. Anal. Biochem. **175:** 67–73.
7. BALDIN, V., A.-M. ROMAN, I. BOSC-BIERNE, F. AMALRIC & G. BOUCHE. 1990. EMBO J. **9**(5): 1511–1517.
8. IMAMURA, T., K. ENGLEKA, X. ZHAN, Y. TOKITA, R. FOROUGH, D. ROEDER, A. JACKSON, J. A. M. MAIER, T. HLA & T. MACIAG. 1990. Science **249**(4976): 1567–1570.

Subcellular Distribution of Multiple Molecular Weight Forms of bFGF in Hepatoma Cells

DAVID R. BRIGSTOCK AND MICHAEL KLAGSBRUN

Departments of Surgery and Biological Chemistry
Children's Hospital and Harvard Medical School
300 Longwood Avenue
Boston, Massachusetts 02115

INTRODUCTION

It is now known that four translational start sites exist for human basic fibroblast growth factor (bFGF) resulting in the synthesis of a bFGF protein of M_r 18,000 and three N-terminally extended forms of M_r 22,500, 23,100, and 24,000.[1-3] All human bFGF proteins appear to lack classical secretory signal peptides[1,2] and are principally cell associated.[3,4] The lack of bFGF secretion has raised the possibility that bFGF may be an intracrine factor.[5] Since there may be differences in the subcellular localization of the individual molecular weight forms of bFGF,[2] we have used cell fractionation techniques to study the intracellular distribution of bFGF proteins in the human hepatoma cell line SK Hep-1.

METHODS

Subcellular fractions were prepared from cultured SK Hep-1 cells as described.[6,7] Cytosol and NaCl extracts of nuclei and membranes were individually subjected to Biorex-70 cation-exchange chromatography followed by heparin-affinity chromatography.[3] Growth factor activity in column fractions was monitored using a Balb/c 3T3 DNA synthesis assay.[8] Active fractions from the heparin-affinity step were subjected to sodium dodecyl sulfate-polyacrylamide gel electrophoresis (SDS-PAGE) and electroblotting.[3] Nitrocellulose blots were probed with antisera to either the peptide bFGF[33-43][3] or the peptide RGRAAERVGGRGRGAAAPRAAPGARG-PRQGLG-Cys which is 90% homologous to residues −35 to −2 of the N-terminal extension sequence of human bFGF.[3]

RESULTS AND DISCUSSION

The distribution of subcellular markers among the fractions (i.e., lactate dehydrogenase for cytosol, DNA for nuclei, 5'-nucleotidase for membranes) was essentially as predicted. However, it was found that bFGF was present in each of the hepatoma cell subcellular compartments and that there were both quantitative and qualitative differences in bFGF subcellular distribution. As shown in FIGURE 1, growth factor activity in the subcellular fractions was retained on Biorex-70 columns and heparin-affinity columns and required, respectively, 0.6 M and 1.5 M NaCl for elution,

427

characteristic of the elution position of bFGF. Correcting for the losses of the relevant subcellular marker, it was calculated that, as a proportion of the total bFGF purified from all three subcellular fractions by heparin-affinity fast protein liquid chromatography (FPLC) alone, 65% of biologically active bFGF was present in cytosol, 17% in nuclei, and 18% in membranes. When Western blots were probed with an antiserum raised against bFGF[33-43] (FIGURE 2A), 18-kDa bFGF was detected in all subcellular fractions whereas 22.5/23.1-kDa bFGF and 24-kDa bFGF were mainly present in nuclei and membranes. To confirm that the 22.5 to 24-kDa proteins were authentic forms of bFGF, we also probed blots with an antiserum that

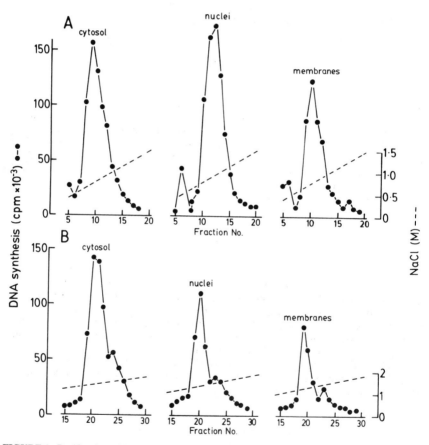

FIGURE 1. Purification of bFGF from subcellular fractions. **A:** Cytosol and 1 M NaCl extracts of nuclei or membranes were diluted and applied to individual Biorex-70 columns that had been equilibrated with 10 mM Tris HCl (pH 7.4) containing 0.1 M NaCl. After sample application, the columns were washed with equilibration buffer and bound proteins were subsequently eluted with a 0.1–3 M NaCl gradient in 10 mM Tris HCl (pH 7.4). Fractions of approximately 2 ml were collected and assayed for their ability to stimulate [³H]thymidine incorporation into the DNA of quiescent Balb/c 3T3 cells.[8] **B:** The active fractions from the Biorex-70 step were pooled and applied directly to a TSK heparin-affinity FPLC column. The column was washed and eluted with a 0.6–2 M NaCl gradient as described.[3] Fractions of 1 ml were collected during gradient formation and assayed for their stimulation of DNA synthesis in 3T3 cells.

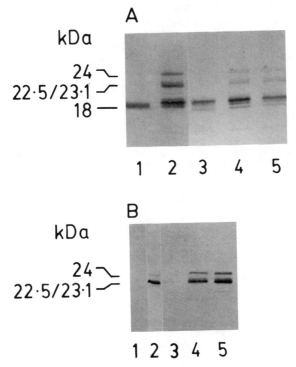

FIGURE 2. Western blot analysis of bFGF in subcellular fractions. Basic FGF purified from subcellular fractions by Biorex cation exchange chromatography and heparin-affinity chromatography (see FIGURE 1) was subjected to SDS-PAGE and Western blotting. Nitrocellulose blots were incubated with rabbit antisera raised against either (A) bFGF[33-43] or (B) a bFGF N-terminal extension peptide sequence. Immunoreactive bands were subsequently visualized by incubation of the blots with alkaline phosphatase–conjugated goat antirabbit immunoglobulin G followed by BCIP/NBT chromogenic substrates.[3] **A** and **B:** Lane 1, recombinant 18-kDa bFGF; lane 2, whole cell extract; lane 3, cytosol; lane 4, nuclei; lane 5, membranes.

specifically recognizes the N-terminal extension of human bFGF. Using this antibody, it was possible to verify that high molecular weight bFGFs were present at high levels in nuclei and membranes but absent from cytosol (FIGURE 2B).

Our results regarding the intracellular distribution of multiple bFGF proteins are very similar to those recently reported by Renko *et al.*[9] and are suggestive of possible functional differences between the various molecular weight forms of bFGF. Since an N-terminally extended form of the FGF oncogene *int*-2 is localized principally in the nucleus,[10] the highly cationic N-terminal extensions[3] of both bFGF and *int*-2 may be important in the intracellular translocation of these proteins.

REFERENCES

1. PRATS, H., M. KAGHAD, A. C. PRATS, M. KLAGSBRUN, J. M. LELIAS, P. LIAUZAN, P. CHALON, J. P. TAUBER, F. AMALRIC, J. A. SMITH & D. CAPUT. 1989. Proc. Nat. Acad. Sci. USA **86:** 1836–1840.

2. FLORKIEWICZ, R. & A. SOMMER. 1989. Proc. Nat. Acad. Sci. USA **86:** 3978–3981.
3. BRIGSTOCK, D. R., M. KLAGSBRUN, J. SASSE, P. A. FARBER & N. IBERG. 1990. Growth Factors **4:** 42–52.
4. VLODAVSKY, I., R. FRIDMAN, R. SULLIVAN, J. SASSE & M. KLAGSBRUN. 1987. J. Cell. Physiol. **131:** 402–408.
5. LOGAN, A. 1990. J. Endocrinol. **125:** 339–343.
6. EVANS, W. H. 1979. *In* Laboratory Techniques in Biochemistry and Molecular Biology. T. S. Work & E. Work, Eds. **7:** 1–266. Elsevier/North Holland Biomedical Press. Amsterdam, the Netherlands.
7. MURAMATSU, M., T. HAYASHI, M. ONISHI, K. SAKAI & T. KASHIMAYA. 1974. Exp. Cell. Res. **88:** 345–351.
8. KLAGSBRUN, M. & Y. SHING. 1985. Proc. Nat. Acad. Sci. USA **82:** 805–809.
9. RENKO, M., N. QUARTO, T. MORIMOTO & D. RIFKIN. 1990. J. Cell. Physiol. **144:** 108–114.
10. ACLAND, P., M. DIXON, G. PETERS & C. DICKSON. 1990. Nature **343:** 662–665.

Internalization of Basic Fibroblast Growth Factor by CCL39 Fibroblast Cells

Involvement of Heparan Sulfate Glycosaminoglycans

LEILA GANNOUN-ZAKI, ISABELLE PIERI,
JOSETTE BADET, MICHEL MOENNER, AND
DENIS BARRITAULT

Biotechnology Laboratory of Eucaryotic Cells
University of Paris XII
Avenue du Général de Gaulle
94010 Créteil, France

Basic fibroblast growth factor (bFGF) is a cationic polypeptide that has been purified from various tissues by taking advantage of its affinity for heparin. Basic FGF interacts with a wide variety of cells as a mitogenic and differentiating agent.[1,2] These cellular responses appear to be mediated via specific high-affinity cell-surface receptors whose molecular weights have been estimated at 130,000–150,000 Daltons by cross-linking experiments. In addition to its receptors, bFGF recognizes, with lower affinity, distinct and more abundant pericellular sites.[2,3]

We have previously shown that bFGF, consecutively to its interaction with CCL39 cells, at 37°C, was rapidly internalized via high- and low-affinity binding sites according to the concentration of the growth factor used.[4]

To further study these different processes, bFGF was derivatized with a photocatalyzable cross-linker, N-Hydroxysuccinimidyl-4-azidobenzoate (HSAB) in order to visualize and characterize the bFGF binding site complexes internalized after interaction of the growth factor with the surface of CCL39 cells. According to the bFGF/HSAB molar ratio used in the coupling reaction, the resulting HSAB-bFGF displayed different growth activity on CCL39 cells. ^{125}I-HSAB-bFGF 1 was found mitogenic on CCL39 cells while ^{125}I-HSAB-bFGF 2 was not (FIGURE 1). Scatchard analyses of binding data at 4°C on bovine brain membrane preparations[5] evidenced the existence of both high- and low-affinity sites for ^{125}I-HSAB-bFGF 1, while only low-affinity sites for ^{125}I-HSAB-bFGF 2 were detected.

Cross-linking experiments with ^{125}I-HSAB-bFGF 1 at 4°C visualized high- and low-affinity complexes with M_rs ranging from 80,000 to 250,000 daltons. After internalization at 37°C, high-affinity complexes from 115,000 to 180,000 daltons could not be detected while low-affinity complexes of M_r 230,000–250,000 and 80,000–90,000 daltons were present in the cellular extracts.

Although ^{125}I-HSAB-bFGF 2 kept its affinity for heparin-Sepharose, it did not induce DNA synthesis in CCL39 cells but was internalized. This process decreased partially after pretreatment of the cells with heparinase II, whereas chondroitinase ABC had no effect. Affinity labeling experiments after internalization, at 37°C, of ^{125}I-HSAB-bFGF 2 by CCL39 cells showed, among other bands, the presence of intracellular high molecular radioactive complexes (230,000–250,000 daltons). These complexes were found to be sensitive to heparinase II pretreatment (FIGURE 2).

These results suggest that biologically active bFGF internalizes via both high-

431

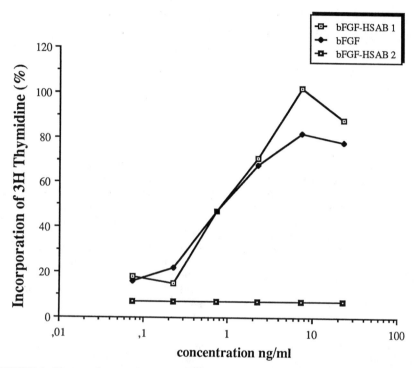

FIGURE 1. Bioassay for growth activity of ^{125}I-HSAB-bFGF. Basic FGF was first coupled to HSAB and then iodinated with [125]INa using two experimental conditions: ^{125}I-HSAB-bFGF 1: 1.5 µM of bFGF were incubated with 100 µM of HSAB for 30 minutes and then iodinated with 0.6 mCi of [125]INa. ^{125}I-HSAB-bFGF 2: 0.8 µM of bFGF was incubated with 100 µM of HSAB for 30 minutes and then iodinated with 1 mCi of [125]INa. ^{125}I-HSAB-bFGF 1 and 2 were then purified on a heparin-Sepharose column. Subconfluent cultures of CCL39 cells in 48-well plates were arrested in G_0/G_1 by a 24-hour incubation in serum-free Dulbecco's minimal essential medium. bFGF and ^{125}I-HSAB-bFGF were then added for 24 hours, and ^{3}H-thymidine (0.5 µCi/well) was incubated for the last 4 hours. ^{3}H-thymidine incorporation was determined after TCA precipitation and cell solubilization. The radioactivity was counted by liquid scintillation in a β counter.

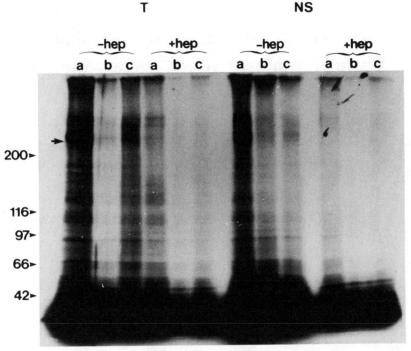

FIGURE 2. Cross-linking experiment of ^{125}I-HSAB-bFGF 2 on heparinase-pretreated CCL39 cells. Cells were pretreated or not with 0.5 U/ml of heparinase II for 1 hour at 37°C, washed with binding buffer, and incubated with 1 ng/ml of ^{125}I-HSAB-bFGF 2 for 3 hours at 4°C in the absence (T) or in the presence (NS) of a 100-fold molar excess of bFGF. **a:** Cells were submitted to UV flash for 7 minutes at 254 nm, then scraped from culture dishes. Cell-associated radioactivity was analyzed by a 3–10% gradient polyacrylamide gel electrophoresis and autora-diography. **b:** Cells were washed with acetic acid buffer for 5 minutes at 4°C,[6] submitted to UV flash, and analyzed as described above. **c:** Cells were transferred at 37°C for 30 minutes, rinsed with binding buffer, and washed with acetic acid buffer. After UV irradiation, the internalized radioactivity was analyzed by electrophoresis.

and low-affinity sites. The internalization process via low-affinity sites might involve interactions with heparan sulfate glycosaminoglycans.

REFERENCES

1. GOSPODAROWICZ, D., G. NEUFELD & L. SCHWEIGERER. 1986. Cell Diff. **19:** 1–17.
2. BURGESS, W. H. & T. MACIAG. 1989. Annu. Rev. Biochem. **58:** 575–606.
3. SAKSELA, O. & D. B. RIFKIN. 1990. J. Cell Biol. **110:** 767–775.
4. GANNOUN-ZAKI, L., I. PIERI, J. BADET, M. MOENNER & D. BARRITAULT. (Submitted for publication.)
5. COURTY, J., M. C. DAUCHEL, A. MEREAU, J. BADET & D. BARRITAULT. 1988. J. Biol. Chem. **263:** 11217–11220.
6. HAIGLER, H. T., F. R. MAXFIELD, M. C. WILLINGHAM & I. PASTAN. 1980. J. Biol. Chem. **255:** 1239–1241.

Expression and Characterization of a Basic Fibroblast Growth Factor–Saporin Fusion Protein in *Escherichia coli*[a]

IGNACIO PRIETO, DOUGLAS A. LAPPI, MICHAEL ONG,
RISË MATSUNAMI, LUCA BENATTI,[b]
RICARDO VILLARES,[c] MARCO SORIA,[b,d]
PAOLO SARMIENTOS,[b] AND ANDREW BAIRD

Department of Molecular and Cellular Growth Biology
The Whittier Institute
9894 Genesee Avenue
La Jolla, California 92037

[b]*Farmitalia Carlo Erba*
Viale Bezzi, 24
20146 Milano, Italy

[c]*Antibioticos Farma S.A.*
Madrid Spain

In an ongoing effort to develop antagonists against basic fibroblast growth factor (bFGF), we have used its high affinity for its receptor to develop reagents that can specifically eliminate cells that express a functional receptor on their surface. In previous studies, we chemically conjugated basic FGF to saporin (SAP), a ribosome-inactivating protein from the plant *Saponaria officinalis*.[1,2] The conjugate binds to the basic FGF receptor through the basic FGF moiety, and penetrates the cell. The saporin moiety of the conjugate then induces ribosome damage that leads to cell death. In the studies here, we have constructed genes that encode basic FGF and SAP fusion proteins and expressed them in *Escherichia coli*.

A cDNA encoding SAP[3] was inserted into a plasmid (pFC80) containing a human basic FGF gene (wt bFGF). The SAP coding sequence was ligated to the 3′ end of the basic FGF gene, and expression of the fusion proteins regulated by the *trp* promoter.

The first set of constructions were performed with a basic FGF gene in which a synthetic polylinker sequence was inserted in the 5′ end of the gene. The new restriction sites in the polylinker were then used to produce basic-FGF-containing fusion genes. This construction (pFKB1) expresses a modified basic FGF (ibFGF) of 180 aa. Using this plasmid, the cDNA coding for SAP protein was inserted,

[a]Research was funded by grants from the National Institutes of Health (DK-18811, NS-28121) and from Erbamont.

generating a fused gene (ΔFGF-ΔSAP) coding for a 412 amino acid protein that contains the first 137 aa of basic FGF and the mature SAP protein.

We have also constructed fused genes containing the whole basic FGF sequence attached to the SAP gene. Mutagenesis removed the stop codon of the basic FGF gene, thus generating a DNA sequence encoding a 22-aa extended basic FGF (1bFGF). Using this construction, a basic FGF-ΔSAP gene was obtained, containing the complete wt basic FGF sequence, a 2 amino acid long connecting peptide, and the SAP protein. FIGURE 1 shows the basic FGF mutant proteins and the basic FGF-SAP fusion molecules.

We have expressed all the basic-FGF-containing constructs in *E. coli* using the constitutive expression system of the *trp* promoter. FIGURE 2 shows a Western blot of

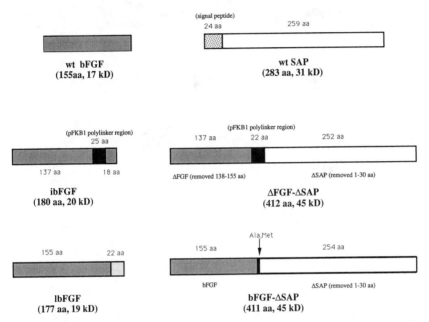

FIGURE 1. Constructions used or obtained in this work. The SAP protein has an amino terminal exporting signal that is removed in all of the FGF-SAP fused genes, along with 6 aa of the mature protein. All the basic-FGF-containing constructions have a constitutive expression in M9 minimal medium, using an *E. coli* B derived strain (Fice 2) as host. All diagrams are on the same scale.

the bacterial extracts containing the ibFGF, ΔFGF-ΔSAP, 1bFGF, and bFGF-ΔSAP proteins: using anti-SAP (A) or anti-basic FGF (B) antibodies.

The expressed proteins are currently purified using a combination of ion exchange and heparin-Sepharose affinity chromatography in an effort to obtain material for radioreceptor binding, ribosomal inhibiting activity, and mitotoxin experiments on basic FGF target cells.

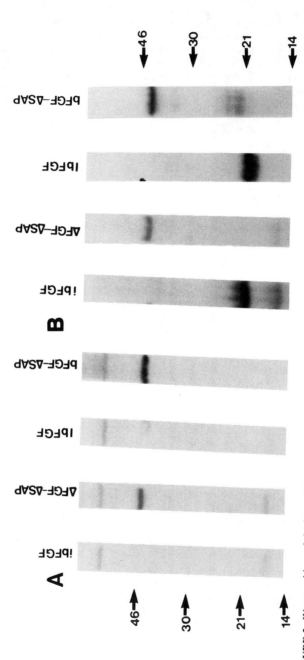

FIGURE 2. Western blots of the bacterial extracts that contain the ibFGF, ΔFGF-ΔSAP, 1bFGF, and bFGF-ΔSAP proteins: using anti-SAP (A) or anti–basic FGF (B) antibodies. The numbers and horizontal arrows on the left and right indicate molecular weight values of markers, expressed in kDa. Controls of pure SAP and basic FGF were run in parallel lanes (not shown).

REFERENCES

1. STIRPE, F., A. GASPERI-CAMPANI, L. BARBIERI, A. FALASCA, A. ABBONDANZA & W. A. STEVENS. 1983. Ribosome-inactivating proteins from the seeds of *Saponaria officinalis* L. (soapwort), of *Agrostemma githago* L. (corn cockle) and of *Asparagus officinalis* L. (asparagus), and from the latex of *Hura crepitans* L. (sandbox tree). Biochem. J. **216:** 617–625.
2. LAPPI, D. A., D. MARTINEAU & A. BAIRD. 1989. Biological and chemical characterization of basic FGF-saporin mitotoxin. Biochem. Biophys. Res. Commun. **160:** 917–923.
3. BENATTI, L., M. B. SACCARDO, M. DANI, G. NITTI, M. SASSANO, R. LORENZETTI, D. A. LAPPI & M. SORIA. 1989. Nucleotide sequence of cDNA coding for saporin-6, a type-1 ribosome-inactivating protein from *Saponaria officinalis*. Eur J. Biochem. **183:** 465–470.

Basic Fibroblast Growth Factor-Saporin Mitotoxin

In Vitro Studies of Its Cell-Killing Activity and of Substances that Alter that Activity

DARLENE MARTINEAU, DOUGLAS A. LAPPI,
EMELIE AMBURN, AND ANDREW BAIRD

Department of Molecular and Cellular Growth Biology
The Whittier Institute for Diabetes and Endocrinology
at Scripps Memorial Hospital
9894 Genesee Avenue
La Jolla, California 92037

We have examined the effect on a number of cell types of a mitotoxin[1] (bFGF-SAP) synthesized by the chemical conjugation of saporin[2] (SAP), a powerful ribosome-inactivating protein from the plant *Saponaria officinalis* and basic fibroblast growth factor (bFGF). The conjugate enters target cells via the bFGF receptor, inhibits protein synthesis, and elicits cell death. The results suggest that the mitotoxin acts by binding to cells in the same way as bFGF and that its cytotoxicity is dependent upon this interaction. Further evidence indicates that bFGF-SAP exerts its cytotoxic effect through the bFGF receptor.

In the first experiment, we attempted to inhibit the cytotoxic effect of the mitotoxin on melanoma cell line SK-MEL-28 by addition of bFGF 30 minutes before the addition of bFGF-SAP. The cytotoxic effect of 200 pM bFGF-SAP was successfully competed by the prior addition of 2 μM bFGF. The addition of TGFα, IGF-I, IGF-II, and NGF in the same manner and at the same concentration had no effect on the conjugate's cell-killing activity (see FIGURE 1). Because pretreatment with bFGF results in a complete inhibition of the cytotoxic effect, and the other growth factors have no effect, it appears that excess bFGF blocks the access of bFGF-SAP to the cells.

In a second series of experiments we tested the possibility that heparin administration might rescue the target cells. As shown in FIGURE 2, washing of baby hamster kidney (BHK) cells with heparin at times ranging from 5 to 180 minutes after treatment with 10 nM bFGF-SAP results in inhibition of the conjugate's cell-killing activity.

The inhibitory effect on bFGF-SAP's cytotoxicity is very strong at 5 minutes, but decreases as the time of washing increases until the 180-minute time point, where little effect of heparin is observed. When treated cells are washed with phosphate-buffered saline (PBS) instead of heparin, the full cytotoxic effect of the bFGF-SAP conjugate is, as expected, retained. Washing with 50 μg/ml heparin alone has no effect on these cells. The results suggest that the bFGF-SAP binds to low-affinity binding sites when it first contacts cells and that the heparin washes successfully compete with its binding to glycosaminoglycans (GAGs) on the cells. By the 180-minute time point, enough of the conjugate is internalized to kill the target cell. Accordingly, the heparin wash has little effect. Because the PBS wash is unable to provide any protection from the cytotoxic activity of bFGF-SAP, the results strongly support the hypothesis that the binding of bFGF to GAGs precedes and is perhaps required for activity.[3]

The hypothesis that the mitotoxin exerts its cytotoxic effect through specific interaction with the bFGF receptor is supported by other experiments conducted in

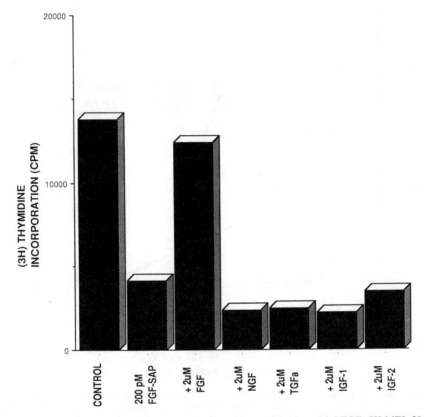

FIGURE 1. Inhibition of cytotoxicity of bFGF-SAP by competition with bFGF. SK-MEL-28 cells in 100 μl of growth media were added to the wells of a 96-well plate at a concentration of 5000 cells per well. Sixteen hours later, the growth factors, followed 30 minutes later by the mitotoxin, were added in quadruplicate. After 22 hours of incubation the samples were removed by aspiration, replaced with growth media, and the incubation was continued for an additional 48 hours. ^{3}H-thymidine was added to the wells for four hours, and the plate was frozen overnight. The plate was then thawed and well contents were collected with a Skatron cell harvester and radioactivity determined by scintillation counting.

our laboratory.[4] As an example, bFGF-SAP stimulates tyrosine phosphorylation of the 90-kDa substrate that characterizes the initial cellular response to basic FGF. Cross-linking experiments show that radiolabeled bFGF-SAP binds to the known high-affinity receptor and suramin, an inhibitor of growth factor receptor binding, inhibits the cytotoxicity of bFGF-SAP. In a study of four different cell types, there is

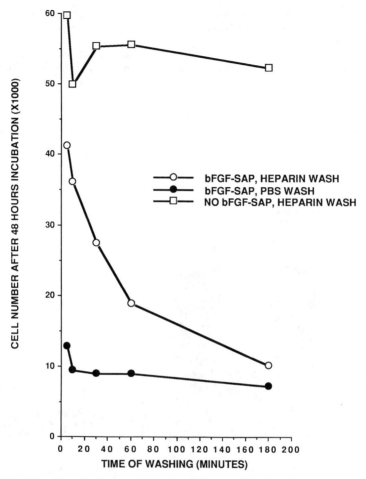

FIGURE 2. Heparin wash of cells at various times after treatment with bFGF-SAP. BHK cells in 1 ml of growth media were added to the wells of 24-well plates at a concentration of 7000 cells per well. Sixteen hours later, 10 nM bFGF-SAP was added to the wells. At 5, 10, 30, 60, and 180 minutes after this addition, the media were aspirated from the wells and the cells were washed 3 times with PBS (filled circles) or twice with 50 μg/ml heparin in PBS and a third time with PBS alone (open circles); 1 ml of growth media was added to the wells, and incubation was continued for 48 hours. In some of the wells no bFGF-SAP was added but the wells were washed with heparin as described above (squares). At the end of the 48-hour incubation, the cells were removed using trypsin, diluted, and counted with a coulter counter.

also a decrease in the ED_{50} of the mitotoxin as the receptor number per cell increases. Accordingly, the potential therapeutic use of bFGF-SAP and related receptor specific mitotoxins may be dependent on a combination of the presence of high-affinity receptors, their relative concentration on the cell surface, and the delivery of the factors to the high-affinity receptors by cell-associated GAGs.

REFERENCES

1. LAPPI, D. A., D. MARTINEAU & A. BAIRD. 1989. Biological and chemical characterization of basic FGF-saporin mitotoxin. Biochem. Biophys. Res. Commun. **160:** 917–923.
2. STIRPE, F., A. GASPERI-CAMPANI, L. BARBIERI, A. FALASCA, A. ABBONDANZA & W. A. STEVENS. 1983. Ribosome-inactivating proteins from the seeds of *Saponaria officinalis* L. (soapwort), of Agrostemma githago L. (corn cockle) and of *Asparagus officinalis* L. (asparagus), and from the latex of *Hura crepitans* L. (sandbox tree). Biochem. J. **216:** 617–625.
3. KIEFER, M. C., J. C. STEPHANS, K. CRAWFORD, K. OKINO & P. J. BARR. 1990. Ligand-affinity cloning and structure of a cell surface heparan sulfate proteoglycan that binds basic fibroblast growth factor. Proc. Nat. Acad. Sci. USA **87:** 6985–6989.
4. LAPPI, D. A., P. A. MAHER, D. MARTINEAU & A. BAIRD. 1991. The basic fibroblast growth factor–saporin mitotoxin acts through the basic fibroblast growth factor receptor. J. Cell. Physiol. **147:** 17–26.

Basic FGF-SAP Mitotoxin in the Hippocampus

Specific Lethal Effect on Cells Expressing the Basic FGF Receptor

ANA-MARIA GONZALEZ, DOUGLAS A. LAPPI,
MARINO L. BUSCAGLIA, LAURIE S. CARMAN,[a]
FRED H. GAGE,[a] AND ANDREW BAIRD

Department of Molecular and Cellular Growth Biology
The Whittier Institute for Diabetes and Endocrinology
9894 Genesee Avenue
La Jolla, California 92037

[a]Department of Neuroscience
University of California at San Diego
La Jolla, California 92037

Basic fibroblast growth factor (FGF), which was first isolated from brain and pituitary extracts and then from several other tissues, has been shown to be a potent neurotrophic agent that can promote the survival of neurons *in vivo* and *in vitro*.[1] *In situ* hybridization studies have shown that while basic FGF mRNA is expressed in the CA2 field of the hippocampus,[2] the basic FGF receptor (*flg*) mRNA is expressed at higher levels in the CA3 field and in the granular layer of the dentate gyrus.[3,4]

Lappi *et al.* have developed a new potent antagonist for basic FGF, the mitotoxin FGF-SAP,[5] which we have used to study the physiological role of basic FGF. Basic FGF is conjugated to SAP, a ribosome-inactivating protein, and the mitotoxin acts through the functional basic FGF high-affinity receptors to become a potent cytotoxic agent. Accordingly, its potency is proportional to the number of high-affinity receptors expressed on the cell surface.[6]

Based on the demonstration that basic FGF acts through a high-affinity receptor on target cells,[6] we injected the basic FGF-SAP mitotoxin (250 ng) into the hippocampus, in order to establish if neurons in the CA3 field are targets for basic FGF. Seven days after injection of the mitotoxin, the neurons of the CA3 field were destroyed (FIGURE 1A). In the other layers of the hippocampus, no detectable cytotoxicity was seen. Injection of a nonconjugated mixture of SAP and basic FGF (FIGURE 1B), SAP, or basic FGF alone showed no damage in any area of the hippocampus. Furthermore, when the dose of mitotoxin was increased to 1 μg, it had a toxic effect in all the fields of the hippocampus (FIGURE 1C) except in the contralateral hemisphere, where only the CA3 region of the hippocampus was affected (FIGURE 1D).

As basic FGF acts specifically on cells expressing high-affinity receptors, these results strongly suggest that the neurons of the CA3 field and granular layer of the hippocampus are actively synthesizing a functional receptor protein. They also support the hypothesis that there exists a functional communication between the two hemispheres in the brain. Furthermore, they clearly show the value of the basic

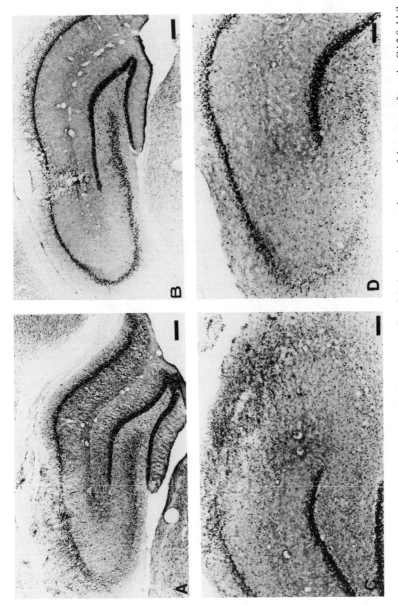

FIGURE 1. Toxicity of basic FGF-SAP mitotoxin in the hippocampus. Panel A shows the severe damage of the neurons from the CA3 field (bar = 200 μm); Panel B is a control section after the injection of the mixture of FGF and SAP (bar = 200 μm). Increased amounts of the mitotoxin (1 μg) completely destroy the ipsilateral hippocampus (Panel C) (bar = 100 μm), while in the contralateral hemisphere, only the CA3 field of the hippocampus is affected (Panel D) (bar = 100 μm).

FGF-SAP conjugate as a potential tool in understanding the physiological role of basic FGF and basic FGF responsive cells in specific areas of the brain.

REFERENCES

1. BAIRD, A. & P. BÖHLEN. 1990. Fibroblast growth factors. *In* Growth Factors. M. Sporn & A. Roberts, Eds.: 369–417. Springer-Verlag. New York, N.Y.
2. EMOTO, N., A. M. GONZALEZ, P. A. WALICKE, E. WADA, D. M. SIMMONS, S. SHIMASAKI & A. BAIRD. 1989. Identification of specific loci of basic fibroblast growth factor synthesis in the rat brain. Growth Factors 2: 21–29.
3. GONZALEZ, A. M., S. A. FRAUTSCHY, J. FARRIS & A. BAIRD. Distribution of basic FGF and its receptor's mRNA in the adult rat brain. Unpublished results.
4. WANAKA, A., E. M. JOHNSON, JR. & J. MILBRANDT. 1990. Localization of FGF receptor mRNA in the adult rat central nervous system by *in situ* hybridization. Neuron 5: 267–281.
5. LAPPI, D. A., D. MARTINEAU & A. BAIRD. 1989. Biological and chemical characterization of basic FGF–saporin mitotoxin. Biochem. Biophys. Res. Commun. 160: 917–923.
6. LAPPI, D., P. MAHER, D. MARTINEAU & A. BAIRD. 1991. Basic fibroblast growth factor–saporin mitotoxin acts through the basic fibroblast growth factor receptor. J. Cell. Physiol. 147: 17–26.

Basic Fibroblast Growth Factor Is Phosphorylated by an Ecto–Protein Kinase Associated with the Surface of SK-Hep Cells

ISABELLE VILGRAIN AND ANDREW BAIRD

Department of Molecular and Cellular Growth Biology
The Whittier Institute for Diabetes and Endocrinology
9894 Genesee Avenue
La Jolla, California 92037

Previous studies from this laboratory[1,2] have shown that basic fibroblast growth factor (FGF) is a good substrate for phosphorylation by the purified catalytic subunit of the cAMP-dependent protein kinase (PK-A) and that the phosphorylation reaction is modulated by components of the extracellular matrix (i.e., laminin, fibronectin, heparin). Moreover, a phosphorylated form of basic FGF can be immunoprecipitated with specific anti-basic-FGF antibodies from a human hepatoma cell line (SK-Hep cells) that is known to synthesize an endothelial mitogen that is structurally related to basic FGF.[3] For these reasons, we have turned our attention to the possibility that posttranslational modifications of basic FGF could regulate its biological activity.

SK-Hep cells (125,000/well) are incubated in the presence of $[\gamma\text{-}^{32}P]$-ATP and increasing concentrations of basic FGF. The products of protein phosphorylation are shown in FIGURE 1A. After 15 minutes of incubation, a phosphorylated form of basic FGF is readily detected. The addition of increasing concentrations of basic FGF to the cells results in a dose-dependent phosphorylation of the growth factor. The linear part of the phosphorylation reaction is consistent with an apparent $K_m \sim 170$ nM. The time course of basic FGF phosphorylation by the ecto–protein kinase is linear between 0 and 10 minutes of incubation and reaches a plateau at 10 to 20 minutes (FIGURE 1B). While 12-O-tetradecanoylphorbol 13-acetate (TPA) is a weak activator of the reaction, the addition of cAMP to the incubation medium increases the incorporation of phosphate into exogenous basic FGF in a dose-dependent manner. Accordingly, we conclude that basic FGF is phosphorylated by a cell surface protein kinase.

In an attempt to further characterize the specific requirements for basic FGF phosphorylation by SK-Hep-cell-associated ecto–protein kinase, we investigated the role of the interaction of basic FGF with the cell surface. The phosphorylation of basic FGF by SK-Hep cells was performed in the presence of increasing concentrations of heparin (1 to 100 μg/ml). As shown in FIGURE 2, basic FGF phosphorylation is decreased in a dose-dependent manner by the addition of heparin. This result suggests that the strong affinity of basic FGF for heparin inhibits its interaction with its low-affinity receptors as well as with the cell-surface-associated PK. Accordingly, basic FGF must be bound to its high- and/or low-affinity receptors to be phosphorylated.

This study reports the first evidence that basic FGF is phosphorylated outside the cells by an ecto–protein kinase activity associated with the surface of a human

445

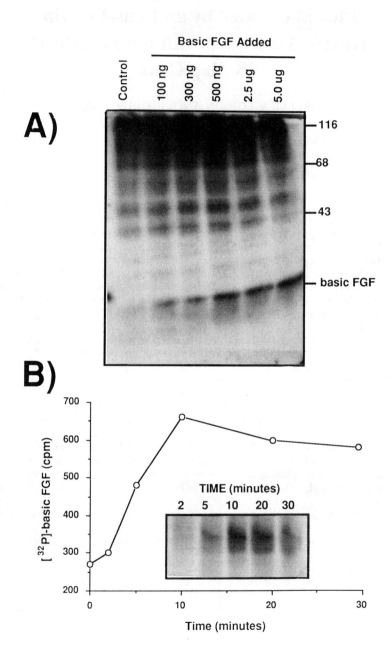

Heparin Added

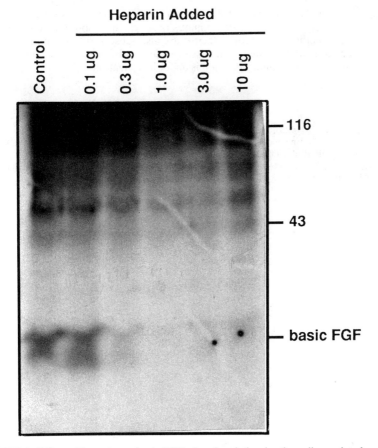

FIGURE 2. Effect of heparin on basic FGF phosphorylation by the cell-associated protein kinase. The phosphorylation assay was performed with 150,000 cells as described in FIGURE 1 and in the presence of 0.5 μg of basic FGF, 10 μM cAMP, and increasing concentrations of heparin from 0 to 10 μg/ml. After 20 minutes of incubation, phosphoproteins were analyzed by SDS-PAGE and autoradiography. Molecular weight markers (in kDa) and the mobility of basic FGF are indicated on the right.

FIGURE 1. Basic FGF phosphorylation by intact SK-Hep cells. **Panel A.** SK-Hep cells were washed 3 times with assay buffer (25 mM HEPES, pH 7.4 containing NaCl (145 mM), KCl (5 mM), $MgCl_2$ (0.8 mM), $CaCl_2$ (1.8 mM), and glucose (20 mM) and incubated in 0.1 ml/well of this solution. Human recombinant basic FGF, at the concentrations indicated, was added to the cells in the presence of 10 μM cAMP. The reaction was started with the addition of [γ^{32}P]-ATP (5 μM, 10 μCi/well) and terminated by removal of the cell supernatant, two washes of the cells with assay buffer and the addition of 100 μl of RIPA buffer containing 50 mM NaF. Phosphorylated proteins were analyzed by sodium dodecyl sulfate-polyacrylamide gel electrophoresis (SDS-PAGE) followed by autoradiography. Molecular weight markers (in kDa) and the mobility of basic FGF are indicated on the right. **Panel B.** The time course of [^{32}P] incorporation into basic FGF was performed with 0.5 μg of protein. At the indicated times, the incubation buffer was removed and the cells processed. After autoradiography, the radioactive band corresponding to basic FGF was cut off from the gel and [^{32}P] incorporated was quantitated by counting each sample in a β counter.

hepatoma cell line (SK-Hep cells). The discovery that a growth factor like basic FGF is a potential target for ecto–protein kinases raises the question of its physiological significance. Whether other growth factors are also substrates for ectophosphorylation remains to be determined. The possible role of this reaction in ligand binding, signal transduction, and internalization is under investigation.

REFERENCES

1. FEIGE, J. J. & A. BAIRD. 1989. Proc. Nat. Acad. Sci. USA **86:** 3174–3178.
2. FEIGE, J. J., J. D. BRADLEY, K. FRYBURG, J. FARRIS, L. C. COUSENS, P. J. BARR & A. BAIRD. 1989. J. Cell. Biol. **109:** 3105–3114.
3. KLAGSBRUN, M., J. SASSE, R. SULLIVAN & J. A. SMITH. 1986. Proc. Nat. Acad. Sci. USA **83:** 2448–2452.

Stimulation of DNA and Protein Synthesis in Epiphyseal Growth Plate Chondrocytes by Fibroblast Growth Factors

Interactions with Other Peptide Growth Factors

D. J. HILL, A. LOGAN,[a] AND D. DE SOUSA

MRC Group in Fetal and Neonatal Health and Development
Lawson Research Institute
St. Joseph's Health Center
London, Ontario N6A 4V2, Canada

[a]Department of Clinical Chemistry
University of Birmingham
Birmingham B15 2TT, United Kingdom

INTRODUCTION

Within epiphyseal growth plate cartilage, chondrocytes undergo a number of sequential steps of maturation including a phase of rapid proliferation, a period of intense matrix protein synthesis, and, ultimately, hypertrophy and growth arrest. This sequence of events is necessary for the continued maintenance of growth plate size and structure, the initial calcification of cartilage, and subsequent replacement with bone in a longitudinal axis. Basic fibroblast growth factor (bFGF) has been shown to be a potent mitogen for both isolated articular and growth plate chondrocytes, and also enhances the production of extracellular matrix molecules. Further, bFGF was reported to inhibit the terminal differentiation of isolated growth plate chondrocytes measured by the appearance of alkaline phosphatase. In fetal life, epiphyseal growth plate chondrocytes and adjacent bone synthesize several growth factors including insulinlike growth factor II (IGF II) and transforming growth factor β (TGF β). This study investigates the interactions between bFGF and other potentially available growth factors during chondrocyte proliferation and protein synthesis.

MATERIALS AND METHODS

Growth plate cartilage was dissected from the proximal tibia taken from lamb fetuses of 35–130 days gestation. Similar results were obtained regardless of fetal age. Chondrocytes were isolated from diced cartilage by incubation with 0.2% (w/v) bacterial collagenase for 2 hours at 37°C before cells were washed with phosphate-buffered saline and plated into tissue culture grade plastic flasks. Cells were cultured in Dulbecco's modified Eagle's medium (DMEM) supplemented with 10% (v/v) fetal calf serum (FCS) and antibiotics and used between third and sixth passage when still expressing type II collagen. For estimation of DNA synthesis chondrocytes were growth restricted when approximately 80% confluent in DMEM + 0.1% FCS for 48 hours, before transfer to serum-free medium with exogenous growth factors for 24 hours. [³H]thymidine (1 μCi/ml) was added for the final 4 hours of culture, the cells washed with DMEM, DNA precipitated with 10% (w/v) trichloroacetic acid and solubilized in 1 M sodium hydroxide, and the incorporated isotope estimated by

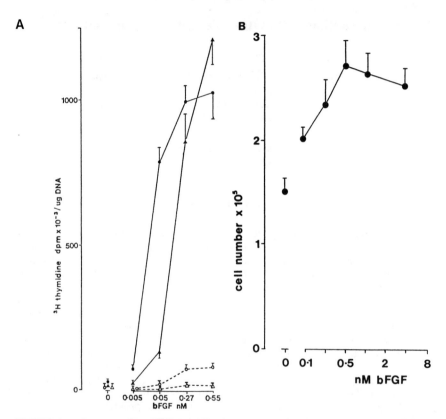

FIGURE 1. **A:** Incorporation of [³H]thymidine into DNA by chondrocytes following incubation with increasing amounts of basic (circles) or acidic (triangles) FGF in the absence (open symbols) or presence (filled symbols) of 70 μg/ml heparin. **B:** Increase in cell number over 5 days of chondrocytes incubated with increasing amounts of bFGF. Figures show mean + standard error of the mean (SEM) ($n = 6$).

liquid scintillation counting. Similar protocols were used for the estimation of total protein synthesis using 5 μCi/ml [³H]leucine. Increase in cell number was estimated following the plating of cells at a density of 0.5×10^5 per well, and their culture for between 5 and 7 days in DMEM + 2% FCS and exogenous growth factors. Cells were released from the culture dish with 0.25% trypsin/EDTA, and cell number

A

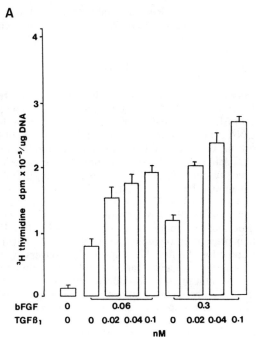

B

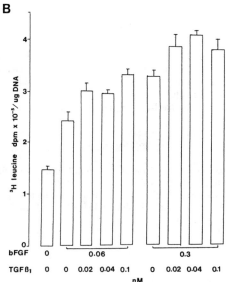

FIGURE 2. Incorporation of [³H]thymidine into DNA (A) or [³H]leucine into protein (B) by chondrocytes incubated with increasing amounts of bFGF with or without TGF β. Figures show mean + SEM (*n* = 6).

estimated with a Coulter counter. Recombinant human bFGF was a gift from Dr. P. Barr, Chiron Corporation, and acidic FGF from Dr. M. Fox, Amgen Corporation.

RESULTS

When chondrocytes were incubated with increasing concentrations of bFGF alone, a fivefold increase in DNA synthetic rate was seen compared to control incubations, with an ED_{50} of approximately 100 pM. However in the presence of 70 µg/ml heparin, DNA synthetic rate was elevated 50-fold with an ED_{50} of approximately 10 pM (FIGURE 1). Acidic FGF was without effect in the absence of heparin and was 10-fold less active than bFGF in the presence of heparin. In longer term studies, bFGF, in the presence of 2% FCS, caused a doubling in cell number over five days of culture. On a molar basis basic FGF was 300 times more potent than IGF I, 800 times more potent than IGF II, and 100 times more active than insulin as a mitogen for chondrocytes. Cells were growth restricted and exposed to bFGF or insulin for varying periods of G_1 phase of the cell cycle. While bFGF was most potent as a mitogen when present throughout the incubation period, it was next most effective when present for the first 12 hours of G_1. Conversely, insulin was most effective in the latter part of G_1. Coincubation of bFGF with suboptimal or maximally effective amounts of insulin or IGF I produced additive effects on DNA synthetic rate. Basic FGF caused a dose-related increase in chondrocyte total protein synthesis. TGF β_1 alone had no effect on either chondrocyte DNA or protein synthesis, but was synergistic in the presence of bFGF for both parameters (FIGURE 2). However, TGF β_1 inhibited the mitogenic or protein synthetic actions of insulin or IGF I.

CONCLUSIONS

Results shown in this volume by us suggest that bFGF is synthesized by fetal growth plate chondrocytes. This study shows that bFGF is an extremely potent mitogen for growth plate chondrocytes which can synergise with other growth factors known to be expressed locally. These interactions are likely to control the rate of both chondrogenesis and extracellular matrix synthesis.

REFERENCES

1. KATO, Y. & M. IWAMATO. 1990. Fibroblast growth factor is an inhibitor of chondrocyte terminal differentiation. J. Biol. Chem. **265:** 5903–5909.
2. HILL, D. J. & D. DE SOUSA. 1990. Insulin is a mitogen for epiphyseal growth plate chondrocytes from the fetal lamb. Endocrinology **126:** 2661–2670.

Basic Fibroblast Growth Factor Is an Autocrine Factor for Rat Thyroid Follicular Cells

A. LOGAN,[a] A. M. GONZALEZ,[c] M. L. BUSCAGLIA,[c]
E. G. BLACK,[b] AND M. C. SHEPPARD[b]

[a]Department of Clinical Chemistry
[b]Department of Medicine
University of Birmingham
Birmingham B15 2TT, United Kingdom

[c]Department of Molecular and Cellular Growth Biology
The Whittier Institute
La Jolla, California 92037

INTRODUCTION

Basic fibroblast growth factor (basic FGF) has been implicated in the regulation of a number of highly vascular endocrine tissues. Clearly its angiogenic activities may be of importance to these organs helping to maintain the integrity of the complex capillary network. There is also good evidence that basic FGF can regulate growth, differentiation, and physiological function of endocrine cells. Basic FGF has been localized in pituitary thyrotroph cells, and it potentiates thyrotropin releasing factor stimulated prolactin and thyrotropin secretion from cultured anterior pituitary cells.[1] Very little is known of its *in vivo* relevance to the pituitary-thyroid axis. In order to link basic FGF to thyroid function, we have investigated the distribution and synthesis of basic FGF and its *flg* receptor in rat thyroid follicular cells. Furthermore we have examined some of its biological activities on the rat thyroid follicular cell line FRTL-5.

MATERIALS AND METHODS

Adult rat thyroids were perfusion fixed and processed for immunocytochemistry and *in situ* hybridization using established methods and antibodies and [^{35}S]-labeled cRNA probes specific for basic FGF and *flg*.[2,3] Cultures of FRTL-5[4] cells were grown on chamber slides, fixed *in situ,* and similarly processed. Cell proliferation in response to exogenous basic FGF was assessed in cultures of FRTL-5 cells by measuring [^{3}H]thymidine incorporation into precipitated DNA.[4] Total inositol phosphate production after a one-hour exposure to basic FGF was assessed by prelabeling cells with [^{3}H]inositol followed by poststimulation Dowex anion chromatographic extraction.[4] Cyclic AMP levels were measured using a radioimmunoassay.[4]

RESULTS

Immunoreactive basic FGF is seen in the follicular cell cytoplasm of the normal rat thyroid. The protein is also seen strongly in the surrounding cell membrane, suggesting a site of storage, probably associated with the glycosaminoglycans (see FIGURE 1). In addition basic FGF and *flg* mRNA and protein are localized in the cytosol of rat thyroid FRTL-5 cells. These results suggest that basic FGF is localized and can be synthesized by rat thyroid follicular cells.

Rat thyroid FRTL-5 cells also respond to exogenous basic FGF. On a molar basis *in vitro*, basic FGF is a more potent mitogen than thyrotropin-stimulating hormone (TSH), stimulating proliferation in a dose-dependent manner (see FIGURE 2). Moreover, anti-basic-FGF antibody can inhibit basal cell proliferation rates. For example, antibody at a concentration of 12.5 μg protein/ml reduces cell proliferation in unstimulated FRTL-5 cultures to $83 \pm 5\%$ of control [mean ± standard error of the mean (SEM)]. The stimulation of proliferation by exogenous basic FGF is accompanied by phosphoinositide hydrolysis ($292 \pm 30\%$ of controls at 0.4 ng basic FGF/ml) and modulation of cAMP levels ($80 \pm 5\%$ of control at 4 ng basic FGF/ml).

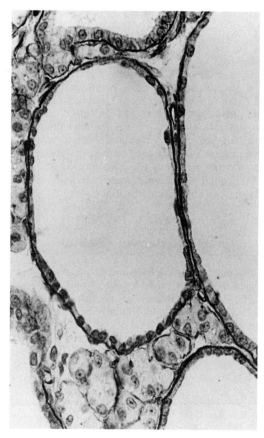

FIGURE 1. Staining of basic FGF in the rat thyroid. Specific staining for basic FGF is associated with thyroid follicular cells and the basement membrane.

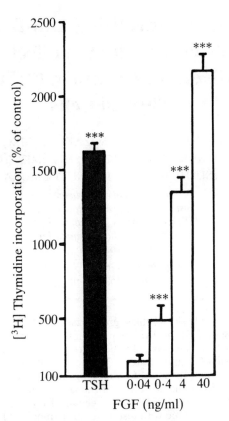

FIGURE 2. [³H]thymidine incorporation in FRTL-5 cells in response to basic FGF. FRTL-5 cells showed a concentration related increase in [³H]thymidine incorporation in response to basic FGF. The shaded bar indicates [³H]thymidine incorporation when 150 μU TSH/ml was added to each well. Values are means ± SEM expressed as a percentage of control wells measured in the same experiments ($n = 7$). ***$p < 0.001$ compared to control (Students t-test).

CONCLUSIONS

Thus, rat thyroid follicular cells can synthesize, localize, and respond to basic FGF. Taken together these results strongly implicate basic FGF as an autocrine factor that may regulate physiological and pathophysiological function of the thyroid. Further studies are under way to define the roles of basic FGF in the normal and diseased thyroid.

REFERENCES

1. BAIRD, A. 1985. Biochemistry **24:** 7855–7860.
2. GONZALEZ, A. M., M. BUSCAGLIA, M. ONG & A. BAIRD. 1990. J. Cell Biol. **110:** 753–765.
3. EMOTO, N., A. M. GONZALEZ, P. A. WALICKE, E. WADA, D. M. SIMMONS, S. SHIMASAKI & A. BAIRD. 1989. Growth Factors **2:** 21–29.
4. BLACK, E. G., A. LOGAN, J. R. E. DAVIS & M. C. SHEPPARD. 1990. J. Endocrinol. **127:** 39–46.

An Immunoneutralizing Anti-Basic-FGF Antibody Potentiates the Effect of Basic FGF on the Growth of FRTL-5 Thyroid Cells

NAOYA EMOTO, OSAMU ISOZAKI, MARIKO ARAI,[a]
HITOMI MURAKAMI, KAZUO SHIZUME,[a] TOSHIO
TSUSHIMA, AND HIROSHI DEMURA

Department of Medicine
Institute of Clinical Endocrinology
Tokyo Women's Medical College
8-1 Kawada-cho, Shinjuku-ku
Tokyo 162, Japan

[a] Institute of Growth Science
Tokyo 162, Japan

Recently, we found that basic fibroblast growth factor (FGF) exists in normal adult porcine thyroids. Basic FGF in thyroids promotes angiogenesis, stimulates thyroid cell growth, and inhibits the uptake of iodide *in vitro*.[1] In the course of this study, we found that anti-basic-FGF antiserum (773; generous gift from Dr. Andrew Baird, The Whittier Institute, La Jolla, Calif.), raised against synthetic basic FGF peptide, FGF-(1-24)-NH_2,[2,3] potentiates the effect of human recombinant basic FGF on the growth of the functional rat thyroid follicular cell line FRTL-5.

FRTL-5 cells were cultured with test materials in modified Ham's *F*-12 medium supplemented with 5% calf serum and a five-hormone preparation (5H medium) consisting of insulin, transferrin, glycyl-L-lysine acetate, somatostatin. The cultures were labeled with ^{3}H-thymidine (TdR) for 3 days. The mitogenic effect was estimated by the amount of ^{3}H-TdR incorporated into trichloroacetic acid (TCA) insoluble cellular fractions. Recombinant human basic FGF at a concentration of 0.1–10 ng/ml increased the incorporation of ^{3}H-TdR into FRTL-5 cells. 773 serum at a dilution of $1:10^5$ potentiated significantly the mitogenic effect of basic FGF (10 ng/ml) on FRTL-5 cells, and at a dilution of $1:10^4$ maximally enhanced it with a twofold increase. The effect of 773 was concentration dependent; it potentiated the effect of basic FGF at concentrations of 1.0–10.0 ng/ml (FIGURE 1, top panel). Other antisera—716, 757, 810, and 926 (also generous gifts from Dr. Andrew Baird) raised against synthetic basic FGF fragments [Tyr^{10}]FGF-(1-10), FGF-(73-87)-NH_2, FGF-(30-50)-NH_2, and FGF-(106-120)-NH_2—had no effect on the growth of FRTL-5 cells stimulated by basic FGF (FIGURE 2).

In contrast to these findings, 773 reduced the mitogenic effect of basic FGF on cultured normal porcine thyroid cells (FIGURE 1, bottom panel). The exact mechanism of this discrepancy is unknown. However, it should be noted that porcine thyroid cells in culture appear to produce basic FGF, but FRTL-5 cells do not. Accordingly, basic FGF mRNA is detected in Northern blot analysis of poly A$^+$ RNA isolated from porcine thyroid cells in culture but not from FRTL-5 cells. Although further studies are necessary, there may be a possibility that the chemical characters

456

of the basic FGF receptors are different between two types of thyroid cells; one producing basic FGF and the other not. Therefore, antibody binding may modify the effects of basic FGF differently.

The analysis of this difference may be important to an understanding of the mechanism of action of basic FGF. Furthermore, these data suggest the possible involvement of an anti-basic-FGF antibody in the modification of thyroid cell growth in autoimmune thyroiditis.

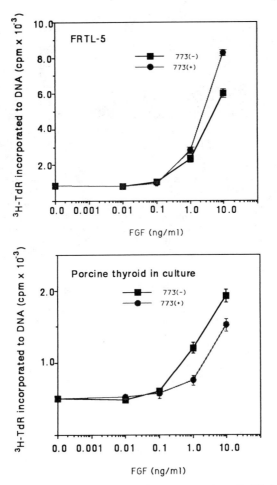

FIGURE 1. Effects of serum 773 on the growth of bFGF-stimulated FRTL-5 rat thyroid cells (top panel) and porcine thyroid cells in culture (bottom panel). FRTL-5 cells were cultured in 5H medium supplemented with 5% calf serum (FCS). Porcine thyroid cells were cultured in F-12 medium containing 1% FCS. Both cells were cultured for 3 days with various concentrations of recombinant human basic FGF in the presence or the absence of serum 773 (1:1000) and labeled with ^{3}H-thymidine (^{3}H-TdR). The data are expressed as counts per minute of ^{3}H-TdR incorporated into TCA-insoluble cellular fractions.

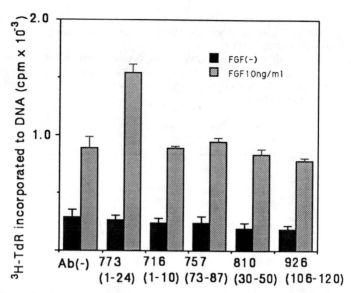

FIGURE 2. Effects of various anti-bFGF antisera on the growth of bFGF-stimulated FRTL-5 cells. FRTL-5 cells were cultured for three days in 5H medium supplemented with 5% calf serum with various anti-bFGF antisera (serum no. 773, no. 716, no. 757, no. 810, and no. 926, final dilution = 1:1000) in the presence or the absence of bFGF (10 ng/ml). The numbers in parentheses indicate the peptide sequence number of the complete primary structure of bFGF (1–146). Antisera were prepared against the synthetic bFGF fragments indicated in the figure. The data are expressed as the amount of ^{3}H-TdR incorporated (cpm) into TCA-insoluble material.

REFERENCES

1. EMOTO, N., O. ISOZAKI, M. ARAI, H. MURAKAMI, K. SHIZUME, A. BAIRD, T. TSUSHIMA & H. DEMURA. 1991. Identification and characterization of basic fibroblast growth factor in porcine thyroids. Endocrinology **128:** 58–64.
2. ESCH, F., A. BAIRD, N. LING, N. UENO, F. HILL, L. DENOROY, R. KLEPPER, D. GOSPODARO-WICZ, P. BOHLEN, & R. GUILLEMIN. 1985. Primary structure of bovine pituitary basic fibroblast growth factor (FGF) and comparison with the amino-terminal sequence of bovine brain acidic FGF. Proc. Nat. Acad. Sci. USA **82:** 6507–6511.
3. BAIRD, A., D. SHUBERT, N. LING & R. GUILLEMIN. 1988. Receptor- and heparin-binding domains of basic fibroblast growth factor. Proc. Nat. Acad. Sci. USA **85:** 2324–2328.

Expression and Release of Basic Fibroblast Growth Factor by Epiphyseal Growth Plate Chondrocytes

A. LOGAN,[a] D. J. HILL,[b] AND A. M. GONZALEZ[c]

[a]Department of Clinical Chemistry
University of Birmingham
Birmingham B15 2TT, United Kingdom

[b]MRC Group in Fetal and Neonatal Health and Development
Lawson Research Institute
St. Joseph's Health Center
London, Ontario N6A 4V2, Canada

[c]Department of Molecular and Cellular Growth Biology
The Whittier Institute
La Jolla, California 92037

INTRODUCTION

Chondrocytes from the epiphyseal growth plates express a number of peptide growth factors including the insulinlike growth factors and transforming growth factor B_1 (TGF B_1) which are likely to act as autocrine or paracrine agents in the regulation of chondrocyte proliferation and differentiation. Previous studies have identified basic fibroblast growth factor (bFGF) as a potent mitogen for isolated chondrocytes derived from both the articular and epiphyseal cartilage. The purpose of this study was to determine whether growth plate chondrocytes *in vitro* synthesized bFGF, and if so did this have the potential to act as an autocrine growth factor during fetal development.

MATERIALS AND METHODS

Cartilage was dissected from proximal tibial growth plates taken from fetal lambs between 35 and 130 days gestation. Chondrocytes were isolated from diced cartilage by incubation with 0.2% w/v collagenase for 2 hours at 37°C before plating onto tissue culture grade plastic and culture in Dulbecco's modified Eagle's medium (DMEM) supplemented with 10% v/v fetal calf serum (FCS) and antibiotics. Cells were used between third and sixth passage when type II collagen was still uniformly expressed as determined by immunohistochemistry. For the immunohistochemical localization of bFGF in cell monolayers, the cells were grown to approximately 80% confluency in 8-well chamber slides before washing three times in serum-free DMEM and culture in this medium for 24 hours. Cells were fixed in 2% w/v paraformaldehyde in phosphate-buffered saline (pH 7.6), and immunohistochemistry performed using the avidin biotin peroxidase technique. The primary antiserum (Ab 773, a rabbit polyclonal antibody against 1–24 synthetic bovine bFGF) was used at 1.25 µg/ml final concentration and color developed using diaminobenzidine. For *in situ* hybridization of mRNA encoding the *flg* receptor, the fixed cells were

459

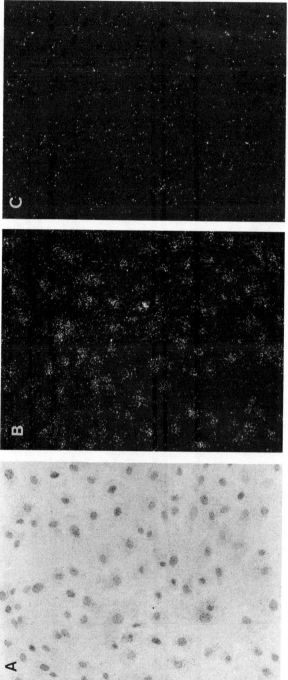

FIGURE 1. Localization of *flg* receptor mRNA within monolayer cultures of growth plate chondrocytes using *in situ* hybridization. **A:** Light-field microscopy of cell monolayer. **B:** Dark-field illumination to demonstrate positive hybridization with an ^{35}S-labeled antisense RNA on all cells. **C:** Dark-field illumination to demonstrate nonspecific hybridization using ^{35}S-labeled sense strand RNA.

incubated with proteinase K prior to hybridization with a 1000 base pair (bp) ^{35}S-labeled riboprobe derived from a rat *flg* receptor cDNA. Immunoreactive bFGF within and released from chondrocyte cultures was quantified using a double antibody radioimmunoassay. Chondrocytes were grown to approximately 80% confluency before washing in serum-free DMEM and culture for 48 hours in DMEM + 0.1% FCS with or without exogenous growth factors. Medium was collected for estimation of bFGF content, and for Western blot analysis following separation on

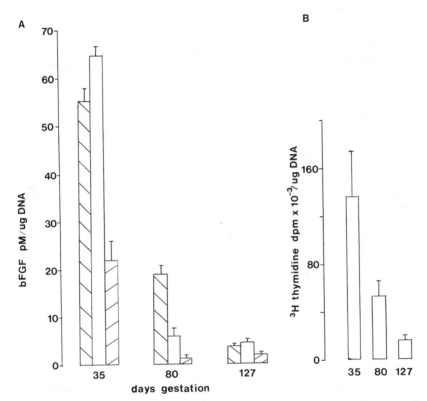

FIGURE 2. A: Radioimmunoassay of bFGF present in conditioned medium (▨), in the extracellular matrix fraction (□), and within the cells (▧) from monolayers of growth plate chondrocytes derived from fetuses at various gestational ages. B: Basal DNA synthetic rate assessed by [³H]thymidine incorporation into DNA of chondrocytes at each gestational age. Figures represent mean values ± standard error of the mean (n = 3–5).

8% sodium dodecyl sulfate-polyacrylamide gel electrophoresis (SDS-PAGE) under nonreducing conditions and transfer to nitrocellulose. Cells were released with 0.25% trypsin/EDTA and aprotinin immediately added prior to collection of cells for centrifugation. The supernatant was assayed for that bFGF associated with extracellular matrix proteins. Finally, cells were centrifuged at 20,000 × *g* for 30 minutes at 4°C and aprotinin added to the supernatant. This was assayed for intracellular bFGF. For the estimation of DNA synthetic rate, chondrocytes were growth restricted in DMEM + 0.1% FCS for 48 hours, before transfer to serum-free medium with or without exogenous bFGF or anti-bFGF antiserum for 24 hours. [^{3}H]thymidine (1 μCi/ml) was added for the final 4 hours of incubation, the cells washed, DNA precipitated with 10% trichloroacetic acid and solubilized in 1 M sodium hydroxide, and the incorporated isotope estimated by liquid scintillation counting. Recombinant human bFGF was a gift from Dr. P. Barr, Chiron Corporation.

RESULTS

Immunohistochemistry revealed that growth plate chondrocytes contained bFGF. In most cells diffuse staining was seen in the cytoplasm, and in a subpopulation this was accompanied by dense nuclear staining. When the primary antiserum was preincubated with 280 nM bFGF before application to the cells, all staining was abolished. Using *in situ* hybridization, *flg* receptor mRNA was localized to all cells within the cultures (FIGURE 1). Immunoreactive bFGF was released into conditioned medium, which contained as much activity as that recovered from the extracellular matrix and activity three times greater than that present in the cells (FIGURE 2). Western blot analysis of conditioned medium revealed a single immunoreactive bFGF species of approximately 18 kDa. No change in the bFGF content of each fraction was seen when the glucose content of the medium was altered between 2.7 mM and 16.7 mM, but this decreased with advancing gestational age (FIGURE 2). A parallel decline was noted in basal DNA synthetic rate of the cells. The bFGF content of conditioned medium, matrix, and cytoplasm showed a twofold increase following incubation with 200 pM TGF B$_1$. Anti-bFGF antiserum (1:50 Ab 773, 25 μg/ml) caused a 20% decrease in chondrocyte basal DNA synthetic rate, and blocked the mitogenic actions of exogenous bFGF.

CONCLUSIONS

Cultures of ovine fetal epiphyseal growth plate chondrocytes contained and released immunoreactive bFGF, which, via an abundant expression of *flg* receptor, has the potential to function as an autocrine mitogen. The synthesis and release of bFGF declined with gestational age, but was stimulated by exogenous TGF B$_1$. The cellular mechanisms therefore exist for bFGF to act locally during chondrogenesis.

Acidic Fibroblast Growth Factor, Heart Development, and Capillary Angiogenesis[a]

G. L. ENGELMANN,[b] C. A. DIONNE,[c] AND M. C. JAYE[c]

[b]Department of Medicine
Loyola University of Chicago
2160 South First Avenue
Maywood, Illinois 60153

[c]Rhône-Poulenc Rorer
King of Prussia, Pennsylvania

Evidence suggests that the fibroblast growth factor (FGF) family of polypeptides plays a critical role in the general proliferative capacity of a majority of mesoderm- and neuroectoderm-derived cells of the embryo, and thereby plays an essential role in early development.[2] Mammalian ventricular myocytes (cardiomyocytes) are of mesodermal origin, and their proliferative growth is almost entirely limited to embryonic and fetal stages of development. Although compelling, causal relationships between specific growth factors and their receptors such as acidic FGF (aFGF), and the development of distinct components of the cardiovascular system remain speculative.

Autocrine and/or paracrine regulation of tissue growth and differentiation of mammalian development may even influence embryo implantation.[6] Therefore, because embryonic and fetal heart development represent the most critical periods of cardiac morphogenesis and growth, aberrations and malformations that occur have life-threatening consequences.[14] Although the embryonic-fetal myocardium is a contractile mass generating systemic circulation and fetal-placental exchange, its growth represents a uniquely brief period of cardiomyocyte proliferation and rapid cellular commitment and differentiation. Cardiomyocyte maturation and increases in tissue mass of the heart during postnatal development are the result of hypertrophic myocyte growth. An additional facet of heart development is angiogenesis. Greater than 90% of the ventricular capillaries are formed during neonatal development.[7]

Evidence has been presented that fetal rodent[9] and avian ventricles[8] and cultured neonatal rat cardiomyocytes[13] contain and/or release FGFs into the extracellular matrix. Adult rat myocytes contain both a and bFGF.[3] The fetal heart contains transcripts for several members of the FGF family.[5] Therefore, FGFs may play a vital role in both proliferative myocyte growth and tissue maturation of the heart.

Analysis of the expression of aFGF and its receptor, flg, at the protein level were examined in fetal, neonatal, and adult myocardium. Immunohistochemical analysis for aFGF peptide and flg receptor (not shown) in ventricular tissue indicated extensive staining of the myocytes of the embryonic and fetal heart. Cell culture studies of fetal myocytes have confirmed their aFGF immunoreactivity and that exogenous aFGF will stimulate thymidine incorporation (not shown). As the heart matures, the 3-week-old and mature myocardium showed limited, small foci of

[a]Supported in part by HL42218 (GLE).

staining for aFGF. In contrast, the mature myocardium shows limited staining of the myocytes with anti-flg antisera, yet strong staining of the vasculature. Further study of aFGF immunoreactivity in the maturing myocardium by immunoelectron micros- copy localized the majority of the aFGF to the extracellular matrix (ECM) surround- ing the lightly stained myocytes (M). Immediately adjacent to the M-ECM were vascular endothelial cells (E) (not shown). This suggests that the fetal and neonatal myocyte may produce aFGF and deposit it into the ECM of the growing ventricle wherein it may facilitate the growth of the capillary bed during structural and functional maturation of the heart.

An essential component of an FGF-mediated heart development paradigm is demonstration of the expression of cognate FGF receptors.[4,11] The FGF receptors (flg, bek, cek-2) are related to the *fms*-like gene (FLG) which encodes a protein tyrosine kinase.[12] High affinity FGF receptors are lost permanently during mouse skeletal muscle cell terminal differentiation *in vitro*[10] and may also be relevant to the final proliferation of the cardiomyocytes during late fetal development. If cardiomy- ocyte produced FGFs are angiogenic peptides facilitating vasculature growth in a local milieu, cell-type-specific FGF receptor expression must be theorized as part of a paracrine model.

Expression of the FGF receptor gene, flg, during heart development is shown in FIGURE 1A. In two different rat strains, the general expression levels suggested a biphasic trend directly related to both myocyte proliferation and angiogenesis. During fetal heart development (E14–E20), when the ventricle contains primarily actively proliferating myocytes, the level of flg is readily detected. As the myocytes cease proliferating immediately after birth (N1–N7), the level of flg declines to near

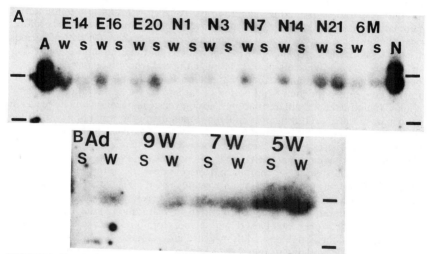

FIGURE 1. Expression of *flg* transcripts during heart development. **A:** Northern blot hybridiza- tion of total ventricular RNA (10 μg/lane hearts, 2 μg/lane N & A) separated under denaturing conditions, transferred to nylon membrane, and hybridized to random prime labeled *flg* cDNA insert. Days of age: E = embryonic, N = neonatal, or 6M = 6 months old. Samples used: N = NT2 teratocarcinoma cell line, A = A204 rhabdomyosarcoma cell line, W = Wistar Kyoto rat, S = spontaneously hypertensive rat. A/N are over-expressing cell lines. Bars indicate ribosomal RNA migrations. **B:** Northern blot hybridization of total ventricular RNA (20 μg/lane) from 5- 7- and 9-week and 12-month-old adult (Ad) rat hearts.

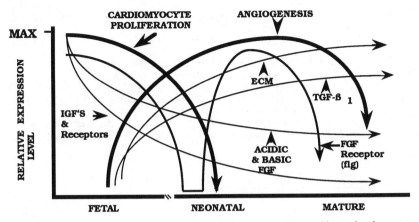

FIGURE 2. Hypothesis of growth factor/receptor gene expression and heart development.

nondetectable levels and may mimic the skeletal muscle myoblast-to-myotube loss of FGF receptors. As the heart grows in the neonate (N14–N21), flg expression returns to high levels while the myocytes are postmitotic and as the capillaries begin to increase markedly in number. Because capillary angiogenesis is only beginning at 3 weeks of age, we examined additional ages (5–9 weeks of age) to determine the expression level of flg (FIGURE 1B). In agreement with morphometry data,[1] the expression of flg coincides with the major period (5–7 weeks) of capillary angiogenesis of the two strains. This has been substantiated by *in situ* hybridization indicating that flg expression at fetal and neonatal stages of heart development is generally localized to the myocyte and nonmyocyte (i.e., endothelium) populations, respectively (not shown).

In conclusion, we have formed a working model of heart development (FIGURE 2) that indicates the general expression level of several growth factor families and their receptors in relationship to the processes of myocyte proliferation and capillary angiogenesis. We propose that fetal myocytes produce aFGF and respond to it as a mitogen when expressing flg receptors. During maturation, aFGF production declines, myocytes lose their FGF receptors, yet the growing vasculature remains responsive in a paracrine model.

REFERENCES

1. ANVERSA, P., M. MELISSARI, CESAREBEGHI & G. OLIVETTI. 1984. Am. J. Physiol. **246:** H739–H746.
2. BURGESS, W. H. & T. MACIAG. 1989. Annu. Rev. Biochem. **58:** 575–606.
3. CASSCELLS, W., E. SPEIR, J. SASSE, M. KLAGSBURN, P. ALLEN, M. LEE, B. CALVO, M. CHIBA, L. HAGGROTH, J. FOLKMAN & S. E. EPSTEIN. 1990. J. Clin. Invest. **85:** 433–441.
4. DIONNE, C. A., G. CRUMLEY, F. BELLOT, J. M. KAPLOW, G. SEARFOSS, M. RUTA, W. H. BURGESS, M. JAYE & J. SCHLESSINGER. 1990. EMBO J. **9:** 2685–2692.
5. HÉBERT, J. M., C. BASILICO, M. GOLDFARB, O. HAUB & G. R. MARTIN. 1990. Dev. Biol. **138:** 454–463.
6. HEYNER, S., B. A. MATTSON, R. M. SMITH & I. Y. ROSENBLUM. 1989. Growth Factors in Mammalian Development. I. Y. Rosenblum & S. Heyner, Eds.: CRC Press. 91–112. Boca Raton, Fla.

7. HUDLICKA, O. & K. R. TYLER. 1987. Angiogenesis: the Growth of the Vascular System: 41–66. Academic Press. New York, N.Y.
8. JOSEPH-SILVERSTEIN, J., S. A. CONSIGLI, K. M. LYSER & C. V. PAULT. 1989. J. Cell Biol. **108:** 2459–2466.
9. KARDAMI, E. & R. R. FANDRICH. 1989. J. Cell Biol. **109:** 1865–1875.
10. OLWIN, B. B. & S. D. HAUSCHKA. 1988. J. Cell Biol. **107:** 761–769.
11. RUTA, M., W. BURGESS, D. GIVOL, J. EPSTEIN, N. NEIGER, J. KAPLOW, G. CRUMLEY, C. DIONNE, M. JAYE & J. SCHLESSINGER. 1989. Proc. Nat. Acad. Sci. USA **86:** 8722–8726.
12. ULLRICH, A. & J. SCHLESSINGER. 1990. Cell **61:** 203–212.
13. WEINER, H. & J. L. SWAIN. 1989. Proc. Nat. Acad. Sci. USA **86:** 2683–2687.
14. ZAK, R. 1984. Growth of the Heart in Health and Disease. Raven Press. New York, N.Y.

FGF Mediation of Coronary Angiogenesis

M. E. SCHELLING

Program in Genetics and Cell Biology
Washington State University
Pullman, Washington 99164-4234

The fibroblast growth factors (FGFs) have been found to modulate coronary endothelial cell proliferation and differentiation, and will therefore perhaps be used therapeutically to increase coronary angiogenesis and collateral formation following myocardial infarction. We isolated coronary venular endothelial cells (CVECs) as a relevant cell line with which to study coronary angiogenesis, blood vessel formation in the heart.[1] I have previously reported the characterization of the 110 kDa acidic and basic fibroblast growth factor receptor from CVECs.[2] Heparin was found to modulate the binding of a and b FGF to the CVEC receptor and a and bFGF-induced endothelial cell proliferation.[3] Over a concentration range of heparin from 0.1 to 8.0 μg/ml, heparin potentiated the binding of aFGF and aFGF-induced cell proliferation. Over the same concentration range, heparin decreased the binding of bFGF to the cells and bFGF-induced cell proliferation. We have also previously

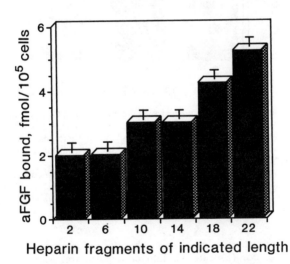

FIGURE 1. The effect of heparin fragment size on the binding of aFGF to the coronary venular endothelial cell (CVEC) FGF receptor. The binding of a and bFGF to CVECs was determined in the presence of a series of heparin fragments of various lengths at heparin fragment concentrations of 10 μg/ml, 1 μg/ml, 0.1 μg/ml, 10 ng/ml, and 1 ng/ml. Hexasaccharides and smaller heparin fragments, regardless of level of sulfation, did not increase the level of aFGF binding to the CVEC FGF receptor above the level of the no-heparin control. Octasaccharides, decasaccharides, dodecasaccharides, and larger fragments increasingly potentiated the binding of aFGF to the CVEC receptor. A high level of sulfation (data not shown) further potentiated binding of aFGF to the CVEC receptor.

reported some of the molecular controls involved in "coronary *in vitro* angiogenesis."[4] Additionally, we have prepared antiidiotypic and other monoclonal antibodies against the CVEC a and bFGF receptor for use as probes in studying receptor structure/function and signal transduction.[5]

Commercially available heparin is heterogeneous both in fragment length and level of sulfation. It is therefore not suitable to further characterize the interactions between heparin, a and bFGF, and CVECs. In collaborative studies with Dr. Carl M. Svahn, Kabi Cardiovascular, Stockholm, Sweden, who produced heparin fragments defined in length and level of sulfation, we found heparin fragment length and level of sulfation important in the regulation of the binding of a and bFGF to the CVEC

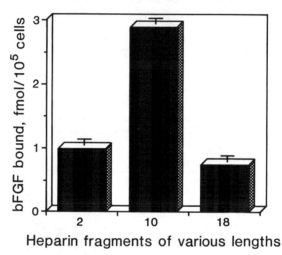

Heparin fragments of various lengths

FIGURE 2. The effect of heparin fragment size on the binding of bFGF to the coronary venular endothelial cell (CVEC) FGF receptor. The effect of heparin fragment size on the binding of bFGF to CVECs was markedly different from the effect of heparin fragment size on the binding of aFGF to CVECs. Binding in the presence of any of the heparin fragments tested inhibited binding to the receptor as compared to the no-heparin control. Binding was moderately inhibited in the presence of the decasaccharide at any of the heparin concentrations tested. Binding was markedly inhibited by shorter and longer fragments: the disaccharide and octadecasaccharide fragments markedly inhibited binding. Increased sulfation (data not shown) further decreased binding of bFGF to the receptor.

FGF receptor.[6] Disaccharides and hexasaccharides did not increase the binding of aFGF, to the receptor (FIGURE 1). Longer heparin fragments potentiated the binding of aFGF, increasingly so when highly sulfated. The effect of heparin fragment length on the binding of bFGF was markedly different (FIGURE 2). All fragments inhibited the binding of bFGF. Disaccharides and octadecasaccharides markedly inhibited the binding of bFGF to the CVEC receptor. Decasaccharides inhibited the binding of bFGF much less. The effect of sulfation on the binding of aFGF differed from the effect of sulfation on the binding of bFGF. Sulfation was found to decrease the binding of bFGF. Sudhalter *et al.* previously reported that heparin fragment size affected the aFGF-induced proliferation of adrenal capillary endothelial cells.[7]

REFERENCES

1. SCHELLING, M. E., C. J. MEININGER, J. R. HAWKER & H. J. GRANGER. 1988. Venular endothelial cells from bovine heart. Am. J. Physiol. **254:** H1211–H1217.
2. SCHELLING, M. E., J. R. HAWKER & H. J. GRANGER. 1987. Characterization of the endothelial cell growth factor (acidic fibroblast growth factor) receptor of coronary venular endothelial cells. J. Cell Biol. **105:** 110a.
3. SCHELLING, M. E., J. R. HAWKER & H. J. GRANGER. 1988. Heparin modulation of the binding of acidic and basic fibroblast growth factor to their receptor on coronary venular endothelial cells. FASEB J. **2**(6): A1714.
4. SCHELLING, M. E., J. R. HAWKER, C. J. MEININGER & H. J. GRANGER. 1987. In vitro angiogenesis by coronary venular endothelial cells. Fed. Proc. **46:** 533.
5. SCHELLING, M. E. 1989. Production of monoclonal antibodies to the coronary venular endothelial A and B FGF receptor. J. Cell Biol. **109:** 247a.
6. SCHELLING, M. E., S. E. CHANTLER & C. M. SVAHN. 1990. Effect of heparin fragment size and level of sulfation on acidic and basic fibroblast growth factor binding to the coronary venular endothelial FGF receptor. J. Cell Biol. **111**(5): 223a.
7. SUDHALTER, J., J. FOLKMAN, C. M. SVAHN, K. BERGENDAL & P. A. D'AMORE. 1989. Importance of size, sulfation, and anticoagulant activity in the potentiation of acidic fibroblast growth factor by heparin. J. Biol. Chem. **264:** 6892–6897.

Expression of Multiple Forms of bFGF in Early Avian Embryos and Their Possible Role in Neural Crest Cell Commitment

LARRY SHERMAN, KATE M. STOCKER, SEAN REES,
RICHARD S. MORRISON,[a] AND GARY CIMENT[b]

Department of Cell Biology and Anatomy
Oregon Health Sciences University
3181 Southwest Sam Jackson Park Road
Portland, Oregon 97201

[a]R.S. Dow Neurological Sciences Institute
1120 Northwest 20th Avenue
Portland, Oregon 97209

The neural crest (NC) is a population of cells that arise during neurulation in vertebrate embryos. These cells migrate from the dorsal neural tube and differentiate into a variety of cell types, including melanocytes of the integument and iris, Schwann cells, sensory and autonomic neurons, various endocrine and paracrine cells, and mesenchymal cells of the head and face.[1]

The mechanisms by which NC cell diversity is established are unclear. It is likely, however, that at least some NC cells are multipotent and that their commitment is influenced by regionally specific environmental cues.[2] Several lines of evidence suggest, for example, the existence of a bipotent progenitor cell that can give rise to melanocytes and Schwann cells.[3] One observation supporting this notion is that cultures of peripheral nerves from avian embryos, which consist largely of Schwann cell precursors and normally do not contain melanocytes, undergo pigmentation in response to the phorbol ester 12-*O*-tetradecanoyl phorbol-13-acetate (TPA).[4] These data suggest that TPA can reverse the developmental restriction of melanogenesis that normally occurs during commitment to the Schwann cell lineage, causing Schwann cell precursors to become melanocytes.

To determine whether various peptide growth factors play a role in the commitment of this putative melanocyte/Schwann cell progenitor, we investigated whether several growth factors could influence melanogenesis in NC derivatives.[5] Dorsal root ganglia (DRG) from 7 day quail (*Coturnix coturnix japonica*) embryos were cultured as explants in a culture medium ("CM") containing 10% E10 chick embryo extract and 20% fetal bovine serum in the presence or absence of various peptide growth factors. Live cultures were examined via microscopy at various times, and were scored as positive when they contained one or more pigmented cells. FIGURE 1A contains data from one such experiment and shows that basic fibroblast growth factor (bFGF), but none of the other growth factors tested, induced pigmentation in 49% of these cultures. In a separate experiment using peripheral nerve explants, bFGF induced pigmentation in 63% of the cultures.[5] These data suggest that at least some of the bFGF-responsive cells would normally have given rise to Schwann cells, and

---

[b]To whom correspondence should be addressed.

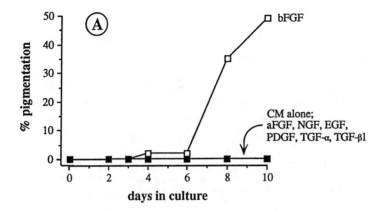

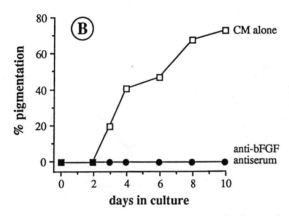

FIGURE 1. Effects of bFGF and bFGF-neutralizing antibodies on pigmentation of DRG cells. **A:** DRG explants from E7 quail embryos were cultured in either CM alone or in the presence of 10 ng/ml of either basic fibroblast growth factor (bFGF), acidic FGF (aFGF), nerve growth factor (NGF), epidermal growth factor (EGF), platelet-derived growth factor (PDGF), transforming growth factor-beta (TGF-β_1) or TGF-alpha (TGF-α), and then cultures were examined at various times for the presence of pigment cells. Each data point represents 48 culture wells. Note that bFGF, but not other growth factors, induced adventitious pigmentation. This experiment was repeated four times with qualitatively similar results. **B:** DRG explants from E5 quail embryos were cultured in CM in the presence (filled circles) and absence (open squares) of a 1:500 dilution of a bFGF-neutralizing antiserum. Each data point represents 24 culture wells. Note that the bFGF-neutralizing antiserum completely inhibited the spontaneous pigmentation of E5 DRG cells. This experiment was repeated three times with qualitatively similar results.

that bFGF can reverse the developmental restriction of melanogenesis at this stage of NC development.

It has been shown that DRG cultures from embryos younger than E7 readily undergo pigmentation in CM alone.[4] To determine whether bFGF that is present in CM (presumably in the chick embryo extract)[6] was responsible for inducing this pigmentation, E5 quail DRG were cultured in CM in the presence of a bFGF-

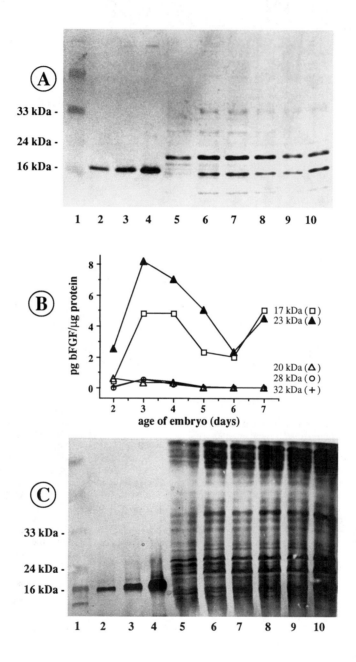

neutralizing antiserum (American Diagnostica).[5] As shown in FIGURE 1B, this antiserum completely blocked pigmentation in these cultures at a dilution of 1:500. This inhibition could be overcome, moreover, by adding excess recombinant human bFGF (rh-bFGF).[5] These data suggest that the factor promoting melanocyte commitment in CM is likely to be bFGF or a closely related growth factor.

To determine whether bFGF is expressed in avian embryos at times when Schwann cell and melanocyte progenitors differentiate, homogenates were made from both quail and chick (*Gallus domesticus*) embryos from 2 to 7 days of age, and analyzed by Western blot analysis. FIGURE 2A is one such Western blot, and shows that 5 distinct forms of bFGF are recognized. It is likely that all of these bands correspond to forms of FGF, since each binds to heparin[7] (data not shown). FIGURE 2B shows that the 17- and 23-kDa forms are predominant bFGF species, while the 20, 28, and 32 kDa forms are expressed at comparatively low levels. All five forms are maximally expressed between E3 and E4. Interestingly, bFGF expression is low at E6, after which time melanogenesis becomes developmentally restricted. FIGURE 2C is a Coomassie-blue–stained blot of a duplicate gel of FIGURE 2A, and shows that the lanes contained approximately equal amounts of protein.

The data presented above indicate that multiple forms of bFGF are present in early avian embryos during NC cell migration, and support the idea that the fate of the putative melanocyte/Schwann cell precursor may be influenced by bFGF or other members of this growth factor family. The localization of bFGF and its various forms along neural crest migratory pathways will provide further support for a role of this family of growth factors in the process of cell commitment.

REFERENCES

1. Le Douarin, N. M. 1982. The neural crest. Cambridge University Press. Cambridge, England.
2. Weston, J. A. 1970. Adv. Morphog. **8:** 41–114.
3. Ciment, G. 1990. Comments Dev. Neurobiol. **1:** 207–223.
4. Ciment, G., B. Glimelius, D. Nelson & J. A. Weston. 1986. Dev. Biol. **118:** 392–398.
5. Stocker, K., L. Sherman, S. Rees & G. Ciment. 1990. Development **111:** 635–645.
6. Kimura, I., Y. Gotoh & E. Ozawa. 1989. In Vitro Cell Dev. Biol. **25:** 236–242.
7. Haynes, L. W. 1988. Mol. Neurobiol. **2:** 263–289.

FIGURE 2. Expression of multiple forms of bFGF during early avian development. **A:** Western blot analysis of chick embryo extracts using monoclonal anti–human recombinant bFGF (DE6) primary antibody and alkaline phosphatase–conjugated secondary antibody. **Lane 1:** Molecular weight markers. **Lanes 2, 3, and 4:** 2.5, 5.0, and 10.0 ng/ml rh-bFGF. **Lanes 5–10:** 600 μg of protein from E2–E7 embryos, extracted in 10 mM Tris, 2 M NaCl, 0.1% CHAPS, 0.02 mM leupeptin, and 0.2 mM PMSF. Note that the 18-kDa and 23-kDa forms of bFGF are predominant at each of the stages assayed. **B:** Graphic representation of the changes in bFGF expression based on relative densitometry values from the Western blot above. Values were converted to pg bFGF/μg protein based on a standard curve generated from lanes containing known amounts of rh-bFGF. Note that all bFGF forms reached maximum expression between E3 and E4, and minimum expression by E6. **C:** Coomassie-blue-stained duplicate nitrocellulose blot of the Western blot in A following separation by sodium dodecyl sulfate-polyacrylamide gel electrophoresis. Note that lanes containing embryo extracts contain approximately equal amounts of total protein.

Basic Fibroblast Growth Factor and Central Nervous System Injury

A. LOGAN,[a] S. A. FRAUTSCHY,[b] AND A. BAIRD[b]

[a]Department of Clinical Chemistry
University of Birmingham
Birmingham B15 2TT, United Kingdom

[b]Department of Molecular and Cellular Growth Biology
The Whittier Institute
La Jolla, California 92037

INTRODUCTION

Following a penetrating injury to the central nervous system (CNS), transient unsuccessful regeneration is seen correlated with the production of a dense permanent scar. In the brain, basic fibroblast growth factor (basic FGF) is present in neurones, in the vascular basement membrane of blood vessels, and in the ependymal cells of the ventricles. *In vitro* studies suggest its potential as a neurotrophic factor and a growth factor promoting glial proliferation, angiogenesis, and fibrotic scar production after injury.[1] In this study we have examined changes in basic FGF levels, distribution, and synthesis in the injured rat brain in order to further implicate it in the CNS wounding response.

MATERIALS AND METHODS

Groups of matched adult female rats had stereotactically defined lesions made unilaterally in the cerebrum. Between 4 hours and 14 days later they were terminated and processed. Some brains were frozen for later heparin-Sepharose basic FGF extraction and radioimmunoassay or RNA extraction and quantitative analysis by northern/dot blot hybridization. The amount of basic FGF mRNA in each sample was quantified by autoradiography of dot blots followed by densitometric scanning. Each sample was normalized to oligo dT and the results expressed as basic FGF mRNA as a percent of the 0 day control, which was taken as 100%. Standard errors of the mean (SEM) were calculated for each group ($n = 4$). Other animals were perfusion fixed and 20-μm cryostat sections were cut for *in situ* hybridization and immunocytochemistry. These analyses used established methods[2,3] and a rabbit polyclonal antibody raised against the 1–24 synthetic fragment of bovine basic FGF(1– 146), a cDNA containing the coding region [0.5 kilobase (kb)] of rat basic FGF cDNA clone RobFGF103 and a cRNA probe that was made from the XhoI-XhoI fragment of the RobFGF103 clone which was subcloned into pBluescript SK$^+$ and linearized with NcoI. The cRNA probe encoding the antisense strand of coding sequence was transcribed using T7 RNA polymerase and [^{35}S]UTP. An RNA probe encoding the sense strand of noncoding sequence was used on control tissue sections.

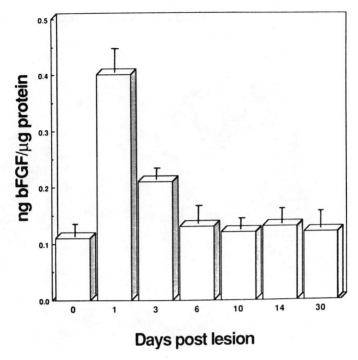

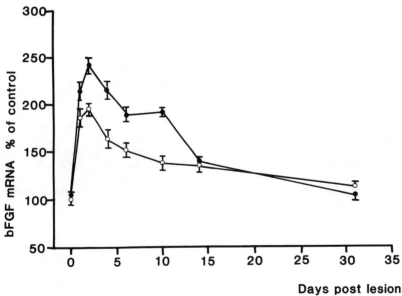

FIGURE 1. Top: Postinjury levels of immunoreactive basic FGF (expressed as ng basic FGF/μg protein) in the lesioned rat cerebral hemisphere. **Bottom:** Basic FGF mRNA levels in the lesioned (filled circles) and contralateral unlesioned (open circles) hemisphere of adult rats. The amount of specific basic FGF mRNA was quantified by autoradiography followed by densitometric scanning. Values represent means ± SEM expressed as a percentage of 0 day control animals.

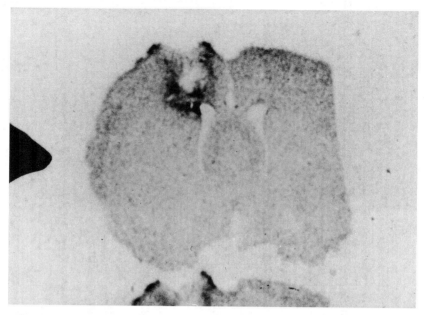

FIGURE 2. *In situ* hybridization of basic FGF mRNA. Focal increases in basic FGF mRNA are seen in neural tissue around the lesion site at 3 days following injury.

RESULTS

In the normal rat brain, multiple basic FGF mRNAs are seen by northern analysis, the major species being 1.7 and 5.9 kb. Following injury, basic FGF mRNA and protein levels rose rapidly and significantly in the lesioned cerebral hemisphere to peak between 1 and 2 days postlesion (see FIGURE 1). The increase in both had virtually disappeared 14 days after injury. At a cellular level both the intensity and the number of cells staining for basic FGF was increased focally around the wound when compared to controls as did the density of mRNA hybridization (see FIGURE 2). Moreover, there was a shift in the type of cells that contain basic FGF mRNA and protein. Early in the response (4 hours to 3 days) macrophagelike cells are clearly immunoreactive. By 7 days, basic FGF appears to be contained mostly in reactive astrocytes and this response subsided by 14 days.

CONCLUSIONS

We conclude that basic FGF is translocated and synthesized locally as a consequence of injury to the CNS. *In vivo* studies are under way to define the role of this growth factor in the injured brain.

REFERENCES

1. LOGAN, A. 1990. Trends Endocrinol. Metab. **1:** 149–154.
2. GONZALEZ, A. M., M. BUSCAGLIA, M. ONG & A. BAIRD. 1990. J. Cell Biol. **110:** 753–765.
3. EMOTO, N., A. M. GONZALEZ, P. A. WALICKE, E. WADA, D. M. SIMMONS, S. SHIMASAKI & A. BAIRD. 1989. Growth Factors **2:** 21–29.

Acidic and Basic Fibroblast Growth Factors Are Present in Glioblastoma Multiforme and Normal Brain[a]

DAVID F. STEFANIK,[b] LAILA R. RIZKALLA,[c]
ANURADHA SOI,[c] SIDNEY A. GOLDBLATT,[c]
AND WAHEEB M. RIZKALLA[c]

[b]Department of Radiation Oncology
Laurel Highlands Cancer Program
1086 Franklin Street
Johnstown, Pennsylvania 15905

[c]Department of Pathology
Conemaugh Valley Memorial Hospital
Johnstown, Pennsylvania

Glioblastoma multiforme is a uniformly fatal brain tumor comprised of pleomorphic astrocytes. Abundant neovascularization takes place.[1] Acidic and Basic fibroblast growth factors (FGFs) are mitogenic for glial and endothelial cells.[2] They have been isolated from normal brain tissue.[3] Immunohistochemistry was utilized to determine the distribution of FGFs in normal human brain. A series of glioblastomas was then evaluated.

Immunohistochemistry was performed on paraffin-embedded tissue. Neutralizing polyclonal antibodies to acidic and basic fibroblast growth factors were employed. The antibodies were prepared in rabbits against bovine antigen and purified by the manufacturer using protein A sepharose chromatography. As a control, slides were probed with purified polyclonal rabbit immunoglobulin from nonimmunized animals. Twelve tumors were probed. Brain tissues from five victims of trauma served as a normal tissue control. The Vectastain ABC peroxidase kit was used for detection (Vector Labs, Inc., Burlingame, Calif.). Ten micrograms of each polyclonal antibody was applied per slide.

The distribution of acidic and basic FGF in normal human brain is presented in FIGURE 1, and in glioblastoma multiforme in FIGURE 2.

Acidic fibroblast growth factor was abundantly present in astrocytes from all glioblastomas studied. Basic fibroblast growth factor was found in the matrix surrounding proliferating blood vessels in most of the glioblastomas. In contrast, normal brain tissue had relatively few astrocytes containing acidic FGF, and perivascular matrix staining was not demonstrated for basic FGF in the normal brain. Both growth factors could be demonstrated in neurons, Purkinje cells, capillary endothelium, and arterial walls in the normal brain. This study implicates both growth factors in the pathogenesis of malignant glioma. Both may be significant mediators of angiogenesis in glioblastoma.

[a]This work was supported by the Conemaugh Valley Memorial and Lee Hospitals, both of which are members of the Laurel Highlands Cancer Program. The Laurel Highlands Cancer Program is a member of the Pittsburgh Cancer Institute's Community Network.

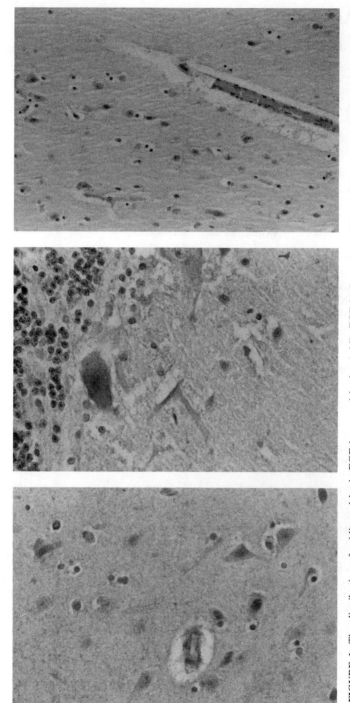

FIGURE 1. The distribution of acidic and basic FGF in normal brain. Acidic FGF is demonstrated in cortical neurons, Purkinje cells, capillary endothelium, some astrocytes, and the muscular and adventitial layers of large arteries (left panel). Basic FGF is present in Purkinje cells, rare astrocytes, and the musclar, but not adventitial, layer of arteries (center panel). The control slides demonstrate no staining (right panel).

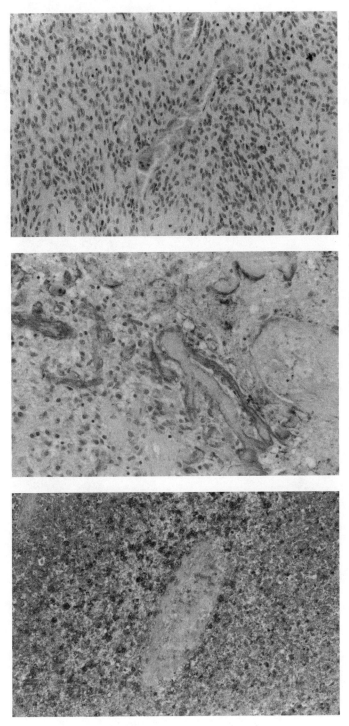

FIGURE 2. Distribution of acidic FGF and basic FGF in glioblastoma (GBM). Acidic FGF was found abundantly in pleomorphic astrocytes from all 12 tumors (left panel). Basic FGF was found in the perivascular matrix in 9 of 12 tumors (center panel). Acidic FGF was distributed diffusely about the tumor, whereas basic FGF was found in focal areas. The control slides demonstrate no staining (right panel).

REFERENCES

1. KORNBLITH, P. L. & M. D. WALKER. 1985. Neoplasms of the central nervous system. *In* Principles and Practices of Oncology. V. T. Davita, S. Hellman & S. A. Rosenberg, Eds. 2nd Edit. **2:** 1437–1510. J.B. Lippincott Co. Philadelphia, Pa.
2. FOLKMAN, J. & M. KLAGSBRUN. 1987. Angiogenic factors. Science **235:** 442–447.
3. GOSPODAROWICZ, D., H. BIALECKI & G. GREENSURG. 1978. Fibroblast growth factor activity from bovine brain. J. Biochem. **253:** 3736–3743.

Acidic Fibroblast Growth Factor in Normal, Injured, and Jimpy Mutant Developing Mouse Brain

DIDIER THOMAS,[a,d] JEAN-PIERRE CARUELLE,[b]
FRANÇOIS LACHAPELLE,[c] DENIS BARRITAULT,[b]
AND BENONI BOILLY[a]

[a]Laboratory of the Biology of Growth Factors
Lille, France

[b]Biotechnology Laboratory of Eucaryotic Cells
Créteil France

[c]Laboratory of Cellular, Molecular, and Clinical Neurobiology
Paris, France

Several growth factors have been reported to be mitogenic for various brain-derived cells while others display neurotrophic activities. Among these, the fibroblast growth factors (FGFs) have been well characterized[1] and have an unusual characteristic since they stimulate oligodendrocytes,[2,3] astrocytes,[4,5] and neuroblasts[6] to proliferate and/or to differenciate *in vitro*. They also show neurotrophic activities on neurons from various origins.[7] In addition, FGFs are potent angiogenic molecules.[8] As FGFs are the most representative growth factors of the central nervous system (CNS), they are presumed to play a key role in brain development. However, the physiological function of FGFs in the CNS is still currently unknown. To assess the possible role of acidic FGF (aFGF) in the induction of specific phases of brain development, we quantified aFGF, aFGF mRNAs, and mRNAs of the Bek form of FGF receptors in embryonic and postnatal mouse brain, at different stages of development. Moreover, we examined the effect of a mechanical lesion and jimpy mutation on these different parameters.

aFGF measurements were performed on heparin-purified brain extracts with an enzymoimmunoassay (EIA) using a specific polyclonal antibody directed against bovine aFGF.[9] Poly A$^+$ mRNA was hybridized on Northern blots with ^{32}P cDNA probes for bovine aFGF[10] and for the human Bek form of FGF receptor.[11]

NORMAL DEVELOPMENT

During the embryonic period studied (E10 to E18), aFGF level was constant at 0.2 ng per mg of extracted proteins; this level increased after birth to reach a first peak at postnatal day 5 (1.7 ng/mg). A plateau was observed from P9 to P11; from P14 to P18, a highly significant increase in aFGF was detected (5 ng/mg). From day 30 through adulthood, the amount of aFGF remained constant (2.5 ng/mg). Using the ^{32}P cDNA probe for aFGF, a 4.5-kilobase (kb) transcript was detected only for

---

[d]Address correspondence to Dr. Thomas at Centre de Neurochimie du CNRS, LNMIC, 5 rue Blaise Pascal, 67084, Strasbourg Cedex, France.

postnatal stages with a similar profile to that observed for protein levels: weak at P5, the signal markedly increased at P11 and apparently reached a plateau from P14 onwards. A shorter minor transcript (2.7 kb) could also be detected from P11 to adult stage but at a lower level. With the Bek probe, a 4.5-kb transcript was detected at every stages studied with no marked variation of signal. Another 1.5-kb transcript was also detected but only from P11 to adulthood.

INJURED BRAIN

Mechanical lesions consisted in a disruption with an angled needle, of about 60 mm^3 of the right hemisphere of 3-day-old mice. With such a treatment, our results indicated a 2.5-fold decrease of aFGF level at P14 and P18. However, aFGF mRNA and Bek mRNA levels remained unchanged after the lesion.

JIMPY MUTANT BRAIN

Jimpy mutation affects only the CNS and is associated with myelin destruction, oligodendrocyte cycle abnormalities, neuronal degeneration, and astrocyte abnormalities.[12] aFGF levels were measured on crude brain extracts at P5 and P14. At P5, the difference observed between jimpy animals and controls was not significant. On the other hand, at P14, the difference between jimpy and control was significant: 17 ng/mg vs. 7.5 ng/mg. Detection of mRNA revealed that in jimpy brains aFGF mRNAs level was twofold higher than in control at P18 and P30 and that there was no difference for Bek mRNA level.

CONCLUSION

In the mouse brain, aFGF is highly expressed during the late maturation phases of development. Bek mRNA (4.5 kb) is present in the brain at all stages studied, suggesting that this Bek form of FGF receptors is present during all the phases of brain development. In contrast, the 1.5-kb messenger appeared at later stages of development, which may suggest a different regulation of its transcription. These results may indicate that aFGF acts during postnatal events of mouse brain maturation. However, our results cannot rule out the possibility that low levels of aFGF could act locally during early phases of CNS development. Regarding injured brain and jimpy mutant brain, we cannot propose a coherent interpretation of our results and, moreover, these two models of CNS lesions did not give additional clarity on the precise function of aFGF in the CNS.

REFERENCES

1. BAIRD, A. & P. BÖHLEN. 1990. *In* Handbook of Experimental Pharmacology. Peptide Growth Factors and Their Receptors. M. B. Sporn & A. B. Roberts, Eds. 95-1: 369–418. Springer Verlag. Berlin, Heidelberg & New York.
2. ECCLESTON, P. A. & D. H. SILBERBERG. 1985. Dev. Brain Res. 21: 315–318.
3. SANETO, R. P. & J. DE VELLIS. 1985. Proc. Nat. Acad. Sci. USA. 82: 3509–3513.
4. MORRISON, R. S. & J. DE VELLIS. 1981. Proc. Nat. Acad. Sci. USA. 78: 7205–7209.
5. PETTMANN, B., M. WEIBEL & M. SENSENBRENNER. 1985. FEBS Lett. 189: 102–108.

6. GENSBURGER, C., G. LABOURDETTE & M. SENSENBRENNER. 1987. FEBS Lett. **217:** 1–5.
7. WALICKE, P. A. 1988. Exp. Neurol. **102:** 1124–1128.
8. FOLKMAN, J. & M. KLAGSBRUN. 1987. Science **235:** 442–447.
9. CARUELLE, D., J. GRASSI, J. COURTY, B. GROUX, D. BARRITAULT & J.-P. CARUELLE. 1988.
    Anal. Biochem. **173:** 117–128.
10. ABRAHAM, J. A., A. MERGIA, J. L. WHANG, A. TUMOLO, J. FRIEDMAN, K. A. HJERRILD, D.
    GOSPODAROWICZ & J. C. FIDDES. 1986. Science. **233:** 545–548.
11. DIONNE, G. A., G. CRUMLEY, F. BELLOT, J. M. KAPLOW, G. SEARFOSS, M. RUTA, W. H.
    BURGESS, M. JAYE & J. SCHLESSINGER. 1990. EMBO J. **9:** 2685–2692.
12. CAMPAGNONI, A. T. & W. B. MACKLIN. 1988. Mol. Neurobiol. **2:** 41–89.

Fibroblast Growth Factors Influence Central Nervous System Development

Evidence from Intraocular Grafts

M. M. J. GIACOBINI, B. HOFFER,[a] AND L. OLSON

Department of Histology and Neurobiology
Karolinska Institute
Box 60400
10401 Stockholm, Sweden

[a] *Department of Pharmacology*
University of Colorado Health Sciences Center
Denver, Colorado 80262

Grafting small defined areas of the developing rat central nervous system (CNS) to the anterior eye chamber of adult rat hosts permits detailed studies of graft survival and growth.[1,2] The intraocularly grafted tissue readily adheres to and becomes vascularized from the host iris. *In oculo* growth and development can be followed by stereomicroscopical observations in the living host animal and the grafts become accessible for morphological and electrophysiological studies. This system has previously been used to monitor effects of nerve growth factor (NGF),[3] truncated insulinlike growth factor-1 (tIGF-1),[4] epidermal growth factor (EGF), and platelet-derived growth factor (PDGF) on a variety of grafted developing brain areas. In the present study, the *in vivo* model of intraocular transplantation has been used to study several defined areas of the developing CNS under the influence of acidic or basic fibroblast growth factor (aFGF, bFGF).

Treatment with growth factors began with incubation of the fetal brain tissue pieces prior to grafting and was followed by 5-μl intraocular injections of the appropriate factor on days 5, 10, and 15 postgrafting. Growth of grafts was monitored by direct observation and measurement through the cornea of the living host rats.

Acidic FGF significantly enhanced growth of transplanted E17-E20 parietal cortex, E20 hippocampus, but not E14 spinal cord when compared to bovine serum albumin (BSA) vehicle alone. Parietal cortex grafts increased by approximately 200% and hippocampus grafts by 100% when stimulated with aFGF. Effective doses were seen in the range of 25 μg/ml. Basic FGF was found to be even more potent than aFGF, enhancing growth of intraocularly transplanted E17-E18 parietal cortex (see FIGURE 1), E16-E17 hippocampus, and E14 spinal cord by approximately 400%, 100%, and 50% respectively when compared to vehicle alone. Effects on all areas were seen using concentrations as low as 2.5 μg/ml.

Histochemical and immunohistochemical studies were carried out on cryostat sectioned grafts looking at markers for overall morphological organization (cresyl violet), vascularization (laminin), glial (GFAP), and neuronal populations (neurofilament; NPY). The results from these experiments are summarized in TABLE 1 and indicate that neither basic nor acidic FGF alters the morphological organization of the grafted areas. Basic FGF seems to increase the vascularization in a nonpathological manner, and aFGF seems to exert an overall normalization of the grafted tissue as determined by the immunohistochemical markers listed above.

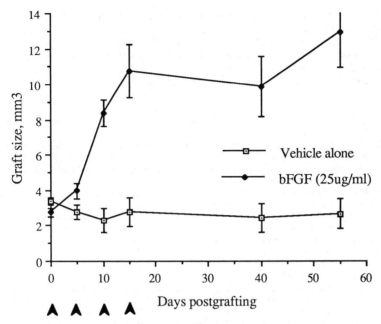

FIGURE 1. Volume increase of intraocularly grafted E18 fetal parietal cortex; $n = 12$ for each group. Arrowheads indicate times of bFGF treatment. On day 0, tissue was incubated in bFGF or vehicle alone prior to grafting. On days 5, 10, and 15, 5-µl injections of either bFGF or vehicle alone were given using a Hamilton syringe and a scleral approach into the anterior chamber of the eye containing the graft. The effects were significant ($p < 0.01$) using a multivariate ANOVA repeated measures design.

TABLE 1. Immunohistochemical Changes Induced by Fibroblast Growth Factors on Fetal Brain Tissue Grafts[a]

Area	GFA	Laminin	NF	NPY
Cortex				
aFGF	(Increase)	NS	Increase	Decrease
bFGF	NS	Increase	Increase	Not tested
Spinal cord				
aFGF	Decrease	NS	Increase	Not tested
bFGF	(Increase)	NS	(Increase)	Not tested
Hippocampus				
aFGF	Increase	NS	Decrease	Not tested
bFGF	Increase	(Increase)	(Decrease)	Decrease

[a]Summary of immunohistochemical changes induced by aFGF and bFGF. Semiquantitative comparison of density of GFA immunoreactivity, vascular appearance, density of NF immunoreactivity, and density of NPY positive cells and fibers of intraocular transplants treated with either aFGF, bFGF, or vehicle solution alone as compared to *in situ* were carried out. Based on these findings, we have indicated whether the various immunohistochemical parameters have increased or decreased. NS is not significant.

The results from our study suggest that both aFGF and bFGF augment the growth of developing brain regions *in oculo,* retaining an organotypic organization. This is in accordance with studies from several laboratories which have reported neuronal survival in primary cultures of several CNS regions.[5,6] In summary, our data taken together with data from other laboratories suggest that FGF may play an important role in neuronal maturation and graft survival.

REFERENCES

1. OLSON, L., Å. SEIGER & I. STRÖMBERG. 1983. Intraocular transplantation in rodents. A detailed account of the procedure and examples of its use in neurobiology with special reference to brain tissue grafting. *In* Advances in Cellular Neurobiology. S. Fedoroff, Ed. **4:** 407–442. Academic Press. New York, N.Y.
2. OLSON, L., H. BJÖRKLUND & B. HOFFER. 1984. Camera bulbi anterior: new vistas on a classical locus for neural tissue transplantation. *In* Neural Transplants, Development and Function. J. Sladek & D. Gash, Eds.: 125–165. Plenum. New York, N.Y.
3. ERIKSDOTTER-NILSSON, M., S. SKIRBOLL, T. EBENDAL & L. OLSON. 1989. Nerve growth factor can influence growth of cortex cerebri and hippocampus: evidence from intraocular grafts. Neuroscience **30**(3): 755–766.
4. GIACOBINI, M., L. OLSON, B. HOFFER & V. SARA. 1990. Truncated IGF-1 exerts trophic effects on fetal brain tissue grafts. Exp. Neurol. **108**(1): 33–37.
5. MORRISON, R. S., A. SHARMA, J. DEVELLIS & R. BRADSHAW. 1986. Basic fibroblast growth factor supports the survival of cerebral cortical neurons in primary culture. Proc. Nat. Acad. Sci. USA **83**(19): 7537–7541.
6. WALICKE, P. A. 1988. Basic and acidic fibroblast growth factors have trophic effects on neurons from multiple CNS regions. J. Neurosci. **8**(7): 2618–2627.

Localization of bFGF in Human Transplant Coronary Atherosclerosis[a]

F. F. ISIK, H. A. VALENTINE,[b] T. O. McDONALD,[c]
A. BAIRD,[c] AND D. GORDON

Department of Pathology
University of Washington Medical Center
Seattle, Washington 98195

[b]*Stanford University*
Stanford, California

[c]*The Whittier Institute*
La Jolla, California

INTRODUCTION

Among heart transplant patients, accelerated transplant arteriosclerosis is the major obstacle to long-term survival.[1] The prominent finding in these vessels is an excessive proliferation of smooth muscle cells (SMCs) and collagen synthesis in the intima.[1,2] Since basic fibroblast growth factor (bFGF) is known to be a potent mitogen for SMCs in cell culture systems[3] and *in vivo*,[4] we looked for its presence in the coronary arteries of human transplanted hearts.

MATERIALS AND METHODS

Segments of coronary artery were taken from hearts removed at 10 days, 1 year, 2 years, and 3 years after transplantation, either at the time of retransplantation or at autopsy. All tissues were fixed in methyl Carnoy's fixative and paraffin embedded. Serial 5-μ sections were subjected to immunocytochemistry, as previously described.[5] Two bFGF specific antibodies were used: a monoclonal anti–bovine brain bFGF antibody (no. 05118, Upstate Biotechnology Inc.)[6] and a rabbit polyclonal antibody to bovine bFGF (773B11, courtesy of Andrew Baird). A proliferating cell nuclear antigen (PCNA) antibody (Coulter Diagnostics) was also used to indicate proliferative activity, with the additional use of cell-type-specific antibodies, as previously described.[5]

RESULTS

With either bFGF antibody, only nuclear staining was observed. A segment of normal coronary artery revealed minimal-to-absent staining in the intima and media, with the exception of an occasional luminal surface cell (presumably endothelial). The 10-day posttransplant arteries showed strong nuclear staining for bFGF diffusely throughout the intima as well as in occasional luminal surface cells. Most of

[a]This study was supported by National Institutes of Health grant HL43322.

the bFGF staining was towards the medial half of the intima, whereas the luminal half of the intima was markedly positive with the anti-PCNA antibody, and with antibodies to monocyte-macrophages and lymphocytes. The 1-, 2-, and 3-year-old transplanted heart vessels showed only focal bFGF reactions, scattered throughout mostly the medial half of the coronary artery intima. Focal, minimal medial staining was also seen of lesser intensity. Serial sections stained with anti-PCNA showed some minimal proliferative activity which was not spatially correlated with areas of bFGF presence. Anti-muscle-actin antibodies revealed that the predominant intimal cell was smooth muscle in all specimens.

DISCUSSION AND CONCLUSIONS

Using two different antibodies on the same tissues, we have found bFGF staining in human coronary arteries of transplanted hearts. However associated proliferative activity in the intima (indicated by anti-PCNA staining) was high only for the 10-day-old sample. In general, no tight spatial association was observed between bFGF and PCNA reactivity. These preliminary findings suggest at least two possibilities: (a) intimal bFGF may support cell proliferation only early in the development of transplant arteriosclerosis, or (b) proliferation is better correlated with an early inflammatory response, and possibly with other growth factors, distinct from bFGF.

REFERENCES

1. BILLINGHAM, M. E. 1987. Transplantation 19(Suppl. 5): 19–25.
2. URETSKY, B. F., S. MURALI, P. S. REDDY, B. RABIN, A. LEE, B. P. GRIFFITH, R. L. HARDESTY, A. TRENTO & H. T. BAHNSON. 1987. Circulation 76: 827–834.
3. KLAGSBRUN, M. & E. R. EDELMAN. 1989. Arteriosclerosis 9: 269–278.
4. LINDNER, V., D. LAPPI, A. BAIRD, R. MAJACK & M. A. REIDY. 1991. Circ. Res. 68: 106–113.
5. GORDON, D., M. A. REIDY, E. P. BENDITT & S. M. SCHWARTZ. 1990. Proc. Nat. Acad. Sci. USA 87: 4600–4604.
6. MATSUZAKI, K., Y. YOSHITAKE, Y. MATUO, H. SASAKI & K. NISHIKAWA. 1989. Proc. Nat. Acad. Sci. USA 86: 9911–9915.

Distribution of Fibroblast Growth Factor Immunoreactivity in Skin during Wool Follicle Development

D. L. du CROS,[a] K. ISAACS, AND G. P. M. MOORE

CSIRO Division of Animal Production
Post Office Box 239
Blacktown, New South Wales 2148, Australia

Basic fibroblast growth factor (bFGF) is a member of a large family of related growth and differentiation factors originally considered as mitogenic for cells of mesenchymal origin. However, bFGF is now recognized as a potent mitogen for derivatives of both the ectoderm and mesoderm. Epidermal keratinocytes and dermal fibroblasts in culture proliferate in response to bFGF.[1] The growth factor has also been detected in cultured fibroblasts[2] and human keratinocytes,[3] and a bFGF-like peptide has been detected in immunoblots of ovine skin keratinocytes.[1] The growth factor has a strong affinity for heparin[4] and may therefore act as a cell proliferation factor following sequestration by the extracellular matrix. Furthermore, bFGF has been implicated in early embryonic induction in amphibia,[5] suggesting that it may also participate in mammalian development.

Relatively little is known about the distribution of bFGF in the skin. A study by Gonzalez, which surveyed tissues of the 18-day rat fetus, identified bFGF predominantly as a component of the basal lamina during follicle development.[6] In order to obtain more information on the distribution of this growth factor during the whole phase of follicle development, we carried out an immunohistochemical study with fetal sheepskin using antibodies to bFGF.

Primary follicles begin their initiation in midside skin at about 60 days of gestation and are mature by 110 days, well before birth at 148 days. Other follicle populations begin to appear at 85–100 days and mature later. Prior to the appearance of follicle primordia, bFGF was evident in the distal regions of the epidermis: the intermediate layer and the periderm. Immunofluorescent label was also present at the epidermal-mesenchymal junction. During early primary follicle formation (70–76 days), immunoreactivity persisted in the upper epidermis and at the epidermal-mesenchymal junction, particularly in the regions of cell condensation. Immunofluorescence was seen at the periphery of the follicle at the plug stage, and later became distributed among the follicle cells. The mesenchyme showed very little binding activity. bFGF in the follicle increased during further development, at 90 days being predominantly associated with the more distal cells of the elongating epithelial column and the region of the basal lamina. Little fluorescence was detected in the surrounding mesenchyme or the developing dermal papilla. The distribution of immunoreactivity in the developing sweat gland was similar to that in the follicle: first the periphery of the gland bud was stained, followed by the appearance of a more widespread reaction among the component cells as the gland elongated. In the mature follicle, bFGF became localized in the outer root sheath (ORS). Fluores-

---

[a] Present affiliation: School of Medicine SM-20, University of Washington, Seattle, Washington 98195.

cence appeared around the cells and at the periphery of the follicle between the ORS and the mesenchyme. This extended proximally to include the zone adjacent to the bulb matrix. Occasionally, a reaction was observed at the junction between the papilla and the matrix of the bulb.

The presence of bFGF-immunoreactive material in the epidermis and associated with the epithelial component of the follicle perhaps suggests more than one role for the growth factor in morphogenesis. Certainly, its association with epithelial proliferation during this period and in the mature follicle is consistent with its established role as an epidermal mitogen. For example, the localization of bFGF at the epidermal-mesenchymal junction and its presence in the cells of the growing plug and developing sweat gland bud indicates the maintenance of a proliferative effect on the follicle. Finally, its presence in the ORS and, more particularly, adjacent to the proliferative zone of the follicle bulb at maturity suggests a similar growth-promoting activity. If bound to the basal lamina in an active form, bFGF could provide a continuous proliferative stimulus to the matrix cells and thus directly influence fiber growth.

REFERENCES

1. PISANSARAKIT, P., D. L. DU CROS & G. P. M. MOORE. 1990. Cultivation of keratinocytes derived from epidermal explants of sheep skin and the roles of growth factors in the regulation of proliferation. Arch. Dermatol. Res. **281:** 530–535.
2. STORY, M. T. 1989. Cultured human foreskin fibroblasts produce a factor that stimulates their growth with properties similar to basic fibroblast growth factor. In Vitro Cell. Dev. Biol. **25:** 402–408.
3. HALABAN, R., R. LANGDON, N. BIRCHALL, C. CUONO, A. BAIRD, G. SCOTT, G. MOELLMAN & J. McGUIRE. 1988. Basic fibroblast growth factor from human keratinocytes is a natural mitogen for melanocytes. J. Cell Biol. **107:** 1611–1619.
4. VLODAVSKY, I., J. FOLKMAN, R. SULLIVAN, R. FRIDMAN, R. ISHAI-MICHAELI, J. SASSE & M. KLAGSBRUN. 1987. Endothelial cell–derived basic fibroblast growth factor: synthesis and deposition into subendothelial extracellular matrix. Proc. Nat. Acad. Sci. USA **84:** 2292–2296.
5. SLACK, J. M. W., B. G. DARLINGTON, J. K. HEATH & S. F. GODSAVE. 1987. Mesoderm induction in early *Xenopus* embryos by heparin-binding growth factors. Nature **326:** 197–200.
6. GONZALEZ, A-M., M. BUSCAGLIA, M. ONG & A. BAIRD. 1990. Distribution of basic fibroblast growth factor in the 18-day rat fetus: localization in the basement membranes of diverse tissues. J. Cell Biol. **110:** 753–765.

Changes in the Immunolocalization of bFGF in the Mouse Endometrium during Implantation[a]

ROBERT J. WORDINGER, AMY E. MOSS, I-FEN CHEN
CHANG, AND TONUIA JACKSON

Department of Anatomy and Cell Biology
University of North Texas
Texas College of Osteopathic Medicine
3500 Camp Bowie Boulevard
Fort Worth, Texas 76107

INTRODUCTION

Basic FGF is a potentially critical regulatory factor which may modulate various physiological responses within the uterus (1) during the ovulatory cycle, (2) in response to steroid hormone stimulation, or (3) during implantation and decidualization. Fibroblast growth factors have been purified and characterized from porcine[1] and bovine[2] uteri. We have demonstrated the immunolocalization of bFGF within the endometrium of the mouse and guinea pig.[3] Basic FGF was observed within the endometrial extracellular matrix, myometrium, and various basal laminae but was not influenced by the stage of the ovarian cycle. The objective of this study was to examine the localization of bFGF during implantation in the mouse.

MATERIALS AND METHODS

Female CD-1 mice were placed with individual males and inspected daily for a vaginal plug which was considered day 1 of pregnancy. Pregnant mice were sacrificed on day 5 of gestation. Implantation sites were dissected, quickly frozen in liquid nitrogen, and sectioned at 10 μm. Sections were allowed to dry at room temperature before fixation in cold (4°C) absolute methanol (10 minutes). Following rinsing in phosphate-buffered saline (PBS), sections were covered with 10% goat serum in PBS for 10 minutes. The goat serum was removed and the sections incubated for 24 hours at 4°C with rabbit anti–human recombinant bFGF (Biomedical Technologies Inc.) at either 1:10 or 1:20 in PBS. Following several rinses in PBS, all tissue sections were incubated in goat anti–rabbit immunoglobulin G (IgG) conjugated to the phycobili-protein, phycoerythrin (PE) (1:35) (Biomeda Corp., Foster City, Calif.). The use of phycobiliproteins decreased background staining and yielded a greater fluorescent signal. Some sections were counterstained with propidium iodide (1:10) in 0.1% trisodium citrate for 10 minutes. Specificity was assessed by (1) incubation of tissue sections with goat anti–rabbit IgG conjugated to PE in the absence of the primary antisera, (2) incubation of tissue sections in pooled rabbit sera followed by goat anti–rabbit IgG-PE, and (3) extraction of tissue sections with 2.0 M NaCl for 60

[a]Supported by National Institutes of Health Grant HD-25638 to RJW.

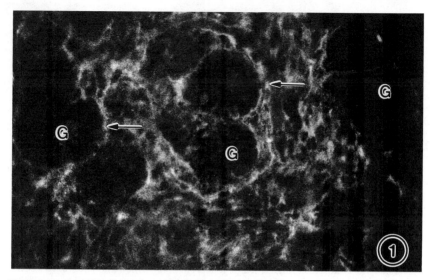

FIGURE 1. Immunolocalization of bFGF in the nondecidualized mouse endometrium. Note the localization (arrows) within the extracellular matrix and the basal laminae associated with the uterine glands (G). The epithelial cells of the uterine glands and surface epithelium (not shown) are negative. (25×)

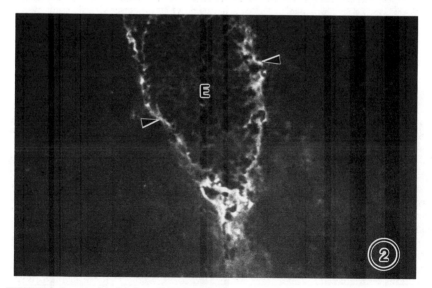

FIGURE 2. Immunolocalization of bFGF during implantation. The embryo (E) contains bFGF (arrows). The trophoblast cells and the decidual tissue in immediate approximation to the embryo were negative for bFGF. Decidual cells further removed from the embryo in the antimesometrial zone (not shown) were encircled by bFGF. (25×)

minutes at 37°C prior to the immunolocalization of bFGF which removes extracellular bFGF.[4]

RESULTS AND DISCUSSION

No immunofluorescence was observed with any of the specificity control sections. Immunofluorescence indicative of bFGF localization was seen in the extracellular matrix of nondecidualized tissue (FIGURE 1). Localization was observed associated with the basal laminae of uterine glandular epithelial cells, surface epithelial cells, and endothelial cells of blood vessels.

During implantation, immunohistochemical localization of bFGF was observed within the embryo (FIGURE 2). Trophoblast cells and the immediate primary decidual zone were negative for bFGF (FIGURE 2). There was a gradual increase in bFGF localization within antimesometrial decidual tissue away from the implanting embryo. Large, ovoid decidual cells towards the periphery of the antimesometrial decidual tissue were surrounded by bFGF. There was also localization associated with the basal laminae of newly formed blood vessels within this region. The myometrium was also positive for the presence of bFGF.

REFERENCES

1. BRIGSTOCK, D. R., R. B. HEAP, P. J. BARKER & K. D. BROWN. 1990. Purification and characterization of heparin-binding growth factors from porcine uterus. Biochem. J. **266:** 273–282.
2. MILNER, P. G., Y-S. LI, R. M. HOFFMAN, C. M. KODNER, N. R. SIEGEL & T. F. DEUEL. 1989. A novel 17 kD heparin-binding growth factor (HBGF-8) in bovine uterus: purification and N-terminal amino acid sequence. Biochem. Biophys. Res. Commun. **165**(3): 1096–1103.
3. WORDINGER, R. J., A. E. MOSS, I-F. C. CHANG & T. L. JACKSON. 1990. Immunohistochemical localization of bFGF in the mouse and guinea pig endometrium. J. Cell Biol. **111**(5): 224a.
4. GONZALEZ, A. M., M. BUSCAGLIA, M. ONG & A. BAIRD. 1990. Distribution of basic fibroblast growth factor in the 18-day rat fetus: localization in the basement membranes of diverse tissues. J. Cell Biol. **110**: 753–765.

Inhibition of Pregnancy in the Passively and Actively Immunized Mammals Rabbit, Rat, and Mouse

MARINO L. BUSCAGLIA, MICHAEL ONG, JOHN FULLER,
ANA-MARIA GONZALEZ, AND ANDREW BAIRD

Department of Molecular and Cellular Growth Biology
The Whittier Institute for Diabetes and Endocrinology
9894 Genesee Avenue
La Jolla, California 92037

Basic fibroblast growth factor (FGF) is a ubiquitous growth factor that is a potent mitogen for many different cell types, has angiogenic and neurotropic properties *in vivo,* and can modulate tissue regeneration and differentiation.[1] Recent immunocytochemical studies have established a wide distribution of immunoreactive basic FGF in the mammalian fetus, and the presence of basic FGF in the gonads, placenta, and pituitary suggests that basic FGF could play a role during reproduction.[2] Using active and passive immunoneutralization, we show that antibodies directed against basic FGF interfere with fertility.

All otherwise untreated but immunized animals used in the experiments described here were unaffected by the active or passive immunization. They also exhibit normal wound healing. In a first experiment, 6 rabbits (3 males, 3 females) were immunized using the synthetic peptide fragment of bovine basic FGF (1–24). Immunized males were mated with normal females and immunized females were mated with normal males. In controls, both normal groups were mated. In immunized animals, the mating behavior was not modified in any of the experimental groups, but fertility was strongly impaired in females. In contrast, fertility in immunized males and pregnancy of control females were normal (FIGURE 1A).

In a second series of experiments, 14 three-month-old Balb/c mice (5 males, 9 females) were immunized against a synthetic peptide fragment of rat basic FGF (1–23). Mating between immunized males and normal females resulted in normal pregnancy. All males were fertile; however, 72% of the matings between immunized females and normal males did not result in normal pregnancy (FIGURE 1B). Indeed, even the immunized females that successfully delivered pups proved to have low titers of the anti-basic-FGF antibodies as compared with the infertile females. For both of these experiments, the antiserum generated was characterized by its titer and its cross-reactivity to endogenous basic FGF in Western blot analysis. We thus concluded that immunoneutralization of basic FGF affects critical steps necessary in female reproduction before and after implantation.

In order to better understand this phenomenon, we injected (intraperitoneally) an immunoglobulin G fraction of a rabbit antiserum raised against the synthetic fragment peptide of bovine basic FGF (1–24) to pregnant female Sprague-Dawley rats. Injections made daily from days 1 to 15 (4 rats) (not shown) and days 4 to 19 of pregnancy resulted in a 50% decrease in successful pregnancy in both cases (FIGURE 1C). Narrowing the time frame of the experiment revealed that a significant impairment of pregnancy could only be observed when the injections were made

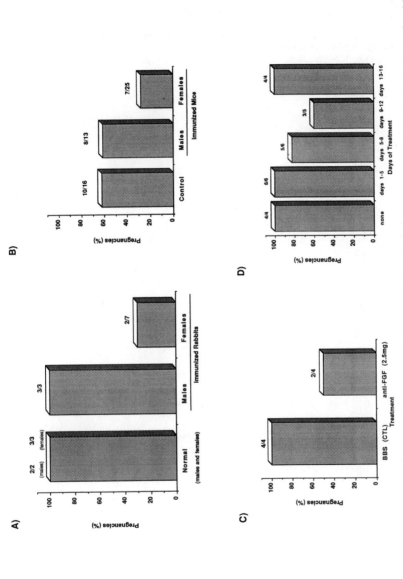

FIGURE 1. A: Number of successful pregnancies in normal and immunized rabbits. **B:** Number of successful pregnancies in normal and immunized mice. **C:** Effect of passive immunization on rat pregnancy (days 4–19). **D:** Effect of the time of anti-FGF administration on rat pregnancy.

between days 5 to 8 or 9 to 12 of pregnancy (FIGURE 1D). Thus, the results presented here imply that basic FGF is involved in fertility and normal pregnancy in mammals.

REFERENCES

1. BAIRD, A. & P. BÖHLEN. 1990. Handbook of Experimental Pharmacology. **95/I.** Springer Verlag. Berlin & Heidelberg.
2. GONZALEZ, A. M., M. BUSCAGLIA, M. ONG & A. BAIRD. 1990. J. Cell Biol. **110:** 753–765.

The Biodistribution of ^{111}In-Anti-bFGF in a Variety of Tumors[a]

Correlation with Cell-Binding Assays

S. J. KNOX, J. M. BROWN, J. McGANN, W. SUTHERLAND,
M. L. GORIS,[b] W. F. HERBLIN,[c] AND J. L. GROSS[c]

Department of Radiation Oncology
[b]*Department of Diagnostic Radiology and Nuclear Medicine*
Stanford University Medical Center
Stanford, California 94305

[c]*DuPont Merck Corporation*
Wilmington, Delaware

INTRODUCTION

Basic fibroblastic growth factor (bFGF) is a factor produced by a variety of tumors. This factor promotes tumor growth by (1) autocrine stimulation mediated by binding of bFGF to receptors on the surface of tumor cells, and (2) stimulation of angiogenesis or neovascularization via binding of bFGF to receptors on capillary endothelial cells. Anti-bFGF (DG2) is a monoclonal antibody (MAB) that should inhibit tumor growth by depriving tumor cells of autocrine stimulation as well as angiogenesis. Studies have shown that this MAB can inhibit (1) binding of bFGF to its receptor, (2) bFGF stimulation of fibroblast proliferation, as measured by ^{3}H-thymidine incorporation, and (3) bFGF-induced angiogenesis in a rat kidney capsule assay.[1] The ability of DG2 to neutralize bFGF-induced angiogenesis in an *in vivo* model distinguished it from other anti-bFGF MABs. Furthermore, anti-bFGF was able to partially inhibit the proliferation of rat C6 glioma cells in culture and C6 glioma tumors in nude mice, and appeared to decrease the invasiveness of the treated tumors as well[2] (Gross *et al,* unpublished data).

Whereas most MABs studied to date are directed to tumor-specific antigens on a particular tumor type, anti-bFGF is directed to bFGF which is produced by many different kinds of tumors, and may therefore be effective for treating a wide variety of tumors. This antibody has two properties that make it ideal for therapy: (1) unlabeled, it effectively binds to bFGF, and (2) it has both antiangiogenesis and antitumorigenic effects. In theory, radiolabeled DG-2 should specifically deliver radioactivity to tumor cells, and increase its tumor-killing capability, if the MAB is able to bind to bound bFGF in the bFGF receptor. In the series of biodistribution and cell-binding assays described here, the potential of DG2 for radioimmunotherapy was studied.

[a]This research was supported in part by an American Cancer Society Institutional Research Grant.

MATERIALS AND METHODS

Mice

Twelve to 21-week-old C3H and SCID mice were obtained from the Stanford University Radiation Biology Mouse Colony. Mice were tested and found to be negative for antibodies to mouse hepatitis virus, Sendai virus, and *Mycoplasma pulmonis*. Groups of C3H mice were injected subcutaneously in the flank with 2×10^5 RIF 1 cells or SCCVII cells in a 0.1-ml volume. Similarly, SCID mice were injected with 1×10^6 HT-1080 cells, 1×10^7 C6, or 1×10^6 LS174T cells. Approximately 1–4 weeks later animals had discrete palpable tumors with a mean diameter of approximately 10 mm.

Antibody

Monoclonal anti-bFGF (DG2) was produced and purified by DuPont Merck Corp (Wilmington, Del.) as previously described.[1] It is a murine (immunoglobulin G) MAB directed against human recombinant bFGF, which also recognizes bFGF of mouse, rat, and human origin. An isotype-matched irrelevant MAB (MB-1; IDEC, Mt. View, Calif.) was used as a control.

Chelation, Labeling, and Radioimmunoactivity

DG2 and MB-1 were chelated with diethylenetriaminepentaacetic acid (Sigma, St. Louis, Mo.) and labeled with ^{111}In (Amersham Corp, Arlington Heights, Ill.).[3] The ^{111}In-MAB was separated from the reactants using a gel filtration column (Pharmacia, Sweden). The specific activity was 1 μCi/μg for both ^{111}In-DG2 and ^{111}In-MB-1 in 4 experiments and 0.62 μCi/μg in one experiment. The labeling efficiency of the MAB was 90–94% and 75–93%, respectively. The immunoreactivity of the eluant was determined with a solid-phase immunoadsorption technique as previously described[4] using cyanogen bromide–activated sepharose 4B beads (Sigma) coupled to purified bFGF (Dupont Merck Corp.). Using this method, the immunoreactivity of ^{111}In-labeled DG2 was $\geq 67.5\%$ and that of ^{111}In-MB-1 was negligible. Mice were given injections shortly thereafter with 4–18 μCi ^{111}In-MAB intravenously through the tail vein or intraperitoneally.

Biodistribution Experiments

Biodistribution studies were performed by sacrificing animals at 24, 48, 72, and 120 hours after injection of ^{111}In-MAB. The weight and activity of the entire mouse as well as dissected organs were measured and the percentage of injected dose/g was calculated.

Cell-Binding Assays

Binding of DG2 to three cell lines (C6, A159, SNB-75) known to produce and express high levels of bFGF and bFGF receptors respectively (Dupont Merck Corp.) was measured using two methods. Confluent monolayers (2 cm^2/well) or single-cell

TABLE 1. Percent Injected Dose/Gram[a]

		Liver	Spleen	Kidney	Stomach	Intestine	Lung	Heart	Femur	Muscle	Blood	Tumor	Total
RIF 1													
Day 2	S	2.7	2.1	3.8	0.6	1.1	1.5	1.5	0.7	0.4	4.1	3.5	22.0
	NS	3.1	3.3	8.4	1.0	1.8	7.1	1.9	1.0	0.5	6.0	5.0	39.1
	S/NS	0.87	0.64	0.45	0.60	0.61	0.21	0.79	0.70	0.80	0.68	0.70	—
Day 3	S	2.4	1.9	3.5	0.7	1.1	3.4	1.1	0.4	0.3	3.1	3.6	21.5
	NS	2.7	2.9	9.4	0.8	1.5	3.1	1.7	0.9	0.5	3.8	4.6	31.9
	S/NS	0.89	0.66	0.37	0.86	0.73	1.09	0.65	0.44	0.60	0.82	0.78	—
SCC VII													
Day 1	S	11.7	4.8	9.1	2.5	2.7	9.0	5.7	1.4	1.1	11.7	6.4	66.1
	NS	4.0	4.0	18.8	1.2	1.5	5.3	3.2	0.8	0.7	6.1	4.7	50.3
	S/NS	2.93	1.20	0.48	2.08	1.80	1.70	1.78	1.75	1.57	1.92	1.36	—
Day 3	S	8.4	4.5	8.3	1.1	1.4	4.2	3.5	1.2	0.9	5.6	6.2	45.3
	NS	3.8	4.6	13.2	1.0	1.4	3.4	2.0	1.4	0.6	2.9	4.7	39.0
	S/NS	2.21	0.99	0.63	1.10	1.00	1.23	1.75	0.86	1.50	1.93	1.32	—
Day 5	S	5.3	3.5	8.3	0.7	1.1	2.6	1.4	0.8	0.5	3.8	5.0	33.1
	NS	ND	ND	ND	ND	ND	ND	ND	ND	ND	ND	ND	ND

[a]S, specific MAB; NS, nonspecific MAB; ND, not done.

TABLE 2. Percent Injected Dose/Gram[a]

		Liver	Spleen	Kidney	Stomach	Intestine	Lung	Heart	Femur	Muscle	Blood	Tumor	Total
HT1080 Day 1	S	10.3	6.4	5.6	1.0	1.4	8.8	4.8	1.4	0.9	9.5	6.5	56.6
	NS	7.2	8.4	16.5	1.3	1.9	7.9	4.8	1.8	0.9	9.8	6.5	67.0
	S/NS	1.43	0.76	0.34	0.77	0.74	1.11	1.00	0.78	1.00	0.97	1.00	—
Day 3	S	9.0	6.6	5.5	1.1	1.6	5.3	2.8	1.3	0.7	6.4	4.6	44.9
	NS	6.5	7.0	9.6	1.0	2.4	4.3	2.2	2.2	0.8	5.2	4.9	46.1
	S/NS	1.38	0.94	0.57	1.10	0.67	1.23	1.27	0.59	0.88	1.23	0.94	—
C6 Day 1	S	12.1	9.6	7.3	1.0	1.4	6.5	3.8	1.7	0.8	9.6	10.3	64.1
	NS	7.0	8.4	16.9	1.0	1.9	7.2	3.0	1.8	1.0	8.6	10.4	67.2
	S/NS	1.73	1.14	0.43	1.00	0.74	0.90	1.27	0.94	0.80	1.12	0.99	—
Day 3	S	12.0	6.7	6.9	1.0	1.7	5.1	2.6	1.6	0.8	8.3	10.5	57.2
	NS	5.6	5.3	11.7	1.0	1.8	3.6	2.1	1.6	0.6	4.4	8.1	45.8
	S/NS	2.14	1.26	0.59	1.00	0.94	1.42	1.24	1.00	1.33	1.89	1.30	—
LS174T Day 1	S	14.9	6.1	5.8	2.3	1.7	5.9	3.2	2.0	0.9	7.4	4.2	54.4
	NS	8.3	7.8	17.2	1.9	2.9	8.3	4.2	2.8	1.5	10.7	5.7	71.3
	S/NS	1.80	0.78	0.34	1.21	0.59	0.71	0.76	0.71	0.60	0.69	0.74	—
Day 3	S	10.3	3.0	5.5	1.5	1.5	6.2	3.8	0.7	0.8	5.8	2.9	42.0
	NS	8.6	8.7	13.7	1.3	2.5	8.3	4.1	2.4	1.7	6.4	4.2	61.9
	S/NS	1.20	0.34	0.40	1.15	0.60	0.75	0.93	0.29	0.47	0.91	0.69	—
Day 5	S	9.6	10.3	5.1	1.4	1.4	4.0	2.5	1.7	0.8	3.8	3.5	44.1
	NS	5.8	7.2	10.6	1.1	1.8	4.7	3.2	0.2	1.3	3.6	3.7	43.2
	S/NS	1.66	1.43	0.48	1.27	0.77	0.85	0.78	8.50	0.62	1.06	0.95	—

[a]S, specific MAB; NS, nonspecific MAB.

suspensions (7.5 × 10⁶ cells; prepared by incubation of cells with 50 mmol EDTA for 20 min at 37°C) were incubated for 30 minutes on ice with and without 50 μg of unlabeled unchelated DG2. Then 250 μl ¹¹¹In-DG2 at 50 μg/ml was added in triplicate to wells containing 500 μl of serially diluted cells. Following incubation and mixing at room temperature for 2 hours, cell suspensions were centrifuged and supernatants were removed. Supernatants were collected from wells that were washed 3 times with phosphate-buffered saline (PBS). The PBS, supernatants, and pellets were counted in a gamma counter, and cell binding was calculated.

RESULTS

The results of the biodistribution experiments using five different tumor systems are summarized in TABLES 1 and 2. The mean percent injected dose/g of ¹¹¹In-DG2 (specific) and ¹¹¹In-MB-1 (nonspecific) MAB, as well as the ratio of uptake of specific/nonspecific MAB, is shown for tumor and normal tissues for groups of mice, with most groups containing 3–5 mice. In general, the standard error of the mean was ≤20% of the percent injected dose/g. There were no significant differences between the specific and nonspecific MABs in any of the tissues studied in any of the tumor lines. Furthermore, there was no significant binding of ¹¹¹In-DG2 to C6, A159 (human endometrial adenocarcinoma), or SNB-75 (human glioma) cells, confirming previous observations that significant active bFGF was either not extracellularly associated or was inaccessible (Herblin *et al.,* unpublished data).

DISCUSSION

These results are consistent with the hypothesis that DG2 inhibits tumor growth either by removing the source of autocrine stimulation or by interfering with microvasculature angiogenesis by binding to free bFGF. Future studies will utilize another MAB, tumor-specific anti-bFGF receptor, that will soon be available for study. While bFGF receptors are present on the surface of tumors and many cells, sequence differences exist between tumor-associated bFGF receptors and bFGF receptors found on normal tissues (Horlick *et al.,* unpublished data). In theory, tumor-specific anti-bFGF receptor MAB should have an antitumorigenic effect mediated by the inhibition of autocrine stimulation. In addition, this MAB may be an effective vehicle to carry radioactivity specifically to targeted tumor tissue, augmenting its efficacy. Because the receptor is a cell-associated antigen, it may be a better target for RIT than bFGF, since dose deposition is determined in part by the retention time of the radiolabeled MAB at the tumor site.

ACKNOWLEDGMENTS

The authors thank Laurie Russell for her secretarial services in preparing this manuscript.

REFERENCES

1. REILLY, T. M., D. S. TAYLOR, W. F. HERBLIN, M. J. THOOLEN, A. T. CHIU, D. W. WATSON & P. B. M. W. M. TIMMERMANS. 1989. Biomed. Biol. Res. Commun. **164:** 736–743.

2. GROSS, J. L., W. F. HERBLIN, B. A. DUSAK, P. CZERNIAK, M. D. DIAMOND & D. L. DEXTER. 1990. Proc. Am. Assoc. Cancer Res. **31:** 79.
3. HNATOWICH, D. J., W. W. LAYNE, R. L. CHILDS, D. LANTEIGNE, M. A. DAVIS, T. W. GRIFFIN & P. W. DOHERTY. 1983. Science **220:** 613–615.
4. KNOX, S. J., R. LEVY, R. A. MILLER, W. UHLAND, J. SCHIELE, W. RUEHL, R. FINSTON, P. DAY-LOLLINI & M. L. GORIS. 1990. Cancer Res. **50:** 4935–4940.

A Model for the Role of Basic Fibroblast Growth Factor in Pituitary Tumorigenesis[a]

JOEL SCHECHTER[b] AND RICHARD WEINER[c]

[b]Department of Anatomy and Cell Biology
University of Southern California
Los Angeles, California 90033

[c]Department of Obstetrics and Gynecology
University of California
San Francisco, California

It is well established that prolonged estrogen stimulation can induce prolactimona formation in rats[1,2] and that different strains of rats have different susceptibilities to estrogen.[3,4] In previous studies comparing the effects of estradiol (E_2) treatment on ovariectomized adult Fischer 344 and Sprague Dawley (SD) rats, we reported on a dramatic response of the vasculature of the anterior pituitary (AP) of F344 rats.[5,6] After 20 days of E_2 treatment, well-formed arteries were found within the AP of F344 rats, as well as disrupted capillaries contributing to the formation of hemorrhagic lakes. An additional study demonstrated that folliculostellate cells within the AP became activated as phagocytes in response to E_2, and that degradation of matrix materials of the capillary bed might be a component of this phagocytic activation.[7] The dramatic changes noted in the vasculature, including arteriogenesis, suggested that angiogenic growth factors, e.g., fibroblast growth factor (FGF), might play an important role in tumorigenesis in F344 rats.

We now report intense focal concentrations of FGF-positive cells in E_2-treated F344 rats after 10–20 days predominantly near the posterolateral edge of the AP, i.e., a region rich in gonadotrophs and lactotrophs.[8,9] Immunostaining of serial sections as well as electron microscopy reveals that FGF-positive cells are gonadotrophs, and that the immunoprecipitate is cytosolic (FIGURE 1). Immunostaining for matrix-associated FGF reveals foci of positivity also at posterolateral AP sites. E_2-treated SD rats do not reveal comparable localization patterns. Immunostaining for collagenase activity was demonstrated only in E_2-treated F344 rats, frequently in cells whose morphological features suggested that they are folliculostellate cells.

We propose that E_2-induced prolactinomas in F344 rats result from increased enzymatic degradation of matrix materials and heightened synthesis and release of FGF from gonadotrophs at discrete peripheral foci of the AP which then stimulates angiogenesis from systemic blood vessels in the adjacent meninges.

[a]Supported by DK 40945.

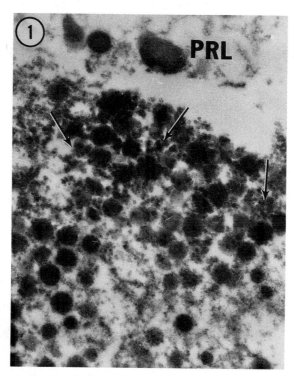

FIGURE 1. Immunolocalization of FGF in an E_2-treated F344 rat (cryoultrathin section without counterstain). Immunoprecipitate (arrows) is most evident in the cytosol between secretory granules of gonadotrophs. Part of a lactotroph is seen (at top) and consistently lacked any FGF immunoprecipitate. Characteristic secretory granules of lactotroph (PRL). 28,800×.

REFERENCES

1. ZONDEK, B. 1936. Lancet **1:** 776–778.
2. FURTH, J. & K. H. CLIFTON. 1966. The Pituitary Gland. G. W. Harris & B. T. Donovan, Eds. **2:** 460–497. Butterworths. London, England.
3. DUNNING, W. F., M. R. CURTIS & A. SEGALOFF. 1947. Cancer Res. **7:** 511–521.
4. WIKLUND, J., N. WERTZ & J. GORSKI. 1981. Endocrinology **109:** 1700–1707.
5. ELIAS, K. A. & R. I. WEINER. 1984. Proc. Nat. Acad. Sci. USA **81:** 4549–4553.
6. SCHECHTER, J., N. AHMAD, K. ELIAS & R. WEINER. 1987. Am. J. Anat. **179:** 315–323.
7. SCHECHTER, J., N. AHMAD & R. WEINER. 1988. Neuroendocrinology **48:** 569–576.
8. BAKER, B. L. 1974. The Pituitary Gland and Its Neuroendocrine Control. E. Knobil and W. H. Sawyer, Eds.: 45. American Physiological Society. Washington, D.C.
9. BAKER, B. L., J. G. PIERCE & J. S. CORNELL. 1972. Am. J. Anat. **135:** 251–267.

Mechanism of Herpesviral Entry into Cells

Role of Fibroblast Growth Factor Receptor

ROBERT J. KANER[a]

Pulmonary Service
Department of Medicine
Memorial Sloan-Kettering Cancer Center
1275 York Avenue
New York, New York 10021

INTRODUCTION

Herpes simplex viral infections are widespread in the general population. By 10 years of age, over 50% of children demonstrate antibodies to herpes simplex type 1 (HSV-1).[1] The incidence of herpes simplex virus type 2 is more difficult to estimate, but various studies cite a prevalence of 0.3%–22%.[2] In the 35 to 44-year-old age group, well over 10% demonstrate antibodies to HSV type 2.[1] A more recent study concluded the incidence is 16.4% in the population of age 15 to 74.[3] Localized infections occur in most normal people, whereas dissemination can occur in the immunocompromised host.

The mechanism whereby the herpes virion enters the cell to initiate infection is undefined. The initial interaction of the virion with the cell was recently shown to occur via cell-surface heparan sulfate,[4] presumably interacting with virally encoded envelope glycoproteins. Subsequent interactions between viral glycoproteins and the cell surface are thought to occur.

RESULTS

We observed that basic fibroblast growth factor (bFGF) inhibits herpes simplex virus type 1 infection of bovine smooth muscle cells[5] (FIGURE 1). We noted that the inhibition of plaque formation (or uptake of radiolabeled virions) occurred only if the growth factor was present before or simultaneously with the introduction of virus. Growth factor added two hours after virus had no effect on viral infection (FIGURE 2). These effects were specific for basic FGF: other growth factors (PDGF, TGF-β, NGF) had no effect on uptake of radiolabeled virus or infectivity. Similar inhibitory effects were observed when infection was studied in human arterial smooth muscle or umbilical vein endothelial cells (not shown) or rat arterial smooth muscle cells (FIGURE 3). Later, it was found that infection in VERO cells (African green monkey kidney) is not inhibited by basic FGF (M. Muggeridge, personal communication and unpublished observations).

To determine if an FGF receptor was involved in HSV penetration, we used polypeptides analogous to various regions of the basic FGF molecule to inhibit

[a]Dalsemar Research Scholar of the American Lung Association.

infection.[6] We observed an approximate 70% block in HSV uptake using bFGF peptides from the 103–120 amino acid region, which is thought to mediate binding of basic FGF to its high-affinity cell surface receptor (FIGURE 4). Polypeptides that did not contain this region were ineffective in blocking infection (FIGURE 4). These initial results suggested a role for a high-affinity basic FGF receptor in HSV-1 uptake and infectivity.

To further define the role of high-affinity FGF receptors in cellular HSV

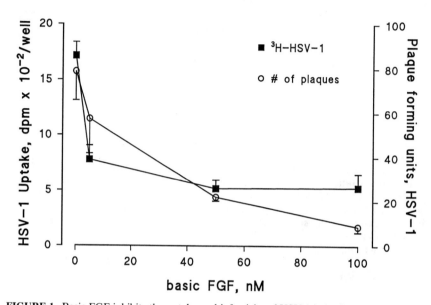

FIGURE 1. Basic FGF inhibits the uptake and infectivity of HSV-1 in bovine smooth muscle cells. For infectivity (plaque) assays, equal volumes of a diluted crude HSV-1 preparation were added to confluent monolayers of bovine smooth muscle cells in the presence of increasing concentrations of basic FGF. After a 2-hour incubation, the monolayers were washed 3 times with phosphate-buffered saline (PBS) and overlaid with M-199/0.1% agarose. Viral plaques were counted in the light microscope after 72 hours incubation at 37°C. Results are expressed as the mean ± standard deviation for triplicate wells. For assays of cellular uptake of radiolabeled virus, ³H-thymidine-labeled HSV-1 was prepared as described in Reference 5. Bovine smooth muscle cells were plated at 50,000 cells/well in 24-well plates. The following day, 30,000 dpm/well of ³H-thymidine-labeled HSV-1 was added in the presence of increasing concentrations of bFGF. After 2 hours at 37°C, the cells were washed extensively with PBS and dissolved in 0.1 N NaOH/0.2% sodium dodecyl sulfate (SDS). Uptake of radiolabel was determined by scintillation counting. Results are expressed as the mean ± standard deviation for quadruplicate wells.

penetration, we studied the uptake of radiolabeled HSV-1 into CHO cells that were transfected with a mouse analogue of a chicken high-affinity FGF receptor (**flg**).[7] These cells overexpressed high-affinity basic FGF receptors with two immunoglobulinlike extracellular domains and demonstrated high-affinity binding of labeled basic FGF.[6] While these cells did not develop typical cytopathological effects of HSV-1 infection and do not support viral replication, our results showed a greater than 10-fold increase in uptake of radiolabeled virus in the transfected cells compared to

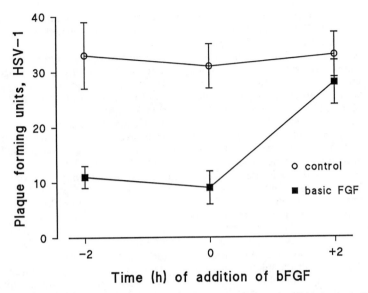

FIGURE 2. Time dependence of the inhibitory effect of bFGF on HSV-1 infection of bovine smooth muscle cells. Viral plaque assays were performed as in FIGURE 1. Basic FGF was added 2 hours before (pre-), simultaneous with, or 2 hours after (post-) the addition of HSV-1 to the monolayer of cells. Basic FGF added after the virion has attached to the cell surface has no effect on infectivity, implicating an early step in the infection cycle.

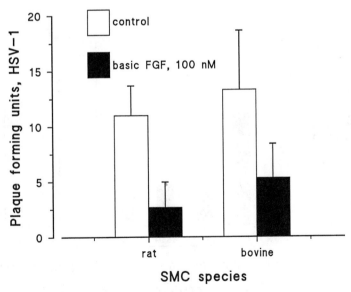

FIGURE 3. Basic FGF inhibits HSV-1 infection of smooth muscle cells of various species. Viral plaque assays were performed as in FIGURE 1. Equal numbers of plaque-forming units (infectious virus) were added to each well.

either parental or nonsense-transfected cell lines (FIGURE 5). In addition, binding experiments at 4°C. showed similar enhanced binding of radiolabeled virions to the cell surface when the **flg** protein was overexpressed.

These results lead to the question of how the herpes virion could recognize an FGF receptor. A search of the gene bank failed to identify any strong homologies at the amino acid level between the basic FGF and the known HSV-1 envelope glycoproteins, particularly glycoproteins B, C, and D, which are thought to participate in the early interactions between the virion and the cell surface during viral attachment and/or entry. We found that basic FGF could be identified in both crude and density gradient purified viral preparations by Western analysis using purified rabbit polyclonal anti–basic FGF antibodies.[8] These same antibodies inhibited HSV-1 infection in bovine smooth muscle cells (FIGURE 6). The HSV-1 preparations could compete with labeled basic FGF for binding to its high-affinity receptor both in a radioreceptor assay and in cross-linking studies.[7] The HSV-1 preparations were also capable of inducing the phosphorylation of the 90 kilodalton (kDa) substrate which indicates activation of the high-affinity basic FGF receptor-specific tyrosine kinase activity.[7] Combining this information with the immunohistochemical localization of a basic FGF-like molecule associated with the nuclear membrane[9] leads us to speculate that the newly forming virion, which is thought to acquire its lipid envelope by budding through the nuclear membrane, might also simultaneously acquire and display on its lipid envelope cellular basic FGF, which allows it to subsequently recognize high-affinity basic receptor(s) at the time of infection of a new cell.

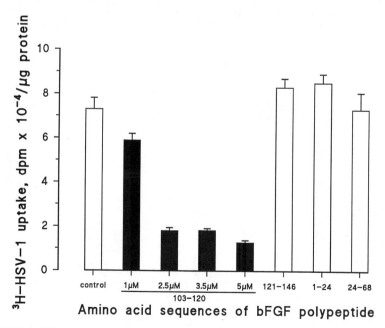

FIGURE 4. Effect of synthetic polypeptide analogues of basic FGF on the uptake of radiolabeled HSV-1 by bovine smooth muscle cells. Uptake of radiolabeled virus was performed as in FIGURE 1. Polypeptides were added at a concentration of 5 μM, except as indicated. The numbering for the amino acids is based on the sequence reported by F. Esch *et al.*[16]

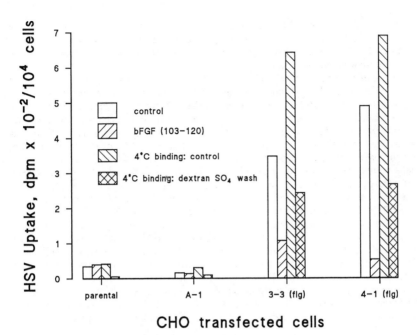

CHO transfected cells

FIGURE 5. Effects upon uptake and binding of radiolabeled HSV-1 by CHO cells transfected with a cDNA encoding a basic FGF receptor (**flg**). CHO = parental CHO cells; A-1 = CHO cells transfected with **flg** in the reverse $3' \rightarrow 5'$ orientation; 3-3, 4-1 = CHO cells transfected with **flg** and overexpressing the receptor. Viral uptake studies were performed as in FIGURE 1. Binding studies were performed at 4°C as in FIGURE 1 except that twice the amount of input virus was used. After a 2-hour incubation with labeled HSV-1, the cells were washed for 30 minutes with PBS alone or PBS containing 1 mg/ml dextran sulfate.

Subsequent experiments established that parental CHO cells are capable of taking up HSV-1.[10] Our own recent data indicate detectable uptake of radiolabeled HSV-1 into parental CHO cells confirmed by Southern blotting for two different HSV-specific genes. Furthermore, they appear to do so in quantities comparable to the transfected cells. There are several possible reasons for this discrepancy from our original results, including, (but not limited to) variability in viral preparations, e.g., in terms of the quantity of basic FGF contained, and the instability of these transfected cells where they lose their capacity for high-affinity binding of labeled basic FGF (unpublished observations). Thus, the receptor density may have decreased with serial passage and a certain critical density of receptors might be necessary to see enhanced viral uptake. This and other possibilities are currently under investigation.

DISCUSSION

The interpretation of these experiments in terms of proving or disproving the hypothesis that a high-affinity basic FGF receptor is involved in cellular infection is complicated in a number of other ways. While transfected CHO cells are clearly capable of overexpressing high-affinity basic FGF binding sites, binding studies with radiolabeled FGF have difficulty establishing whether the parental cells express low numbers vs. zero high-affinity sites. The parental cells do *not* respond mitogenically to basic FGF. However, **flg**-specific mRNA is detectable in the parental cells (A. Baird, personal communication).

In addition, data presented at this meeting[11–13] support the concept that low-affinity (heparan sulfate) binding of basic FGF is required for high-affinity binding. This makes interpretation of experiments that attempt to distinguish between low- and high-affinity responses all the more ambiguous, since experimental maneuvers that inhibit low-affinity binding will inhibit high-affinity binding as well. One approach is to investigate the properties of truncated soluble forms of the extracellular portion of the **flg** gene as potential competitive inhibitors. Conditioned media from cells transfected with this gene inhibited HSV-1 infection by 30–50% in NIH3T3 fibroblasts, and in arterial smooth muscle cells (FIGURE 7), respectively. Future experiments will focus on the antiviral effects of purified soluble receptor protein and the ability of HSV-1 to bind to this protein when it is immobilized.

While the suggestive evidence outlined above is intriguing, definitive proof that high-affinity basic FGF receptors are implicated in HSV-1 infection remains elusive. If high-affinity basic FGF receptors are one of the cellular proteins participating in HSV-1 infection of smooth muscle cells *in vitro,* it remains to be seen whether this effect is limited to allowing HSV-1 to extend its range of infectable cells *in vitro* or whether these receptors are important participants in HSV infection *in vivo.* Another important possibility is that this pathway may be only one of several that herpes simplex virus uses to enter cells. Recent data from other investigators support the

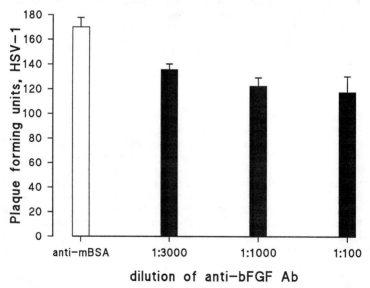

FIGURE 6. Effect of polyclonal rabbit anti-bFGF antibodies on HSV-1 infection of bovine smooth muscle cells. Purified anti-bFGF polyclonal antibodies were mixed with equal volumes of HSV-1 at the indicated dilutions. The antibodies were incubated with the virus at 37°C. for 30 minutes prior to infecting the cell monolayer. Viral plaque assays were then completed as in FIGURE 1. Purified rabbit polyclonal anti-mBSA antibodies (1:100) were used as a control.

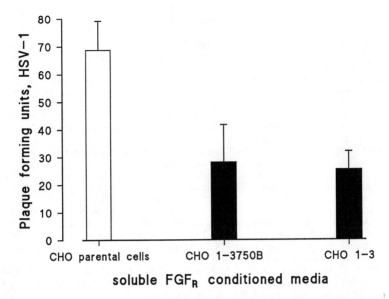

FIGURE 7. Effect of conditioned media from cell lines transfected with a gene encoding a soluble truncated form of the extracellular domain of **flg** on HSV-1 infection of bovine smooth muscle cells. Conditioned media (Dulbecco's modified Eagle medium/0.1% fetal calf serum) from two different cell lines (1-3750B; 1-3), each transfected with a truncated **flg** gene encoding the extracellular domain of the receptor were collected after 24 hours. Conditioned medium was mixed 1:1 with the viral preparation diluted in fresh medium, and incubated at 37°C for 1 hour. Viral plaque assays were then performed as in FIGURE 1. Conditioned medium from parental CHO cells was used as a control.

concepts that there are cell surface receptors that mediate HSV penetration into cells, that there is more than one cell surface receptor pathway involved in HSV-1 infection,[14,15] and that separate cell surface receptors are recognized by different proteins on the virion surface.[15] This multiplicity of different receptors could potentially explain why basic FGF is unable to completely inhibit infection,[15] since other potential portals of entry may co-exist in the same cells. Furthermore, cell lines that do not exhibit high-affinity bFGF binding and are receptor negative, but are infectable with HSV-1 (D. Mirda, personal communication), indicate that high-affinity bFGF receptors are not required for HSV infection in all mammalian cells. The fact that infection in VERO cells is not inhibited by bFGF, whereas heparin in high concentrations can inhibit infection in all cells studied to date (although heparin is not as effective an inhibitor of HSV-2), is consistent with the idea that more than one pathway of infection may exist in different cell types.

ACKNOWLEDGMENTS

I thank A. Baird, C. Basilico, R. Florkiewicz, D. Hajjar, P. Maher, A. Mansukhani, and B. Summers for their essential contributions to this work. I thank E. Cha, M. Ong, J. Sharkey, B. Ursea and R. Ursea for technical assistance.

REFERENCES

1. NAHMIAS, J. K. 1978. Epidemiology of herpes simplex virus 1 and 2. *In* Viral Infections of Humans—Epidemiology and Control. A. S. Evans, Ed.: 253–271. Plenum Medical Book Co. New York, N.Y.

2. ROONEY, J. F. 1985. Epidemiology of herpes simplex. Straus SE, Moderator. Herpes simplex virus infection: biology, treatment and prevention. Ann. Intern. Med. **103:** 404–419.

3. JOHNSON, R. E., A. J. NAHMIAS, L. S. MAGDER, F. K. LEE, C. A. BROOKS & M. A. SNOWDEN. 1986. A seroepidemiologic survey of the prevalence of herpes simplex virus type 2 infection in the United States. N. Engl. J. Med. **321:** 7–12.

4. WUDUNN, D. & P. G. SPEAR. 1989. Initial interaction of herpes simplex virus with cells is binding to heparan sulfate. J. Virol. **63:** 52–58.

5. KANER, R. J., A. BAIRD, A. MANSUKHANI, C. BASILICO, B. D. SUMMERS, R. Z. FLORKIEWICZ & D. P. HAJJAR. 1990. Fibroblast growth factor receptor is a portal of cellular entry for herpes simplex virus type 1. Science **248:** 1410–1413.

6. BAIRD, A., D. SCHUBERT, N. LING & R. GUILLEMIN. 1988. Receptor- and heparin-binding domains of basic fibroblast growth factor. Proc. Nat. Acad. Sci. USA **85:** 2324–2328.

7. MANSUKHANI, A., D. MOSCATELLI, D. TALARICO, V. LEVYTSKA & C. BASILICO. 1990. A murine fibroblast growth factor (FGF) receptor expressed in CHO cells is activated by basic FGF and Kaposi FGF. Proc. Nat. Acad. Sci. USA **87:** 4378–4382.

8. BAIRD, A., R. Z. FLORKIEWICZ, P. MAHER, R. J. KANER & D. P. HAJJAR. 1990. Mediation of virion penetration into vascular cells by association of basic fibroblast growth factor with herpes simplex virus 1. Nature **348:** 344–346.

9. FLORKIEWICZ, R. Z., A. BAIRD & A-M. GONZALEZ. 1991. Multiple forms of bFGF: differential nuclear and cell surface localization. Growth Factors **4:** 265–275.

10. SHIEH, M. L. & P. G. SPEAR. 1991. Fibroblast growth factor receptor: does it have a role in the binding of HSV-1? Science **253:** 209–210.

11. YAYON, A., M. KLAGSBRUN, J. ESKO, P. LEDER & D. ORNITZ. 1991. Cell surface, heparin-like molecules are required for binding of basic fibroblast growth factor to its high affinity receptor. Cell **64:** 841–848.

12. GANNOUN-ZAKI, L., I. PIERI, J. BADET, M. MOENNER & D. BARRITAULT. Internalization of basic fibroblast growth factor by CCL39 fibroblast cells: involvement of heparan sulfate glycosaminoglycans. Ann. N.Y. Acad. Sci. (This volume.)

13. RAPRAEGER, A. C., A. KRUFKA & B. B. OLWIN. 1991. Requirement of heparan sulfate for bFGF-mediated fibroblast growth and myoblast differentiation. Science **252:** 1705–1708.

14. CAMPADELLI-FIUME, G., D. STRIPE, A. BOSCARO, E. AVITABILE, L. FOA-TOMASI, D. BARKER & R. ROIZMAN. 1990. Glycoprotein C–dependent attachment of herpes simplex virus to susceptible cells leading to productive infection. Virology **178:** 213–222.

15. SEARS, A. E., B. S. McGWIRE & B. ROIZMAN. 1991. Infection of polarized MDCK cells with herpes simplex virus 1: two asymmetrically distributed cell receptors interact with different viral proteins. Proc. Nat. Acad. Sci. USA **88:** 5087–5091.

16. ESCH, F., A. BAIRD, N. LING, N. UENO, F. HILL, L. DENOROY, R. KLEPPER, D. GOSPDAROWICZ, P. BÖHLEN & R. GUILLEMIN. 1991. Primary structure of bovine pituitary fibroblast growth factor (FGF) and comparison with the amino-terminal sequence of bovine acidic FGF. Proc. Nat. Acad. Sci. USA **82:** 6507–6511.

Index of Contributors